Topics in Fluorescence Spectroscopy

Volume 4

Probe Design and Chemical Sensing

Topics in Fluorescence Spectroscopy

Edited by JOSEPH R. LAKOWICZ

Volume 1: Techniques
Volume 2: Principles
Volume 3: Biochemical Applications
Volume 4: Probe Design and Chemical Sensing

Topics in Fluorescence Spectroscopy

Volume 4
Probe Design and Chemical Sensing

Edited by

JOSEPH R. LAKOWICZ

Center for Fluorescence Spectroscopy and
Department of Biological Chemistry
University of Maryland School of Medicine
Baltimore, Maryland

SPRINGER SCIENCE+BUSINESS MEDIA, LLC

Library of Congress Cataloging-in-Publication Data

On file

ISBN 978-1-4757-8808-2 ISBN 978-0-306-47060-8 (eBook)
DOI 10.1007/978-0-306-47060-8

Originally published by Plenum Press, New York in 1994
Softcover reprint of the hardcover 1st edition 1994

Contributors

J. Ricardo Alcala • Department of Biomedical Engineering, Case Western Reserve University, Cleveland, Ohio 44106

Shabbir B. Bambot • Department of Chemical and Biochemical Engineering, University of Maryland Baltimore County, Baltimore, Maryland 21228; and Medical Biotechnology Center of the Maryland Biotechnology Institute, University of Maryland at Baltimore, Baltimore, Maryland 21201

David J. S. Birch • Department of Physics and Applied Physics, University of Strathclyde, Glasgow G4 0NG, Scotland, United Kingdom

Gary Carter • Department of Electrical Engineering, University of Maryland Baltimore County, Baltimore, Maryland 21228

Guillermo A. Casay • Department of Chemistry, Georgia State University, Atlanta, Georgia 30303

Anthony W. Czarnik • Department of Chemistry, Ohio State University, Columbus, Ohio 43210

B. A. DeGraff • Department of Chemistry, James Madison University, Harrisonburg, Virginia 22807

J. N. Demas • Department of Chemistry and Biophysics, University of Virginia, Charlottesville, Virginia 22908

K. T. V. Grattan • Department of Physics, Measurement and Instrumentation Centre, School of Electrical Engineering and Applied Physics, City University, London ECl V0HB, United Kingdom

Raja Holavanahali • Department of Electrical Engineering, University of Maryland Baltimore County, Baltimore, Maryland 21228

Graham Hungerford • Department of Physics and Applied Physics, University of Strathclyde, Glasgow G4 0NG, Scotland, United Kingdom

Simon C. W. Kwong • Department of Chemical and Biochemical Engineering, University of Maryland Baltimore County, Baltimore, Maryland 21228; and Medical Biotechnology Center of the Maryland Biotechnology Institute, University of Maryland at Baltimore, Baltimore, Maryland 21201

Joseph R. Lakowicz • Center for Fluorescence Spectroscopy and Department of Biological Chemistry, University of Maryland School of Medicine, Baltimore, Maryland 21201

René Lapouyade • Photophysique et Photochimie Moléculaire, URA (CNRS) No. 348, Université de Bordeaux I, F-33405 Talence, France.

Dieter Oelkrug • Institute for Physical and Theoretical Chemistry, University of Tübingen, DW-7400 Tübingen, Germany

Alvydas J. Ozinskas • Becton Dickinson Diagnostic Instrument Systems, Sparks, Maryland 21152

Gabor Patonay • Department of Chemistry, Georgia State University, Atlanta, Georgia 30303

Govind Rao • Department of Chemical and Biochemical Engineering, University of Maryland Baltimore County, Baltimore, Maryland 21228; and Medical Biotechnology Center of the Maryland Biotechnology Institute, University of Maryland at Baltimore, Baltimore, Maryland 21201

W. Rettig • W. Nernst Institute for Physical and Theoretical Chemistry, Humboldt University at Berlin, D-10117 Berlin, Germany

Dana B. Shealy • Department of Chemistry, Georgia State University, Atlanta, Georgia 30303

Jeffrey Sipior • Medical Biotechnology Center of the Maryland Biotechnology Institute, University of Maryland at Baltimore, Baltimore, Maryland 21201

Henryk Szmacinski • Center for Fluorescence Spectroscopy and Department of Biological Chemistry, University of Maryland School of Medicine, and Medical Biotechnology Center of the Maryland Biotechnology Institute, University of Maryland at Baltimore, Baltimore, Maryland 21201

Richard B. Thompson • Department of Biological Chemistry, University of Maryland School of Medicine, Baltimore, Maryland 21201

Bernard Valeur • Laboratoire de Chimie Générale, Conservatoire National des Arts et Métiers 75141 Paris Cedex 03, France

Z. Y. Zhang • Department of Physics, Measurement and Instrumentation Centre, School of Electrical Engineering and Applied Physics, City University, London ECl V0HB, United Kingdom

Preface

Time-resolved fluorescence spectroscopy is widely used as a research tool in biochemistry and biophysics. These uses of fluorescence have resulted in extensive knowledge of the structure and dynamics of biological macromolecules. This information has been gained by studies of phenomena that affect the excited state, such as the local environment, quenching processes, and energy transfer.

Topics in Fluorescence Spectroscopy, Volume 4: Probe Design and Chemical Sensing reflects a new trend, which is the use of time-resolved fluorescence in analytical and clinical chemistry. These emerging applications of time-resolved fluorescence are the result of continued advances in laser detector and computer technology. For instance, photomultiplier tubes (PMT) were previously bulky devices. Miniature PMTs are now available, and the performance of simpler detectors is continually improving. There is also considerable effort to develop fluorophores that can be excited with the red/near-infrared (NIR) output of laser diodes. Using such probes, one can readily imagine small time-resolved fluorometers, even hand-held devices, being used for doctor's office or home health care.

Volume 4 is intended to summarize the principles required for these biomedical applications of time-resolved fluorescence spectroscopy. For this reason, many of the chapters describe the development of red/NIR probes and the mechanisms by which analytes interact with the probes and produce spectral changes. Other chapters describe the unique opportunities of red/NIR fluorescence and the types of instruments suitable for such measurements. Also included is a description of the principles of chemical sensing based on lifetimes, and an overview of the ever-important topic of immunoassays.

Additional volumes in this series will be published to reflect further advances in fluorescence spectroscopy and its many applications. I welcome your suggestions for future topics or volumes, offers to contribute chapters on specific topics, or comments on the present volumes.

Finally, I thank all the authors for their excellent contributions, and for their patience with the inevitable delays incurred in release of this volume.

Joseph R. Lakowicz

Baltimore, Maryland

Contents

5. Fluorescence Probes Based on Twisted Intramolecular Charge Transfer (TICT) States and Other Adiabatic Photoreactions

W. Rettig and René Lapouyade

6. Red and Near-Infrared Fluorometry
Richard B. Thompson

7. Near-Infrared Fluorescence Probes
Guillermo A. Casay, Dana B. Shealy, and Gabor Patonay

8. Fluorescence Spectroscopy in Turbid Media and Tissues

Dieter Oelkrug

9. Real-Time Chemical Sensing Employing Luminescence Techniques

J. Ricardo Alcala

1

Emerging Biomedical Applications of Time-Resolved Fluorescence Spectroscopy

Joseph R. Lakowicz

1.1. Introduction

Time-resolved fluorescence spectroscopy is presently regarded as a research tool in biochemistry, biophysics, and chemical physics. Advances in laser technology, the development of long-wavelength probes, and the use of lifetime-based methods, are resulting in the rapid migration of time-resolved fluorescence to the clinical chemistry lab, to the patient's bedside, to flow cytometers, and even to the doctor's office and home health care. Additionally, time-resolved imaging is now a reality in fluorescence microscopy, and will provide chemical imaging of a variety of intracellular analytes and/or cellular phenomena. In this introductory chapter we attempt to describe some of the opportunities available using chemical sensing based on fluorescence lifetimes. In fact, it was the rapid migration of time-resolved fluorescence to biomedical applications that resulted in the present volume on probe design and chemical sensing.

Time-resolved fluorescence spectroscopy has resulted in significant advances in our understanding of the structure and dynamics of biological macromolecules.[1–3] There can be no doubt that such experimentation has contributed immensely to our present understanding of biological macromolecules and their assemblies. At present, time-resolved measurements require relatively complex instrumentation, resulting in a number of monographs on this topic.[4–6]

In addition to these research applications of fluorescence, there is a continuing use of fluorescence detection to replace analytical methods based on radioactivity, as can be judged from the recent books and conferences on fluorescence sensing methods.[7–11] These emerging applications of fluorescence can be seen by the growth and introduction of improved methods for immunoassays, enzyme-linked immunoassays

Joseph R. Lakowicz • Center for Fluorescence Spectroscopy and Department of Biological Chemistry, University of Maryland School of Medicine, Baltimore, Maryland 21201.

Topics in Fluorescence Spectroscopy, Volume 4: Probe Design and Chemical Sensing, edited by Joseph R. Lakowicz. Plenum Press, New York, 1994.

(ELISA), protein and DNA staining, and protein and DNA sequencing. A major driving force in this evolution is the introduction of long-wavelength probes which allow excitation with simple and robust light sources such as laser diodes. Importantly, the use of red-NIR excitation improves detection limits because of the lower levels of autofluorescence observed with these long excitation wavelengths. Consequently, several laboratories are now directing their efforts to develop long-wavelength probes which can be excited with laser diodes from 635 to 820 nm. In our opinion, long-wavelength sensing probes, when combined with laser diode sensing and time-resolved methods, will result in a new generation of clinical assays and medical devices.

1.2. Schemes for Fluorescence Sensing

The various possible schemes for fluorescence sensing are summarized in Figure 1.1. At present, most fluorescence assays are based on the standard intensity-based methods, in which the intensity of the probe molecule changes in response to the analyte of interest. However, there has been the realization that lifetime-based methods possess intrinsic advantages for chemical sensing. (A more detailed description of

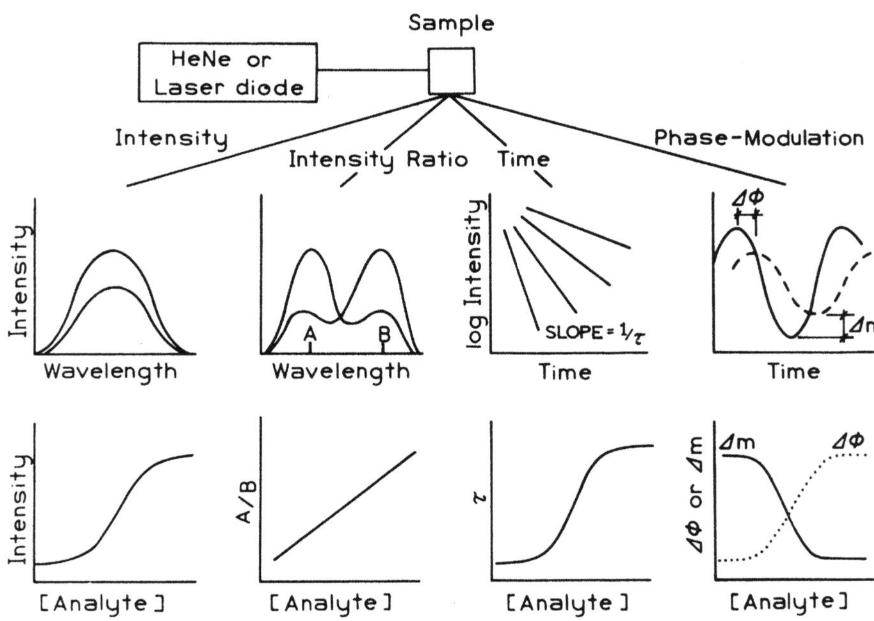

Figure 1.1. Schemes for fluorescence sensing: intensity, intensity ratio, time-domain, and phase-modulation, from *left* to *right*.

lifetime-based sensing is given in Chapter 10 of this volume.) If the intensity of a probe varies in response to an analyte, or if the amount of signal is proportional to the analyte, then it appears simple and straightforward to relate this intensity to the analyte concentration (*left*). Intensity-based methods are initially the easiest to implement because many probe fluorophores change intensity and/or quantum yield in response to analytes. Additionally, collisional quenching processes, such as quenching by oxygen, iodide, and chloride, etc., result in changes in intensity without significant shifts to the emission spectrum. While intensity measurements are simple and accurate in the laboratory, these are often inadequate in real-world situations. This is because the sample may be turbid, the optical surfaces may be misaligned or imperfect, and the probe concentration may vary from sample to sample, as summarized in Table 1.1. In the case of fluorescence microscopy, it is often impossible to know the probe concentration at each point in the image because the intensity changes continually due to photobleaching, phototransformation, and/or diffusive processes.

In principle, the problems of intensity-based sensing can be avoided using wavelength-ratiometric probes, i.e., fluorophores that display spectral changes in the absorption or emission spectrum on binding or interaction with the analytes (Figure 1.1). In this case, the analyte concentration can be determined independently of the probe concentration by the ratio of intensities at two excitation or two emission wavelengths.

Table 1.1. Characteristics of Intensity, Intensity-Ratio, and Time-Resolved Sensing Methods

Methods	Comments
Intensity-based sensing	Many probes are available
	Dependent on probe concentration
	Sensitive to absorption, scatter, and sample autofluorescence
	Sensitive to light lasers, photobleaching, etc.
	Difficult to measure small changes in intensity
Wavelength-ratiometric sensing	Limited number of probes and analytes
	Requires UV excitation with most present probes, except pH probes
Lifetime-based sensing	Independent of probe concentration
Time-domain and phase-modulation	
	Insensitive to simple absorption and scatter
	Insensitive to light losses and optical imperfections
Phase-modulated sensing	Simple and robust instrumentation
	Easy to measure small changes
	High accuracy
	Probe development simpler than wavelength ratiometric probes

Wavelength-ratiometric probes provide a straightforward means of avoiding the difficulties of intensity-based sensing. However, few such probes are available, and it is clear that they are difficult to create.[12] For instance, in spite of the enormous interest in measurements of intracellular Ca^{2+} concentration, there appears to be no practical wavelength-ratiometric indicator for Ca^{2+} that allows visible wavelength excitation. A recently synthesized visible wavelength ratiometric Ca^{2+} probe[13] remains to be tested in a fluorescence microscope. The two most widely used probes, Fura-2 and Indo-1, both require ultraviolet (UV) excitation with the associated problems of complex UV laser sources, and the high amounts of autofluorescence that are excited at these wavelengths. Attempts to make long-wavelength Ca^{2+} probes have resulted in probes which may change intensity, but do not display spectral shifts in either the excitation or emission spectra, such as Rhod-2, Fluo-3 and Calcium Green.[14] Wavelength-ratiometric probes for pH have recently become available.[14–15]

The difficulties of intensity-based measurements and of the scarcity of probes may be circumvented by the use of time-resolved or lifetime-based sensing. Several years ago we decided that it would probably be easier to identify and/or synthesize probes that display changes in lifetime in response to analytes, rather than to design and synthesize probes that display spectral shifts. Our opinion was based on the knowledge that a wide variety of quenchers and/or molecular interactions result in changes in the lifetimes of fluorophores, while changes in spectral shape were the exception rather than the rule. This prediction proved to be correct, as we now know that probes such as the Calcium Green series,[16] and the analogous Mg^{2+} probes,[17] all display changes in lifetime in response to binding their specific cations. Additionally, the pH probes of the seminaphtofluorescein (SNAFL) and seminaphtophodafluor (SNARF) series also display changes in lifetime on pH-induced ionization.[18] Of course, collisional quenchers like Cl^-, O_2, etc. also cause changes in lifetimes, as summarized in Lakowicz and Szmacinski.[19]

It is important to notice that a change in lifetime is not a necessary result of a change in fluorescence intensity. For instance, the Ca^{2+} probe Fluo-3 displays a large increase in intensity on binding Ca^{2+}, but there is no change in lifetime. This is because the Ca-free form of the probe is effectively nonfluorescent, and its emission does not contribute to the lifetime measurement. In order to obtain a change in lifetime, the probe must display detectable emission from both the free and cation-bound forms. Then the lifetime reflects the fraction of the probe complexed with cations. Of course, this consideration does not apply to collisional quenching, when the intensity decay of the entire ensemble of fluorophores is decreased by diffusive encounters with the quencher.

1.2.1. Instrument Complexity, Measurement Scheme, and the Spectral Properties of Fluorophores

It is well known that the lifetime of fluorophores is typically in the range of 1–10 nsec, and that it is easily possible to spend $50,000–$500,000 for lifetime instrumentation. How then can one rationally propose such measurements at the patient's bedside? Such instruments and measurements are possible if we reverse the usual paradigm of designing the instrument to suit the spectral properties of the probe molecules. When using this approach, one may be forced to use complex and expensive laser sources, as well as sophisticated schemes for generation of pulsed or amplitude-modulated light. Simple and robust instrumentation can readily be designed if we first decide on the laser source, the probe molecule, and the measurement scheme. For instance, laser diodes provide an ideal source of light from 635 to 800 nm. Importantly, the output of the laser diodes can be amplitude-modulated at any desired frequency up to several GHz,[20] and these devices have been used in phase-modulation fluorometry.[21–23] A bedside or even a hand-held lifetime instrument can be readily design if we synthesize and develop functional and specific probes which can be excited with laser diode sources. At present, there are many dyes in this range of wavelength, but only a handful which can be covalently attached to macromolecules, and to the best of our knowledge, none which are specifically sensitive to Ca^{2+}, Mg^{2+}, or other analytes.

Prior to describing the possible applications of laser-diode fluorometry, it is important to understand the two methods now used to measure fluorescence lifetimes; these being the time-domain (TD)[4, 5, 24] and frequency-domain (FD) or phase-modulation methods.[25] In TD fluorometry, the sample is excited by a pulse of light followed by measurement of the time-dependent intensity. In FD fluorometry, the sample is excited with amplitude-modulated light. The lifetime can be found from the phase angle delay and demodulation of the emission relative to the modulated incident light. We do not wish to fuel the debate of TD versus FD methods, but it is clear that phase and modulation measurements can be performed with simple and low cost instrumentation, and can provide excellent accuracy with short data acquisition times.

1.2.2. Lifetime-Based Sensing

Why can we expect lifetime-based sensing to be superior to intensity-based sensing? We feel this is the case because real-world sensing applications occur in environments that are not equivalent to optically clear and clean cuvettes. Instead, there are numerous factors that can affect the intensity values, such as imperfections or misalignment of surfaces and light losses in optical fibers. Additionally, many desired applications, such as homogeneous immunoassays or trans-dermal sensing measurements, require quantitative measurements in highly turbid or absorbing media (Figure 1.2, *top*). Such factors preclude quantitative measurements of intensities, or even intensity ratios.

Lifetime-based sensing can be mostly insensitive to these real-world effects. This is because these factors are not expected to alter the rate at which the intensity decays (Figure 1.2, *middle*). In our opinion, phase-modulation sensing provides additional advantages (Figure 1.2, *bottom*). The instruments take advantage of radio-frequency

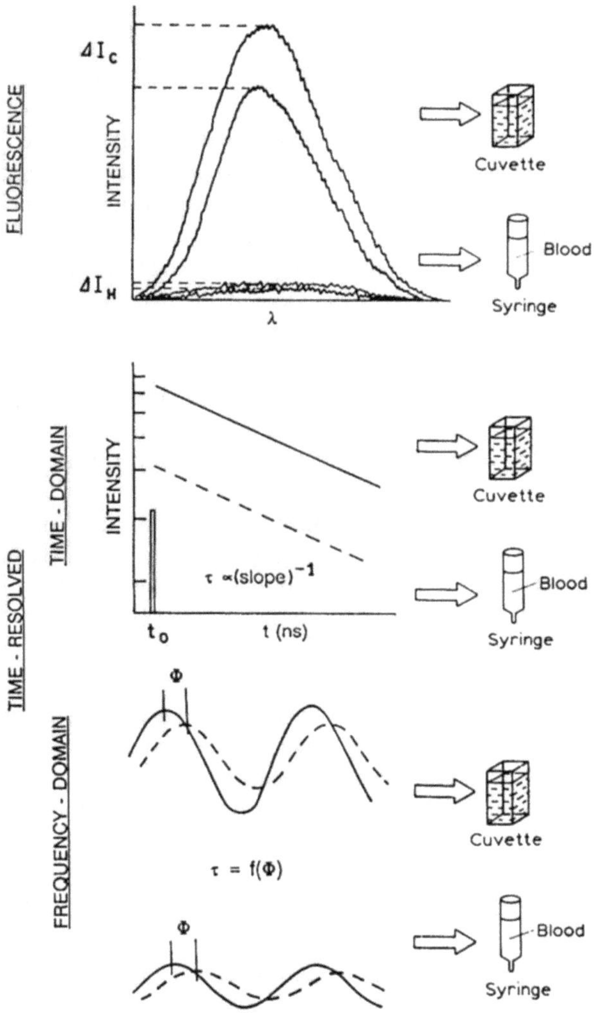

Figure 1.2. Intensity, time-domain, and frequency-domain sensing, as applied in the laboratory, a cuvette, and blood sample in a clinical setting.

methods to reject noise and filter signals, resulting in reliable data even in electrically noisy environments. Standard phase-modulation instruments provide 50 psec resolution with just seconds of data acquisition, so that small changes in lifetime can be easily measured. The merits and disadvantages of various sensing schemes are summarized in Table 1.1.

1.3. Applications of Fluorescence to Clinical Sensing

1.3.1. Phase-Modulation Sensing of Blood Gases and/or Blood Septicemia

Optical detection of blood gases [pH, partial pressure of carbon dioxide (pCO_2), and partial pressure of oxygen (pO_2)] is the Holy Grail of optical sensing. This is because current methods do not fully satisfy the needs of the intensive care patient. In these unfortunate cases the blood gases change on the timescale of minutes in response to the patient's physiological status. Measuring a blood gas requires taking a sample of arterial blood, placing it on ice, transporting it to a central laboratory, and measuring the pH using an electrode, and O_2 and CO_2 by the Clark and Severinghous electrodes, respectively (Table 1.2). Even for a stat request, it is difficult to obtain the blood gas report in less than 30 minutes, by which time the patient's status is often quite different. Additionally, handling of blood by health-care workers is undesirable with regard to the risk of acquired immunodeficiency syndrome (AIDS) and other infectious diseases. At present, determination of blood gases is time-consuming and expensive, with a cost of at least $400,000,000 per year in the United States.

How can phase-modulation fluorometry contribute to this health-care need? It now seems possible to construct a lifetime-based blood gas catheter (Figure 1.3), or alternatively, an apparatus to read the blood gas in the freshly drawn blood at the patient's bedside. To be specific, fluorophores are presently known to accomplish the task using a 543-nm Green Helium–Neon laser,[18,19] and it seems likely that the chemistries will be identified for a laser diode source. The use of longer wavelengths should minimize the problems of light absorption and autofluorescence of the samples, and the use of phase or modulation sensing should provide the robustness needed in a clinical environment. For the more technically oriented researcher, we note that the

Table 1.2. Blood Gases (pO_2, pCO_2, and pH) of Critically Ill Patients

Methods	Comments
Present status	Requires arterial blood
	Performed in a central clinical lab
	Time delays, changing patient status
	Complex instrumentation
	Requires handling of blood (AIDS)
	Estimated U.S. market $400,000,000 per year
Optical measurements of blood gases	Available at bedside in the CCU
	Continuous readout of pO_2, pCO_2, and pH
	Rapid response and improved patient care
	No handling of blood by health care workers

Abbreviations: CCU, critical care unit.

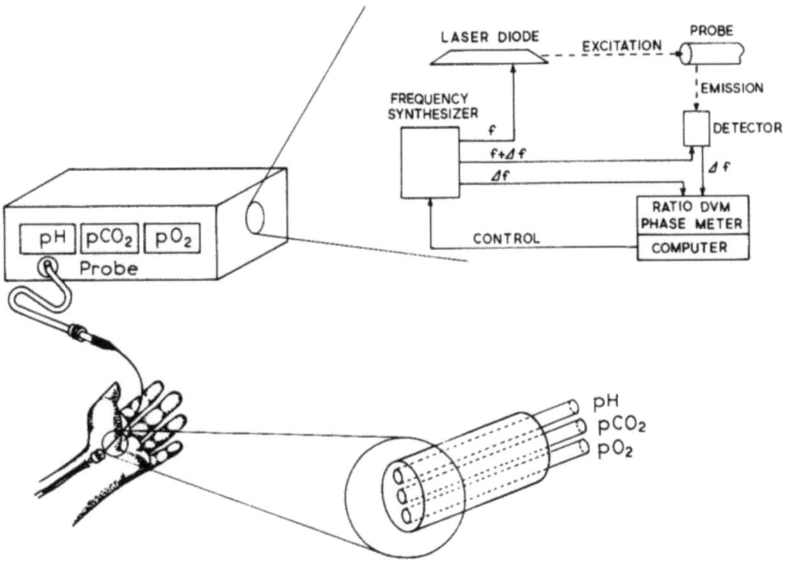

Figure 1.3. Phase-modulation blood gas instrumentation based on a laser diode light source. Blood gas catheter (*top*); direct reading blood gas syringe (*bottom*).

use of both phase angle and modulation measurements, which are simultaneously available, can provide error checking in critical applications.

One can also imagine a blood septicemia assay based on phase-modulation fluorometry (Figure 1.4). It is known that for certain long-lived fluorophores it is possible to use a simple electroluminescent device as the amplitude-modulated light source.[26] In this case, the probe chemistry is available for sensing of O_2, but is not yet completely developed for pH and pCO_2. Nonetheless, construction of the apparatus shown in Figure 1.4 seems straightforward and we have identified preliminary probes for this purpose. Importantly, such a blood septicemia assay (Figure 1.4) would allow for the simultaneous measurement of pH and pO_2, as well as pCO_2, and should be insensitive to optical alignment of the sample vials. Additionally, one can imagine the addition of other affinity-based assays for glucose[27] or antigens,[28–29] should they be clinically informative in such an assay.

1.3.2. Noninvasive Transdermal Glucose Sensing

Noninvasive glucose measurements can potentially be performed with phase-modulation fluorometry. The blood gas application described above requires drawing the blood, i.e., an invasive as well as an unpleasant procedure. Similarly, present measurements of blood glucose also require fresh blood. Insulin-dependent diabetics

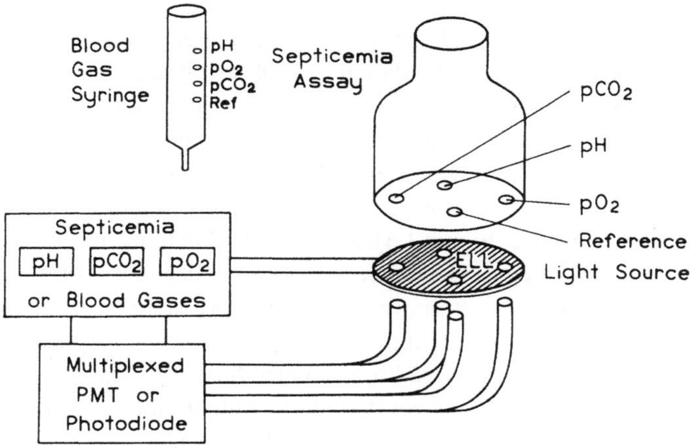

Figure 1.4. Blood gas septicemia apparatus based on an electroluminescent light source PMT = photomultiplier tube.

often require blood glucose to be measured five times per day. The unpleasantness and pain of this procedure results in the minimum number of blood glucose measurements by diabetics. However, it is known that erratic blood glucose control is responsible for the adverse long-term health effects of blindness and heart disease, apparently due to the irreversible glycosylation and modification of blood proteins and blood vessels. Continuous noninvasive monitoring of glucose can provide the input needed for continuous insulin injection, i.e., the "insulin pump," or improved information to the diabetics of the effects of food intake in their glucose levels (Table 1.3).

Table 1.3. Blood Glucose Determination in Diabetics

Methods	Comments
Present status	Invasive, i.e., requires withdrawing of blood
	Infrequent measurements result in blood glucose variations
	Long-term health consequences result from inadequate control of blood sugar
Noninvasive optical measurement of blood glucose	Does not require withdrawing of blood
	Can provide continuous monitoring
	Improved control of blood glucose
	Fewer long-term health effects of poorly regulated blood glucose
	Can be based on laser diode lifetime measurements

Noninvasive monitoring of glucose now appears possible based on our current understanding of lifetime-based sensing and the optical properties of tissues. Recall as a child when you placed a flashlight (a white light source) behind your hand and noticed the red transmitted light. We should have all recognized that this observation enables noninvasive sensing using long-wavelength light sources and time-resolved detection. In modern terms, the red wavelength of laser diodes is only weakly absorbed by skin, but of course are highly scattered. Consider the implantation of a glucose sensing patch below the skin, in which the decay time of the laser diode-excitable probe is sensitive to glucose (Figure 1.5). Because the skin transmits the red light, the sensor will be excited. Because the decay times are not dependent on total intensity, they can be measured in this scattering medium, most probably by the phase-modulation method. The times required for light migration in tissues are typically on the 200 psec timescale,[30] and thus can be readily accounted for when measuring ns lifetimes. Additionally, tissue glucose levels are thought to follow blood glucose to within a 30-minute delay,[31] so that the patch can possibly be under the skin and need not penetrate the venous system. Also, there are probably better locations for this glucose patch than on the forearm (Figure 1.5).

What mechanisms can be used to create a lifetime-based glucose sensor? In our opinion, the mechanism should be fluorescence resonance energy transfer (FRET). The phenomenon of FRET results in transfer of the excitation from a donor fluorophore to an acceptor chromophore, which need not itself be fluorescent. FRET is a through-space interactor which occurs over distances of 20–60 Å.

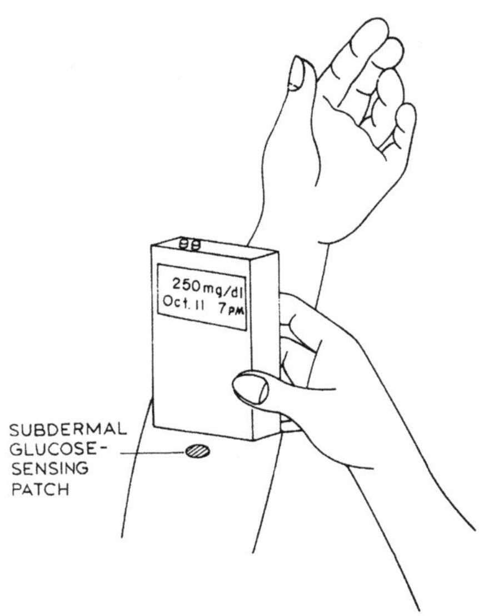

Figure 1.5. Noninvasive sensing of glucose. Within the optical window (650 nm–1300 nm), lifetime-based sensing by fluoresence is possible through the human skin.

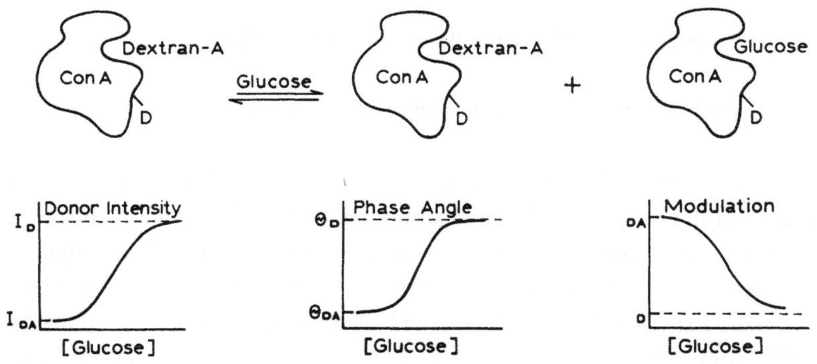

Figure 1.6. Glucose sensing by fluorescence resonance energy transfer (FRET).

The characteristic distances for FRET can be reliably calculated from the spectral properties of the donor (D) and acceptor (A). Importantly, FRET can be reliably predicted to occur for any D-A pair, so that the system can be wavelength-adjusted to match the wavelengths of laser diode sources. The extent of FRET depends on the proximity of the donor and acceptor.

Based on the above considerations, a glucose lifetime sensor can be based on a protein that reversibly binds glucose (Figure 1.6), such as Concanavalin A (Con A). The acceptor (A) should be attached on a polymeric media like dextran which will bind the Con A, but not diffuse out of the semipermeable patch. Glucose will competitively display the Con A from the dextran acceptor, resulting in a monotonic increase in donor lifetime with increasing glucose concentration. It should be noted that this sensor would not require any external connections, would not consume glucose, and can potentially be replenished if needed by injection rather than removal. Implantable devices have now been accepted as a means of birth control. Hence, it seems that individuals with diabetes are likely to accept such an implant if it results in improved or more convenient control of his or her blood glucose.

The glucose and blood gas sensing applications should not be regarded as a "Star Wars" approach, which will only increase the cost of health care without significant benefit. In these two cases, the costs of the new technology will probably be less than existing methods. More importantly, improved monitoring of blood gas is likely to decrease the time spent in the intensive care unit (ICU), and control of blood glucose should reduce the long-term consequence of diabetes. In both cases, the improved care should decrease the total cost of health care and maintenance. Also, the technology for these applications is available today, and require only that these concepts be developed into the actual applications.

1.4. Applications to Cell Biology and Physiology

1.4.1. Intracellular Chemical Analysis and Flow Cytometry

Flow cytometry and/or fluorescence activated cell sorting (FACS) is presently widely used in the diagnosis of cancer and other diseases.[32, 33] Most applications of flow cytometry are based on the presence or absence of cell-surface antigens, or the presence of one or two copies of the DNA, as determined by measurement of the fluorescence intensity of cells labeled with fluorescent antibodies or nucleic acid stains. The immunological or cell-division emphasis of flow cytometry may be a consequence of the difficulty in measuring the precise intensity values during the passage of the cell through the laser beam in 10 μsec or less (Figure 1.7). Also, there is considerable cell-to-cell variations in the extent of staining or uptake of probe molecules.

The difficulties of intensity-based flow cytometry are illustrated by the present difficulties of cell-by-cell measurements of intracellular calcium. This can be accomplished using the calcium probe Indo-1,[34–38] but requires an ultraviolet (UV) laser source which is not routinely available in flow cytometry (Indo-1 is an emission wavelength ratiometric probe). Flow cytometers routinely have argon ion laser sources with outputs of 488 or 514 nm. Measurement of intracellular ions other than Ca^{2+} is nearly impossible. (The SNAFL and SNARF probes should allow pH measurement from the wavelength-ratiometric data.[15])

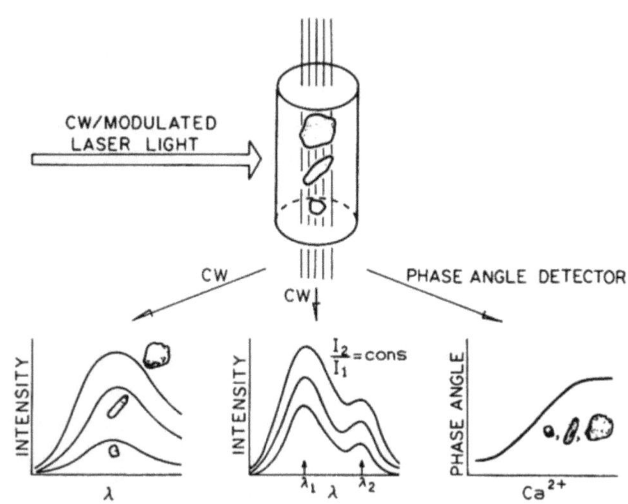

Figure 1.7. Cell-by-cell lifetime measurements in flow cytometry. CW = Continuous detection.

The advantages of lifetime-based sensing can be particularly attractive to flow cytometry, when the size, shape, and degree of labeling can vary between these cells. For instance, the probe Calcium Green displays a lifetime change from 1 to 4 nsec on binding Ca^{2+}, and Calcium Green can be excited with an argon ion laser.[16] Consequently, intracellular Ca^{2+} measurements could be readily accomplished, if cell-by-cell lifetime measurements were possible in flow cytometry. At first, this task seemed nearly impossible in that lifetime measurements almost always require continuous data acquisition from minutes to hours, and even lifetime measurements in one second would not be adequate for the 10 μsec flow cytometry signal from each cell.

The problem has now been solved, and it is possible to measure the phase angle of the probe as the cells pass through the laser beam.[39, 40] While these measurements have not yet been applied to Ca^{2+}, the method has been validated with standard beads and stained cells. In our opinion, this new flow cytometry parameter, and our increasing knowledge of lifetimes of probes, will result in the increased use of flow cytometry for studies of intracellular physiology, in addition to the current emphasis in immunology, cell activation, and ploidy.

1.4.2. Fluorescence Lifetime Imaging Microscopy (FLIM)

Fluorescence microscopy is routinely used to study the location and movement of intracellular species. In general, the fluorescence image reflects the location and concentration of the probe, or that amount of probe remaining in a photobleached sample (Figure 1.8, *lower left*). Consequently, quantitative fluorescence microscopy

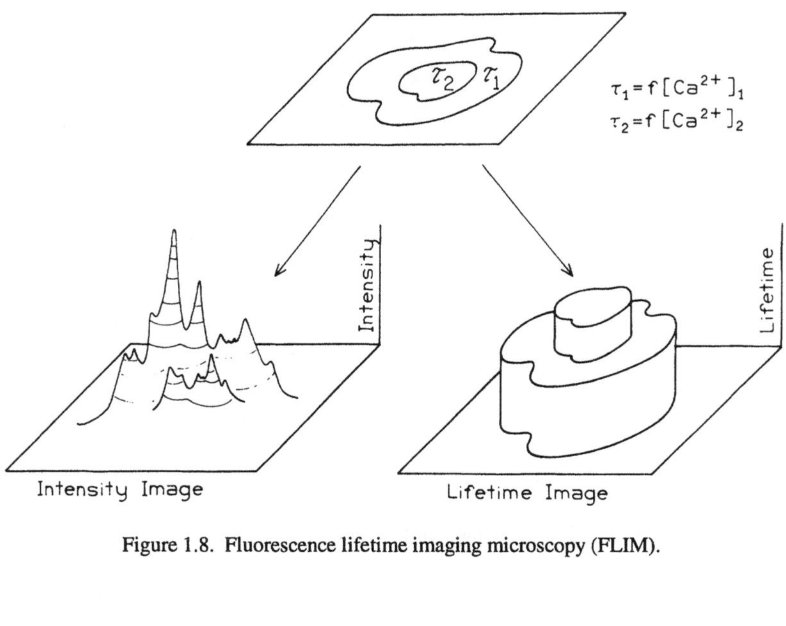

Figure 1.8. Fluorescence lifetime imaging microscopy (FLIM).

is very difficult, except for those cases where wavelength-ratiometric probes are available.

Consider now that the lifetime of the probe is different in the two regions of the cell (Figure 1.8, *top*). If one could create a contrast based on the lifetime at each point in the image, one would resolve two regions of the cell, each with an analyte (Ca^{2+}) concentration which was revealed by the lifetime image. While this type of imaging may seem exotic, in fact, a type of "lifetime imaging" is now routinely used in medical imaging. In magnetic resonance imaging (MRI) the contrast, or black-gray-white scale, is based on the proton relaxation times, which are analogous to a fluorescence lifetime. MRI images are not routinely based on the total signal, which is analogous to the local intensity in the fluorescence microscopic images. MRI provides useful medical images because the contrast reflects the different chemical and physical properties of the organ. In the same way, the contrast in FLIM can provide chemical images of cells based on the local lifetime, which can be affected by cations, anions, pH, O_2, temperature, viscosity, or polarity. In this sense, FLIM is the microscopic analogue of MRI.

The creation of such fluorescence lifetime images, in which the contrast is based on lifetimes, appeared to be a daunting challenge. Imagine performing 2.62×10^5 lifetime measurements for a typical 512×512 image. Given the difficulties of measuring even a single lifetime in a cuvette, such a task seems nearly impossible. However, image intensifiers and charged-coupled device camera technology now makes this possible.[41, 42] Figure 1.9 (*right*) shows the Ca^{2+} lifetime image of COS cells based on the probe Quin-2,[43] along with the intensity image (*left*). The intensity images show the expected spatial variations due to probe localization, and the Ca^{2+}

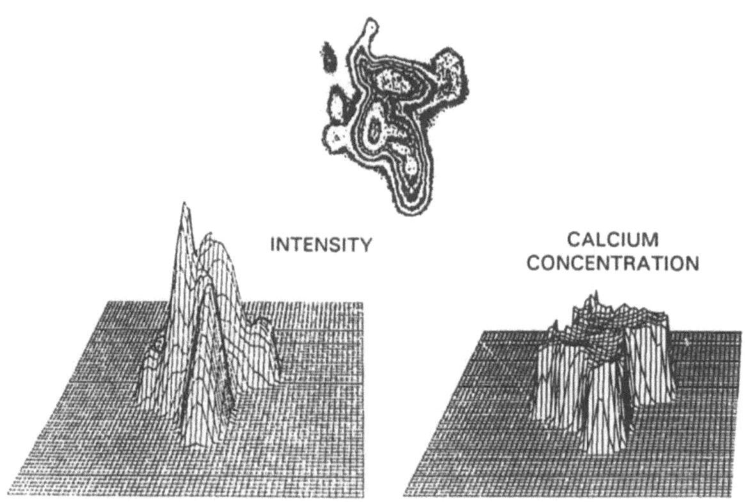

Figure 1.9. Intensity (*left*) and calcium images (*right*) of Quin-2 fluorescence in COS cells.

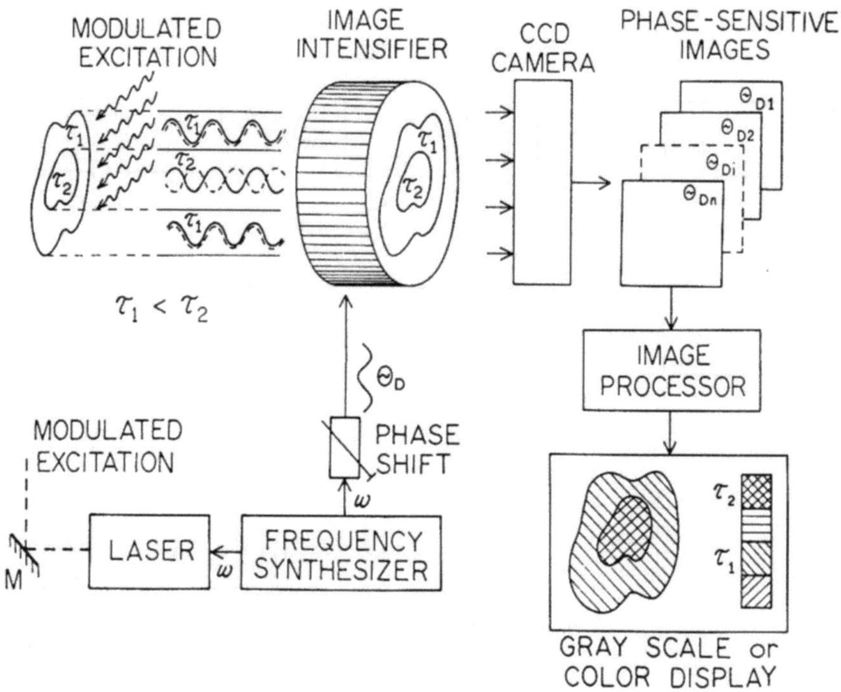

Figure 1.10. Apparatus for fluorescence lifetime imaging microscopy (FLIM).

(phase angle) image shows the expected uniform concentration of intracellular calcium. As predicted, the lifetime imaging provides chemical imaging, which within limits is insensitive to the local probe concentration.

The cellular FLIM images in Figure 1.9 were obtained using moderately complex instrumentation, which consists of a picosecond dye laser, a gain-modulated image intensifier, and a slow-scan scientific-grade CCD camera (Figure 1.10). However, the FLIM instruments in the future can be compact, even mostly a solid-state device. This possibility is shown in Figure 1.11, where we show that the light source can be a laser diode, assuming the probes are available. The image intensifier is a moderately simple device, but is delicate and requires high voltages. Reports have appeared on gatable CCD detectors.[44] Present gatable CCDs are too slow (50 nsec gating time). This time response is likely to improve, and probes can be developed with longer decay times. Then the FLIM apparatus will consist of only modest additions to a standard fluorescence microscope.

What type of chemical imaging will be possible using FLIM technology? Based on our current understanding of FLIM, and factors that affect fluorescence lifetimes,

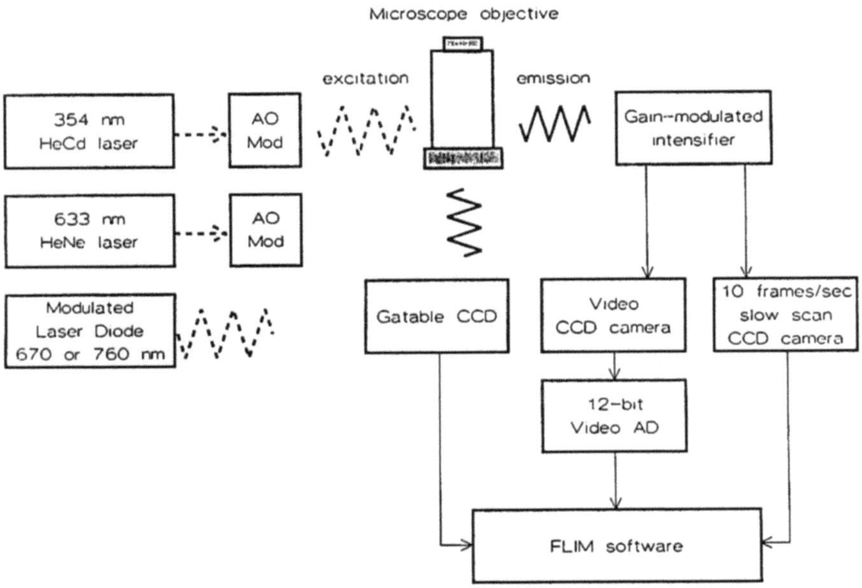

Figure 1.11. Future apparatus for FLIM.

we can predict that lifetime imaging will allow images of a variety of cellular properties. These include imaging of ions, cofactor binding, probe binding, chromosomes, microviscosity and proximity imaging of associating macromolecules (Table 1.4). We also believe that FLIM technology can play an important role in biomedical imaging, process control, and engineering research (Table 1.5). These applications are possible because the lifetime of luminescent paints can be sensitive to the pO_2, and temperature is known to affect the lifetime of many flurophores.

Table 1.4. Biomedical Applications of FLIM

Type of imaging	Analyte or property
Chemical imaging	Ca^{2+}, Mg^{2+}, Cl^-, pH, O_2, Na^+, and K^+
Ligand binding to proteins	NADH, TNS
Chromosome imaging	Acridine lifetimes depend on DNA base composition
Microviscosity imaging	Identify viscosity-lifetime probes
Proximity imaging by energy transfer	Protein–protein binding
	Protein–membrane association

Abbreviations: NADH, nicotinamide-adenine dinucleotide phosphate; TNS, 2-(*p*-toluidino)naphthalene-6-sulfonic acid.

Table 1.5. Additional Applications of FLIM

Type of imaging	Application
Photon migration imaging	Optical tomography
	Clinical sensing
	Mammography
Imaging of air flow by O_2 partial pressure	Aerodynamics research
Temperature imaging	Semiconductor manufacturing
Remote sensing	Lifetime-based LIDAR

Abbreviations: LIDAR, light detection and ranging.

1.5. Conclusion: The Need for Development of New Fluorescence Probes

In my opinion, the application of fluorescence to analytical chemistry, clinical chemistry, flow cytometry, and imaging is limited not by the instrument technology, but by the available probes. There are only a limited number of conjugatable long wavelength probes, and none which display specific analyte sensitivity. What is needed is an arsenal of probes, all of which can be excited with laser diodes or light-emitting diodes (LEDs), and which are specifically sensitive to cations, anions, and other analytes. While several laboratories are working in this topic, the total effort is minor in comparison to the number of scientists engaged in instrument development, technology development, theory or applications. The development of this arsenal of probes is crucial for the practical application of fluorescence to real-world sensing applications. As these new probes are developed, one can predict a number of health care products (Table 1.6). During the next several years we can expect the rapid transfer of technologies from the research laboratory to a variety of health care applications.

Table 1.6. Product Predictions Based on Advanced Optical Devices

- Optical monitoring of bioreactors
- A bedside blood gas analyzer, Green HeNe, then laser diode
- Doctor's office and/or bedside clinical chemistry and diagnostics
- Noninvasive optical transdermal glucose sensor
- Noninvasive diagnostics based on laser diodes
- Fluorescence microscope analogous to MRI images
- Optical mammography
- Optical tomography and imaging

Acknowledgments

The concepts and results described in this chapter were developed over a number of years. I wish to thank the National Institutes of Health (RR-08119, RR-07510, GM 39617, GM 35154) and the National Science Foundation (MCB 8804931, BIR 9319032, DIR 8710401) for their continued support.

References

1. T. G. Dewey, ed., *Biophysical and Biochemical Aspects of Fluorescence Spectroscopy*, Plenum Press, New York (1991).
2. J. R. Lakowicz, ed., *Time-Resolved Laser Spectroscopy in Biochemistry III, SPIE (The Society of Photo-Optical Instrumentation Engineers) 1640,* Billingham, Washington (1992).
3. D. M. Jameson and G. D. Reinhart, eds., *Fluorescent Biomolecules: Methodologies and Applications*, Plenum Press, New York (1989).
4. D. V. O'Connor and D. Phillips, *Time-Correlated Single Photon Counting*, Academic Press, London (1984).
5. J. N. Demas, *Excited State Lifetime Measurements*, Academic Press, New York (1983).
6. J. R. Lakowicz, ed., *Topics in Fluorescence Spectroscopy*, Plenum Press, New York (1991).
7. O. S. Wolfbeis, ed., *Fluorescence Spectroscopy: New Methods and Applications*, Springer-Verlag, New York (1993).
8. S. G. Schulman, *Molecular Luminescence Spectroscopy Methods and Applications*: *Part 1*, John Wiley & Sons, New York (1985).
9. J. R. Lakowicz and R. B. Thompson, *Advances in Fluorescence Sensing Technology, SPIE 1885*, Billingham Washington (1993).
10. K. Van Dyke and R. Van Dyke, eds., *Luminescence Immunoassay and Molecular Applications*, CRC Press, Boca Raton, Florida (1990).
11. O. S. Wolfbeis, ed., Proc. of 1st Euro. Conf. on Optical Chemical Sensors and Biosensors, *Sensors and Actuators B 11*, 1–3 (1993).
12. G. Grynkiewicz, M. Poenie, and R. Y. Tsien, A new generation of Ca^{2+} indicators with greatly improved fluorescence properties, *J. Biol. Chem. 260*, 3440–3450 (1985).
13. E. U. Akkaya and J. R. Lakowicz, Styryl-based wavelength ratiometric probes: A new class of fluorescent calcium probes with long wavelength emission and a large stokes' shift, *Anal. Biochem. 213*, 285–289 (1993).
14. *Bioprobes 13*, 3–4 (1991).
15. J. E. Whitaker, R. P. Haugland, and F. G. Prendergast, Spectral and photophysical studies of benzo[c]xanthene dyes: Dual emission pH sensors, *Anal. Biochem. 194*, 330–344 (1991).
16. J. R. Lakowicz, H. Szmacinski, and M. L. Johnson, Calcium imaging using fluorescence lifetimes and long-wavelength probes, *J. Fluorescence* 2(1), 47–62 (1992).
17. H. Szmacinski and J. R. Lakowicz, Lifetime-based sensing of magnesium (submitted for publication).
18. H. Szmacinski and J. R. Lakowicz, Optical measurements of pH using fluorescence lifetimes and phase-modulation fluorometry, *Anal. Chem. 65*, 1668–1674 (1993).
19. J. R. Lakowicz and H. Szmacinski, Fluorescence lifetime-based sensing of pH, Ca^{2+}, K^+ and glucose, *Sensors and Actuators B* 11, 133–143 (1993).
20. I. P. Kaminow, *An Introduction to Electrooptic Devices*, Academic Press, New York, 1974.
21. R. B. Thompson, J. K. Frisoli, and J. R. Lakowicz, Phase fluorometry using a continuously modulated laser diode, *Anal. Chem.* 64, 2075–2078 (1992).

22. R. B. Thompson and J. R. Lakowicz, Fiber optic pH sensor based on phase fluorescence lifetimes, *Anal. Chem.* 65, 853–856 (1993).
23. K. W. Berndt, I. Gryczynski, and J. R. Lakowicz, Phase-modulation fluorometry using a frequency-doubled pulsed laser diode light source, *Rev. Sci. Instrum. 61*, 2331–2337 (1990).
24. D. J. S. Birch and R. E. Imhof, Time-domain fluorescence spectroscopy using time-correlated single-photon counting, in: *Topics in Fluorescence Spectroscopy* (J. R. Lakowicz, ed.), Vol. 1, pp. 1–95, Plenum Press, New York (1991).
25. J. R. Lakowicz and I. Gryczynski, Frequency-domain fluorescence spectroscopy, in: *Topics in Fluorescence Spectroscopy* (J. R. Lakowicz, ed.), Vol. 1, pp. 293–355, Plenum Press, New York (1991).
26. K. W. Berndt and J. R. Lakowicz, Electroluminscent lamp-based phase fluorometer and oxygen sensor, *Anal. Biochem. 201*, 319–325 (1992).
27. J. R. Lakowicz and B. P. Maliwal, Optical sensing of glucose using phase-modulation fluorometry, *Anal. Chim. Acta 271*, 155–164 (1993).
28. J. R. Lakowicz and B. P. Maliwal, Fluorescence lifetime energy transfer immunoassay quantified by phase-modulation fluorometry, *Sensors and Actuators B 12*, 65–70 (1993).
29. A. J. Ozinskas, H. Malak, J. Joshi, H. Szmacinski, J. Britz, R. B. Thompson, P. A. Koen, and J. R. Lakowicz, Homogeneous model immunoassay of thyroxine by phase-modulation fluorescence spectroscopy, *Anal. Biochem. 213*, 264–270 (1993).
30. B. Chance, J. Leigh, H. Miyake, D. Smith, S. Nioka, R. Greenfield, M. Finlander, K. Kaufmann, W. Levy, M. Young, P. Cohen, H. Yoshioka, and R. Boretsky, Comparison of time-resolved and -unresolved measurements of deoxyhemoglobin in brain, *Proc. Natl. Acad. Sci. 85*, 4971–4975 (1988).
31. G. Velho, P. Froguel, D. R. Thevenot, and G. Reach, *In vivo* calibration of a subcutaneous glucose sensor for determination of subcutaneous glucose kinetics, *Diabetes Nutr. Metab. Clin. Exp. 1*, 227–233 (1988).
32. A. L. Givan, *Flow Cytometry First Principles*, Wiley-Liss, New York (1992).
33. W. McL. Grogan and J. M. Collins, *Guide to Flow Cytometry*, Marcel Dekker, New York (1990).
34. L. K. Jennings, M. E. Dockter, C. D. Wall, C. F. Fox, and D. M. Kennedy, Calcium mobilization in human platelets using Indo-1 and flow cytometry, *Blood 74*(8), 2674–2680 (1989).
35. J. T. Ransom, D. L. DiGiusto, and J. Cambier, Flow cytometric analysis of intracellular calcium mobilization, *Methods Enzymol. 141*, 53–63 (1987).
36. G. L. Rossi, D. J. Young, S. I. Wasserman, and K. E. Barrett, Calcium mobilization in activated mast cells monitored by flow cytometric analysis. *Agents Actions 31*, 257–262 (1990).
37. R. B. Alexander, E. S. Bolton, S. Koenig, G. M. Jones, S. L. Topalian, C. H. June, and S. A. Rosenberg, Detection of antigen specific T lymphocytes by determination of intracellular calcium concentration using flow cytometry, *J. Immunol. Methods 148*, 131–141 (1992).
38. B. Goller and M. Kubbies, UV Lasers for flow cytometric analysis: HeCd versus argon laser excitation, *J. Histochem. and Cytochem. 40*(4), 451–456 (1992).
39. B. G. Pinsky, J. J. Ladasky, J. R. Lakowicz, K. Berndt, and R. A. Hoffman, Phase-resolved fluorescence lifetime measurements for flow cytometry, *Cytometry 14*, 123–135 (1993).
40. J. A. Steinkamp and H. A. Crissman, Resolution of fluorescence signals from cells labeled with fluorochromes having different lifetimes by phase-sensitive flow cytometry, *Cytometry 14*, 210–216 (1993).
41. J. R. Lakowicz, H. Szmacinski, K. Nowaczyk, K. Berndt, and M. L. Johnson, Fluorescence lifetime imaging, *Anal. Biochem. 202*, 316–330 (1992).
42. J. R. Lakowicz, H. Szmacinski, K. Nowaczyk, and M. L. Johnson, Fluorescence lifetime imaging of calcium using Quin-2, *Cell Calcium 13*, 131–147 (1992).
43. J. R. Lakowicz, H. Szmacinski, K. Nowaczyk, W. J. Lederer, and M. L. Johnson, Fluorescence lifetime imaging of intracellular calcium in COS cells using Quin-2, *Cell Calcium 15*, 1–21 (1994).
44. R. K. Riech, R. W. Mountain, W. H. McGonagle, C. M. Huang, J. C. Twichell, B. B. Kosicki, and E. D. Savoye, An integrated electronic shutter for back-illuminated charge-coupled devices, *Proc. IEEE 91*, 171–174 (1991).

2

Principles of Fluorescent Probe Design for Ion Recognition

Bernard Valeur

2.1. Introduction

Fluorescence probing of the structure and dynamics of matter or living systems at a molecular or supramolecular level has been the object of numerous investigations in various fields such as polymers, solid surfaces, surfactant solutions, biological membranes, vesicles, proteins, nucleic acids, living cells, fluoroimmunochemistry, clinical diagnosis, etc. In fact, owing to the sensitivity of fluorescent molecules to their microenvironment, information can be obtained on local physical and structural parameters(1) (polarity, fluidity, order parameters, molecular mobility, distances at a supramolecular level) as well as local chemical parameters(2, 3) (pH, ion concentration). Such a local information is seldom accessible by other techniques.

The increasing interest of researchers for fluorescent probes can be explained by the great improvement of the sensitivity and the spatial or temporal resolution of instruments, and by the development of a wide choice of commercially available probes for particular applications (Molecular Probes, Inc., United States; Lambda Fluoreszenztechnologie Ges.m.b.H., Austria). However, there is still a need for probes with improved specific response and minimum perturbation of the microenvironment, in particular in the field of ion recognition which is the object of this chapter.

Ion recognition is a subject of considerable interest because of its implications in many fields: chemistry, biology, medicine (clinical biochemistry), environment, etc. In particular, selective detection of metal cations involved in biological processes (e.g., sodium, potassium, calcium, magnesium), in clinical diagnosis (e.g., lithium, potassium, aluminum) or in pollution (e.g., lead, mercury, cadmium) has received much attention. Among the various methods available for detection of ions, and more

---

Bernard Valeur • Laboratoire de Chimie Générale, Conservatoire National des Art et Métiers, 75141 Paris Cedex 03, France.

Topics in Fluorescence Spectroscopy, Volume 4: Probe Design and Chemical Sensing, edited by Joseph R. Lakowicz. Plenum Press, New York, 1994.

generally organic and inorganic species, those based on fluorescent sensors[4] offer distinct advantages in terms of sensitivity, selectivity, response time, local observation (e.g., by fluorescence imaging spectroscopy). Moreover, remote sensing is possible by using optical fibers.[5, 6]

Recognition of ions requires special care in the design of fluorescent probes because attention should be paid to both recognition and signaling moieties. The former is responsible for selectivity and efficiency of binding, which is relevant to the field of supramolecular chemistry[7–14] and the latter converts the information into an optical signal which should be as selective as possible of the species to be probed. Therefore, selectivity must be viewed in terms of both selectivity of binding and selectivity of photophysical effects. Furthermore, it should be emphasized that the medium in which recognition takes place is of major importance: parameters such as the nature of the solvent (polarity, hydrogen-bonding ability, protic or aprotic character), pH, ionic strength, etc. play indeed a great role because they can affect not only the efficiency and selectivity of binding, but also the photophysical characteristics of the fluorophore (for instance, protonation may compete with cation binding). In many practical cases, and of course for biological samples, aqueous solutions are mostly considered and water soluble probes are desirable, but in some analytical applications (e.g., based on extraction) the probe can be in an organic phase.

The photophysical changes of a fluorescent probe on ion binding can involve various photoinduced processes: electron transfer, charge transfer (with or without concomitant internal rotation), energy transfer, excimer or exciplex formation or disappearance, etc. These changes on recognition should be of course as marked as possible. Probes undergoing shifts of emission and/or excitation spectra (or appearance or disappearance of bands) are preferable to those that undergo only changes in fluorescence intensity: indeed, after calibration, the ratio of the fluorescence intensities at two appropriate emission or excitation wavelengths provides a measure of the ion or molecule concentration which is independent of the probe concentration (provided that the ion is in excess) and is insensitive to intensity of incident light, scattering, inner-filter effects, and photobleaching.

It is not the aim of this chapter to describe the particular conditions of application of fluorescent probes for ion recognition; the reader is referred to the relevant reviews and papers. Rather, this chapter intends to give readers a comprehensive overview of the two major aspects involved in ion recognition by fluorescent probes: the structure of the ionophore, and the ion-induced photophysical changes (many papers on fluorescence sensing of ions pay little attention to the origin of photophysical changes). A better understanding of both these aspects should help the user and the designer of this kind of fluorescent probe. Proton sensors, i.e., pH probes, will not be discussed is this chapter not only because of space limitations, but also because they are generally not based on a recognition process.

2.2. Fluorescent Signaling Receptors of Cations

2.2.1. Fundamental Aspects

Colorimetric determination of cations based on changes in color on complexation by dye reagents started to be popular a long time ago, especially in the case of alkaline-earth metal ions which are efficiently chelated by agents of the ethylenediamine-tetraacetate (EDTA) type. Fluorimetric techniques being more sensitive than photometric ones, numerous fluorogenic chelating reagents were studied and applied to practical cases.[15] Among them, oxine (8-hydroxyquinoline) and many of its derivatives occupy an important place in analytical chemistry and are still the object of new applications but they are not very specific.[16] In contrast, fluorescent sensors of the EDTA type exhibits high selectivity for calcium with respect to the other ions present in living cells.[17, 18] Examples will be given below.

The discovery of crown ethers and cryptands in the late sixties opened new possibilities of cation recognition with improvement of selectivity, especially for alkali metal ions for which there is a lack of selective chelators. Then, the idea of coupling these ionophores to chromophores or fluorophores, leading to so-called chromoionophores and fluoroionophores, respectively, emerged some years later.[19] As only fluorescent probes are considered in this chapter, chromoionophores will not be described.

In the design of a fluoroionophore, much attention is to be paid to the characteristics of the ionophore moiety and to the expected changes in fluorescence characteristics of the fluorophore moiety on binding. The complexing ability of the ionophore will be considered first.

It should be first emphasized that the cation concentration ranges of interest are very different according to the field. For instance, the concentration of calcium ion in a living cell is in the micromolar range, whereas in blood plasma and urine it is in the millimolar range. The ionophore moiety of calcium probes to be used in cellular biology and in clinical diagnosis should thus be different. Therefore, the range of cation concentration to be measured is an important parameter. The dissociation constant of the complex under the practical conditions (solvent, pH, ionic strength, etc.) of detection of a given cation should match the expected range of cation concentration. Assuming formation of a 1:1 complex between an ion (I) and its receptor (R), the dissociation equilibrium

$$\mathrm{RI} \rightleftharpoons \mathrm{R} + \mathrm{I}$$

is characterized by the dissociation constant K_d defined as

$$K_d = \frac{[\mathrm{R}][\mathrm{I}]}{[\mathrm{RI}]}$$

Stability (or binding) constants K_s are often used instead of dissociation constants ($K_s = 1/K_d$). These equilibrium constants are concentration quotients as the corresponding activity coefficients are given the value 1. However, in many practical situations, other species like inorganic salts are often present in the solution; they may induce changes in activity coefficients and water-structure variations. Nevertheless, concentration quotients can still be used by considering that the standard state is not pure solvent but the solution of these other species.[20]

As for any indicator, the concentration range that is appropriate for optimal measurements is such that $0.1 < [RI]/[R] < 10$, or $0.1\ K_d < [I] < 10\ K_d$, i.e., two decades of concentration, or two units of $\log_{10}[I]$ around pK_d or pK_s.

The dissociation constant of a complex between a given ligand and a cation depends on many factors: nature of the cation, nature of the solvent, temperature, ionic strength, and pH in some cases. In ion recognition, complex selectivity (i.e., the preferred complexation of a certain cation when other cations are present) is of major importance. In this regard, the characteristics of the ionophore, i.e., the ligand topology and the number and nature of the complexing heteroatoms or groups, should match the characteristics of the cation, i.e., radius, charge, coordination number, intrinsic nature (e.g., hardness of metal cations, nature and structure of organic cations, etc.) according to the general principles of supramolecular chemistry.[7–14] The ionophore can be a chelator, an open-chain structure (podand), a macrocycle (coronand, e.g., crown ether), or a macrobicycle (cryptand). The relevant complexes are called chelates, podates, coronates, and cryptates, respectively. The stability of chelates can be extremely different according to the structure of the chelator. Regarding the other ones, the stability of complexes with alkali and alkaline earth metal ions increases in the following order: podates << coronates << cryptates. The high stability of the latter results from the three-dimensional encapsulation, and the complexation selectivity is also usually higher because of their small ability to be deformed. In the case of coronates and cryptates, the most stable complexes are formed with ions having an ionic diameter close to that of the ligand cavity. Another principle generally applicable in chemistry predicts that hard oxygen centers combine with hard alkali metal ions, and soft sulfur or nitrogen centers with soft transition metal ions. For aqueous solutions, an excellent review is devoted to the ligand design for selective complexation of metal ions.[21]

Medium effects play a great role on both stabilities and selectivities of complexation with cations. The main factors are (1) the difference between ligand coordination energy and solvation energy, i.e., the solvating power of the ligand as compared to that of the solvent; and (2) the difference of interaction with the ligand shell and the dielectric medium outside the first solvation shell. Generally, the stability of coronates and cryptates increases markedly from water to methanol and to less polar solvents.

The connection between the ionophore and the fluorophore is a very important aspect of probe design with in mind the search for the strongest perturbation of the photophysical properties of the fluorophore by the cation. The ionophore may be linked to the fluorophore via a spacer, but in many cases some atoms or groups participating in the complexation belong to the fluorophore. More than one ionophore and/or more than one fluorophore may be involved in the structure. Figure 2.1 illustrates some of the structures that have been designed.

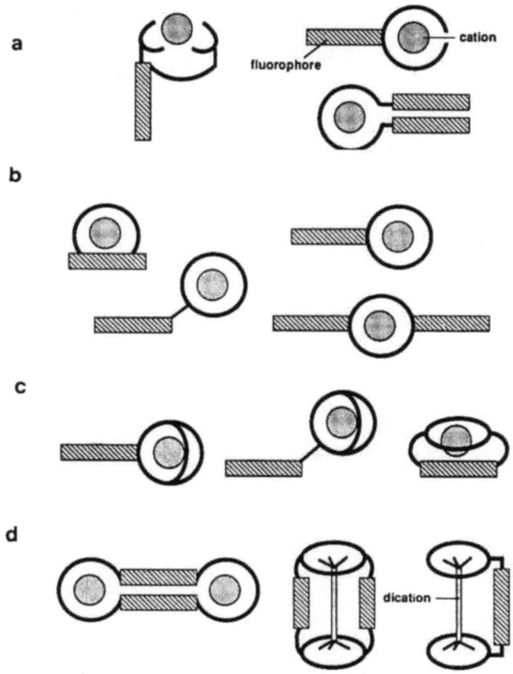

Figure 2.1. Schematic illustration of various structures of fluoroionophores. (a) Chelators and podands; (b) coronands; (c) cryptands; (d) other structures.

Fluorescent signaling receptors of cations are generally classed according to the nature of the cation to be probed, but in this chapter devoted to the principles of design, they will be classed according to the involved photophysical processes.

2.2.2. Recognition Based on Cation Control of Photoinduced Electron Transfer in Nonconjugated Donor–Acceptor Systems

This type of probe, often called fluorescent photoinduced electron transfer (PET) sensors, has been extensively studied (for reviews, see Refs. 22 and 23). Figure 2.2 illustrates how a cation can control the photoinduced charge transfer in a fluoroionophore in which the cation receptor is an electron donor (e.g., amino group) and the fluorophore (e.g., anthracene) plays the role of an acceptor. On excitation of the fluorophore, an electron of the highest occupied molecular orbital (HOMO) is promoted to the lowest unoccupied molecular orbital (LUMO), which enables photoinduced electron transfer from the HOMO of the donor (belonging to the free cation receptor) to that of the fluorophore, causing fluorescence quenching of the latter. On

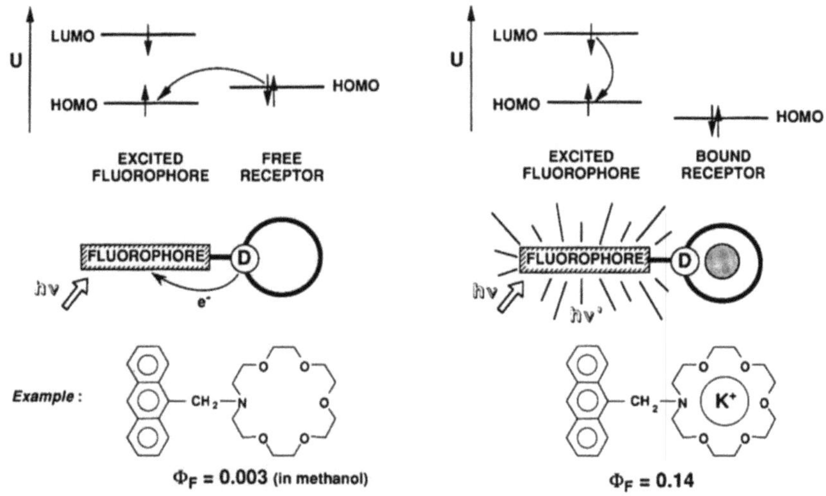

Figure 2.2. Principle of cation recognition based on cation control of photoinduced electron transfer in nonconjugated donor–acceptor systems. (Example from Ref. 26.)

cation binding, the redox potential of the donor is raised so that the relevant HOMO becomes lower in energy than that of the fluorophore; consequently, photoinduced electron transfer is not possible any more and fluorescence quenching is suppressed. In other words, fluorescence intensity is enhanced upon cation binding.

It has long been discovered that the fluorescence of aromatic hydrocarbons is quenched by aliphatic or aromatic amines because of photoinduced electron transfer from the latter to the former. Various systems have been designed in which anthracene is linked, via a spacer, to an ionophore containing a nitrogen atom. Figure 2.3 gives some examples: chelators **1**[24], **2a** and **2b**[25], **2c**, **2d** and **2e**[71], coronands **3**[26] and **4**[27], cryptands **5**[28–30] and **6**[31]. Anthracene fluorescence, which is efficiently quenched by the lone pair of the nitrogen atom, is recovered on binding of a cation which suppresses electron transfer. It can be anticipated in the other fluoroionophores, i.e., **2b**, **2c**, **2d**, **2e**, and the coumaro-cryptand **7**[32] the observed increase in fluorescence intensity on cation binding can be explained along the same line. Note that in **2c**, **2d**, and **2e**, the spacer is –CO–NH–.

Very large enhancements of fluorescence intensity can be observed. For instance, the fluorescence of a solution of **1** in acetonitrile is multiplied by a factor of 1000 on addition of $ZnCl_2$.[24]

It should be noted that the quantum yield is affected by cation binding but the absence of shift of the emission or excitation spectra precludes the possibility of intensity-ratio measurements at two wavelengths. An exception is provided by the anthraceno-cryptands **5** which in some cases may lead to disappearance or appearance of exciplexes characterized by an additional band at higher wavelengths.[28–30]

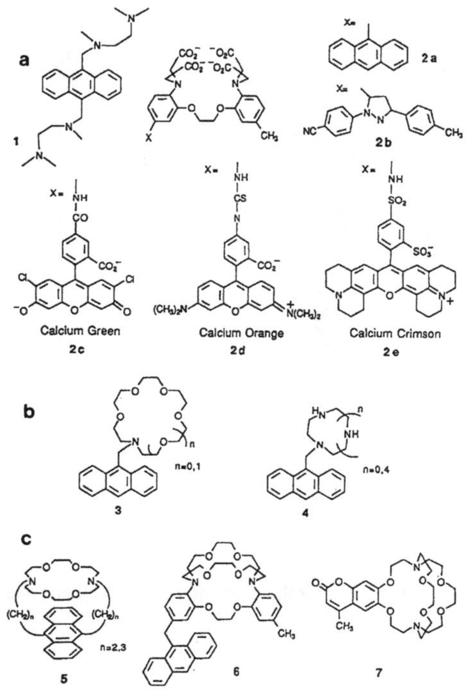

Figure 2.3. Examples of fluoroionophores based on cation control of photoinduced electron transfer. (a) Chelators; (b) coronands; (c) cryptands.

Attention should be paid to possible competitive protonation of the nitrogen atom of the ionophore, which suppresses electron transfer (even more efficiently than cations) and proton may thus have a larger effect on quantum yield than other cations. Aromatic nitrogen, as in **2** and **6**, have lower pK_a than aliphatic amines and thus can be used in neutral media.

Calcium Green (**2c**), Calcium Orange (**2d**), and Calcium Crimson (**2e**) have been developed by Molecular Probes, Inc.: they are suitable for fluorescence lifetime imaging of calcium.[71]

Ditopic receptors **8**[33,34] and **9**[35] (Figure 2.4) are good examples of special design for the recognition of α,ω-alkane diammonium ions. Compound **8** is of particular interest because it exhibits dual fluorescence in the absence of cation owing to excimer (excited anthracene dimer) formation. On addition of the dication $H_3N^+(CH_2)_7N^+H_3,2Cl^-$ to a solution of **8** in methanol the "monomer"-like emission is enhanced by a factor of 10 because fluorescence quenching is hindered, and the excimer band vanishes because the two anthracene rings are parted from each other on complexation. Other examples of cation control of excimer formation will be given in Section 2.2.4.

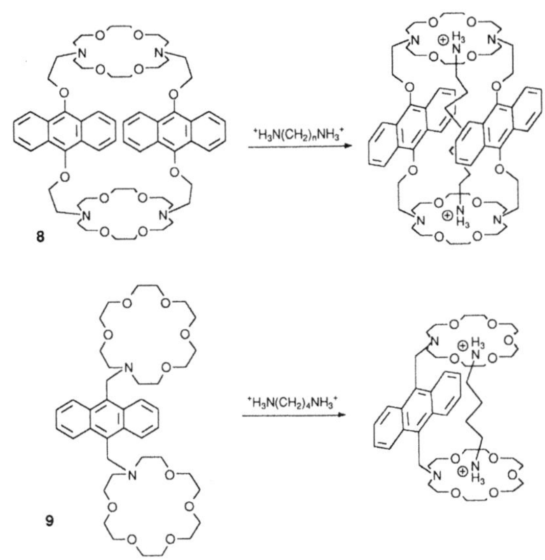

Figure 2.4. Ditopic receptors for the recognition of α,ω-alkane diammonium ions.

2.2.3. Recognition Based on Cation Control of Photoinduced Charge Transfer in Conjugated Donor–Acceptor Systems

Many fluorophores contain an electron-donating group (often an amino group) conjugated to an electron-withdrawing group so that they undergo intramolecular charge transfer from the donor to the acceptor on excitation by light. The consequent change in dipole moment results in a Stokes shift that depends of the microenvironment of the fluorophore; polarity probes have been designed on this basis (For a discussion on polarity probes, see Ref. 1). It can thus be anticipated that cations in close interaction with the donor or the acceptor moiety will change the photophysical properties of the fluorophore because the complexed cation affects the efficiency of intramolecular charge transfer.[36]

When a group (like an amino group) playing the role of an electron donor within the fluorophore interacts with a cation, the latter reduces the electron-donating character of this group; owing to the resulting reduction of conjugation, a blue shift of the absorption spectrum is expected together with a decrease of the extinction coefficient. Conversely, a cation interacting with the acceptor group enhances the electron-withdrawing character of this group; the absorption spectrum is thus red-shifted and the extinction coefficient is increased. The fluorescence spectra are in principle shifted in the same direction as those of the absorption spectra. In addition to these shifts, changes in quantum yields and lifetimes are often observed. All these photophysical effects are

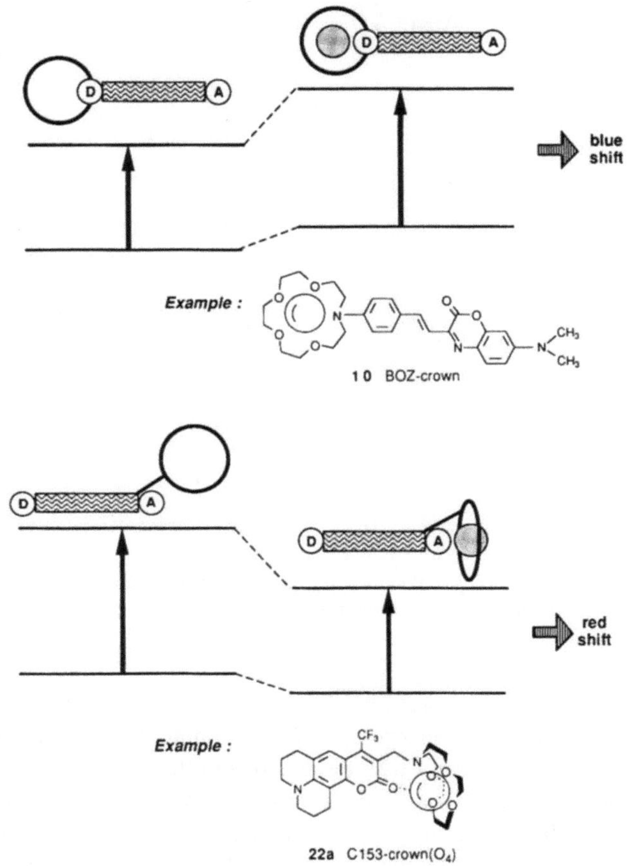

Figure 2.5. Spectral displacement of fluoroionophores based on cation control of photoinduced charge transfer in conjugated donor–acceptor systems.

obviously dependent on the charge and the size of the cation, and selectivity of these effects is expected.

The photophysical changes on cation binding can also be described in terms of charge dipole interaction[29] keeping in mind that the dipole moment in the excited state is larger than that in the ground state. When the cation interacts with the donor group, the excited state is more strongly destabilized by the cation than the ground state, and a blue shift of the absorption and emission spectra is expected. Conversely, when the cation interacts with the acceptor group, the excited state is more stabilized by the cation than the ground state, and this leads to a red shift of the absorption and emission spectra (Figure 2.5).

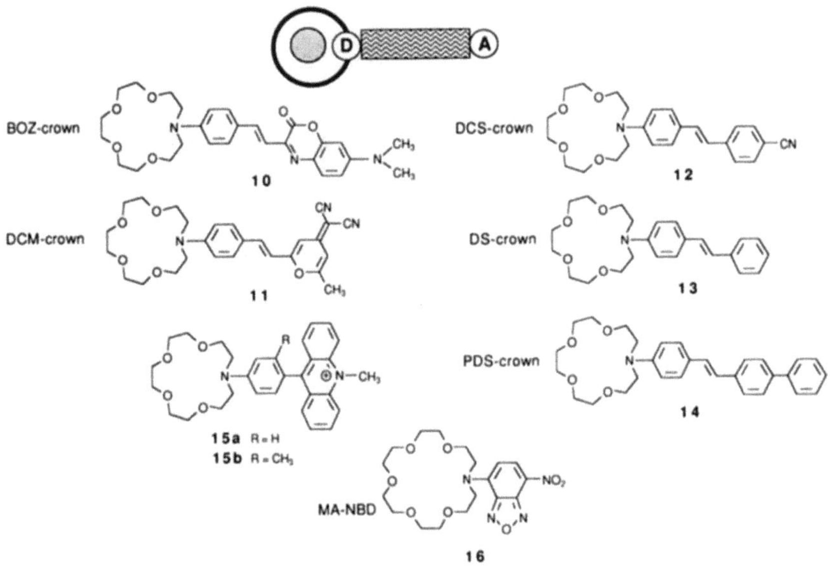

Figure 2.6. Examples of donor–acceptor fluoroionophores in which the electron-donating character of the donor (nitrogen atom of the crown) is cation-controlled.

In Figure 2.6, fluoroionophores **10**,[37–39] **11**,[39, 40], **12–14**,[41] **15**[42, 43] contain the same macrocycle (monoaza-15-crown-5) as the ionophore; the nitrogen of the crown plays the role of an electron donor with respect to both fluorophore and cation. These compounds undergo a more or less drastic blue shift of their absorption spectrum upon cation binding, the alkaline-earth metal ions inducing much larger effects than alkali metal ions, as expected for the double charge of the former. The fact that the blue shift of the fluorescence spectrum is very small in some cases will be discussed below. The largest effects observed for calcium ion in compounds **10–15** are explained by the fact that it fits well into the cavity of monoaza-15-crown-5.

The above-described photophysical changes in terms of cation control of photoinduced intramolecular charge transfer have been oversimplified for the sake of clarity. The phenomena are often more complex because excited-state charge transfer is often accompanied with internal rotation leading to twisted internal charge transfer (TICT) states that can be fluorescent or nonfluorescent.[44] Moreover photoisomerization can occur by rotation around an ethylenic double bond. For donor–acceptor stilbenes (**12–14**), the proposed kinetic scheme[41] contains three states: the planar state E* reached on excitation can lead to state P*(nonfluorescent) by double-bond twist, and to TICT state A* by single-bond twist, the latter being responsible for the main part of emission. The existence of a fluorescent TICT state is also likely to be responsible for the photophysical properties of **10**[45], but there is no photoisomerization in this case. The absence of fluorescence of **15a** and **15b** is due to the formation of a

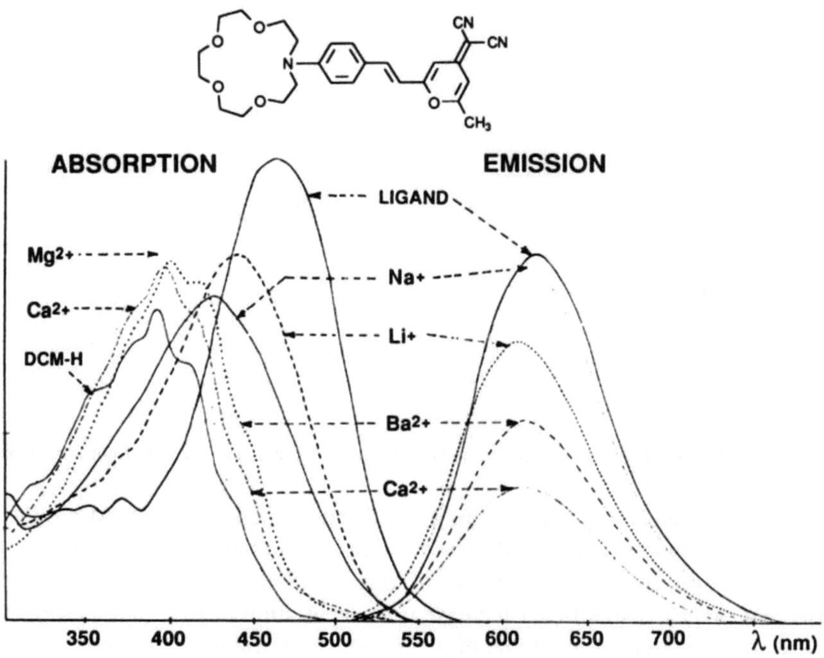

Figure 2.7. Absorption and fluorescence spectra of DCM-crown and its complexes with perchlorate salts in acetonitrile. DCM-H is the compound in which the crown is replaced by a hydrogen atom: this compound cannot undergo intramolecular charge transfer and its fluorescence quantum yield is very low (3×10^{-4}) (adapted from Ref. 40.)

nonfluorescent TICT state, and an acridinium-type fluorescence is recovered on binding of H^+ and Ag^+.[43] The appearance of fluorescence on complexation of **16** with cations[46] is due to the fact that the formation of the nonfluorescent charge transfer state (involving the nitro group as an acceptor) is hindered by the bound cation.

Surprisingly, the emission spectrum of DCM-crown (**11**) undergoes almost no shift whereas the absorption spectrum is drastically affected (Figure 2.7); even more surprising is the fact that the excited-state lifetime is almost unaffected by complexation whereas the fluorescence quantum yield is reduced.[40] In order to understand this peculiar behavior, the photophysics of DCM-crown and its complexes with lithium and calcium in acetonitrile has been studied by picosecond pump-probe spectroscopy.[47] The results provide evidence for disruption of the interaction between a cation and the crown nitrogen atom which takes place in less than 5 and 20 psec for the Li^+ and Ca^{2+} complexes, respectively. Such a disruption arises from the appearance of a partial positive charge on the nitrogen atom of the crown as a result of the photoinduced intramolecular charge transfer from this atom towards the electron-withdrawing group (dicyanomethylene group). Consequently, the stability of the complexes in the excited state is lower than in ten ground state. Then, excitation by an intense light pulse causes some complexed cations to leave the crown and diffuses away, but this can occur as

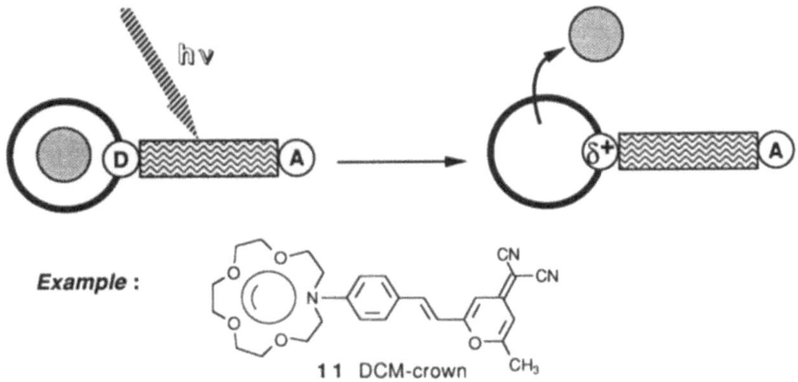

Figure 2.8. Schematic illustration of photoejection of cation from a donor–acceptor fluoroionophore.

long as the crowned compound is in the excited state. In ten case of DCM-crown, the excited-state lifetime is only about 2 nsec which does not allow the cations to move very far. However, by using light pulses of long duration (e.g. in the microsecond or millisecond range), a significant increase in concentration of free cations should be obtained. Alternatively, it would be of great interest to design compounds with long lifetime. Such systems capable of fast and reversible photorelease of cations would be useful to study the response of chemical or biological systems to a cation concentration jump according to the chemical relaxation methodology.

In calcium chelators Indo-1 (**17**) and Fura-3 (**18b**) (Figure 2.9),[18] the fluorophores have donor–acceptor stilbene-like structures rigidified so as to avoid photoisomerization. Based on the same principle, Fura-2 (**18a**)[18] is one of the most popular calcium indicator for microscopy of individual cells because, in contrast to Quin-2 (see Section 2.2.5.), the excitation spectrum is blue shifted on cation binding, thus allowing intensity-ratio measurements. On the other hand, there is almost no shift of the emission spectrum, which can be interpreted along the same line as DCM-crown (see earlier in this section).

By keeping the same fluorophores as in **17** and **18a**, but reducing the "cavity" size of the ionophore, one obtains Mag-Indol (**19**) and Mag-Fura2 (**20**) (Figure 2.9) which are selective for magnesium.[48] All the chelators **17** to **20** are commercially available under the nonfluorescent acetoxymethyl ester form so that they are cell permeant and they recover their fluorescence on hydrolysis by enzymes.[48]

FCryp-2 (**21**)[49] (Figure 2.9) is a nice example of a fluorescent signaling receptor in which the ionophore moiety has been specially designed for determination of intracellular sodium-free concentration. An indole derivative acts as the fluorophore: on sodium binding, the emission maximum shifts from 460 to 395 nm and the fluorescence intensity increases 25-fold. The origin of these photophysical changes have not been studied so far. The large Stokes shift of the free ligand may be accounted for by photoinduced charge transfer with concomitant internal rotation in the excited

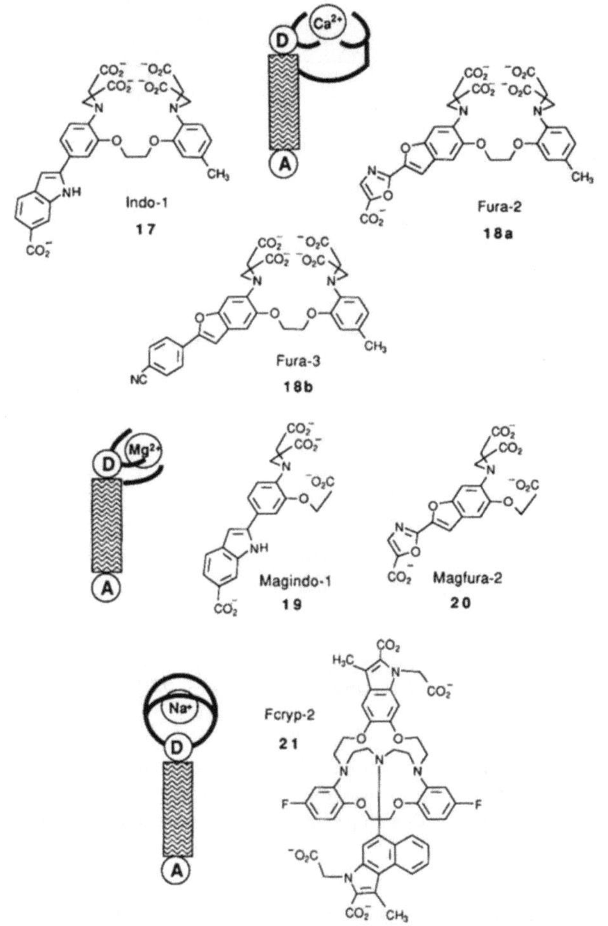

Figure 2.9. Examples of donor–acceptor fluoroionophores designed for the recognition of cytosolic calcium, magnesium, and sodium ions.

state leading to a TICT state. The blue shift of the emission spectrum on sodium binding is likely to be due to the reduction of the electron-donating character of the nitrogen atom of the cryptand which the dye is bound to.

In contrast to the compounds of Figure 2.6, the cation interacts directly with the acceptor moiety (carbonyle group of coumarin 153) in **22** (Figure 2.10).[50] According to the above-described photophysical effects, red shifts of the absorption and emission spectra are observed as expected. These shifts follow exactly the order of increased charge density of the cation. The high stability constants of the complexes observed in acetonitrile and methanol arise from the participation of the carbonyl group of the coumarin into the complexes and from the deconjugation of the nitrogen atom of the azacrown ether from the π-electron system.[50]

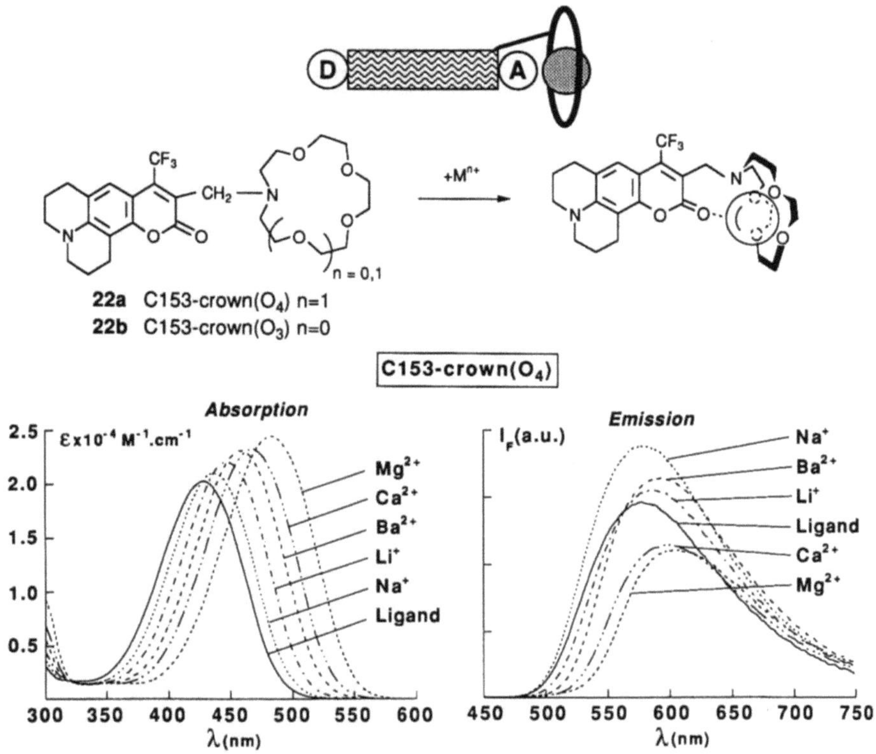

Figure 2.10. Example of a donor–acceptor fluoroionophore in which the electron-withdrawing character of the acceptor (carbonyl group of the coumarine) is cation-controlled. Absorption and fluorescence spectra of C153-crown(O_4) and its complexes with perchlorate salts in acetonitrile. (Adapted from Ref. 50.)

It is worth comparing the stability constants of **10–12**, **15a**, and **22a** under the same conditions because the ionophore moiety is the same (monoaza-15-crown-5). The values determined in acetonitrile are reported in Table 2.1 which clearly shows that the binding constants depend on the fluorophore because the nitrogen atom of the crown is conjugated with the fluorophore except in the case of C153-crown(O_4) whose complex stability constants are three orders of magnitude higher than the others for the reasons explained above. The complex stability of this compound remains high enough in acetonitrile–water mixtures for practical applications to the determination of cation in aqueous samples.[51]

Improvement of the efficiency and selectivity of binding together with an increase in the specificity of photophysical effects can be achieved by linking two fluorophores on the same ionophore as shown in Figure 2.11. The carbonyl groups of the two coumarin moieties in $(C153)_2$-K22 (**23**)[50] participate in the complexes: direct interaction between these groups and the cation explains the high stability constants and the photophysical changes. In addition to the shifts of the absorption and emission

Table 2.1. Effect of the Nature of the Fluorophore on the Stability Constants (K_S = [ML]/[M][L]) of Complexes with Fluoroionophores Containing Monoaza-15-crown-5 in Acetonitrile

Cation	Li^+	Na^+	K^+	Mg^{2+}	Ca^{2+}	Ba^{2+}
Ionic diameter (Å)	1.36	1.94	2.66	1.32	1.98	2.68
BOZ-crown (**10**) $\log_{10} K_S$	2.8	2.2	2.2	2.7	4.1	3.6
DCM-crown (**11**) $\log_{10} K_S$	2.5	2.0	—	—	3.7	3.1
DCS-crown (**12**) $\log_{10} K_S$	1.7	2.0	1.7	2.5	4.2	3.7
15a $\log_{10} K_S$	—	—	—	—	3.1	2.6
C153-crown(O_4) (**22a**) $\log_{10} K_S$	6.3	5.0	4.2	6.2	6.8	6.7

[a]Ref. 40.
[b]Ref. 48.
[c]Ref. 41.
[d]Ref. 47.
[e]Ref. 50.

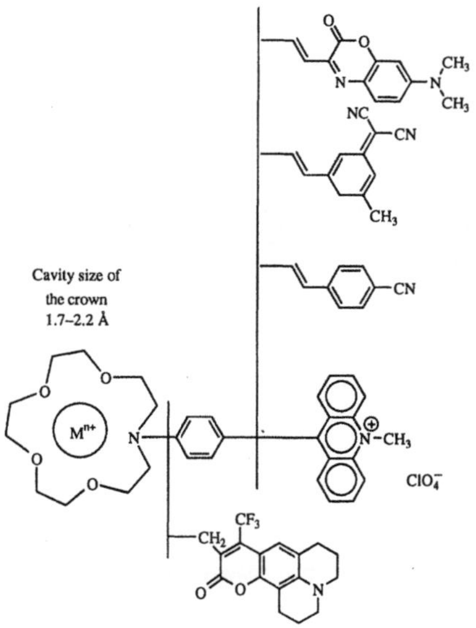

10 11 12 15a 22a

spectra, an interesting specific increase in the fluorescence quantum yield on binding of potassium and barium ions has been observed: in the complexes between $(C153)_2$-K22 and these ions that fit best into the crown cavity, the carbonyl groups of the two coumarins are preferentially on the opposite sides with respect to the cation and self- quenching is thus partially or totally suppressed, whereas with cations smaller than the cavity size of the crown, the preferred conformation of the relevant complexes may be such that the two carbonyls are on the same side and the close approach of the coumarin moieties accounts for static quenching (the two conformations are shown in Figure 2.11).

According to the same principle, in PBFI (**24a**)[52] and SBFI (**24b**)[53] (Figure 2.11), the oxygen atom of the methoxy substituent of the fluorophore can interact with a cation; binding efficiency and selectivity are thus better than that of the crown alone.

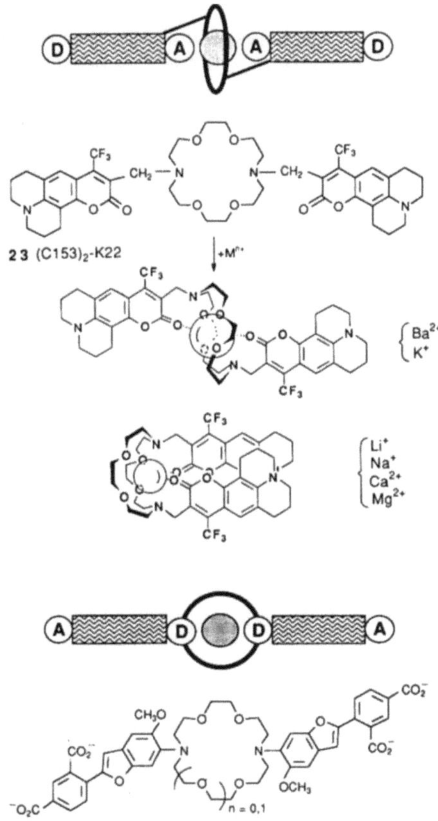

23 (C153)$_2$-K22

24a n=1 : PBFI
24b n=0 : SBFI

Figure 2.11. Examples of fluoroionophores consisting of two donor–acceptor fluorophores.

SBFI has been designed for probing intracellular sodium ions and PBFI for potassium ions. In both compounds, the photophysical changes are likely to be due to the reduction of the electron-donating character of the nitrogen atoms of the diazacrown by the complexed cation.

The importance of solvent effects has been outlined in Section 2.2.1. An illustration with some of the fluoroionophores described in this section is given in Table 2.2. For alkali and alkaline-earth metal ions, the stability constants are higher in acetonitrile than in methanol; these cations are indeed hard and have a stronger affinity for oxygen atoms (hard) than for nitrogen atoms (soft). In contrast, the soft silver atom has a strong affinity for nitrogen atoms and no complexation is observed in acetonitrile, whereas complexes in methanol, ether, and 1,2-dichloromethane are formed.

Table 2.2. Solvent Effect on Stability Constants (K_S = [ML]/[M][L]) of Complexes with Azacrowns Linked to Fluorophores

Azacrown complex	Cation	Solvent	Log_{10} K_s
C153-crown(O4) (**22a**)	Li^+	CH_3CN	6.3[a]
		CH_3OH	2.6[b]
	Na^+	CH_3CN	5.0[a]
		CH_3OH	3.9[b]
	Ca^{2+}	CH_3CN	6.8[a]
		CH_3OH	4.8[b]
	Ba^{2+}	CH_3CN	6.7[a]
		CH_3OH	4.1[b]
$(C153)_2$-K22 (**23a**)	Ba^{2+}	CH_3CN	7.6[a]
		CH_3OH	5.4[b]
DCS-crown (**12**)	Ca^{2+}	CH_3CN	4.2[c]
		CH_2Cl_2	5.2[c]
	Ag^+	CH_3CN	≈0[c]
		CH_3OH	3.1[c]
		Et_2O	4.3[c]
		CH_2Cl_2	5.2[c]

[a]Ref. 50.
[b]J. Bourson, F. Badaoui, and B. Valeur, unpublished results.
[c]Ref. 41.

2.2.4. Recognition Based on Cation Control of the Proximity between Two Fluorophores, or a Fluorophore and a Quencher

The basic principle of this method of recognition is a cation-induced conformational change bringing closer together (or moving away) two moieties able to interact and induce photophysical effects: excimer or exciplex formation (or disappearance), electronic energy transfer and quenching.

2.2.4.1. Cation Control of Excimer Formation

Cation control of excimer formation has already been mentioned in Section 2.2.2. about compound **8**. Figure 2.12 provides two interesting examples of anthraceno-crown ethers. Compound **25**[(54)] exhibits a fluorescence spectrum composed of the characteristic monomer and excimer bands. Gradual addition of sodium perchlorate to a solution of **25** in methanol induces a decrease in the monomer band and an increase in the excimer band. Complexation is indeed expected to bring closer together the two anthracene units which favors excimer formation. Compound **26**[(55)] is based on the same principle but four carbonyl groups have been incorporated because of their known role in complexing divalent ions. Unfortunately, only moderate changes in absorption and emission spectra were found upon addition of barium and strontium ions, the weak complexing ability being due to the rigidity of the system.

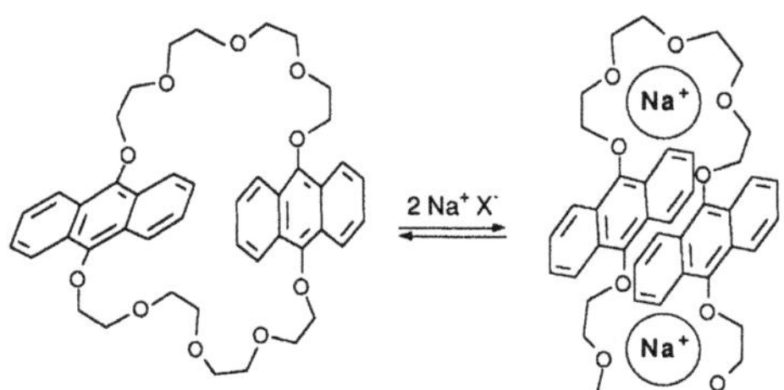

25

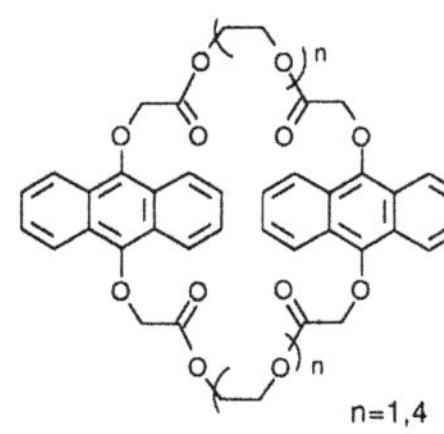

26

Figure 2.12. Cation control of excimer formation in anthraceno-crown ethers (adapted from Refs. 54 and 55.)

Another example of cation control excimer formation will be given in Section 2.2.4.3 dealing with calixarenes.

2.2.4.2. Cation Control of Photoinduced Energy Transfer between Unlike Fluorophores

Bifluorophores consisting of two different fluorescent dyes linked by a flexible spacer containing heteroatoms (oxygen, nitrogen or sulfur atoms) can bind cations. This results in a decrease of the distance between the two fluorophores and, consequently, to an increase in efficiency of photoinduced energy transfer between the two moieties (Figure 2.13) provided that the emission spectrum of the donor (D) overlaps the absorption spectrum of the acceptor (A).[36] The transfer efficiency depends on the distance according to Förster's theory:

$$\Phi_T = \frac{1}{1 + \left[\dfrac{R}{R_0}\right]^6}$$

where R is the distance between the two fluorophores and R_0 is the Förster critical radius. (For a discussion on the use of energy transfer for determination of distances at a supramolecular scale, see Ref. 1). As Φ_T varies drastically around R/R_0 (sixth power dependence), a relatively small variation in distance could lead to a large increase in transfer efficiency provided that the interchromophoric distance is larger than the Förster critical radius in the free ligand and lower in the complex, as illustrated in Figure 2.13.

The first bifluorophore designed according to this principle[56, 57] consists of two coumarins linked by a penta(ethylene oxide) spacer (**27**) (Figure 2.13). Complexation of Pb^{2+} by this compound in acetonitrile is revealed by the changes in absorption, excitation, and emission spectra (whereas alkali and alkaline-earth metal ions have almost no effect). The stability constant K_S of the 1:1 complex was found to be $3.1 \times 10^{-4}\ M^{-1}$. The transfer efficiency increases on complexation from 0.77 to 0.89; this modest variation can be explained as follows. The value for the free ligand is already high because the average interchromophoric distance turns out to be shorter than the Förster critical radius. After complexation the efficiency is not close to 1, as it would be expected, which means that the donor and acceptor moieties are not very close to each other, and helical wrapping of the ligand around one cation may explain the observations (Figure 2.13). Examples of such helical wrapping of acyclic ligands can be found in the literature.[58] An interesting solvent effect has been observed: in propylene carbonate, the stoichiometry of the complex is 1:3, i.e., each ligand binds cooperatively three Pb^{2+} owing to the participation of solvent molecules into the complexes.[56, 57]

For the design of complexing bifluorophores, much attention should be paid to the Förster critical radius of the donor–acceptor pair as compared to the interchromophoric distance (Figure 2.13). This critical radius depends on the donor quantum yield and on the spectral overlap between donor emission and acceptor emission. Complex-

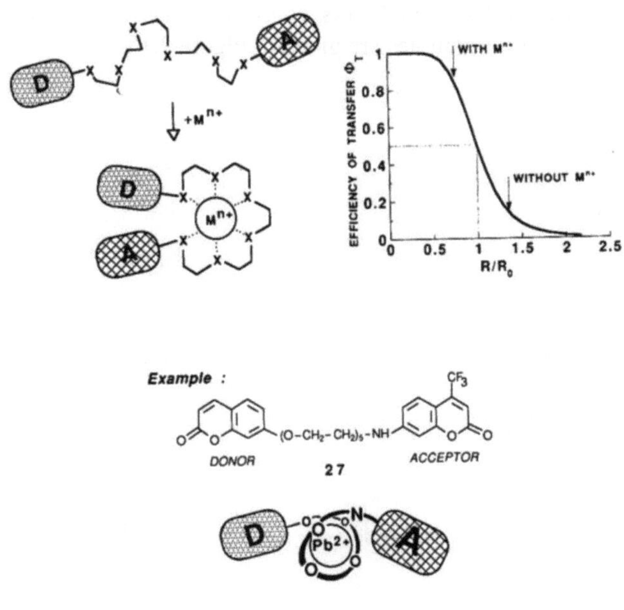

Figure 2.13. Principle of ion recognition based on cation control of excitation energy transfer in a bifluorophore (adapted from Ref. 56.)

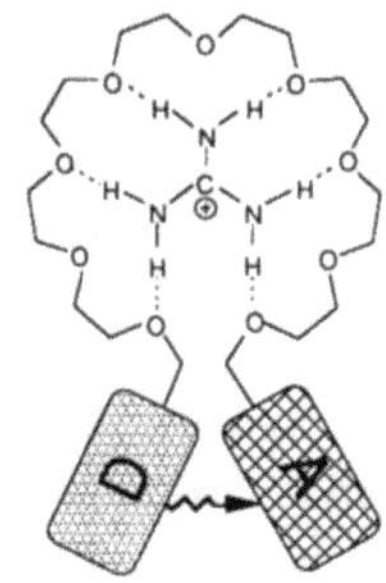

Figure 2.14. Hypothetical structure of a bifluorophore capable of recognition of guanidinium ion.

ing bifluorophores appear to be of limited use for the detection of metal cations for which many other complexing reagents are available, but this principle of design is of potential interest for organic cations like guanidinium ion, as shown in Figure 2.14.(36)

2.2.4.3. Cation-Induced Conformational Changes in Calixarenes Labeled with Fluorophores

Recently it was shown that introduction of fluorescent groups into ionophoric calix[4]arenes provides new systems for recognition of alkali metal ions.(59–62) Because of the mobility of the phenyl units, conformational changes can be induced by solvents and metal ions. Examples are given in Figure 2.15.

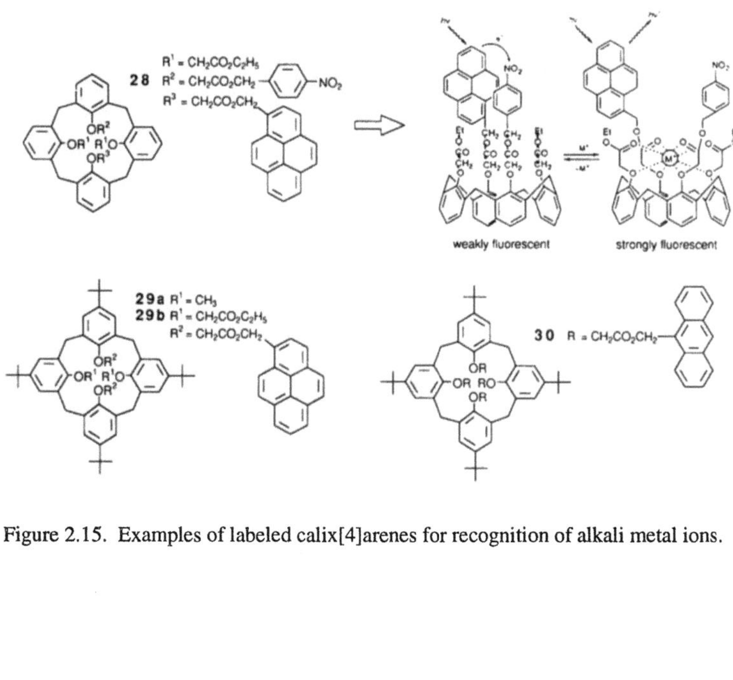

Figure 2.15. Examples of labeled calix[4]arenes for recognition of alkali metal ions.

In compound **28**,[59] the conformational change induced by sodium binding reduces the probability of collision between the pyrene fluorophore and the nitrobenzene quencher, both attached to the calixarene ring.

When the lower rim contains two pyrene residues (**29**), the monomer versus excimer emission is affected by addition of alkali metal ions to solutions of **29a**[60] and **29b**.[61] For the latter, the Na^+/K^+ selectivity, as measured by the ratio of the dissociation constant was found to be 154, while the affinity for Li^+ was too low to be determined.

In calix[4]arenes bearing four anthracene moieties on the lower rim (**30**),[62] changes in fluorescence intensity are observed on binding of alkali metal ions: quenching of fluorescence by Na^+ may arise from interaction of four anthracene residues brought in closer proximity to one another; enhancement of fluorescence by K^+ is difficult to explain.

2.2.5. Miscellaneous

Many fluorescent probes exhibit cation-induced photophysical changes that cannot be explained along the same lines as those described in the preceding sections. Interaction of a cation with a fluorophore often leads to changes in radiative, nonradiative, and/or intersystem crossing rate constants, and consequently, changes in fluorescence quantum yield and lifetime (whereas shift of spectra is not a general rule). In a pioneering work, Sousa *et al.*[63–66] studied the direct interaction between a cation and the π-electron system of naphthalene in compounds **31** (Figure 2.16).

Various other examples that are of practical use are also shown in Figure 2.16. Quin-2 (**32**) was the first practical fluorescent indicator for cytosolic calcium with a simple 6-methoxyquinoline as its fluorophore (For a review, see Ref. 65). Ca^{2+}-binding increases the fluorescence intensity about sixfold (without spectral displacement in contrast to Fura 2). The fluorescence lifetime of Quin-2 is highly sensitive to calcium concentration; Quin-2 can thus be used as a probe in the technique of fluorescence lifetime imaging.[66]

Compounds **33**[67] and **34** (Kryptofix 5, Merck)[68, 69] are also based on methoxyquinoline fluorophore. The geometrical contraints in **33** explain the excellent selectivity for the small lithium ion, and as in the previous example, binding is accompanied with an increase in fluorescence intensity.

Such a fluorescence enhancement in these compounds **32–34** is similar to that observed in metal chelates with 8-hydroxyquinoline (oxine) and its derivatives[16] but is not fully understood yet.

For practical applications it is better to use fluorescent probes excitable at high wavelengths, so that autofluorescence of the sample (especially biological samples) does not interfere. Calcium indicators based on fluorescein and rhodamine fluorophores[70] fulfill this requirement and are thus more convenient for fluorescence

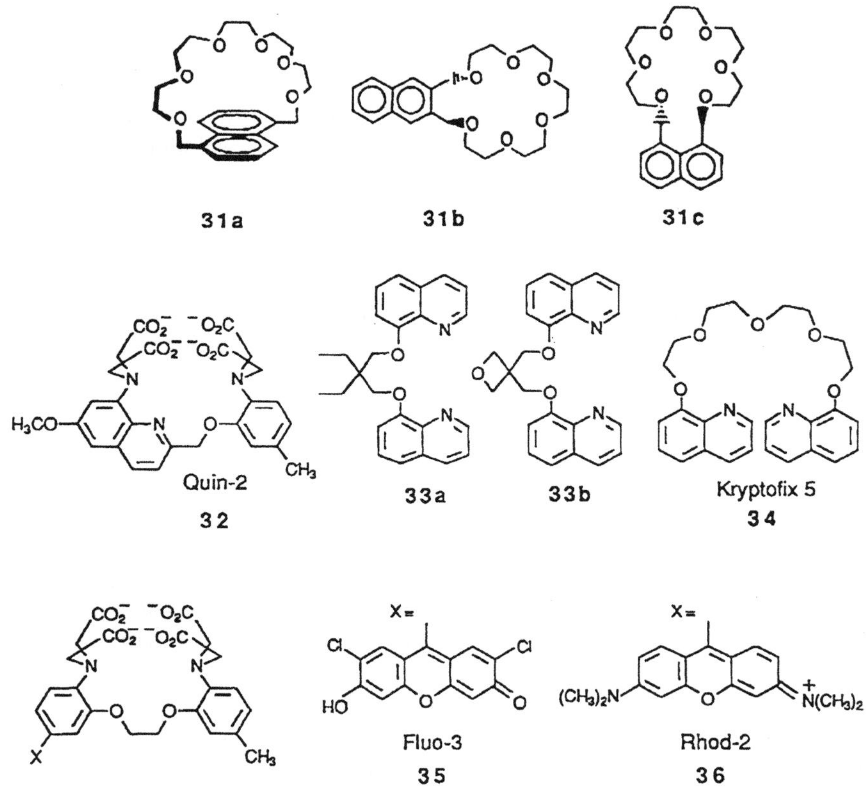

Figure 2.16. Various fluoroionophores undergoing particular changes in some of their photophysical properties on cation binding.

microscopy and flow cytometry, but they do not undergo spectral shifts upon binding. The most interesting dye in this series is Fluo-3 (**35**) and Rhod-2 (**36**).[70]

2.3. Fluorescent Signaling Receptors of Anions

The development of fluorescent probes for anion recognition has been very limited so far in comparison with those for cations. Most of the presently available methods of detection of anions based on fluorescence involve quenching, redox reactions, substitution reactions, ternary complex formation[15] and thus cannot be considered as recognition methods. For instance, the fluorescent sensors that are used for the determination of chloride anions in living cells are based on collisional quenching of a dye by halide ions; 6-methoxy-*N*-(sulfopropyl)quinolinium and

37

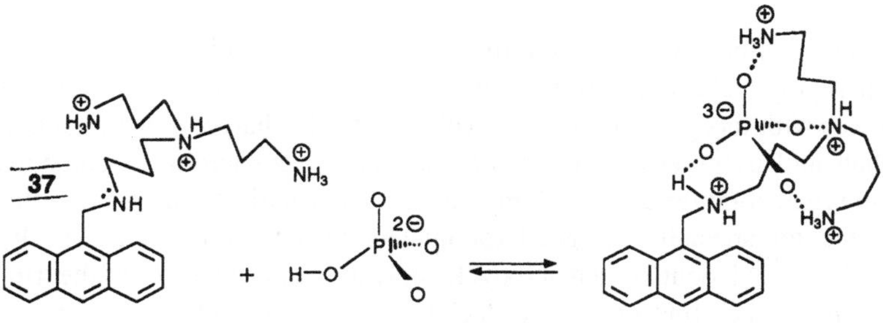

low fluorescence high fluorescence

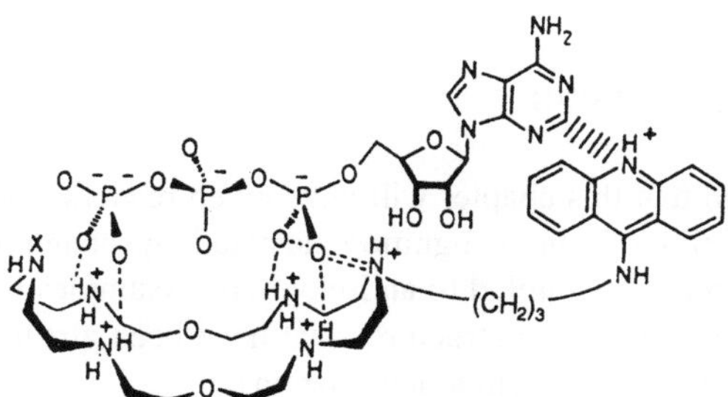

38

Figure 2.17. Examples of fluoroionophores for recognition of anions (Adapted from Refs. 72 and 73.)

N-(6-methoxyquinolyl)acetic acid are examples of such probes.[48] Therefore, one cannot speak of anion recognition because the probe does not associate with the anion but only interacts in the excited state causing a decrease in fluorescence intensity.

There are only few true cases of anion recognition by fluorescent probes. Phosphate groups can be recognized by anthrylpolyamine conjugate probes (**37**)[72] (Figure 2.17). The choice of pH is crucial: at pH 6, a fraction of 70% of **37** exists as a triprotonated form, the nitrogen atom close to the anthracene moiety being not protonated. The very low fluorescence of this compound is due to photoinduced electron transfer from the unprotonated amino group to anthracene (See Section 2.2.2 and Figure 2.2). This trication can bind a complementary structure like monohydrogenophosphate whose three oxygen atoms interact with the three positive charges; the remaining phosphate OH group is in a favorable position to undergo intracomplex proton transfer to the unprotonated amino group, which eliminates intramolecular quenching (according to the scheme of Figure 2.2). Then, binding is accompanied with a drastic enhancement of fluorescence. **37** can also bind adenosine triphosphate (ATP), citrate, and sulfate.[72] This mode of recognition is conceptually very interesting but the stability of the complexes is low.

A better efficiency and selectivity is expected with probes based on macrocyclic and macropolycyclic polycations that are capable of forming strong and selective complexes with inorganic anions and with negatively charged functional groups (i.e., phosphate and carboxylate). In fact, protonated polyazamacrocycles and polyazamacrobicycles can complex anions. Compound **38** is a beautiful example of a fluorogenic anion receptor specially designed for nucleotide recognition and ATP hydrolysis (Figure 2.17)[73]: it contains a macrocyclic polyamine as a receptor of the triphosphate moiety and an acridine group for stacking interaction with the nucleic base. The fluorescence of the acridine group is significantly enhanced on binding. In addition, amino groups of the protonated macrocyclic hexamine catalyzes the hydrolysis of ATP.

2.4. Concluding Remarks

It is to be hoped that this chapter will have given readers a better understanding of the basic principles of ion recognition detected by changes in photophysical properties of a fluorophore coupled to an ionophore. Examples have been chosen to illustrate the immense variety of structures that have been already designed and the inexhaustible possibilities of creating new systems.

At the present time, there is a striking contrast between the extensive development of fluorescent probes for cation recognition and the limited number of available probes for anions notwithstanding the great need for the latter. This is due to the difficulty of the design of selective anion receptors; progress made in the relevant field of supramolecular chemistry will certainly lead in the future to new selective fluorescent signaling receptors of anions.

In the field of cellular biology, numerous fluorescent probes for cation recognition have been designed. Among them calcium probes have been the object of special attention because of the great role played by this cation in biological systems. Highly selective probes with binding constants compatible with intracellular concentration are now available, but there is still a need for calcium probes that fulfill both criteria of high-wavelength excitability and fluorescence spectral shift upon binding. Regarding sodium and potassium, improvement in selectivity is still to be made. The example of a cryptand-linked indole fluorophore[49] confirms that the best way to achieve selectivity for alkali metal ions is to use cryptands, but fluorophores excitable at higher wavelength would be desirable in order to avoid parasitic autofluorescence.

In biomedical applications, the ranges of ion concentration are higher by several orders of magnitude. For instance, the abovementioned calcium probes for living cells cannot be used because the dissociation constant is so low that they would be saturated. Special attention is thus to be paid to the ionophore moiety to achieve proper selectivity and efficiency of binding. For instance, at present there is a need for a selective fluorescent probe for the determination of calcium in blood which could work in the millimolar range in aqueous solutions so that optodes with immobilized probes on the tip could be made for continuous monitoring calcium in blood vessels.

In analytical chemistry, detection of metal ions is of major importance. In particular, the development of simple and reliable methods for continuous control *in situ* of metal ions in the environment is the object of much attention. For instance, the detection of lead, mercury, cadmium, and iron ions in sea water will be performed in the near future by optodes associated with suitable fluoroionophores, thus allowing continuous monitoring by instruments on ships.

It should be emphasized that best design (each application corresponding to a particular design), proper choice, and correct use of fluorescent probes require a thorough knowledge of the basic phenomena involved in ion recognition: medium effect on complexation equilibrium, fundamental photophysical processes, and possible changes from other causes than complexation.

Ion recognition by fluorescent probes is a field in which the word "design" takes its full sense, and there is no doubt that it will remain an active field of research for several decades.

References

1. B. Valeur, Fluorescent probes for evaluation of local physical and structural parameters, in *Molecular Luminescence Spectroscopy, Methods and Applications Part 3* (S. G. Schulman, ed.), pp. 25–84, John Wiley & Sons, New York (1993).
2. R. Y. Tsien, Fluorescent probes of cell signaling, *Ann. Rev. Neurosci. 12*, 227–253 (1989).
3. R. Y. Tsien, Fluorescent indicators of ion concentrations, in *Methods in Cell Biology*, Vol. 30, pp. 127–156, Academic Press, New York (1989).
4. A. W. Czarnik (Ed.), *Fluorescent Chemosensors for Ion and Molecule Recognition*, ACS Symposium Series 538, Washington, D.C. (1993).
5. O. S. Wolfbeis, Fiber optical fluorosensors in analytical and clinical chemistry, in: *Molecular Luminescence Spectroscopy, Methods and Applications: Part 2* (S. G. Schulman, ed.), pp. 129–281, John Wiley & Sons, New York (1988).
6. O. S. Wolfbeis (Ed.), *Fiber Optic Chemical Sensors and Biosensors,* Vols. 1 and 2, CRC Press, Boca Raton, Florida (1991).
7. F. Vögtle, *Host Guest Complex Chemistry II*, Topics in Current Chemistry 101, Springer Verlag, Berlin (1982).
8. F. Vögtle and E. Weber, *Host Guest Complex Chemistry III*, Topics in Current Chemistry 121, Springer Verlag, Berlin (1984).
9. J. M. Lehn, Supramolecular chemistry: Receptors, catalysts, and carriers, *Science* 227, 849–856 (1985).
10. J. M. Lehn, Supramolecular chemistry—scope and perspectives. Molecules, supermolecules, and molecular devices, *Angew. Chemie. Int. Ed. Engl. 27*, 89–112 (1988).
11. Y. Inoue and G. W. Gokel, *Cation Binding by Macrocycles. Complexation of Cationic Species by Crown Ethers*, Marcel Dekker, New York and Basel (1990).
12. F. Vögtle, *Supramolecular Chemistry. An Introduction*, John Wiley & Sons, Chichester, England (1991).
13. B. Dietrich, P. Viout, and J.-M. Lehn, *Aspects de la chimie des composés macrocycliques*, Savoirs Actuels, InterEditions, Paris (1991).
14. G. W. Gokel, *Crown-ethers and Cryptands*, Royal Society of Chemistry, Cambridge, England (1991).
15. A. Fernandez-Gutierrez and A. Muñoz de la Peña, Determination of inorganic substances by luminescence methods, in: *Molecular Luminescence Spectroscopy, Methods and applications: Part*

1 (S. G. Schulman, ed.), pp. 371–546, John Wiley & Sons, New York (1985) and references therein (more than 1400).
16. K. Soroka, R. S. Vithanage, D. A. Phillips, B. Walker, and P. K. Dasgupta, Fluorescence properties of metal complexes of 8-hydroxyquinoline-5-sulfonic acid and chromatographic applications, *Anal. Chem. 59*, 629–636 (1987) and references therein.
17. R. Y. Tsien, New calcium indicators and buffers with high selectivity against magnesium and protons: Design, synthesis, and properties of prototype structures, *Biochemistry 19*, 2396–2404 (1980).
18. G. Grynkiewicz, M. Poenie, and R. Y. Tsien, A new generation of Ca^{2+} indicators with greatly improved fluorescence properties, *J. Biol. Chem. 260*, 3440–3450 (1985).
19. H.-G. Löhr and F. Vögtle, Chromo- and fluoroionophores. A new class of dye reagents, *Acc. Chem. Res. 18*, 65 (1985) and references cited therein.
20. G. Anderegg, The investigation of complex formation equilibrium at constant ionic strength, *Talanta 40*, 243–246 (1993).
21. R. D. Hancock and A. E. Martell, Ligand design for selective complexation of metal ions in aqueous solution, *Chem. Rev. 89*, 1875–1914 (1989).
22. R. A. Bissell, A. Prasanna de Silva, H. Q. Nimal Gunaratne, P. L. Mark Lynch, G. E. M. Maguire, and K. R. A. Samankumara Sandanayake, Molecular fluorescent signaling with 'Fluor-Spacer-Receptor' systems: Approaches to sensing and switching devices via supramolecular photophysics, *Chem. Soc. Rev. 187* (1992).
23. R. A. Bissell, A. Prasanna de Silva, H. Q. Nimal Gunaratne, P. L. Mark Lynch, G. E. M. Maguire, C. P. McCoy, and K. R. A. Samankumara Sandanayake, Fluorescent PET (Photoinduced Electron Transfer) sensors, *Top. Curr. Chem. 168*, 223 (1993).
24. M. E. Huston, K. W. Haider, and A. W. Czarnik, Chelation-enhanced fluorescence in 9,10-bis(TMEDA)anthracene, *J. Am. Chem. Soc. 110*, 4460–4462 (1988).
25. A. Prasanna de Silva, and H. Q. Nimal Gunaratne, Fluorescent PET (Photoinduced Electron Transfer) sensors selectives for submicromolar calcium with quantitatively predictable spectral and ion-binding properties, *J. Chem. Soc., Chem. Commun.* 186 (1990).
26. A. Prasanna de Silva and S. A. de Silva, Fluorescent signalling crown ethers; "switching on" of fluorescence by alkali metal ion recognition and binding in situ, *J. Chem. Soc., Chem. Commun.* 1709 (1986).
27. E. U. Akkaya, M. E. Huston, and A. W. Czarnik, Chelation-enhanced fluorescence of anthrylazamacrocycle conjugate probes in aqueous solution, *J. Am. Chem. Soc. 112*, 3590–3593 (1990).
28. J. P. Konopelski, F. Kotzyba-Hibert, J.-M. Lehn, J.-P. Desvergne, F. Fages, A. Castellan, and H. Bouas-Laurent, Synthesis, cation binding, and photophysical properties of macrobicyclic anthraceno-cryptands, *J. Chem. Soc., Chem. Commun.* 433 (1985).
29. F. Fages, J.-P. Desvergne, and H. Bouas-Laurent, Nonlinear triple exciplexes: Thermodynamic and kinetic aspects of the intramolecular exciplex formation between anthracene and the two anchored nitrogens of an anthraceno-cryptand, *J. Am. Chem. Soc. 111*, 96–102 (1989).
30. F. Fages, J.-P. Desvergne, H. Bouas-Laurent, P. Marsau, J.-M. Lehn, F. Kotzyba-Hibert, A.-M. Albrecht-Gary, and M. Al-Joubbeh, Anthraceno-cryptands: A new class of cation-complexing macrobicyclic fluorophores, *J. Am. Chem. Soc. 111*, 8672–8680 (1989).
31. A. Prasanna de Silva, H. Q. Nimal Gunaratne, and K. R. A. Samankumara Sandanayake, A new benzo-annelated cryptand and a derivative with alkali cation-sensitive fluorescence, *Tetrahedron Lett. 31*, 5193–5196 (1990).
32. K. Golchini, M. Mackovic-Basic, S. A. Gharib, D. Masilamani, M. E. Lucas, and I. Kurtz, Synthesis and characterization of a new fluorescent probe for measuring potassium, *Am. J. Phys. 258*, F438 (1990). See also Ref. 4, Chap. 11, p. 163.
33. F. Fages, J.-P. Desvergne, H. Bouas-Laurent, J.-M. Lehn, J. P. Konopelski, P. Marsau, and Y. Barrans, Synthesis and fluorescence emission properties of a bis-anthracenyl macrotricyclic ditopic receptor. Crystal structure of its dinuclear rubidium cryptate, *J. Chem. Soc., Chem. Commun.* 655 (1990).

34. F. Fages, J. P. Desvergne, K. Kampke, H. Bouas-Laurent, J. M. Lehn, M. Meyer, A. M. Albrecht-Gary, Linear molecular recognition: Spectroscopic, photophysical and complexation studies on α,ω-alkanediyldiammonium ions binding to a bisanthracenyl macrotricyclic receptor, *J. Am. Chem. Soc. 115*, 3658–3664 (1993).
35. A. Prasanna de Silva and K. R. A. Samankumara Sandanayake, Fluorescence 'off-on' signaling upon linear recognition and binding of α,ω-alkane diammonium ions by anthracene 9,10-bis(N-methyl aza-18-crown-6 ether), *Angew. Chemie 29*, 1173 (1990).
36. B. Valeur, J. Bourson, and J. Pouget, Ion recognition detected by changes in photoinduced charge or energy transfer, in: *Fluorescent Chemosensors for Ion and Molecule Recognition*, ACS Symposium Series 538, (A. W. Czarnik, ed.), Chapter 3, Washington, D.C. (1993).
37. S. Fery-Forgues, M.-T. Le Bris, J.-P. Guetté, and B. Valeur, First crown ether derivative of benzoxazinone; a new fluoroionophore for alkaline earth metals recognition, *J. Chem. Soc., Chem. Commun. 5*, 384–385 (1988).
38. S. Fery-Forgues, M.-T. Le Bris, J.-P. Guetté, and B. Valeur, Ion-responsive fluorescent compounds. 1. Effect of cation binding on photophysical properties of a benzoxazinone derivative linked to monoaza-15-crown-5, *J. Phys. Chem. 92*, 6233–6237 (1988).
39. S. Fery-Forgues, J. Bourson, L. Dallery, and B. Valeur, NMR and optical spectroscopy studies of cation binding on chromophores and fluorophores linked to monoaza-15-crown-5, *New J. Chem. 14*, 617–623 (1990).
40. J. Bourson and B. Valeur, Ion-responsive fluorescent compounds. 2. Cation-steered intramolecular charge transfer in a crowned merocyanine, *J. Phys. Chem. 93*, 3871 (1989).
41. J. F. Létard, R. Lapouyade, and W. Rettig, Synthesis and photophysical study of 4-(N-monoaza-15-crown-5) stilbenes forming TICT states and their complexation with cations, *Pure Appl. Chem. 65*, 1705–1712 (1993).
42. S. A. Jonker, F. Ariese, and J. W. Verhoeven, Cation complexation with functionalized 9-arylacridinium ions: Possible applications in the development of cation-selective optical probes, *Recl. Trav. Chim. Pays-Bas 108*, 109–115 (1989).
43. S. A. Jonker, S. I. Van Dijk, K. Goubitz, C. A. Reiss, W. Schuddeboom, and J. W. Verhoeven, Solid-state structure and spectroscopy of chromoionophoric acridinium derivatives, *Mol. Cryst. Liq. Cryst. 183*, 273–282 (1990).
44. W. Rettig, Charge separation in excited states of decoupled systems-TICT compounds and implications regarding the development of new laser dyes and the primary processes of vision and photosynthesis, *Angew. Chem. Int. Ed. Engl. 25*, 971–988 (1986).
45. S. Fery-Forgues, M.-T. Le Bris, J.-C. Mialocq, J. Pouget, W. Rettig, and B. Valeur, Photophysical properties of styryl derivatives of aminobenzoxazinones, *J. Phys. Chem. 96*, 701–710 (1992).
46. K. W. Street, Jr. and S. A. Krause, A new metal sensitive fluorescence reagent, *Anal. Lett. 19*, 735–745 (1986).
47. M. M. Martin, P. Plaza, N. Dai Hung, Y. H. Meyer, J. Bourson, and B. Valeur, Photoejection of cations from complexes with a crown-ether-linked merocyanine evidenced by ultrafast spectroscopy, *Chem. Phys. Lett. 202*, 425 (1993).
48. R. P. Haugland, *Handbook of Fluorescent Probes and Research Chemicals*, Molecular Probes, Inc., Eugene, OR, USA (1992–1994).
49. G. A. Smith, T. R. Hesketh, and J. C. Metcalfe, Design and properties of a fluorescent indicator of intracellular free Na^+ concentration, *Biochem. J. 250*, 227–232 (1988).
50. J. Bourson, J. Pouget, and B. Valeur, Ion-responsive fluorescent compounds. 4. Effect of cation binding on the photophysical properties of a coumarin linked to monoaza- and diazacrown ethers, *J. Phys. Chem. 97*, 4552–4557 (1993).
51. J. Bourson, M.-N. Borrel, and B. Valeur, Ion-responsive fluorescent compounds. 3. Cation complexation with coumarin 153 linked to monoaza-15-crown-5, *Anal. Chim. Acta 257*, 189–193 (1992).
52. S. E. Kasner and M. B. Ganz, Regulation of intracellular potassium in mesangial cells: A fluorescence analysis using the dye, PBFI, *Am. J. Physiol. 262*, F462–F467 (1992).

53. A. Minta and R. Y. Tsien, Fluorescent indicators for cytosolic sodium, *J. Biol. Chem. 264*, 19449–19457 (1989).
54. H. Bouas-Laurent, A. Castellan, M. Daney, J.-P. Desvergne, G. Guinand, P. Marsau, and M.-H. Riffaud, Cation-directed photochemistry of an anthraceno-crown ether, *J. Am. Chem. Soc. 108*, 315–317 (1986).
55. J. Hinschberger, J.-P. Desvergne, H. Bouas-Laurent, and P. Marsau, Synthesis and photophysical properties of fluorescent anthracenophanes incorporating two polyoxadioxoalkane chains, *J. Chem. Soc. Perkin Trans. 2*, 993–1000 (1990).
56. B. Valeur, J. Pouget, J. Bourson, M. Kaschke, and N. P. Ernsting, Tuning of photoinduced energy transfer in a bichromophoric coumarin supermolecule by cation binding, *J. Phys. Chem. 96*, 6545–6549 (1992).
57. B. Valeur, J. Pouget, and J. Bourson, Photoinduced electronic energy transfer in a supramolecular complex between a coumarin bichromophoric molecule and lead(II), *J. Lumin. 52*, 345–347 (1992).
58. F. Vögtle and E. Weber, Multidentate acyclic neutral ligands and their complexation, *Angew. Chemie Int. Ed. Engl. 18*, 753–776 (1979).
59. I. Aoki, T. Sakaki, and S. Shinkai, A new metal sensory system based on intramolecular fluorescence quenching on the ionophoric calix[4]arene ring, *J. Chem. Soc., Chem. Commun.* 730 (1992).
60. I. Aoki, H. Kawabata, K. Nakashima, and S. Shinkai, Fluorescent calix[4]arene which responds to solvent polarity and metal ions, *J. Chem. Soc., Chem. Commun.* 1771 (1991).
61. T. Jin, K. Ichikawa, and T. Koyama, *J. Chem. Soc., Chem. Commun.* 499 (1992).
62. C. Pérez-Jiménez, S. J. Harris, and D. Diamond, A novel calix[4]arene tetraester with fluorescent response to complexation with alkali metal cations, *J. Chem. Soc., Chem. Commun.* 480 (1993).
63. L. R. Sousa and J. M. Larson, Crown ether model systems for the study of photoexcited state response to geometrically oriented perturbers. The effect of alkali metal ions on emission from naphthalene derivatives, *J. Am. Chem. Soc. 99*, 307–309 (1977).
64. J. M. Larson and L. R. Sousa, Crown ether model systems for the study of photoexcited-state response to oriented perturbers. How does a naphthalene derivative respond to an alkali metal cation in its π face?, *J. Am. Chem. Soc. 100*, 1943 (1978).
65. R. Y. Tsien and T. Pozzan, Measurement of cytosolic free Ca^{2+} with Quin-2, in: *Methods in Enzymology*, Vol. 172, pp. 230–262, Academic Press, New York (1989).
66. J. R. Lakowicz, H. Szmacinski, K. Nowaczyk, and M. L. Johnson, Fluorescence lifetime imaging of calcium using Quin-2, *Cell Calcium 13*, 131–147 (1992).
67. K. Hiratani, Drastic change in fluorescence intensity of acyclic polyethers caused by addition of lithium ion, *J. Chem. Soc., Chem. Commun.* 960 (1987).
68. E. Weber and F. M. Vögtle, Kristalline 1:1-alkalimetallcomplexe nichtcyclischer neutralliganden, *Tetrahedron Lett. 29*, 2415–2418 (1975).
69. O. S. Wolfbeis and H. Offenbacher, The effects of alkali cation complexation on the fluorescence properties of crown-ethers, *Monat. Chem. 115*, 647–654 (1984).
70. A. Minta, J. P. Y. Kao, and R. Y. Tsien, Fluorescent indicators for cytosolic calcium based on rhodamine and fluorescein chromophores, *J. Biol. Chem. 264*, 8171–8178 (1989).
71. J. R. Lakowicz, H. Szmacinski, K. Nowaczyk, and M. L. Johnson, Calcium imaging using fluorescence lifetimes and long-wavelength probes, *J. Fluorescence 2*, 47 (1992). See also Ref. 4, Chap. 10, p. 147.
72. M. E. Huston, E. U. Akkaya, and A. W. Czarnik, Chelation enhanced fluorescence detection of non-metal ions, *J. Am. Chem. Soc. 111*, 8735–8737 (1989).
73. M. W. Hosseini, A. J. Blacker, and J.-M. Lehn, Multiple molecular recognition and catalysis, nucleotide binding and ATP hydrolysis by a receptor molecule bearing an anion binding site, an intercalator group, and a catalytic site, *J. Chem. Soc., Chem. Commun.* 596 (1988).

3

Fluorescent Chemosensors for Cations, Anions, and Neutral Analytes

Anthony W. Czarnik

There is a technological driving force for the design of new fluorescent chemosensors. With the use of fluorescent microscopy, analyte fluxes in single living cells can be monitored in real-time. Using fiber optic remote sensing equipment, concentrations in underground water sources and in chemical reactors can likewise be monitored in real-time. There is already a device on the market used to monitor in real-time arterial partial pressure of oxygen (pO_2), partial pressure of carbon dioxide (pCO_2), and pH levels during open heart surgery. Fledgling industries are now developing around each of the above three technologies.

In all three instances, further development principally awaits the synthesis of new molecules that bind analytes from aqueous solution with fluorescent signal transduction. Effective intracellular probes exist for pH and for Ca(II) [and, arguably, for Mg(II)]; it would be highly desirable to be able to sense inorganic phosphate, glucose, and adenosine. Effective remote sensors are being developed for pCO_2 in the deep ocean; we can also see uses for Cu(II) and organic pollutant sensing. While blood gases and pH can be monitored at present, there is an obvious advantage to doing every blood test currently done batchwise instead continuously at the bedside with *in arterio* optical fiber lines.

Each of these goals requires the synthesis of a compound with sufficiently selective binding and concomitant fluorescent signal transduction in water. However, the scientific base on which to start the rational design process does not, at present, exist. We have set as our research agenda the evaluation of new ways in which an ion or molecule recognition event can be transduced into a fluorescence event.

Fluorimetric methods have proven useful for the assay of metal ions in solution[(1)]; e.g., *in vivo* studies of calcium-selective fluorescence probes have been reported by

Anthony W. Czarnik • Department of Chemistry, Ohio State University, Columbus, Ohio 43210.

Topics in Fluorescence Spectroscopy, Volume 4: Probe Design and Chemical Sensing, edited by Joseph R. Lakowicz. Plenum Press, New York, 1994.

Tsien.[2] Most such analytical methods reported to date involve complexation of metal ions with aromatic heterocyclic ligands ("intrinsic" fluoroionophores). In 1977, Sousa described the synthesis of naphthalene-crown ether probes (Figure 3.1) in which the fluorophore π-system is insulated from the donor atoms by at least one methylene group ("conjugate" fluoroionophores).[3a] These compounds demonstrated fluorescence changes on the binding of alkali metal salts in an ethanol glass; the observed changes were attributed to a heavy atom effect (for Cs^+ and Rb^+) and/or a complexation-induced change in triplet energy relative to ground and excited singlet state energies. Subsequent reports by various groups have built on this original premise, in which binding of metals to crowns and azacrowns has been coupled to emission changes of covalently attached fluorophores (Figure 3.1).[3b–m] This chapter primarily describes our group's research on this topic over the past several years.

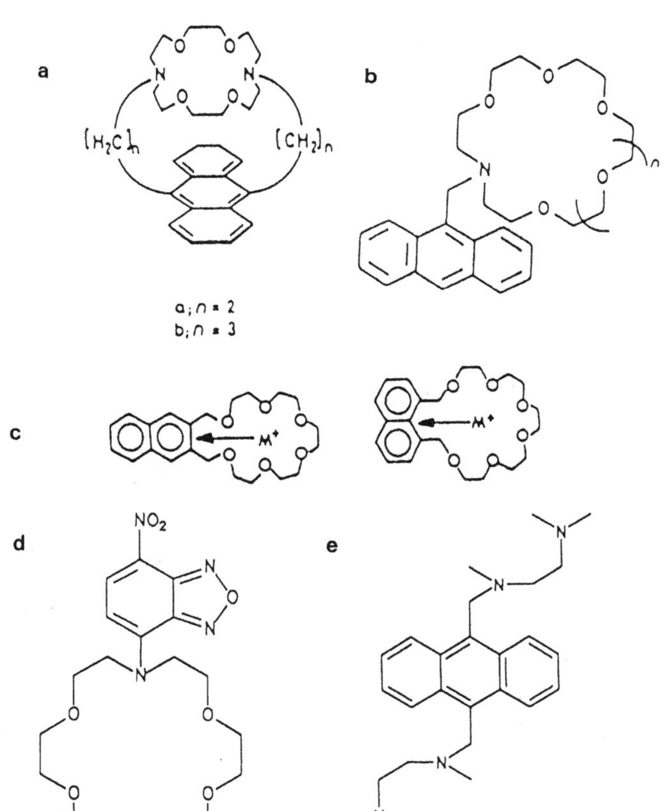

Figure 3.1. Some fluorescent chemosensors for metal ions described by: (a) Konopelski, 1985; (b) de Silva, 1986; (c) Sousa, 1977; (d) Street, 1986; and (e) Czarnik, 1988.

3.1. Chelation-Enhanced Fluorescence in 9,10-Bis(TMEDA)anthracene

Our work on the design of catalysts that associate via reversible covalent bond formation led us in 1987 to report the syntheses (Figure 3.2) of 9,10-bis(TMEDA)anthracene (**1**) and its bis($ZnCl_2$) chelate (**2**).[4] To our surprise, we found that the metal complex was over 1000 times more fluorescent than the free ligand[3i]; the effect is readily apparent without the use of a fluorimeter. A chelation-enhanced fluorescence (CHEF) effect of this magnitude proved unprecedented, and led us to consider both the nature of the fluorescence increase and extensions that might prove useful in providing a generalized "signal" to molecular recognition interactions.

Both the intensity and shape of compound **2**'s emission spectrum closely match that of 9,10-dimethylanthracene (Figure 3.3). What can account for this very large change in fluorescence emission intensity? We believe we are observing fluorescence quenching via exciplex formation, made very efficient in compound **1** by the high effective concentration of the intramolecular amine groups. Fluorescence quenching by inter- and intramolecular amines is, of course, a well-known phenomenon,[5] and the fluorescence of 9,10-dimethylanthracene is quenched by addition of TMEDA as expected. At [DMA] = 0.1 mM in acetonitrile, 4000 equivalents of added TMEDA result in quenching by a factor of 80; the intramolecular quenching at the same concentration of compound **1** with no added TMEDA is roughly 30-times more efficient than even this. Consequently, we explain the observed CHEF in 9,10-di(TMEDA)anthracene by noting that, when chelated to a metal ion, the amine lone pairs become involved in bonding and are unable to donate an electron to the excited state of the anthracene. This explanation is corroborated by the pH profile shown in Figure 3.4 obtained in aqueous solution. At pH 11.7, the amine groups are almost

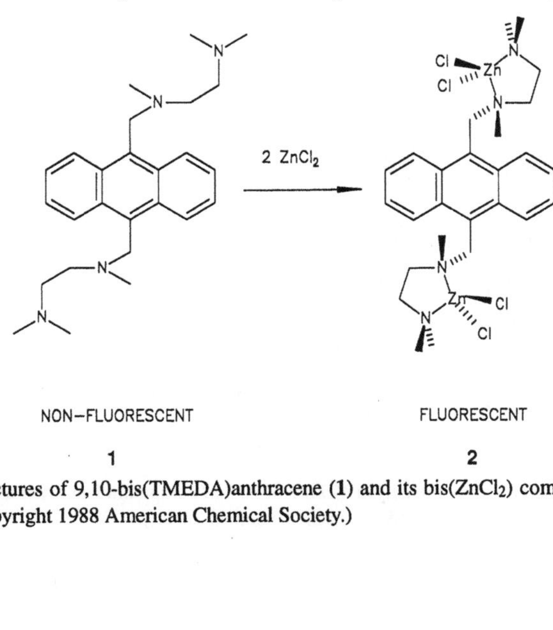

Figure 3.2. Structures of 9,10-bis(TMEDA)anthracene (**1**) and its bis($ZnCl_2$) complex (**2**). (Reproduced from Ref. 3i. Copyright 1988 American Chemical Society.)

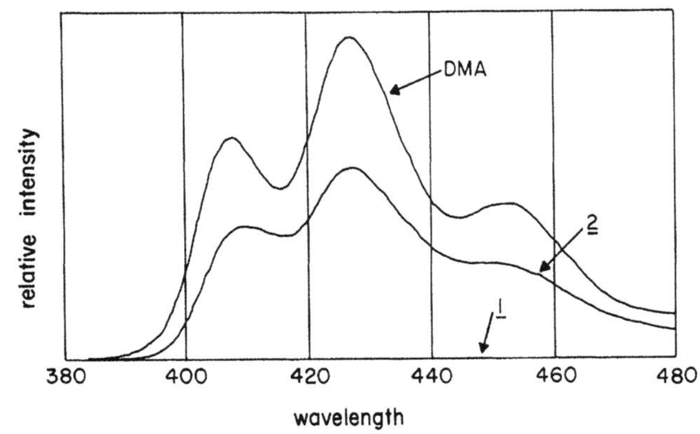

Figure 3.3. Fluorescence emission spectra of compounds **1**, **2**, and 9,10-dimethylanthracene (DMA) in acetonitrile (all 10^{-4} *M* solutions). (Reproduced from Ref. 3i. Copyright 1988 American Chemical Society.)

completely unprotonated and the fluorescence of compound **1** is consequently very low; at pH 1.6, the amines (most or all; one cannot distinguish based on the data in Figure 3.4) are protonated, and the fluorescence increases by a factor of over 300. In acetonitrile, the fluorescence increase may be titrated by addition of metal ion (Figure 3.5). Solubility limits prevented us from further raising the metal ion concentration and therefore determining an asymptotic "intrinsic fluorescence" for complex **2**, which is partially dissociated even in acetonitrile saturated with $ZnCl_2$. The same titration done in aqueous solution fails; it is not surprising that the complexation of zinc ion

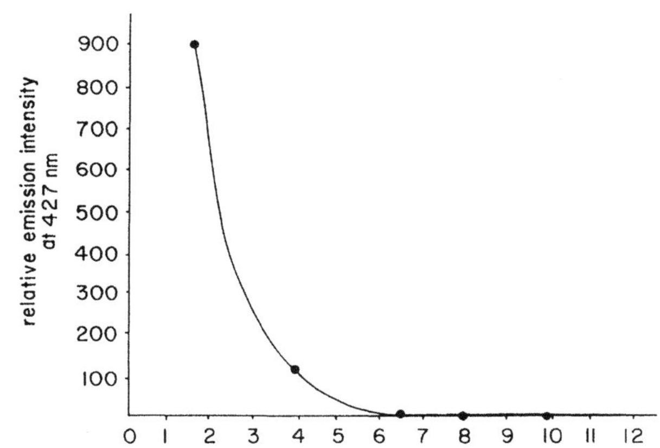

Figure 3.4. Fluorescence emission intensity at 427 nm of 10^{-4} *M* compound **1** in water as a function of pH. (Reproduced from Ref. 3i. Copyright 1988 American Chemical Society.)

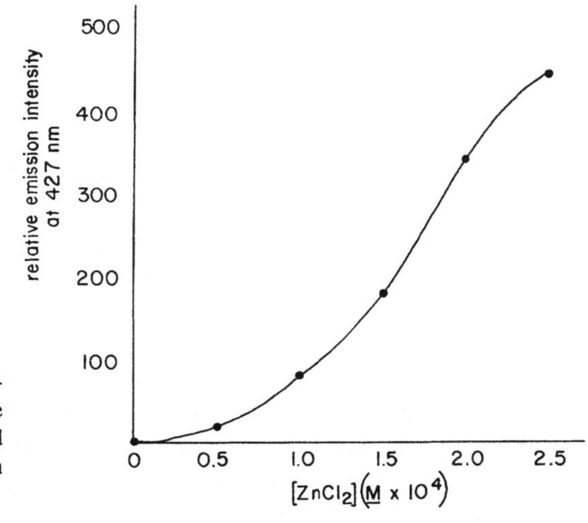

Figure 3.5. Fluorescence emission intensity of 10^{-4} *M* compound **1** in acetonitrile as a function of added $ZnCl_2$. (Reproduced from Ref. 3i. Copyright 1988 American Chemical Society.)

with **1** in water is not complete at equimolar concentrations, and we predicted that cryptands leading to polydentate chelation would provide much larger association constants.

These results suggested that nitrogen-containing ligands with known specificities for metal ions might be utilized as fluorescent chemosensors via a simple, flexible connection to a fluorescent compound such as anthracene. In this scheme, complexation need not change the conformation of the fluorophore as required by an inhibition of vibrational decay mechanism, but it must tie up amine lone pairs. It also seemed possible to tie up amine lone pairs in other interesting ways, such as ion-pairing or strong hydrogen bonding; each interaction suggests potential analytical applications. We have concluded that this approach to the design of fluorescent analytical chemosensors has considerable potential, and have examined the utility of the CHEF method in other media and systems.

3.2. Chelation-Enhanced Fluorescence of Anthlylazamacrocycle Chemosensors in Aqueous Solution

While conjugate probe methods offer considerable potential and flexibility in the design of metal-selective fluoroionophores, large fluorescence changes to date have been seen only in nonaqueous solution. Of course, assays in totally aqueous media are better suited to many potential applications. Accordingly, we have examined the occurrence of CHEF effects in totally aqueous environment using anthrylazamacrocycle chemosensors. Anthrylazamacrocycles **5a–e** were synthesized (Figure 3.6) by the reaction of 9-chloromethylanthracene (**3**) with an excess of the appropriate azamacrocycle (**4, x = 1–4**). The free bases of **5a–e** were obtained as oils after selective

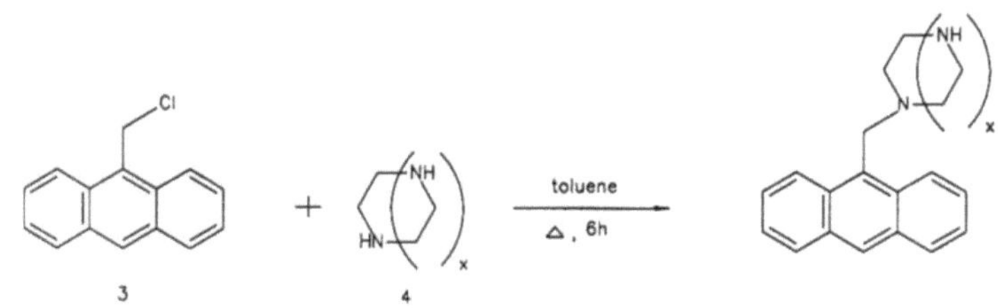

Figure 3.6. Syntheses of anthrylazamacrocycles **5a–e**. **5a**, x = 1; **5b**, x = 2; **5c**, x = 3; **5d**, x = 4; **5e**, x = 5. (Reproduced from Ref. 6. Copyright 1990 American Chemical Society.)

basic extraction. All compounds were subsequently isolated as HCl salts by precipitation from ethanol with concentrated HCl. Integration of the ^{1}H nuclear magnetic resonance (NMR) spectrum in each case demonstrated a 1:1 ratio of anthracene to azamacrocycle protons, confirming that overalkylation had not occurred.[(6)]

The pH-fluorescence profiles for compounds **5a–e** are shown in Figure 3.7. We postulate that protonation at the benzylic nitrogen in all five anthrylazamacrocycles is the key step leading to large fluorescence enhancements; this idea is supported by the studies of Thomas[(7)] and Davidson[(8)] on the distance dependence of exciplex formation in a series of homologous naphthyl- and anthrylalkylamines. While fluorescence quenching is expected at high pH, we also observed some form of quenching under strongly acidic conditions; such a decrease has been observed using other fluorophores, and in these cases is due to an acid-catalyzed, photochemically induced decomposition. Fluorescence titrations with Zn(II) and Hg(II) (net heavy atom quenching)

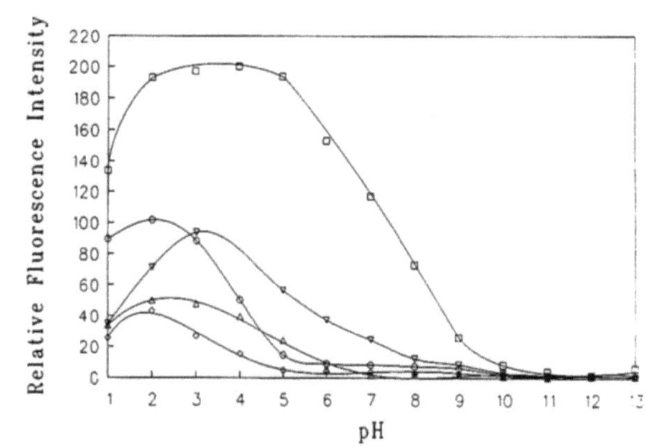

Figure 3.7. pH-Fluorescence profiles for 10 μM solutions of anthrylazamacrocycles **5a** (○), **5b** (△), **5c** (□), **5d** (◇), and **5e** (▽). Excitation was at 335 ± 3 nm; emission was measured at the emission maximum centered near 416 nm. pH's were maintained using the following solutions (all 0.1 *M*): trichloroacetate (pH 1); dichloroacetate (pH 2); chloroacetate (pH 3); acetate (pH 4 and 5); MES (pH 6); HEPES (pH 7 and 8); CHES (pH 9); CAPS (pH 10 and 11), tetrabutylammonium hydroxide (pH 12); NaOH (pH 13). (Reproduced from Ref. 6. Copyright 1990 American Chemical Society.)

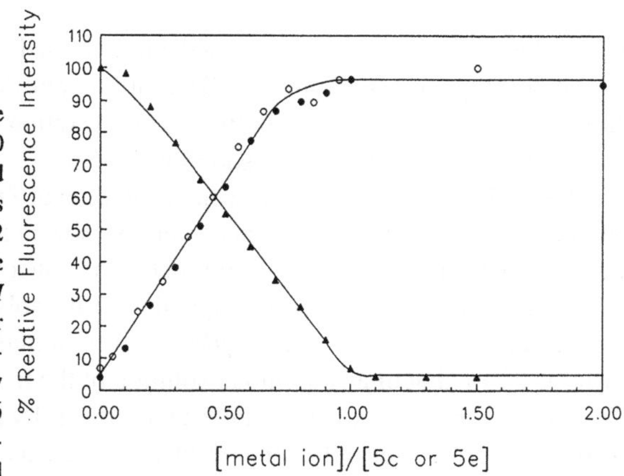

Figure 3.8. Titration of the fluorescence of anthrylazamacrocycles **5c** and **5e** (10 and 100 μM, respectively) by added metal ions. The three experiments shown are: (●) titration of **5c** in pH 12 buffer with $Zn(ClO_4)_2$, (○) titration of **5c** in pH 12 buffer with $Zn(ClO_4)_2$ and 1 M $NaClO_4$; (▲) titration of **5e** in pH 6 buffer with $Hg(ClO_4)_2$. The chelation-enhanced quenching (CHEQ) of **5e** by Hg(II) ion was accomplished at pH 6 because, while the azamacrocycle is sufficiently deprotonated at this pH to bind Hg(II), it is also sufficiently protonated to display some background fluorescence (Figure 3.7). Excitation was at 335 ± 3 nm; emission was measured at the emission maximum centered near 416 nm. (Reproduced from Ref. 6. Copyright 1990 American Chemical Society.)

are shown in Figure 3.8. Both chelation-enhanced fluorescence (CHEF) and chelation-enhanced quenching (CHEQ) are observed for these ions, with overall emission changes of 25- and 20-fold, respectively. At pH 12, the presence of a large (100,000-fold) excess of added $NaClO_4$ does not interfere with the titration of Zn(II) ion. The absorption spectra remain virtually unchanged in the Zn(II) titration experiments. In an effort to test the limits of signal range in aqueous solution, we carried out the titration of **5c** with Cd(II) under strongly basic conditions (pH 13); because of the very low background fluorescence at this pH, a CHEF effect of 190-fold was obtained on addition of a saturating amount (2 equivalents) of $Cd(ClO_4)_2$.

An observed fluorescence dependence on pH is in keeping with an intramolecular amine quenching mechanism. Protonation of an amine group in fluorophore–amine conjugates results in the elimination of photoinduced electron transfer. Therefore, fluorescence is expected to be a function of pH, and pH measurement using anthrylamines has been described by de Silva.[9] The maximal emission intensity of equimolar solutions of **5a–e** at pH 2–3 varies significantly with **5c** clearly the most fluorescent; this interesting but unexplained finding is the focus of current study. In addition, it is noteworthy that compound **5c** functions as a fluorescent pH indicator with a nearly linear response between pH 5 and 10, although its metal binding properties make **5c** nonideal for this purpose.

Our survey of CHEF effects with various anthrylazamacrocycles and metal ions has permitted us to put potential interactions into three categories:

Case 1: Complexation with fluorescence enhancement. If the metal ion itself is not quenching, binds to the azamacrocyle, and does not form a complex capable of absorbing the emitted light, a large CHEF effect (25- to 190-fold, depending on the metal, the anthrylazamacrocycle, and the pH used) is observed. Zn(II) and Cd(II) display this type of behavior. A Zn(II) titration of **5c** demonstrates a linear response as shown in Figure 3.8; nearly complete complexation is observed on addition of one equivalent of Zn(II) to a 10 μM solution of **5c**.

The CHEF effects we observe on the binding of Zn(II) and Cd(II) are significantly smaller than the >1000-fold enhancement we saw in acetonitrile solution.[3i] We have observed that the background fluorescence of noncomplexed anthrylpolyamines increases as the hydrogen-bonding capability of the solvent increases. For example, the background fluorescence of a 10 μM solution of anthrylazamacrocycle **5c** in pH 12 buffer is 230% that of an equimolar solution of the free base of **5c** in acetonitrile. This sensitivity to protic solvent is quite predictable; hydrogen bonding of the benzylic nitrogen will also serve to remove electron density from that position and reduce electron transfer quenching. Thus, fluorescence becomes a potentially useful tool for studying the microenvironment about the benzylic nitrogen.

Case 2: Complexation with intracomplex quenching. If a quenching metal ion (e.g., open-shell, paramagnetic, large or easily reducible cation) binds tightly to the anthrylazamacrocycle derivative, intracomplex quenching takes place.[10] The fluorescence of compounds **5b–e** can be titrated down as shown in Figure 3.8 for **5e** and Hg(II). A higher concentration of the anthrylazamacrocycle is used in this case to provide sufficient emission even in the last 20% of the titration.

Case 3: No complexation. If the binding interaction is not strong enough there is no effect on the fluorescence. The fluorescence intensity of solutions of piperazinyl derivative **5a** does not change on addition of metal ions. Ca(II) and Al(III), both very weak binders to the macrocyclic polyamines,[11] have no effect on the fluorescence of any anthrylazamacrocycle. Na^+ also has no effect, and as shown in Figure 3.8 the titration profile of **5c** with Zn(II) is virtually unaffected even by the presence of a 100,000-fold excess of Na^+ at pH 12.

In summary, we have shown that conjugate fluorescent chemosensors, fluoroionophores in which the metal ligand is not an integral part of the aromatic π-system, demonstrate large (20- to 190-fold) signal changes on transition metal ion binding in 100% aqueous solution. The pH dependence of fluorescence emission intensity is consistent with the elimination of photoinduced electron transfer via amine protonation, and it seems likely that CHEF effects on the binding of Zn(II) and Cd(II) result from a similar mechanism. Under some conditions, the presence of 1 M sodium ion does not interfere with the titration of μM amounts of Zn(II). The binding of inherently quenching metals, such as Cu(II) and Hg(II), results in intracomplex quenching even though both ions are also expected to complex to the quenching amine. In that only a benzylic amine results in a nonemissive exciplex, it seems likely that it is principally the benzylic amine that is responsible for quenching of the uncomplexed anthrylazamacrocycles. Hydrogen bonding to the benzylic amine also results in increased fluorescence, although to a lesser extent than protonation or metal ion complexation.

3.3. Chelatoselective Fluorescence Perturbation in an Anthlylazamacrocycle CHEF Sensor

While fluorimetric methods for the determination of some metal ions in aqueous solution exist using intrinsic probes,[(12)] selective methods for the determination of Zn(II) and Cd(II) do not. As described in the preceding section, anthrylazamacrocycle chemosensors **5b–e** yield large (20- to 190-fold) changes in fluorescence on metal ion complexation in aqueous solution[(6)]; the very large association constants between several transition metals (e.g., Pb(II), Cu(II), Zn(II), Cd(II), Hg(II)) and azamacrocycles[(13)] make the sequestration (and therefore quantitation) of small amounts of such ions possible. Only Zn(II) and Cd(II) bind anthrylazamacrocycles with net CHEF; however, assigning an enhancement to one metal or the other has not been possible heretofore. We have observed that the complexation of Cd(II) and anthrylmethylpentacyclen (**5d**) uniquely demonstrates a perturbation of the fluorophore emission spectrum; the resulting ion discrimination can be utilized directly for simultaneous Zn(II)/Cd(II) analysis.

Normalized emission spectra for the complexes of Zn(II) and Cd(II) perchlorates with anthrylazamacrocycles **5b–e** (Figure 3.6) are shown in Figure 3.9. In each case an anthracenic fluorescence spectrum is observed. However, the Cd(II)-**5d** complex displays an additional broad, red-shifted band yielding the composite spectrum with λ_{max} 446 nm. A typical anthracenic emission is observed for Zn(II) and Cd(II) complexes of (9′-anthrylmethyl)-1,4,7,10,13-pentaazatridecane, a linear analogue of **5d**. The unique fluorescence behavior of the Cd(II)-**5d** complex along with the large binding constants of Zn(II) and Cd(II) to pentacyclen[(13a,b)] allow for the simultaneous determination of each metal ion in aqueous solution. When the total Zn(II) and Cd(II)

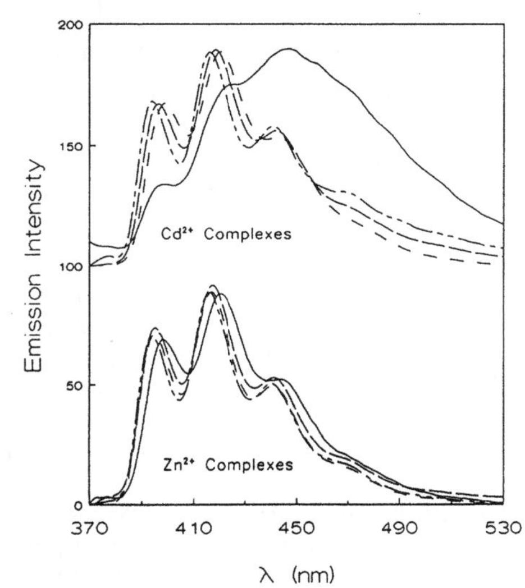

Figure 3.9. Normalized fluorescence spectra of 10 μM solutions of **5b–e** containing $Cd(ClO_4)_2$ (*above*) and $Zn(ClO_4)_2$ (*below*) in 0.1 *M* pH 10 CAPS buffer. Excitation was at 335 ± 3 nm. **5b**:— - -; **5c**: - - - - ; **5d**: ———; **5e**:– – – –. (Reproduced from Ref. 24. Copyright 1990 American Chemical Society.)

concentration is less than that of the probe (10 μM), the concentrations of Zn(II) and Cd(II) can be expressed as in Eqs. (3.1) and (3.2).

$$[Zn(II)] = 0.54\, I_{398} - 0.62\, I_{516} - 1.2 \quad (3.1)$$

$$[Cd(II)] = 1.1\, I_{516} - 0.037\, I_{398} - 0.11 \quad (3.2)$$

In addition to a fluorescence perturbation, the Cd(II)-**5d** combination also uniquely yields a perturbation in the ultraviolet (UV) spectrum. A difference spectrum obtained by subtracting a fractional amount of an uncomplexed **5d** spectrum from the perturbed spectrum is the mirror image of a fluorescence difference spectrum obtained by similar means. Moreover, excitation at 400 nm (where **1–4** are weakly absorbing but where moderate absorption is seen in the difference spectrum) gives rise to an emission spectrum with identical shape and λ_{max} (456 nm) to that of the fluorescence difference spectrum. Thus, evidence points to the existence of two equilibrating ground state species as the physical basis for the chelatoselective emission. Bouas-Laurent has reported a related observation in methanol where a red-shifted CHEF was observed for a Tl(I) π-complex.[14]

The ^{1}H NMR spectrum of the Cd(II)-**5d** complex in D_2O solution reveals the presence of more than one chelate in solution, and H-10 is clearly represented by major and minor singlets (ratio = 6:4). All aromatic peaks are noticeably broadened with respect to a spectrum of **5d** taken in the absence of Cd(II) or in the presence of Zn(II). Stepwise heating of the solution (300–360K) leads to a gradual coalescing of the aromatic resonances. The spectra of all other permutations of **5b–5e** with Zn(II) and Cd(II) show only one species, qualitatively similar to that of the Zn(II)-**5d** complex. Most interestingly, the Cd(II)-**5d** complex uniquely experiences deuterium exchange ($k = 1.5 \times 10^{-3}$ sec^{-1} at 343K and pH 8), occurring only at the 1 and 8 positions.

The observed fluorescence, UV, and ^{1}H NMR perturbations of the Cd(II)-**5d** complex are observed only in water; in methanol, ethanol, and acetonitrile only unperturbed anthracenic spectra are observed.

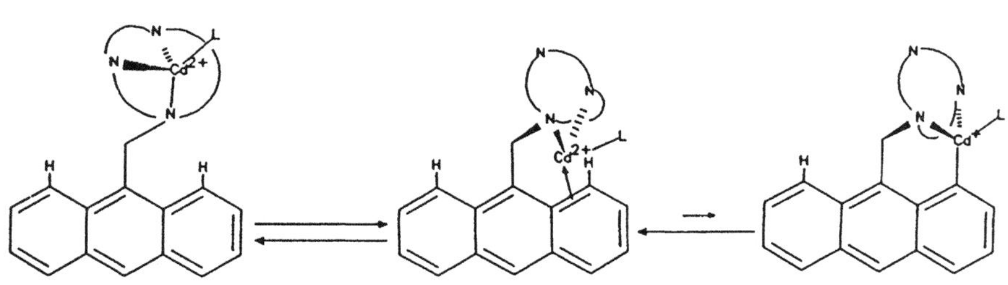

a b c

Figure 3.10. A proposal rationalizing the observed fluorescence perturbation and selective deuteration of the **5d**-Cd(II) complex: (a) n-Ar-Cd(II) complex; (b) Ar-Cd(II) π-complex; (c) Ar-Cd(II) σ-complex. (Reproduced from Ref. 24. Copyright 1990 American Chemical Society.)

These observations are most consistent with the equilibria shown in Figure 3.10. (1) The Cd(II)-**5d** complex uniquely populates a conformer in which an anthracene-Cd(II) π-d orbital interaction is enforced. (2) The π-complex leads to a higher-energy σ-complex that results in deuterium exchange. (3) Solvation strongly influences the position of the conformational equilibrium; the complete selectivity for water versus methanol argues for stringent external ligand steric requirements in the π-complex. Arylcadmium species are well known, but previously restricted to anhydrous environments.[15] In fact, cadmium's position above mercury in the periodic table portends activity as an electrophile toward aromatics. Conclusions regarding the structural basis of chelatoselective fluorescence perturbation suggest variations on the incorporation of such "nonclassical" selectivity into future fluoroionophores.

3.4. Chelation-Enhanced Fluorescence Detection of Nonmetal Ions

Both conjugate and integral fluorescent chemosensors have been applied to date almost exclusively to the detection of metal ions. Because intracomplex protonation or hydrogen bonding at a benzylic nitrogen are each expected to result in CHEF, we examined the interaction of several anions with anthracene-based conjugate fluorescent chemosensors. Anthrylpolyamines **6** and **7** (Figure 3.11) were synthesized by the reaction of tris(3-aminopropyl)amine with 9-chloromethylanthracene and 9,10-bis(chloromethyl)anthracene, respectively; both compounds were isolated and characterized as their HCl salts.[16]

As described previously, examination of the literature leads to the conclusion that a change in protonation or chelation state of a benzylic nitrogen leads to large fluorescence enhancements. Thus, trication **8** (Figure 3.12) is the ionic form of **6** that

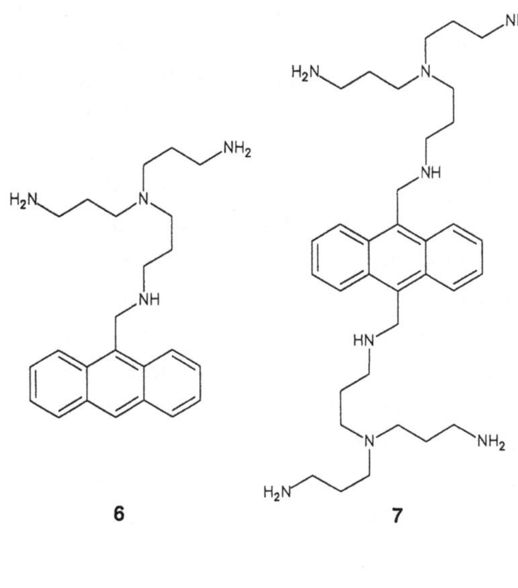

6 **7**

Figure 3.11. Structures of anthrylpolyamines **6** and **7**. (Reproduced from Ref. 16. Copyright 1989 American Chemical Society.)

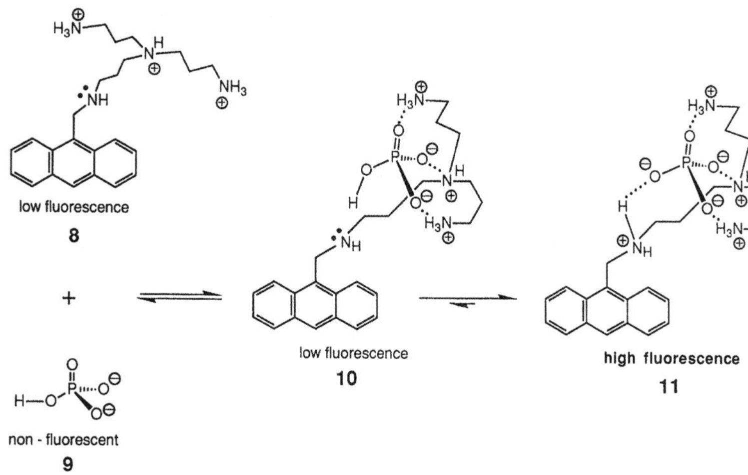

Figure 3.12. A working hypothesis as to how the binding of phosphate (**9**) to anthrylpolyamine **8** can afford a CHEF effect. Intracomplex protonation (or strong hydrogen bonding) of **10** by the HPO_4^{-2} ion leads to loss of the benzylic nitrogen lone pair (**11**), thereby removing this source of intramolecular quenching. (Reproduced from Ref. 16. Copyright 1989 American Chemical Society.)

can act as an anion sensor at pH 6. Complexation of the anionic phosphate oxygens with ammonium ions on **8** places the remaining phosphate OH group in close proximity to the free amine; this species (**10**) will demonstrate low fluorescence due to quenching by the free amine group. However, favorable intracomplex proton transfer will lead to **11** in which intramolecular quenching is eliminated and higher fluorescence is observed. Our analysis using the HPO_4^{-2} ion as shown in Figure 3.12 illustrates the general principle by which CHEF is seen for anions.

Species such as sulfate and acetate, which yield smaller fluorescence enhancements upon binding, are totally dissociated at pH 6 and thus cannot deliver a proton directly; however, the observation of CHEF in these complexations does not refute the mechanism put forward in Figure 3.12. Intracomplex protonation (or strong hydrogen bonding) of **10** by the HPO_4^{-2} ion as shown in Figure 3.12 is indistinguishable from the binding of **8** to the PO_4^{-3} ion with subsequent enhanced benzylic amine protonation from the solvent. Such enhanced protonation could result from either: (1) enhanced basicity of amine groups on ion-pairing of neighboring ammonium ions to the anion, or (2) enhanced amine protonation resulting from intracomplex hydrogen bonding of the ammonium ion to a nearby guest oxygen (e.g., **11**). If a significant concentration of PO_4^{-3} was present at pH 6, one would anticipate a CHEF effect on its binding to **8**. Fluorescence enhancement upon the binding of acetate, sulfate, or dimethyl phosphate can likewise result from effects (a) and (b), which differ from that shown in **11** principally in the basicity of the bound anion.

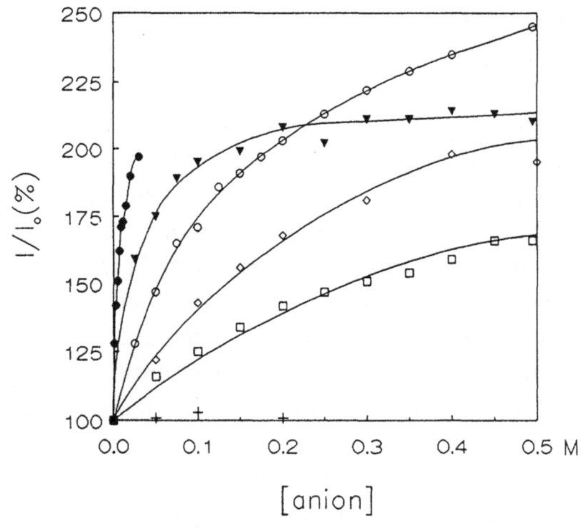

Figure 3.13. Titration of the fluorescence of anthrylpolyamine **6** by various added anions. ○, phosphate ion; ▼, sulfate ion; ◇, acetate ion; □, dimethyl phosphate ion; ●, citrate ion; +, TBACl ion.

Fluorescence spectra were recorded for aqueous solutions (all adjusted to pH 6.0) of **6** in the presence of increasing concentrations of several anions (Figure 3.13). Using the above model the concentration of complex **11** was calculated for each solution of known total anion concentration. The effective binding constant of phosphate (an average of all ionic forms) to **8** (log $K_{eq} = 0.82$) at pH 6 along with the corresponding percent fluorescence intensity increase (>145%) could then be obtained. Similarly, values were determined at pH 6 for the binding of **6** to adenosine triphosphate (ATP)

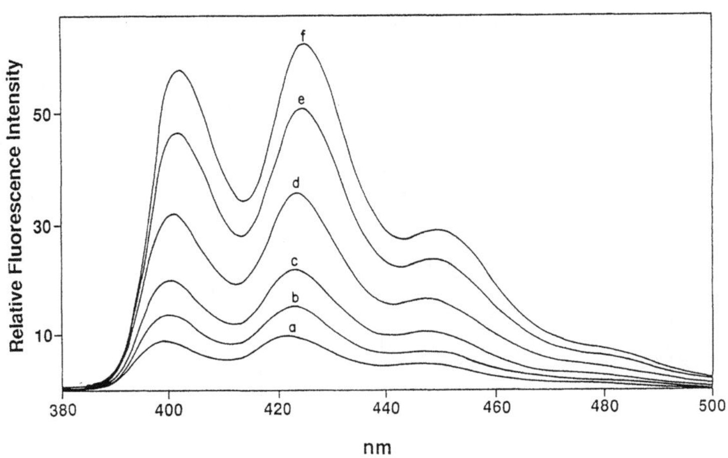

Figure 3.14. Fluorescence spectra for 5 μM solutions of anthrylpolyamine **7** at pH 6 in the presence of increasing concentrations of citrate: (a) none, (b) 0.1 mM, (c) 1 mM, (d) 10 mM, (e) 0.1 M, (f) 0.2 M. Excitation was performed at 335 ± 3 nm. All solutions were adjusted to pH 6. (Reproduced from Ref. 16. Copyright 1990 American Chemical Society.)

(log $K_{eq} = 4.2$, 79%), citrate (log $K_{eq} = 2.3$, 97%), sulfate (log $K_{eq} = 1.6$, 114%), acetate (log $K_{eq} \leq 0.6$, >98%), and dimethyl phosphate (log $K_{eq} \leq 0.5$, >66%). As an indication that even larger fluorescence enhancements are likely with structurally modified conjugate chemosensors, we have observed a sixfold CHEF effect for the binding of citrate to anthrylbispolyamine **7** (Figure 3.14). These results demonstrated that intra-complex protonation of a quenching nitrogen leads to CHEF effects in aqueous solution in the same way that metal ion chelation does. We believe our results suggest a general and heretofore undescribed method for the fluorogenic "signaling" of anion binding. Since the origin of the effect can be rationalized at the molecular level, a structural basis exists for the design of conjugate chemosensors for ionic and hydrogen bonding guests. Given the almost limitless synthetic approaches to nitrogen-containing hosts,[17] the fabrication of useful analytic tools seems likely to result.

3.5. An Assay for Enzyme-Catalyzed Polyanion Hydrolysis Based on Template-Directed Excimer Formation

While the hydrolyses of DNA and RNA can be followed by monitoring changes in their UV absorption spectra, the hydrolyses of biological polyanions lacking a chromophore must be accomplished indirectly. Both radiolabeling techniques and coupled enzyme systems are used frequently as indirect methods for enzyme assay. We have found that the activity of hydrolytic enzymes acting on the polyanions heparin and polyglutamate can be monitored by fluorescence using a template-directed excimer formation effect obtained with anthrylpolyamine conjugate chemosensors.

Anthrylpolyamines **12–15** (Figure 3.15) were prepared via simple substitution reactions of 9-(chloromethyl)anthracene. The full emission spectra of **12–15** (all 1 μM)

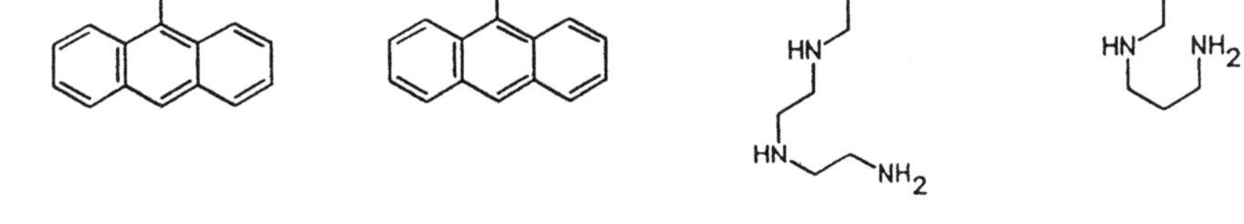

12 **13** **14** **15**

Figure 3.15. Structures of anthrylpolyamines **12–15**. (Reproduced from Ref. 25. Copyright 1990 American Chemical Society.)

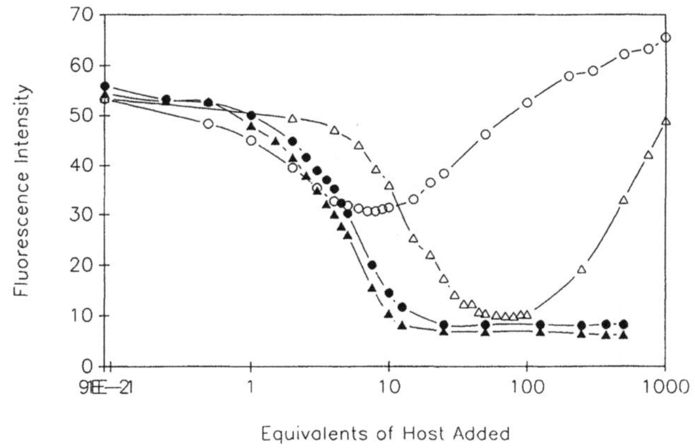

Figure 3.16. Effects on the fluorescence of anthrylpolyamine **14** on titration by four biological polyanions. ○—○, Heparin; △—△. poly-L-glutamate; ●—●, dsDMA; ▲—▲, ssDNA. (Reproduced from Ref. 25. Copyright 1990 American Chemical Society.)

were collected during titration with ds DNA, ss DNA, heparin, and poly-L-glutamate; representative titration data from the monitoring of compound **14** at 422 nm are shown in Figure 3.16. Both disubstituted anthracenes (**14** and **15**) exhibit a 6-nm red shift in λ_{max} when bound either to ds DNA or to ss DNA; likewise, both monosubstituted anthracenes (**12** and **13**) show a 14-nm red shift in their emission spectra in the presence of either ds or ss DNA. Interaction of the nucleotide bases with the anthracene is a likely source of the bathochromic shift and the observed CHEQ effect. Such π-stacking with ss DNA, seldom observed with intercalating compounds, may result from the favorable entropy effect of electrostatic preassociation.[18]

The heparin and poly-L-glutamate titrations show a markedly different behavior than do the DNA titrations. As polyanion is added, the fluorescence of the anthrylpolyamine solution decreases until a well-defined minimum is reached. A new emission at 510 nm, which we assign to the anthracene excimer of **14**, increases and decreases coincidently with the titrated fluorescence minimum. Likewise, the UV spectrum of 10 μM **14** with added heparin shows hypochromism that occurs and disappears coincidently with the fluorescence minimum and a 2-nm red shift. We have proposed template-directed excimer formation as the physical basis for these observations. In the absence of heparin, fluorescence of the unassociated probe is observed. As heparin is added, the fluorescence decreases as a result of heparin-directed interaction between probe molecules. Additional heparin permits the fluorophore population to diffuse over the length of the polyanion, thus avoiding excimer formation and yielding a net CHEF.

The anthrylpolyamine most effective in binding to heparin (**14**) was used to follow the activity of heparinase at pH 5. Samples were prepared containing 1 μM probe **14** and 5 μM heparin in 0.1 M pH 5 NaOAc buffer with 0.05 mM EDTA. Under these

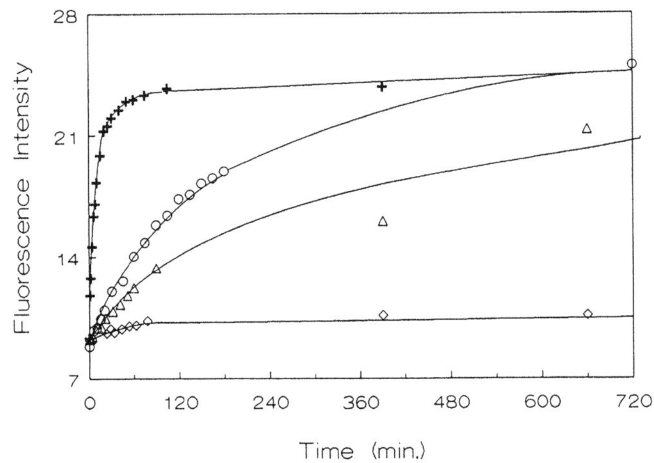

Figure 3.17. Effect of heparinase on the CHEQ of probe **13** with five equivalents of heparin. ◇, No heparinase; △, 10 units heparinase; ○, 25 units heparinase; +, 50 units heparinase. (Reproduced from Ref. 25. Copyright 1990 American Chemical Society.)

conditions, the fluorescence of probe **3** is at its minimum as a result of template-directed excimer formation. Addition of heparinase, an enzyme that hydrolyzes heparin to oligosaccharide units,[19] results in a fluorescence enhancement as shown in Figure

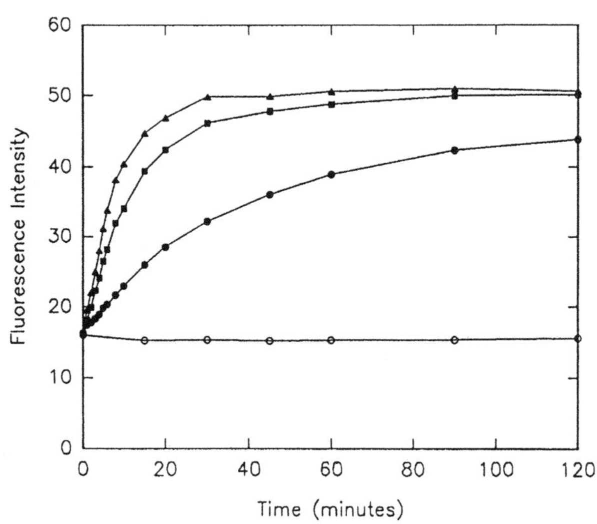

Figure 3.18. Effect of pronase on the CHEQ of probe **13** with 50 equivalents of poly-L-glutamate. ▲, 10 units pronase; ■, 5 units pronase; ●, 1 unit pronase; ○, no pronase.

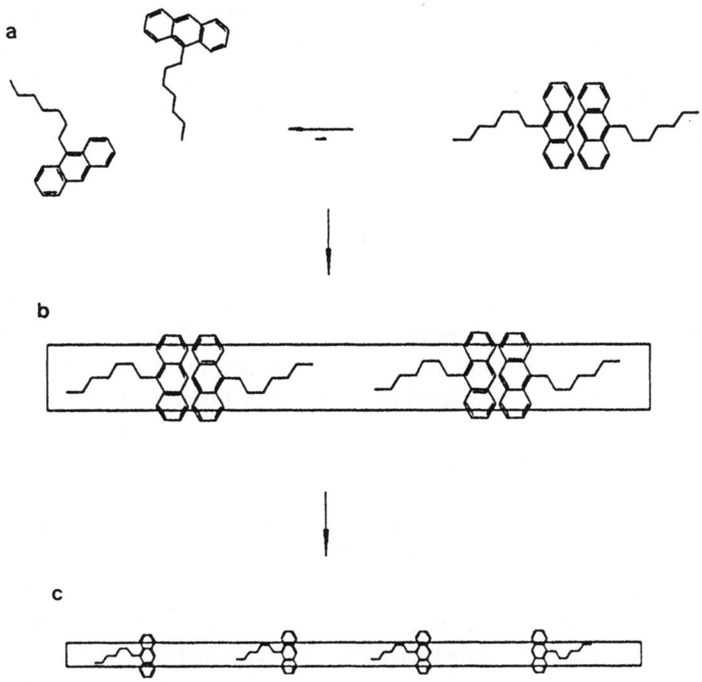

Figure 3.19. Proposed mechanism for template-directed excimer formation. (a) No heparin; (b) minimal heparin (CHEQ); (c) excess heparin (CHEF). (Reproduced from Ref. 25. Copyright 1990 American Chemical Society.)

3.17. One of the most effective polyglutamate binders (**13**) was used to test the activity of pronase at pH 5. Samples were prepared containing 1 μM probe **13** and 90 μM poly-L-glutamate in 0.1 M pH 5 NaOAc buffer with 0.05 mM EDTA. Under these conditions, the fluorescence of probe **13** is also at its minimum. Addition of pronase, a nonselective proteolytic enzyme that hydrolyzes polyglutamate to glutamic acid,[20] results in a fluorescence enhancement (Figure 3.18). Our rationale for the physical basis of these assays is shown in Figure 3.19. The cationic fluoroionophores, which do not associate appreciably in dilute solution, are brought into proximity on binding to the polyanion; excimer formation is thus enforced, and the fluorescence of the probe is quenched. Addition of hydrolytic enzyme causes cleavage of the substrate into fragments that no longer enhance probe aggregation. Consequently, quenching of the fluorescence decreases with time until hydrolysis of the template is complete. Because the binding interaction is principally electrostatic, it should be possible to follow the hydrolyses of other polyanions by this technique as well. Furthermore, it is not even necessary to know the structure of the polyanion for the assay to be useful. For this reason, we believe that the template-directed excimer formation method may prove useful for the assay of other enzymes involved in the anabolism or catabolism of anionic biopolymers.

3.6. Fluorescence Chemosensing of Carbohydrates

Few chemical sensing mechanisms (other than bulk quenching) have been described for neutral molecules; nevertheless, many small analytes of interest are uncharged: glucose, for example. Of course, most carbohydrates are neither fluorescent nor are they fluorescence quenchers; novel signal transduction mechanisms are required.

Anthrylboronic acid **16** in water displays a fluorescence emission centered at 416 nm of similar structure to that displayed by anthracene itself. On addition of base, fluorescence decreases; because the emission can be modulated reversibly, the change is due not to decomposition but to ionization leading to boronate **17** (Figure 3.20). While there are several reasons that the fluorescence of boronate **17** might be quenched compared with that of **16**, the oxidizability of borates suggests electron transfer quenching; photoinduced electron transfer from alkyltriphenylborate salts with resulting fluorescence quenching has been described by Schuster.[21] The pH-fluorescence profile of **16** obtained without buffer is shown in Figure 3.21, from which a pK_a of 8.8 is calculated; this compares favorably to the known phenylboronic acid pK_a of 8.83.[22]

On addition of fructose, the apparent pK_a value decreases, leading to the remaining four curves shown in Figure 3.21. The explanation for this observation lies in the fact that the fructose complex of **16** is a stronger acid than is **16** itself. This result was predicated on the work of Edwards, who reported the same trend in polyol complexes of phenylboronic acid in 1959.[23] As determined in the presence of a near-saturating amount of fructose (100 mM), the apparent pK_a of the fructose-**16** complex is 5.9. The

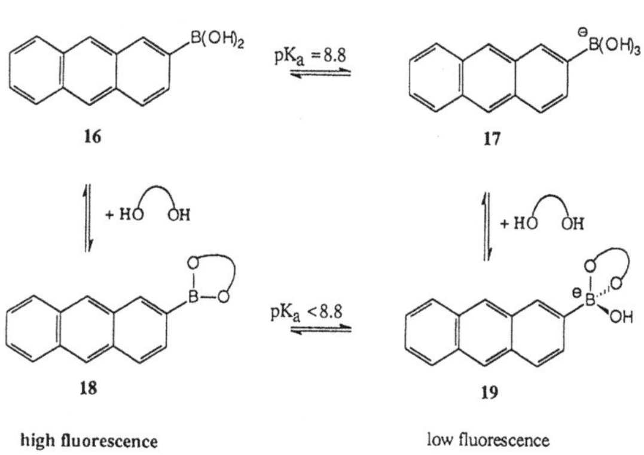

Figure 3.20. Equilibria available to an aqueous solution of anthrylboronic acid **16** in the presence of a generic polyol. Species **16** and **18** display higher fluorescence than do species **17** and **19**. (Reproduced from Ref. 26. Copyright 1992 American Chemical Society.)

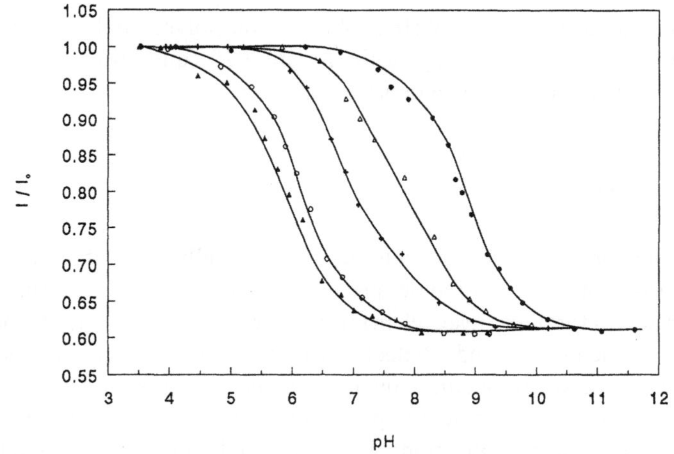

Figure 3.21. pH-Fluorescence titrations of anthrylboronic acid **16** (0.75 μM) as a function of fructose concentration (●, 0 mM; △, 2 mM; +, 5 mM; ○, 50 mM; ▲, 100 mM). All solutions contain 1% (v/v) dimethyl sulfoxide (DMSO). The apparent pK_a's at these concentrations are, respectively, 8.8, 7.8, 7.0, 6.2, and 5.9. (Reproduced from Ref. 26. Copyright 1992 American Chemical Society.)

greatest signal range available is therefore at a pH that is the average of pK_a(**16**) and pK_a(**16×fructose**), or 7.35. Titrations of selected polyols, determined at pH 7.4, are shown in Figure 3.22; the apparent fructose dissociation constant at that pH is 3.7 mM. The stability trends we observe are in good agreement with the reported trends using

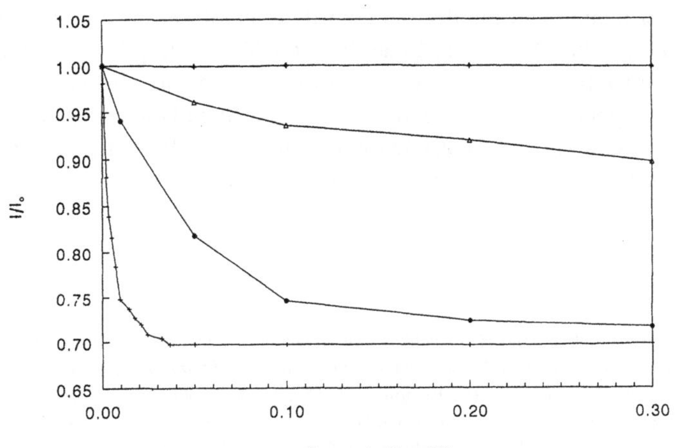

Figure 3.22. Fluorescence titrations of anthrylboronic acid **16** (0.75 μM) at pH 7.4 (20 mM phosphate buffer) as a function of polyol concentration (✚, fructose; ●, 1,1,1-tris(hydroxymethyl)ethane; △, glucose; +, ethylene glycol). All solutions contain 1% (v/v) dimethylsulfoxide (DMSO). (Reproduced from Ref. 26. Copyright 1992 American Chemical Society.)

phenylboronic acid in water.[23] CHEQ obtained on polyol binding at constant pH results, in essence, from a shifting of the Figure 3.20 equilibrium from **16** (higher fluorescence) toward **19** (lower fluorescence).

3.7. Conclusion

Anthrylpolyamines have proven to be a synthetically accessible yet rich source of chemosensors for a variety of ionic species in aqueous solution. The challenges that present themselves for future research in this field are intimately related to the potential applications of such compounds. Selectivity (metal ion versus anion versus proton) will continue to be of the greatest interest. It must be appreciated, however, that absolute selectivity is a theoretical impossibility; thus, useful selectivity ranges will be defined for particular applications. Excitation and emission wavelengths can, in principle, be engineered to avoid background absorption or autofluorescence; again, the range of usefulness must be defined prior to compound design. Down the road, the coupling of chemosensor with fiber optic methods to make remote sensing devices will create a whole new regime of questions relating to sensing on surfaces. All these issues, sitting resolutely at the interfaces of synthesis, coordination chemistry, and photochemistry, await definition and resolution.

Acknowledgments

The experimental and intellectual contributions of Drs. Xavier Cherian, Engin Akkaya, Mike Huston, Mi-Young Chae, Juyoung Yoon, Sung Yeap Hong, and Scott Van Arman are represented by the work described in this chapter. We gratefully acknowledge support for this work from The Ohio State University, the A. P. Sloan Foundation, the Camille and Henry Dreyfus Foundation, the National Science Foundation, Merck & Co., and Eli Lilly and Company.

References

1. (a) G. Schwarzenbach and H. Flaschka, *Complexometric Titrations* (H. Irving, trans.), Methuen, New York (1969); (b) T. S. West, *Complexometry with EDTA and Related Reagents*, BDH, New York (1969); (c) *Indicators* (E. Bishop, ed.), Chapter 6, Pergamon, Oxford (1972); (d) G. G. Guilbault, *Practical Fluorescence*, Chapter 6, Marcel Dekker, Inc., New York (1973); (e) R. J. Clarke, J. H. Coates, and S. F. Lincoln, *Inorg. Chim. Acta, 153*, 21 (1988).
2. (a) G. Grynkiewicz, M. Poenie, and R. Y. Tsien, *J. Biol. Chem. 260*, 3440 (1985); (b) R. Y. Tsien, *Soc. Gen. Physiol. Ser., 40*, 327 (1986).
3. (a) L. R. Sousa and J. M. Larson, *J. Am. Chem. Soc., 99*, 307 (1977); (b) H. Nishida, Y. Katayama, H. Katsuki, H. Nakamura, M. Takagi, and K. Ueno, *Chem. Lett.* 1853 (1982); (c) O. S. Wolfbeis and H. Offenbacher, *Monat. Chem. 115*, 647 (1984); (d) J. P. Konopelski, F. Kotzyba-Hibert, J.-M. Lehn,

J.-P Desvergne, F. Fages, A. Castellan, and H. Bouas-Laurent, *J. Chem. Soc. Chem. Commun.* 433 (1985); (e) A. P. de Silva and S. A. de Silva, *J. Chem. Soc., Chem. Commun.* 1709 (1986); (f) K. W. Street and S. A. Krause, *Anyl. Lett. 19*, 735 (1986); (g) S. Ghosh, M. Petrin, A. H. Maki, and L. R. Sousa, *J. Chem. Phys. 87*, 4315 (1987); (h) *Chemical and Engineering News*, November 9, 1987, p. 26; (i) M. E. Huston, K. W. Haider, and A. W. Czarnik, *J. Am. Chem. Soc. 110*, 4460 (1988); (j) S. Fery-Forgues, M.-T. Le Bris, J.-P. Guette, and B. Valeur, *J. Phys. Chem. 92*, 6233 (1988); (k) S. J. Ganion, R. W. Stevenson, B. Son, C. Nikolakaki, P. L. Bock, and L. R. Sousa, Abstract ORGN 133 from the 197th ACS National Meeting, Dallas, Texas (1989); (l) F. Fages, J.-P. Desvergne, H. Bouas-Laurent, P. Marsau, J.-M. Lehn, F. Kotzyba-Hibert, A.-M. Albrecht-Gary, and M. Al-Joubbeh, *J. Am. Chem. Soc. 111*, 8672 (1989); (m) J. Bourson, and B. Valeur, *J. Phys. Chem. 93*, 3871 (1989).

4. P. Nanjappan and A. W. Czarnik, *J. Amer. Chem. Soc. 109*, 1826 (1987).
5. M. Gordon and W. R. Ware, eds. *The Exciplex*, Academic Press: New York (1975).
6. E. U. Akkaya, M. E. Huston, and A. W. Czarnik, *J. Am. Chem. Soc. 112*, 3590 (1990).
7. E. A. Chandross and H. T. Thomas, *Chem. Phys. Lett. 9*, 393 (1971).
8. D. R. G. Brimage and R. S. Davidson, *J. Chem. Soc., Chem. Comm.* 1385 (1971).
9. A. P. de Silva and R. A. D. D. Ripasinghe, *J. Chem. Soc., Chem. Comm.* 1669 (1985).
10. (a) A. W. Varnes, R. B. Dodson, and E. L. Wehry, *J. Am. Chem. Soc. 94*, 946 (1972); (b) J. A. Kelmo and T. M. Shepherd, *Chem. Phys. Lett. 47*, 158 (1977); (c) S. Formosinho, *J. Mol. Photochem. 7*, 13 (1976).
11. For a review, see N. S. Poonia and A. V. Bajaj, *Chem Rev. 79*, 389 (1979).
12. (a) G. Schwarzenbach and H. Flaschka, *Complexometric Titrations*, (H. Irving, trans.), Methuen, New York (1969); (b) T. S. West, *Complexometry with EDTA and Related Reagents*, BDH, New York (1969); (c) *Indicators* (E. Bishop, ed.), Pergamon, Oxford (1972); (d) G. G. Guilbault, *Practical Fluorescence*, Chapter 6, Marcel Dekker, New York (1973).
13. (a) M. Kodama and E. Kimura, *J. Chem. Soc., Dalton Trans.* 1081 (1978); (b) M. Kodama, E. Kimura, and S. Yamaguchi, *J. Chem. Soc., Dalton Trans.* 2536 (1980); (c) R. D. Hancock, R. Bhavan, C. A. Wagner, and G. D. S. Hosken, *Afr. J. Chem. 39*, 238 (1986).
14. F. Fages, J.-P. Desvergne, H. Bouas-Laurent, P. Marsau, J.-M. Lehn, F. Kotzyba-Hibert, A.-M. Albrecht-Gary, and M. Al-Joubbeh, *J. Am Chem. Soc. 111*, 8672 (1989). An analogous ground state species was also reported for the Ag(I) π-complex; however, it is nonemissive.
15. (a) J. L. Atwood, D. E. Berry, S. R. Stobart, and M. Zorworotko, *J. Inorg. Chem. 2*, 3480 (1983); (b) A. Osman, R. G. Steevensz, D. G. Tuck, H. A. Meinema, and J. G. Noltes, *Can. J. Chem. 62*, 1698 (1984).
16. M. E. Huston, E. U. Akkaya, and A. W. Czarnik, *J. Am. Chem. Soc. 111*, 8735 (1989).
17. For an excellent overview of the great variation achievable in the design of polyammonium receptors, see F. P. Schmidtchen, *Nachr. Chem., Tech. Lab. 8*, 10 (1988).
18. Previous work on polyamine DNA probes has been reported by Gabbay and Ware: (a) E. J. Gabbay, *J. Am. Chem. Soc. 91*, 5136 (1969); (b) E. R. Ware, J. W. Klein, and K. Zero, *Langmuir 4*, 458 (1988).
19. (a) R. J. Linhardt, D. M. Cohen, and K. G. Rice, *Biochemistry 28*, 2888 (1989); (b) R. J. Linhardt, G. L. Fitzgerald, C. L. Cooney, and R. Langer, *Biochim. Biophys. Acta 702*, 197 (1982).
20. D. G. Smyth, *Methods Enzymol. 11*, 214 (1967).
21. (a) S. Chatterjee, P. D. Davis, P. Gottschalk, M. E. Kurz, B. Sauerwein, X. Yang, and G. B. Schuster, *J. Am. Chem. Soc. 112*, 6329 (1990); (b) G. B. Schuster, *Pure Appl. Chem. 62*, 1565 (1990).
22. (a) H. Nakatani, T. Morita, and K. Hiromi, *Biochim. Biophys. Acta 525*, 423 (1978); (b) J. Juillard and N. Geugue, *C. R. Acad. Paris C 264*, 259 (1967).
23. J. P. Lorand and J. D. Edwards, *J. Org. Chem. 24*, 769 (1959).
24. M. E. Huston and A. W. Czarnik, *J. Am. Chem. Soc. 112*, 7054 (1990).
25. S. A. Van Arman and A. W. Czarnik, *J. Am. Chem. Soc. 112*, 5376 (1990).
26. J.-Y. Yoon and A. W. Czarnik, *J. Am. Chem. Soc. 114*, 5874 (1992).

Additional Reading

For the work of several laboratories on fluorescent chemosensors, see: *Fluorescent Chemosensors for Ion and Molecule Recognition* (A. W. Czarnik, ed.), ACS Books, Washington, D.C. (1993).

4

Design and Applications of Highly Luminescent Transition Metal Complexes

J. N. Demas and B. A. DeGraff

4.1. Introduction

Molecular probes and remote sensor technology is currently undergoing a flurry of activity. Particularly promising are devices based on luminescence materials as the indicator. Luminescence probes and sensors are especially attractive because of their potentially high sensitivity and specificity. Currently, sensors include remote fiber-optic-based luminescence systems for measuring O_2, pH, partial pressure of carbon dioxide (pCO_2), temperature, and for immunoassay. Response is monitored by changes in luminescence intensity, lifetime (τ), or spectral distribution.[1–3] Many sensors utilize luminophores in or on a polymer support. There are two major challenges to chemists. First, one must be able to systematically design new materials with responses to specific analytes and possessing specific luminescence properties. Then, since many luminescence sensors are supported in or on some type of polymeric structure, one must have an intimate understanding of the interactions between the sensor or sensitizer and the matrix. In many ways understanding the interactions is more complex than sensor molecule design.

Until relatively recently, most work focused on organic luminophores as sensor-probe materials. However, luminescent transition metal complexes, especially those with platinum metals (Ru(II), Os(II), Re(I), Rh(III), and Ir(III)) have shown considerable promise and are receiving increasing attention. More recently Pt(II) complex have shown promising results.[4] Many of these materials have highly desirable features:

- Long τ's (hundreds of nanoseconds to >100 μsec), which make τ measurements much simpler and less expensive compared to the low-nanosecond lifetimes of most organic fluorophores.

J. N. Demas • Department of Chemistry and Biophysics, University of Virginia, Charlottesville, Virginia 22908. *B. A. DeGraff* • Department of Chemistry, James Madison University, Harrisonburg, Virginia 22807.

Topics in Fluorescence Spectroscopy, Volume 4: Probe Design and Chemical Sensing, edited by Joseph R. Lakowicz. Plenum Press, New York, 1994.

- High-wavelength independent luminescence quantum yields that can exceed 0.5, although 0.04 to 0.2 is more typical.[5–7] However, even these modest values are adequate for a variety of sensor applications.
- Intense visible absorptions; this increases sensitivity, simplifies sensor design, and expands the variety of excitation sources available.
- High thermal, chemical, and photochemical stability, which simplifies sterilization and extends sensor lifetime.

Figure 4.1 shows a typical absorption spectrum ($[Ru(bpy)_3]^{2+}$, bpy = 2,2′-bipyridine), which is the prototype inorganic photosensitizer and sensor. Note in particular the intense blue-ultraviolet (UV) absorptions and the wavelength independence of the luminescence efficiency. In the blue region, a 10^{-4} *M* solution would absorb over 90% of the light in a 1-cm path length.

One of the beauties of this class of inorganic complexes is that molecular engineering permits systematically altering spectroscopic and chemical properties. This chemical flexibility allows design of systems that respond to specific environmental variables, allows ionic or covalent attachment to a support or reagents, and permits tailoring absorption-emission properties to available excitation sources and detectors. While design criteria for making highly luminescent species in homogeneous media are well established,[7] an intimate understanding of the detailed interactions

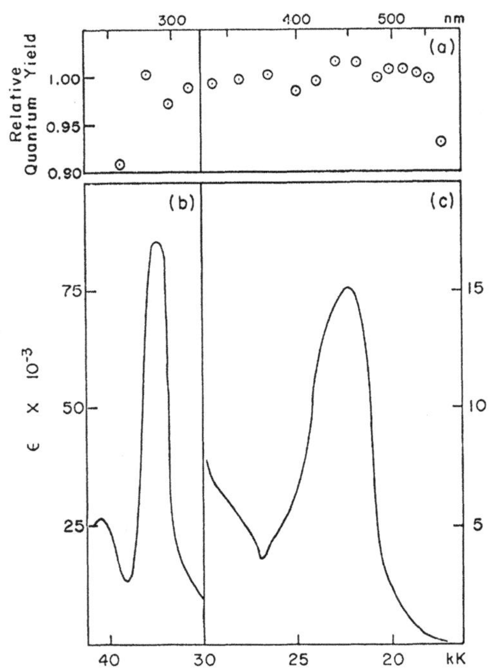

Figure 4.1. Relative luminescence quantum yield (a) and absorption spectrum of $Ru(bpy)_3^{2+}$ in methanol at room temperature (b, c) (7). The decline in luminescence yields at the extremes are experimental artifacts. (Reprinted from Ref. 10 with permission. Copyright 1971 American Chemical Society.)

Table 4.1. Luminescence Properties of d^6 Metal Complexes[a]

Luminescent	Somewhat luminescent	Nonluminescent
$[Ru(bpy)_3]^{2+}$ (RT)	$[Os(bpy)_3]^{2+}$ (RT)	$Fe(bpy)_3]^{2+}$ (RT,77)
		$[Co(bpy)_3]^{3+}$ (RT,77)
$[Rh(bpy)_3]^{3+}$ (77)		$[Rh(bpy)_3]^{3+}$ (RT)
$[Co(CN)_6]^{3-}$ (solid 77)	$[Co(CN)_6]^{3-}$ (solid 77)	$[Co(CN)]_6^{3-}$ (soln. RT)
$[Ru(bpy)_2(py)_2]^{2+}$ (77)		$[Ru(bpy)_2(py)_2]^{2+}$ (RT)
$[Os(phen)_2Cl(CO)^+$ (RT)	$[Os(phen)_3]^{2+}$(RT)	
$[Ru(bpy)_2(CN)_2]$ (RT)		
$[Ru(bpy)_2([CN)_2(H^+)_{1,2}$ (77)	$[Ru(bpy)_2(CN)_2](H^+)_{1,2}$ (RT)	
$Re(bpy)Cl(CO)_3$ (RT)		$[Re(dpk)Cl(CO)_3$(RT)
$[Re(bpy)(py)(CO)_3]^+$ (RT)		
$[Re(bby)(NCR)(CO)]_3^+$ (RT)		
$[Re(bpy)(NCR)(CO)_3]^+$ (RT)		

[a]dpk = 2,2′-dipyridylketone; NCR = alkyl isocyanide; RT = room temperature; 77 = 77K.

responsible for achieving selective binding to specific support sites and for controlling or predicting behavior in different micro environments is still in its infancy.

Considering first the design of the probe molecule, to those unfamiliar with inorganic spectroscopy and photophysics, the pattern of luminescence for inorganic complexes can appear quite random and illogical. Table 4.1 lists representative metal complexes categorized by luminescence efficiency. We will demonstrate that the rational design of luminescent materials with desired properties can be achieved by utilizing a remarkably small set of basic chemical and spectroscopic concepts. We describe briefly the basic electronic structure and spectroscopy of transition metal complexes. Our focus is on systems with d^6 electronic configurations, which is one of the more promising areas. However, most of the principles are generic. We limit ourselves to systems containing at least one α-diimine ligand such as 2,2′-bipyridine (bpy) or 1,10-phenanthroline (phen). The pertinent structure–function relationships are presented and demonstrated by case studies.

However, the rational design of luminescent molecules is only part of the problem. The luminophore must generally be supported on a solid support. Specific supports can radically alter the properties of the probe or sensor and this interaction must be thoroughly understood. Thus, we also explore some of the problems that arise when one goes from homogeneous solvents to generally microheterogeneous supports. We explore the different types of heterogeneity and other support properties and their impact on device design.

The effect of microheterogeneity is so pervasive and its understanding so critical to sensor design and fabrication, that we discuss in some detail the problems associated with unraveling the fundamental nature of real systems from physical measurements. As we will show, this is generally not a trivial problem, and one should exercise considerable caution in using any micromechanistic model based on typical luminescence measurements.

4.2. States of Inorganic Complexes

We first examine the relationships between electron structure and the emission and absorption spectroscopy of metal complexes. Transition metal complexes are characterized by partially filled d orbitals.[8] To a large measure the ordering and occupancy of these orbitals determines emissive properties. Figure 4.2 shows an orbital and state diagram for a representative octahedral MX_6 d^6 metal complex where M is the metal and X is a ligand that coordinates or binds at one site. The octahedral crystal field of the ligands splits the initially degenerate five atomic d-orbitals by an amount Δ into a triply degenerate t level and a doubly degenerate e level. The t and e are group theoretic notations for the symmetry of the states in the specified point group. The splitting arises because the two e orbitals are directed towards the six ligands and the remaining t_2 orbitals point between the ligands. The electrostatic interactions between the filled ligand orbitals and an electron placed in the different d orbitals yields energy differences for the two orbitals. Thus, placing an electron in an e orbital is higher in energy than placing one in a t_2 orbital. The magnitude of the splitting is given by the crystal field splitting, Δ. Δ is a particularly important parameter and its size is determined by the crystal field strength of the ligands and the central metal ion.

The distribution of electrons between the t_2 and e levels and, thus, the state diagram of the system, is profoundly affected by Δ. If Δ is large (i.e., strong field), it is energetically more favorable to pair electrons in the t_2 level than to keep them unpaired by distributing them throughout the t_2 and e levels (weak field). We consider only strong crystal field systems where all six d electrons pair and fill the three t_2 orbitals (Figure 4.2). Weak crystal field d^6 systems are nonemissive.

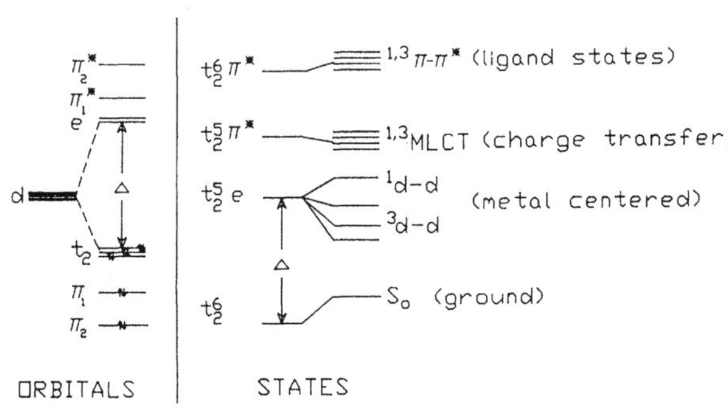

Figure 4.2. Simplified schematic orbital and state diagrams for a d^6 metal in an octahedral environment showing d and π bonding and π antibonding (π^*) orbitals. A strong crystal field is assumed so that the t_2 levels are filled. Ligand to metal charge transfer states are ignored.

The ligands have π and σ orbitals, but only the π orbitals are spectroscopically important for visible and near-UV absorptions and emissions. There are both π bonding and π antibonding (π^*) levels. The π bonding levels are filled.

The spectroscopic states are derived from the different orbital configurations (Figure 4.2). All spins are paired in the ground state making the ground state a singlet state. The excited states arise from an electron promotion to an unoccupied orbital. States are classified by both the original and final orbitals. There are three types of excited states: metal-centered *d–d* states, ligand-based π–π^* states, and charge transfer (CT) states.

The lowest ligand-based excited states are π–π^* ones derived from promoting a bonding π electron to a π^* level. Since the spins can be either paired or unpaired, there are triplets and singlets. The triplet state will always be lower in energy than its corresponding singlet state. These transitions are largely localized on the organic ligands and are spectroscopically very similar to those of the free ligand. The intense 250–300 nm absorptions in $Ru(bpy)_3^{2+}$ (Figure 4.1) are spin-allowed ligand localized π–π^* transitions and are strikingly similar to those of the free bpy ligand. This demonstrates the high degree of separation of the ligand states from the other states.

Similarly, singlet and triplet *d–d* states arise from promoting a t_2 electron to an e level on the metal center (denoted by $t_2^5e^1$) Transitions to *d–d* states are formally forbidden, even for the spin-allowed singlet-singlet ones; this is due to the two *d* orbitals having the same inversion symmetry. Thus, *d–d* emissions are generally characterized by long radiative lifetimes and high susceptibility to environmental quenching. These two factors tend to give very low or negligible luminescence yields for metal-centered emissions, especially at room temperature. A further problem with low-lying *d–d* states in d^6 systems is that the t_2 levels are bonding with respect to the ligands while the *e* orbitals tend to be antibonding. Thus, *d–d* states of $t_2^5e^1$ configuration tend to weaken the metal ligand bonds and make the excited state very reactive with respect to decomposition by ligand dissociation. For these reasons we are aware of no *d–d* emitting d^6 molecules that are likely sensors. There are no visible *d–d* transitions in $Ru(bpy)_3^{2+}$. However, this is not surprising as they are too weak to be observed, being buried under the much more intense charge transfer and π–π^* transitions.

By marrying the metal to an organic ligand with π levels, an entirely new class of transitions can arise. Charge transfer states arise from the exchange of an electron (CT) between the ligand and the metal. An electron can be promoted either from a metal orbital to a ligand orbital ($t_2^5\pi^{*1}$ or more commonly $t_2^5\pi^*$) configuration) or from a ligand to a metal orbital ($\pi^1t_2^6e^1$). The first are metal-to-ligand charge transfer transitions (MLCT) and the second are ligand-to-metal (LMCT). α-diimine ligands are readily reduced and are the primary systems discussed here; they give only MLCT transitions at low-energy. CT transitions tend to be much more strongly allowed than *d–d* transitions with extinction coefficients of 1000–30,000 while *d–d* transitions are in the 1–200 range. Therefore, CT transitions have shorter radiative lifetimes, which makes them much less susceptible to intramolecular and environmental quenching.

The large molar extinction coefficients of the spin-allowed CT transitions make them much easier to pump optically. The intense 454-nm visible band of $Ru(bpy)_3^{2+}$ is an MLCT transition.

An octahedral geometry (O_h symmetry) is, of course, an ideal case, which virtually none of our systems match. All have lower symmetry (typically D_3 or C_2), which further splits the *d* levels. However, the octahedral model is a starting point. Lowering the symmetries does not affect the basic nature of the types of excited states. Further, such important features as the *d* state energies are still dictated by the average Δ of the ligands.

4.3. Design Considerations

As we will now show, knowledge of a few basic rules coupled with the above state information allows synthetic control of the spectroscopy and, thus, luminescence properties. The elements accessible to the chemist are the ligands, geometry, and metal ion. The most important design rule of luminescent transition metal complexes is that the emission always arises from the lowest excited state in the molecule.[9, 10] This is a variation of Kasha's rule for organic molecules that states that luminescence will arise from the lowest state of a given multiplicity. This variation derives from the observation that in metal complexes radiationless decay from upper excited levels is very fast even if it involves a change in multiplicity, and relaxation to the lowest excited level, thus, occurs with nearly 100% efficiency. In organic systems, change of multiplicity can be much more forbidden than in metal-based systems, which means that both fluorescence and phosphorescence can be observed, frequently in the same molecule. The 100% efficient relaxation of the upper excited states is shown for $Ru(bpy)_3^{2+}$ in Figure 4.1 where the luminescence efficiency is independent of whether excitation is into either MLCT or π–π^* excited states. This absence of fluorescences in inorganic systems is a consequence of spin orbit coupling from the metal, which accelerates intersystem crossing to the triplet manifold. This means that all major emission contributions in our systems will arise formally from the lowest triplet state, and the emissions will be forbidden phosphorescences. Thus, control of the luminescence properties of complexes hinges on control of the relative state energies and the nature and energy of the lowest excited state.

Our goal is the rational design of molecules that have a specific set of properties, generally including efficient emissions, long luminescence τ's, simple optical pumping, specific environmental sensitivity, and chemical and photochemical inertness. Unfortunately, some of these aims are mutually antagonistic.

Based on a large body of experimental and theoretical work, the following rules have been found most important in obtaining desirable photophysical and photochemical properties of transition metal complexes.

- The lowest excited state must be either a CT or ligand π–π^*. This minimizes photochemical instability inherent in many *d–d* states.
- Any *d–d* states must be well above the emitting level to prevent their thermal excitation, which results in photochemical instability and rapid excited state decay.
- Spin orbit coupling should be high to increase the allowedness of the emission and permit radiative decay to compete effectively with radiationless decay.
- Pure π–π^* phosphorescences tend to be too long-lived for efficient emission. Either spin orbit coupling or mixing with more allowed CT states must exist to increase the allowedness of the π–π^* phosphorescences so that it can compete with radiationless decay.
- The emitting level must not be at too low an energy. The energy gap law states that radiationless processes become more efficient as the emitting state approaches the ground state.[11, 12]

Generally critical in successful sensors is the removal of the lowest *d–d* state from competition with the emitting level. Controlling the energies of the *d–d* states is accomplished by varying Δ by altering the ligands or central metal ion. Stronger crystal field strength ligands or metals raise *d–d* state energies. Crystal field strength for the ligands increases in the series[8]:

$$\text{Cl} < \text{pyridine} \ll \text{bpy, phen} < \text{CN} < \text{NCR} < \text{CO}$$

For the metal, Δ increases on descending a column in the periodic table. In addition, CT state energies are affected by the ease of oxidation/reduction of the ligands and metal ion. For MLCT transitions, more easily reduced ligands and more easily oxidized metals lower the MLCT states.

The π–π^* energies are largely controlled by the ligands. However, varying the substituents, heteroatoms in the aromatic, or the extent of π conjugation can radically alter the energies and intensities of the π–π^* transitions.

Spin orbit coupling is most effectively increased by using higher Z metals. Heavy atom substituents on the organic ligand have proved of minimal use and may actually make the ligands more photochemically unstable.

To illustrate the very detrimental effect of the energy gap law we show in Figure 4.3 the observed decay rates for a series of $(bpy)ReX(CO)_3^+$ complexes (X = pyridine or substituted pyridine).[13] The nonradiative decay rate, k_{nr} is presented versus the emission maxima of the MLCT state, which is controlled by altering the substituents on the pyridine ligand. The observed τ is essentially equal to $1/k_{nr}$ for this series. With a modest change in emitting state energy of 2.2 kcm^{-1}, τ decreases from 0.96 to 0.11 μsec.[13] Note also the ability to tune the emission over 2.2 kcm^{-1} (red to green) merely by altering the pyridine ligand.

We stress that the above picture of the spectroscopy of metal complexes is an extremely simple-minded view of what can be an exceptionally complicated topic, although we do touch on most of the features important for room temperature sensors.

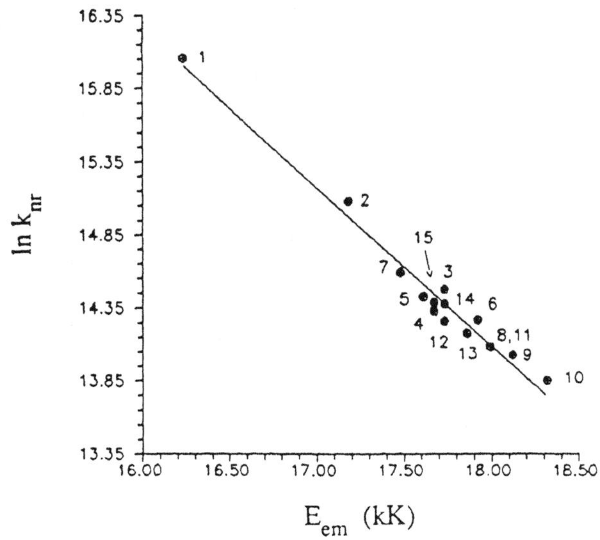

Figure 4.3. Plot of nonradiative decay constants for an homologous series of $Re(bpy)(CO)_3X^+$ (X = pyridine or substituted pyridine). The observed lifetime is close to $1/k_{nr}$. The solid line is the best theoretical fit with the energy gap law. (Reprinted from Ref. 13 with permission. Copyright 1990 American Chemical Society.)

Indeed, for the higher atomic number elements, there is still enormous controversy about the nature of the excited states, which while the problem of the excited state nature is interesting, it is not of direct significance to sensor design. For a view of some facets of this complex issue see the review by Krausz and Ferguson.[14]

4.4. Temperature Effects on Inorganic Sensors

Even when the *d–d* state is at much higher energy than the emitting level, it can still be of paramount importance in the photophysics and photochemistry of the system. Indeed, a major contributor to the temperature-dependent loss of emission intensity in luminescent metal complex based sensor materials is nonradiative decay via high-energy *d–d* excited states.[15] The model for this is shown in Figure 4.4A. The excited state lifetime is given by

$$1/\tau = (k_r + k_{nr}) + k_{dd} \tag{4.1a}$$

$$k_{dd} = k' \exp(-\Delta E / kT) \tag{4.1b}$$

where τ is the observed lifetime, k_r is the radiative rate constant, and k_{nr} is the nonradiative rate for direct depletion of the emitting state to the ground state. k' is the preexponential factor for thermally activated deactivation of the emitting level via a nonemissive *d–d* excited state located ΔE above the emitting state. This model assumes

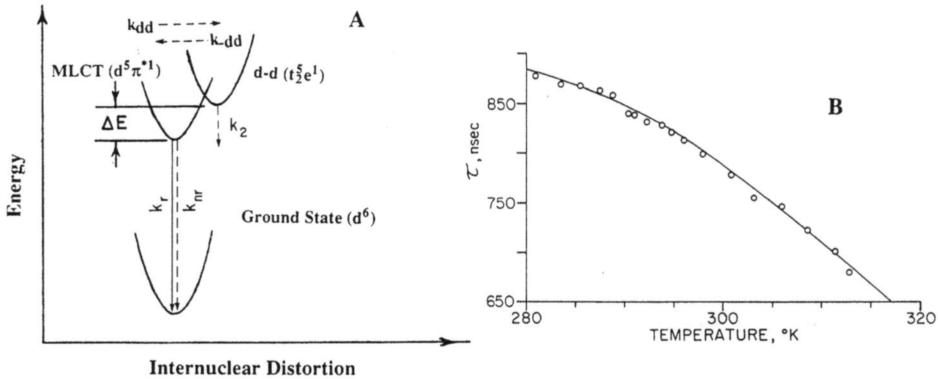

Figure 4.4. (A) State diagram showing the loss of excitation energy via radiationless decay through the *d–d* state. (B) Temperature dependence of the lifetime of $Ru(bpy)_3^{2+}$ in a micellar media. The solid line is the best fit using a thermally activated deactivation via the *d–d* state. (Reprinted from Ref. 15 with permission. Copyright 1986 American Chemical Society.)

that once the *d–d* state is reached quenching is certain; this means that $k_{dd} >> d_{-dd}$. However, the alternative model of an excited state equilibrium between the emitting charge transfer excited state and the *d–d* state yields slightly different equations that are indistinguishable experimentally from Eq. (4.10). A typical experimental curve with a fit to Eq. (4.1) is shown in Figure 4.4B.

Although not plotted, there is a parallel correlation between absolute luminescence efficiency, ϕ, (photons emitted per photon absorbed) and observed lifetime.

$$\phi = \phi_{\text{limit}} (\tau/\tau_{\text{limit}}) \tag{4.2a}$$

$$\tau_{\text{limit}} = 1/(k_r + k_{nr}) \tag{4.2b}$$

$$\phi_{\text{limit}} = k_r/(k_r + k_{nr}) = k_r\tau_{\text{limit}} \tag{4.2c}$$

where τ_{limit} and ϕ_{limit} are the observed limiting lifetime and quantum yield at low temperature, respectively. Thus, if the lifetime is 1% of the low-temperature limiting value, the quantum yield is only 1% of the low-temperature limiting yield, which may be much smaller than unity.

To demonstrate the importance of k' and ΔE in determining excited state lifetime, we plot typical data for τ versus these quantities in Figure 4.4B. For τ_{limit} we select 10 μsec, which is appropriate for the widely used $Ru(Ph_2phen)_3^{2+}$ complex (Ph_2phen = 4,7-diphenyl-1,10-phenanthroline). For k' we use 10^{10} sec^{-1} and 10^{13} sec^{-1}. The higher k' is typical for a rapidly deactivating *d–d* state while the lower value might arise if there were actually an excited state equilibrium between the *d–d* and the emitting CT excited state.[15]

The absolutely pivotal role of the ΔE in controlling τ is shown in Figure 4.5. For $k' = 10^{13}$ sec^{-1}, which is similar to that for $Ru(bpy)_3^{2+}$, usable room temperature

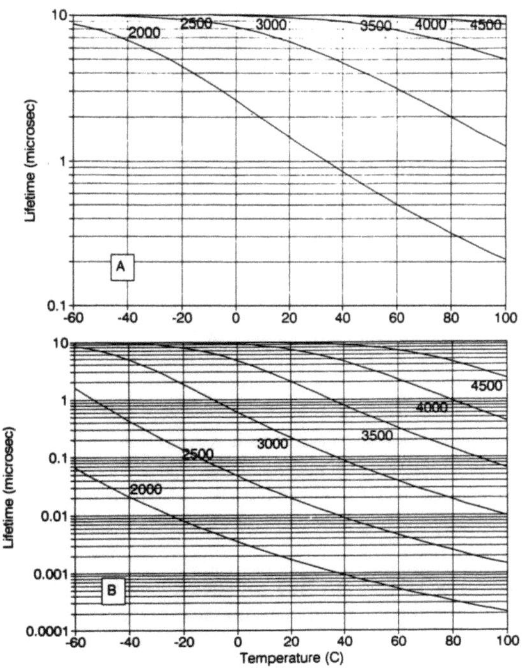

Figure 4.5. Calculated observed lifetimes of the system described by Figure 4.4A as a function of k' and ΔE in cm^{-1} indicated next to each curve. (A) $k' = 1 \times 10^{10}\ sec^{-1}$. (B) $k' = 1 \times 10^{13}\ sec^{-1}$. $\tau_{limit} = 10\ \mu s$, which is similar to that of $Ru(Ph_2phen)_3^{2+}$.

lifetimes are not obtained for $\Delta E < 2500\ cm^{-1}$. For $Ru(bpy)_3^{2+}$ the observed lifetime at 20°C is 0.6 μsec versus τ_{limit} of 5 μsec. The measured ΔE is about 3560 cm^{-1}. Thus, the short lifetime, relatively low quantum yield (4%), and significant temperature dependence of this complex at room temperature arise from the presence of a deactivating *d–d* state that is actually at quite a high energy relative to the emitting state.

We now address the possibility of controlling temperature effects by molecular design. This has potential applications in temperature sensors as well as in the minimization of temperature effects on the response of sensors. Clearly, if one could raise ΔE above 4000 cm^{-1}, then thermal deactivation could be largely eliminated. By suitable choices of strong field ligands and central metal ions, this design feature can probably be realized, although one still must control the energy of the emitting state so as to have absorption spectra that allow easy optical pumping and maintain good luminescence quantum yields. Alternatively, if one could lower k' to $10^{10}\ sec^{-1}$, thermal losses could be largely eliminated even for ΔE down to about 2500 cm^{-1}. At this time it is not clear what molecular design features can be exploited to lower k'. One suggestion is that k' can be reduced by stabilizing the *d–d* state to retard deactivation from it. Since the *d–d* is dissociative and deactivation probably proceeds in part via a partial dissociative mechanism, a possible approach would be to anchor the dissociating ligand by a cage structure so that even partial dissociation via a bond lengthening would be reduced.

There are possible advantages of the temperature dependence of excited state lifetimes. The dependence can be used for luminescence-based temperature sensors, which is an area of considerable interest. Clearly, with a suitable ΔE, complexes that give significant lifetime changes in a wide range of possible temperatures can be designed. We discuss this issue elsewhere.(16)

4.5. Design Examples

Specific examples are now used to demonstrate these concepts. First, consider the group $Ru(bpy)_3^{2+}$ (luminescent), $Os(bpy)_3^{2+}$ (slightly luminescent), and $Fe(bpy)_3^{2+}$ (nonluminescent) (Table 4.1). For $Fe(bpy)_3^{2+}$, despite an exhaustive search no emission has ever been detected even at 77K; we routinely use it as a nonemissive solution filter. All three iso structural d^6 systems are in the same oxidation state with the same electronic configuration (t_2^6). The Fe(II) complex has an intense MLCT band at 510 nm, and the Ru(II) complex at 450 nm; the Os(II) complex has intense MLCT bands that stretch out to 700 nm. The π–π^* transitions are all quite similar in all three complexes with intense absorptions around 290 nm and ligand triplet states at 450 nm (inferred from the free ligand and other emissive complexes and the insensitivity of these states to coordination to different metals).

Despite their chemical similarities, there are radical differences in luminescence properties that can be attributed to their excited state energies and orderings. See Figure 4.6, which shows the positions of their lowest d–d, MLCT, and π–π^* triplet states. The ligands are the same, but Δ increases dramatically on going from iron (Fe) to ruthenium (Ru) to osmium (Os). Thus, the ^3d–d state energies rise pronouncedly on going from Fe to Ru to Os. Further, the ease of oxidation of the metal to the +3 oxidation state is

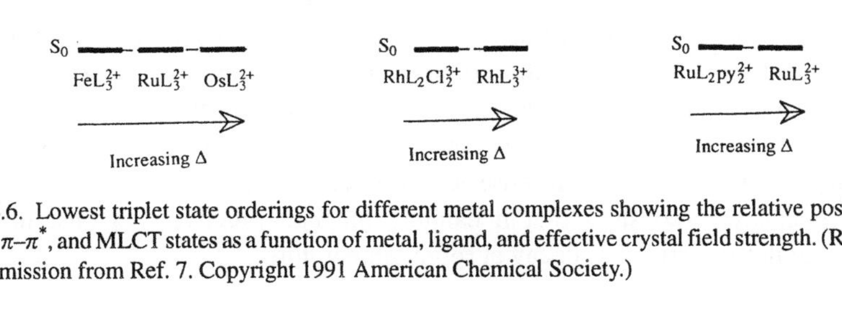

Figure 4.6. Lowest triplet state orderings for different metal complexes showing the relative positions of the d–d, π–π^*, and MLCT states as a function of metal, ligand, and effective crystal field strength. (Reprinted with permission from Ref. 7. Copyright 1991 American Chemical Society.)

easier for Os than for Ru which, in turn, is easier than for Fe; this trend mirrors the MLCT state energies.

The striking absence of luminescence for the Fe(II) complex under any conditions arises from two reasons: The lowest excited state is a *d–d* triplet state and is very close to the ground state. Being so close to the ground state, the d-d depletes the upper levels and quenches emission. The forbiddenness of the potential *d–d* phosphorescence allows radiationless decay to be more effective at depleting the excited state. The forbiddenness of the *d–d* state is exacerbated by the low atomic number of Fe, which minimizes the spin orbit coupling that could impart allowedness to the transition. The greater Δ of Ru relative to Fe raises the *d–d* state above the MLCT and makes the MLCT the potentially emissive state. Since the MLCT state of Ru is well above the ground state, the energy gap law decreases the efficiency of nonradiative decay and permits the more allowed CT emission to compete with radiationless decay. The trend, due to crystal strength, continues in Os with the *d–d* state being well above the MLCT state, which then becomes the expected emitting state. However, the Os complex is much less emissive than the Ru one, because its MLCT state is lower in energy and, thus, quenching to the ground state is more efficient as expected by the energy gap law.

The $Co(CN)_6^{3-}$ is an interesting system. It has everything working against it as a sensor. This pale yellow complex shows no room temperature luminescence in solution, but in the solid state it does give an attractive red luminescence, which is quite bright at 77K. There are no discernible absorptions in the visible; the lowest energy absorptions are in the UV and are *d–d* transitions. The cyanides are very strong field ligands and raise the *d–d* excited states to high enough energy to be emissive. Cyanide has no accessible ligand absorption and CT transitions are of too high an energy to see. At room temperature the complex is quite photosensitive in solution. The solid state lifetime of about 300 μsec at 77K is consistent with a phosphorescence. The broad structureless red–infrared emission leaves no doubt the emission is of *d–d* character. The photochemistry is loss of a CN^-, which is consistent with the weakened ligand bond in the $t_2^5e^1$ excited state.

The $Co(bpy)_3^{3+}$ is nonemissive even at 77K. The bpy brings MLCT states into the near-UV, but the weaker crystal field strength of bpy versus CN^- apparently allows the lowest *d–d* state to fall below the ligand and MLCT states. Indeed, as with $Fe(bpy)_3^{2+}$, the *d–d* state appears to be at too low an energy to give an emission competitive with radiationless decay directly back to the ground state.

$Rh(bpy)_3^{3+}$ is an example of a complex that exhibits an almost pure π–π^* phosphorescence and demonstrates one of the limitations of nearly pure ligand localized emissions. At 77K, the complex is highly emissive with a beautifully structured blue ligand phosphorescence ($\lambda_{max} = 446$ nm for the first peak) having a τ in the tens of msec,[17] but it has no detectable room temperature emission. It is this very long radiative lifetime that causes the absence of room temperature emission. The radiative decay is so slow that it cannot compete effectively against inter- and intramolecular radiationless decay at room temperature.

The remarkable effect of altering crystal field strength is shown merely by replacing one bpy in $Rh(bpy)_3^{3+}$ with two chlorides to yield *cis*-$Rh(bpy)_2Cl_2^+$. The emission character is completely altered. $Rh(bpy)_2Cl_2^+$ emits only at 77K with a red broad structureless emission ($\lambda_{max} = 740$ nm).(9) Refer to Figure 4.6 for the explanation. Chloride is a much weaker field ligand than bpy. Therefore, the average Δ for $Rh(bpy)_2Cl_2^+$ is much lower than for $Rh(bpy)_3^{3+}$ and the *d–d* state for the chloro complex is the lowest state in the molecule giving the structureless emission. The CT states are at too high an energy to emit, because Rh(III) is very difficult to oxidize or reduce. Since the *d–d* state is so forbidden, it is deactivated by radiationless paths at room temperature. In addition, the presence of the low-lying *d–d* makes the complex susceptible to photochemical Cl^- displacement.

Similarly, $Ru(bpy)_2(CN)_2$ emits well at room temperature while $[Ru(bpy)_2(CN)Cl]^+$ does not. The main factor is the much lower effective crystal field strength of Cl versus CN.

Ligand field strength can significantly alter photostability even in apparently closely related complexes. Consider *cis*-$[Ru(bpy)_2(py)_2]^{2+}$ where py has a smaller Δ than bpy versus $Ru(bpy)_3^{2+}$. The absorption spectrum of $[Ru(bpy)_2(py)_2]^{2+}$ is virtually identical to $Ru(bpy)_3^{2+}$. The two bpy ligands can accept an electron from the Ru(II) and the MLCT states are little perturbed from those of $Ru(bpy)_3^{2+}$. At 77K $[Ru(bpy)_2(py)_2]^{2+}$ and $Ru(bpy)_3^{2+}$ give very similar beautifully structured emissions. However, at room temperature, not only is $[Ru(bpy)_2(py)_2]^{2+}$ not emissive, but also it is extraordinarily photounstable; a solution in a cuvette will undergo substantial photodegradation merely by carrying it across a lighted room to the spectrometer. The pyridine ligands are labilized to give $Ru(bpy)_2(py)S^{2+}$ and $Ru(bpy)_2(S)_2^{2+}$ (S = solvent). The 77K results show that the lowest excited state is MLCT, but the room temperature photochemistry shows that the *d–d* is thermally accessible from the excited MLCT state and provides the decomposition pathway.

Figure 4.6 illustrates this more clearly where the relative energies of $[Ru(bpy)_3]^{2+}$ and $[Ru(bpy)_2(py)_2]^{2+}$ are shown. The lower gap between the *d–d* state from the emitting MLCT allows very efficient depopulation-decomposition via the *d–d* state for $[Ru(bpy)_2(py)_2]^{2+}$ but not for $[Ru(bpy)_3]^{2+}$ where the gap is much larger.

Photochemistry and deactivation via *d–d* states can be a problem even for large separations between the *d–d* and the emitting level.(18–20) For example, as pointed out above, even the emissive $Ru(bpy)_3^{2+}$ is deactivated in part via an upper *d–d* state; however, because of the higher Δ, the gap to the *d–d* state is larger and deactivation is less efficient than for the py complex. There is some *d–d* photochemistry with $[Ru(bpy)_3]^{2+}$. However, unlike the pyridine complex where rupture of one bond results in py dissociation, it is more difficult to remove the bidentate bpy ligand. The yield for permanent photochemistry is much smaller than for the pyridine complex; the bpy can anneal by reattaching itself at the vacant site.(18–20) Since the deactivating *d–d* state is thermally activated, there is an appreciable increase in the deactivation rate as the temperature is raised. Thus, $[Ru(bpy)_3]^{2+}$ shows a strong change in τ above room temperature, which limits its utility at higher temperatures. Figure 4.4b shows a typical plot of the emission lifetime versus temperature along with a fit to Eq. (4.1).

Thus, this again stresses that a most desirable design goal is to make the *d–d* states as thermally inaccessible as possible from the emitting MLCT or π–π^* state. This can be done by increasing Δ by using a higher atomic number metal ion of the same oxidation state. This problem is exacerbated by the very high *k*'s which permit *d–d* states with even huge ΔE's (>3000 cm^{-1}) to contribute substantially to deactivation.

While one might conclude that the highest possible higher atomic number is desired for optimum performance, other factors are important. For example, compared to $[Ru(bpy)_3]^{2+}$, $[Os(bpy)_3]^{2+}$ has a much larger splitting between the *d–d* and MLCT states because of the higher Δ of Os relative to Ru and because of the easier oxidation of Os, which lowers the MLCT state (Figure 4.2). Thus, $[Os(bpy)_3]^{2+}$ shows no photochemistry or *d–d* deactivation. Unfortunately, the MLCT state has dropped so low that direct radiationless decay (energy gap law) reduces τ to about 50 nsec versus 600 nsec for $[Ru(bpy)_3]^{2+}$.

Chemical modifications can be used to tune the state energies and enhance properties. CO ligands greatly stabilize the *t* levels, which results in both an increased Δ and a higher-energy MLCT transition. For example, the primary MLCT bands of $Os(phen)_3^{2+}$ and $[Os(phen)_2Cl(CO)]^+$ are at 430 and 365 nm, respectively. The emissions are similarly shifted from 710 to 646 nm. In keeping with the expectations of the energy gap law, the τ of $[Os(phen)_3]^{2+}$ and $[Os(phen)_2Cl(CO)]^+$ are 74 and 234 nsec, respectively.[21]

An interesting point can arise with Re(I) complexes (e.g., fac-$Re^+L(CO)_3X$ where X is a halide, pyridine, nitrile, or isonitrile) where the π–π^* and MLCT states can be very nearly isoenergetic. This near-degeneracy allows the lowest state to be rather easily switched back and forth between π–π^* and MLCT by suitable choices of ligands and in some cases merely by altering the temperature.[22, 23]

Even when the π–π^* state is lowest in Re(I) complexes, the emissions can be very efficient, in marked contrast to $[Rh(bpy)_3]^{3+}$. We attribute this to two causes. First, Re has a higher atomic number than Rh which enhances spin-orbit coupling, and this increases the allowedness of the phosphorescence. Second, because of the proximity of the MLCT and π–π^* states, mixing of the two will occur. Thus, the less allowed π–π^* state steals allowedness from the more allowed MLCT state, which increases the π–π^* state's radiative rate constant and enhances its emission efficiency. We have reported a detailed study on the luminescence and binding properties of a series of $[ReL(CO)_3NCR]^+$ complexes where the π–π^* and MLCT states are nearly degenerate and can be inverted by small changes in ligand structure or temperature.[22]

This illustrates a principle. State mixing is frequently desirable since, when properly done, the best attributes of both states are incorporated into the emitting level. In the Re(I) case just discussed, the MLCT state confers environmental sensitivity and its more efficient radiative rate. Concurrently, the π–π^* confers a longer lifetime, and for reasons that are not obvious slows down radiationless decay. For example, in the $ReL(CO)_3NCR^+$ systems the π–π^* phosphorescences have lifetimes at 77K in the 100–1000 μsec range, which is very short for a ligand-localized phosphorescence.[22, 23, 24] Much of the allowedness must be gained via state mixing with more allowed MLCT states.

Lastly, photochemically unstable ligands should be avoided. $Re(bpy)(CO)_3Cl$ shows a moderately efficient MLCT emission at room temperature (R. M. Ballew, unpublished results from our laboratory). However, the apparently closely related $Re(dpk)(CO)_3Cl$ (dpk = 2,2′-dipyridyl ketone) shows a benzophenone like phosphorescence at 77K indicating that the n–π^* excited state of the ketone in complex is the lowest state of the complex. No luminescence is seen at room temperature, and even at 77K the dpk triplet state is such a powerful hydrogen atom extractor that it removes protons from alcohol glasses as seen by the formation of the intense blue color of the keto free radical. The absence of an MLCT emission is caused by the greater difficulty of reducing dpk relative to bpy, which pushes the MLCT states above the dpk ligand states.

4.6. Sensor Design and Applications

4.6.1. Probe/Sensor Design

We have discussed how to make highly luminescent species, but we have left unaddressed the more difficult question of how to incorporate specific sensitivity into molecular probes. There are two basic problems. First, one must develop a moiety with the desired specific sensitivity. Second, in doing so, one must not violate the basic criteria established earlier and inadvertently turn off the luminescence or introduce unacceptable photochemical sensitivity.

For environmental sensitive probes there is a large body of data associated with organic bioprobes.[25] Generally, large solvent sensitivities are highly desirable with the greatest sensitivity being found with molecules having the largest permanent dipole moment change on going from the ground to the excited state. This dipole change interacts strongly with the solvent dipoles and causes large solvent effects on state energies and spectra. Molecules showing large solvatochromism on their absorption spectra are frequently good environmental reporters.

In general, large solvatochromism is only observed in metal complexes with asymmetric MLCT states. For example, $[Ru(bpy)_3]^{2+}$ is a poor choice since the promoted electron can distribute itself in a roughly spherical fashion with no overall change in dipole moment. In keeping with this, $Ru(bpy)_3^{2+}$ shows small solvent effects on its absorption or emission. On the other hand, *cis*-$Ru(bpy)_2(CN)_2$ has an enormous change in the dipole on MLCT excitation since the charge is moved asymmetrically toward the two bpy. *cis*-$Ru(bpy)_2(CN)_2$ shows large solvatochromism in both emission and absorption as well as a large solvent sensitivity on τ. $Ru(bpy)_2$(oxalate) has a comparably large solvatochromism, but the MLCT state is moved so far to the red that radiationless decay deactivates the complex before it can emit at room temperature.

Solvatochromism is probably a requirement for any system that will demonstrate large environmental sensitivity.

There is a small solvent dependence for $[Ru(L)_3]^{2+}$ complexes which depends on the structure of L. The emission spectra of the $Ru(bpy)_3^{2+}$ is mildly sensitive to media, the $Ru(phen)_3^{2+}$ much less so, and the $Ru(Ph_2phen)_3^{2+}$ is virtually media independent.[26] This decreasing sensitivity to solvent perturbation is a consequence of the excitation being localized in the metal α-diimine portion of the complex (-N=C-C=N).[27] The more extended the complex, the greater the shielding of the excited portion and the smaller the solvent perturbations of the emission spectrum. In particular, the bulky phenyl groups are extremely effective at shielding the excited state from environmental perturbations.

Other special probe properties can be built in by suitable ligand modification. For example, 4,7-dihydroxy-1,10-phenanthroline complexes of Ru(II) show pH sensitivity. Ligands with exposed basic nitrogens also show some promise for pH sensors.[28] fac-$[Re(bpy)(CO)_3NC(CH_2)_nCH_3]^+$ (n = 0–17) and the related $Re(bpy)(CO)_3CN(CH_2)_nCH_3^+$ provide hydrophobic hooks to attach relatively polar complexes to hydrophobic media such as micelles or cyclodextrins. This can be used to anchor the probe to specific structural features of the target.

Temperature sensitivity can be achieved by having two states of different orbital types within kT of each other. Altering populations by temperature will then alter the decay rates, emission intensities, and τ's. Thermal activation of a *d–d* state is one choice, although careful design to minimize permanent decomposition becomes a problem. The temperature dependence of $[Ru(bpy)_3]^{2+}$ of Figure 4.4b is such an example. Crosby *et al.* have reported extraordinarily large changes in τ's versus T for $Ru(II)L_3$ complexes ($T < 40K$) with negligible changes in the quantum yields.[29] They have suggested the use of these complexes as cryogenic thermometers. For room temperature, some of the Re(I) systems with close π–π^* and MLCT states are obvious choices.[22] These systems have proved quite interesting as a two-state model is inadequate. The ligand-localized and MLCT states can be within thermal reach of each other and are both emissive, while the upper *d–d* state can deplete energy via thermal activation.

For quenching-based oxygen sensors the dominant design features appear to be the highest τ's and quantum yield.[30–32] For a given class of complexes the bimolecular quenching constants are relatively insensitive to structure. Ruthenium(II) complexes have proved the most successful to date. Re(I) complexes have lower bimolecular quenching constants, and frequently poorer photochemical stabilities, although their long lifetimes can make them more sensitive than the Ru(II) systems. Figure 4.7 shows a Stern–Volmer quenching plot for $[Ru(Ph_2phen)_3]^{2+}$ and two other Ru(II) complexes in silicone rubber. Figure 4.8 shows the luminescence response of $[Ru(Ph_2phen)_3]^{2+}$ being breathed across.[32] The high sensitivity of the luminescence intensity to $[O_2]$ is quite evident. Note the lower O_2 concentration (higher emission intensity) on the initial exhalation (greater exchange time in the lungs), irregularities of the first few breaths due to the higher CO_2 levels in the blood, and restoration of the equilibrium oxygen concentration.

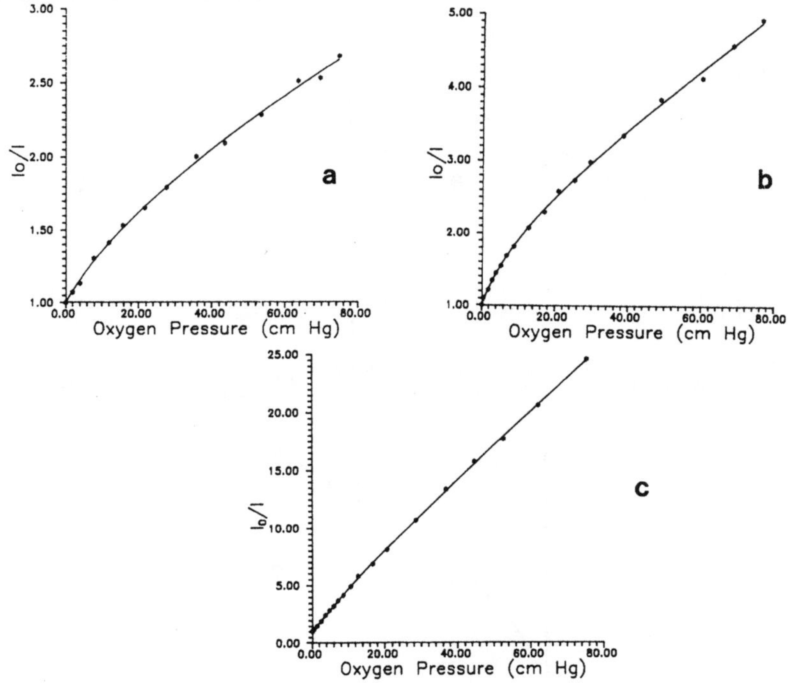

Figure 4.7. Intensity O_2 quenching data for (a) $Ru(bpy)_3^{2+}$, (b) $Ru(phen)_3^{2+}$ (phen = 1,10-phenanthroline), and (c) $Ru(Ph_2phen)_3^{2+}$ in GE RTV 118 silicone rubber. The solid lines are the best fits using the two-site model. (Reprinted from Ref. 26 with permission. Copyright 1987 American Chemical Society.)

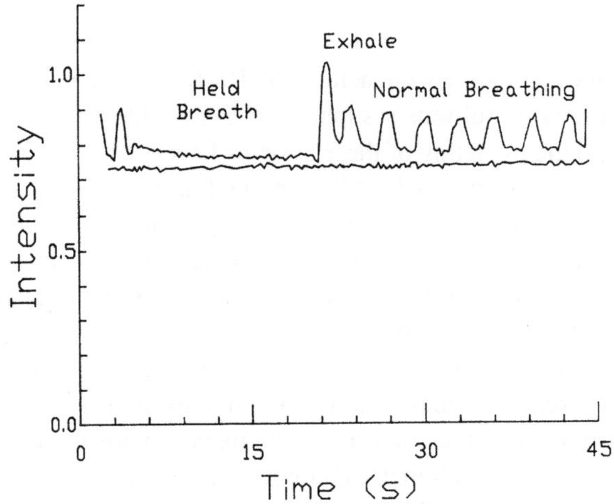

Figure 4.8. Luminescence intensity of $Ru(4,7\text{–}Ph_2phen)_3^{2+}$ in silicone rubber while being breathed over. (Reprinted from Ref. 32 with permission. Copyright 1987 American Chemical Society.)

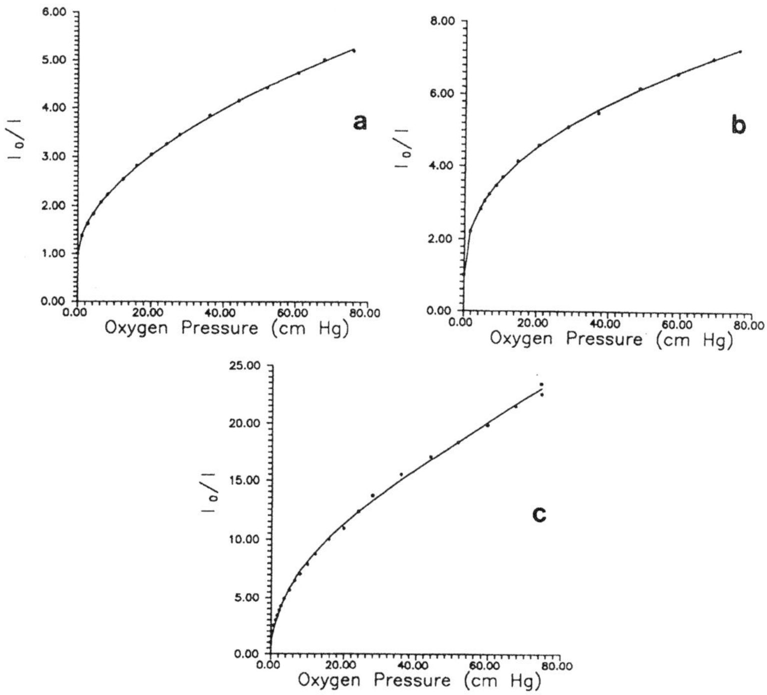

Figure 4.9. Stern–Volmer intensity quenching of (a) $Ru(bpy)_3^{2+}$, (b) $Ru(phen)_3^{2+}$, and (c) $Ru(Ph_2phen)_3^{2+}$ on hydrophilic Cab-O-Sil silica disks. The solid lines are the best fits using a two-parameter Freundlich adsorption model. (Reprinted from Ref. 33 with permission. Copyright 1991, American Chemical Society.)

We also examined oxygen quenching of Ru(II) complexes bound to a silica surface—a potentially useful sensor support.[(33)] Figure 4.9 shows typical quenching curves. We explore the modeling of these systems later.

Design of long-lived probes is still a compelling goal. This will allow greater sensitivity, easier lifetime measurements, and a longer time scale in rotational probes. More recent work of ours shows that a series of Re(I) complexes have great promise as sensors and as molecular probes. Their unquenched τ's can exceed 100 μsec with luminescence quantum yields of >0.5.[(22)]

Trigonal ML_3 metal complexes exist as optically active pairs. The complexes can show enantiomeric selective binding to DNA and in excited state quenching.[(34)] One of the optically active enantiomers of RuL_3 complexes binds more strongly to chiral DNA than does the other enantiomer. In luminescence quenching of racemic mixtures of rare earth complexes, resolved ML_3 complexes stereoselectively quench one of the rare earth species over the other.[(35–39)] Such chiral recognition promises to be a useful fundamental and practical tool in spectroscopy and biochemistry.

4.6.2. Applications

We now present several useful, or potentially useful, systems based on luminescent metal complexes in organic or inorganic supports. The probe molecules were chosen based on the above guidelines. Each probe is supported in some way by a second material to produce a practical device. The results presented show how metal complexes can be applied to diverse problems. These results also identify some problem areas with polymeric supports. We discuss some of our insights into the design and modeling of such systems and describe measurements that help unravel these complex systems.

Our discussion will refer to two different classes of microheterogeneous media. For example, binding of probes to cyclodextrins, micelles, and polymers can yield a probe in several, or a continuum of, spectroscopically different states, but this heterogeneity does not always yield emissions with characteristic heterogeneous behavior (e.g., polyexponential decays, multiple emissions, and non—Stern–Volmer quenching). This dichotomy can frequently be traced to what we will call transient and persistent microheterogeneity. Transient microheterogeneity arises when the lifetimes of the probes' excited states are long compared to the time scale of microscopic rearrangement. Luminescence and quenching then sample a time-averaged environment that is identical for every molecule regardless of the initially formed microenvironment. Many micelle-bound systems satisfy this criteria and show single exponential decay and single emission spectra.

Persistent microheterogeneity exists when the luminescence and quenching decay are fast compared to conformational-environmental rearrangement. Excited molecules can then emit and be quenched in distinctly different chemical environments. Decays can then be nonexponential, multiple emissions may be present, and simple Stern–Volmer kinetics can break down.

4.6.2.1. Luminescent Quantum Counters (QCs)

Common optical detectors have responses that are not constant in energy or photon units with respect to wavelength. Photomultipliers are among the worst offenders. Variation of sensitivity with wavelength makes accurate intensity measurements over a range of wavelengths quite difficult. Luminescence QC's[40–42] overcame some of these shortcomings. An optically dense dye is viewed by a conventional detector. Over the range where the dye absorbs all of the incident light and has a wavelength independent quantum efficiency and emission spectrum, the combination device has a quantum flat response (i.e., equal response for equal number of photons). Rhodamine B in organic solvents is the most widely used QC with a relatively flat response (4.2% maximum spread for 350–600 nm). Its operating range is 250–600 nm.[42] We have shown that traditional organic QC dyes supported in polymer matrices give good performance and avoid solution difficulties such as evaporation, bubbles, and leakage.

We showed that $[Ru(bpy)_3]^{2+}$ in a polyvinyl alcohol matrix has nearly perfect QC properties[41] with a flatness of better than 1% over the measured 360–520 nm range. The response is the flattest we have observed within instrumental resolution. This represents an almost ideal sensor-support system. The polymer appears to function as a nearly homogeneous solvent that minimizes complex–complex interaction by separation. The rigid polymer reduces or eliminates O_2 quenching, which would otherwise reduce sensitivity. As judged by the nearly exponential decays, the local site seen by the complex is nearly homogeneous. However, microcrystal formation at very high concentrations gave rise to a second exponential decay. Similar single exponential decay results have been observed for low concentrations of RuL_3^{2+} complexes in polymethyl methacrylate polymers.[29]

4.6.2.2. Singlet-Oxygen $^1O_2$ Generators

Photochemically generated $^1O_2$ is a powerful synthetic tool.[43] Homogeneous dissolved organic sensitizers are widely used, but separation of dye and products can be bothersome. Solid polymer-supported organic dyes are more convenient. The insoluble polymer-sensitizer is removed by filtration[44–46] and is reusable. We have demonstrated the utility of polymer-supported inorganic sensitizers.[45] $[Ru(bpy)_3]^{2+}$ and related complexes are excellent homogeneous $^1O_2$ generators (efficiencies approaching 85%). We avoided the problems of covalently linking the sensitizer to the polymer by exploiting the essentially irreversible binding of the complexes to strong cation exchangers.

In this case the polymer is a highly solvent and gas-permeable but insoluble support. It is important, however, that the support not greatly quench the luminescence or the $^1O_2$-generation efficiency will suffer. The system shows an interesting triphasic structure. See the original reference for details.[45]

The luminescence decays are somewhat nonexponential for the ionically bound metal complexes; nonexponentiality is exacerbated by the presence of O_2. We suggest that nonexponential decays reflect a persistent microheterogeneity around the complex. This was our first clear evidence of spectroscopically different binding sites. In this case, oxygen enhances heterogeneity detection by differentially quenching different sites.

A persistent nonuniform distribution of ionic binding sites on the ion exchange resin coupled with the variation in local electrostatic charge around the complex seems quite likely given the variable structure of the ionic binding sites. Variations of behavior with changes in ionic environment were observed on ion pairing of metal complexes with different counter ions.[47] Ion pairing caused localization of the excited electron in the antibonding ligand orbitals on a single α-diimine ligand that is furthest from the counter ion. In the ion exchange resins, much larger local fields exist and the magnitude of the effect on the excited state properties would be larger than for a single counter ion.

4.6.2.3. Oxygen Analyzer

An important class of luminescence sensors is O_2 quenching sensors, which are based on the decrease of luminescence intensity and lifetime of the sensor material as a function of O_2 tension.[48–50] We will deal in greater mathematical detail with these luminescence sensors in the next section.

We find that metal complexes immobilized on polymers are excellent O_2 sensors.[32, 33] Figure 4.7 shows the intensity quenching data for three different complexes in silicone rubber. Note the large degree of quenching, especially for $[Ru(Ph_2phen)_3]^{2+}$ as a function of O_2 pressure. The downward curvature is from the presence of different persistent local environments for the complexes, each with a different quenching constant. The greater resistance of some sites to quenching causes the downward-curved plot. The solid lines are the best fit using the model described later. The unquenched lifetime is 5–6 μsec. Precise lifetime measurements at this level are relatively easy and inexpensive. For comparison, the best fluorescent organic sensors are a few hundred nanoseconds, which makes τ measurements much more expensive.

In contrast to the $^1O_2$ generator, the polymer used in the O_2 sensor must be highly gas permeable but solvent impenetrable. Solvent penetration will alter the probe properties and make calibration dependent on environment. Furthermore, good solvents for the probe will leach the probe and destroy the sensor. Again, it is important that the support dissolve the probe well and not greatly quench the luminescence.

In many persistent microheterogeneous systems, the decay curves become multiexponential because of the presence of a range of sites with different lifetimes and quenching constants. In addition to lifetime heterogeneity, we have observed a low-level decomposition of our complexes under intense, protracted irradiation.[26, 32, 33] We have shown that this is not due to $^1O_2$. Further, we have demonstrated from the kinetics of the photochemistry that there are at least two photochemically distinguishable sites: a relatively reactive component and a largely photoinert component. This result demonstrates that both photophysics and photochemistry are strongly affected by persistent microheterogeneity.

Lifetime, as opposed to intensity measurements, are especially appealing. Once the lifetime versus oxygen pressure and temperature has been calibrated, there is no need for an unquenched reference; this is unlike an intensity measurement in which drifts in source intensity, alignment, photodecomposition, and so on can cause spurious shifts in the apparent oxygen concentration. For microheterogeneous systems where there is no single decay time, phase shift measurements are particularly attractive. A phase shift is a single measurement and does not depend on complex data fitting of a nonexponential decay. Ruthenium(II) complexes are especially useful because they can be pumped with blue light emitting diodes (LEDs) or electroluminescent strips.[51]

All three of these devices demonstrate the need for some type of support for the probe molecule. We turn now to some of the problems that can arise when probes are bound to these supposedly passive supports.

4.7. Microheterogenous Systems

While general guidance exists for sensor/complex design, much of this earlier work is based on homogeneous or transient microheterogeneous systems. For sensor design, a polymeric support cannot be treated as a continuous media because it exists in a variety of microdomains. These domains are essentially static during the excited state lifetime of the sensors, and the sensor in different domains experiences different average environments during its decays. This heterogeneity of environments is referred to as microheterogeneity and can complicate the behavior of the sensor, especially if one tries to extrapolate behavior in continuous media to a microheterogeneous one.

In spite of the numerous studies on practical applications of luminescence sensors, the understanding of the fundamental primary processes and underlying photophysics is still in its infancy. Only when details of the types of sites occupied by the sensor molecules, the local environment, and the quenching processes are understood and correlated with a variety of sensor molecules and supports can new materials and supports be rationally designed.

The paucity of fundamental data on microheterogeneous systems arises in part from the difficulty of acquiring data of sufficient quality to even postulate reasonable models. Further, even with very good data, establishing the uniqueness of a model can be extremely difficult; multiple reasonable models can fit the same data equally well. Reliable models are needed not only in generating calibration curves, but in understanding and correcting the behavior of a sensor when there is an inconsistency in the performance of different sensors that should be equivalent.

We focus on luminescence quenching based sensors, but many of the principles are generic. Quenching-based oxygen detectors are probably the most mature. In homogeneous media with only a single component exponential decay the intensity and lifetime forms of the Stern–Volmer equations are:

$$\tau_0/\tau = 1 + K_{SV}[Q] \tag{4.3a}$$

$$K_{SV} = k_2\,\tau_0 \tag{4.3b}$$

$$I_0/I = 1 + (K_{SV} + K_{eq})[Q] + K_{eq}K_{SV}[Q]^2 \tag{4.3c}$$

where [Q] is the quencher concentration, τ's are lifetimes, I's are intensities, K_{SV} and k_2 are the Stern–Volmer and bimolecular quenching constants, respectively, and K_{eq} is the association constant for the binding of the quencher to the luminescent species. This form assumes the possibility of both dynamic (i.e., diffusional) and static (i.e., formation of a nonluminescent complex) quenching. The subscript 0 denotes the value in the absence of quencher. If plots of τ_0 / τ or I_0 / I versus quencher concentration are linear and match, quenching is purely dynamic (i.e., $K_{eq} = 0$). If I_0/I is above τ_0/τ, static quenching is present. Fitting both intensity and lifetime curves determines the static and dynamic quenching contributions.

In many microheterogeneous systems simple homogeneous models fail. The intensity Stern–Volmer plots are generally curved downward, which is the opposite of mixed dynamic-static quenching plots. Further, the decays are invariably multiexponential. Only now are reasonable models of the complexities of these systems being developed. For example, in our work on luminescent metal complexes, we find that a two-site model, with each site having different K_{SV}'s, can fit all of the intensity oxygen quenching data for several metal complexes in a silicone rubber.[26] Also, we have found that for several complexes bound to a silica, a Freundlich adsorption isotherm coupled with surface diffusional quenching modeled intensity quenching data to within experimental error.[32] Very high temporal resolution experiments with organic systems suggest that Gaussian distributions of sites are appropriate for some organics bound to silica.[52]

The problem of assessing the degree of static and dynamic quenching in persistent microheterogeneous systems was largely unanswered until recently. We reported a model-independent method for unambiguously measuring the degree of static quenching even without any knowledge of the lifetime distribution function. Details are given elsewhere.[53] We have yet to find an oxygen quenching system that exhibits any detectable static quenching.

Unfortunately, severe microheterogeneity of luminescent sensors on polymers or inorganic substrates is the rule rather than the exception. It is also clear that one cannot use O_2 quenching in a homogeneous solvent as a reliable predictor of sensor performance in polymers. This microheterogeneity causes exceptional difficulty in reliably fitting experimental data to verifiable models. It is easy to fit data very well even though the model bears little or no resemblance to reality (see below).[54, 55] We discuss several widely used microheterogeneous models and show which experimental techniques and conditions are required to reliably differentiate between different models. Our basic approach will be to numerically simulate experimental data using different models, reduce the data, and determine which models can be differentiated.

4.7.1. Simulations

Numerical simulation provides an especially convenient approach, particularly in some complicated areas such as assessing the consistency of different models with the data, differentiating between different models, and "quick and dirty" ways of getting a feel for the expected behavior of different models. We provide examples of all of these uses. Simulations are much more efficient than trying to infer results from poorly understood real systems. This is especially true given powerful simulation programs such as Mathcad, Maple, Theorist, and Mathematica and powerful microcomputers. We will consider distribution and quenching models based on both intensity and lifetime measurements. Full details of the calculations and additional details are given elsewhere.[54, 55] We merely summarize the more important points here.

Our simplest continuous microheterogeneous model assumes that the luminophore exists in a distribution of spectroscopically different environmental sites. For a tractable, yet plausible, model each site is assumed to be quenched by normal Stern–Volmer quenching kinetics. For luminescence decays each individual component is assumed to give a single exponential decay with the following impulse response:

$$i(t,\tau) = K(\tau)\exp(-t/\tau) \tag{4.4}$$

and the observed decay $D(t)$ for the distribution is $i(t, \tau)$ integrated over all possible lifetimes.

$$D(t) = \int_0^\infty K(\tau)\exp(-t/\tau)d\tau \tag{4.5}$$

Each concentration profile of sites as a function of their unquenched lifetime was assumed to be a Gaussian distribution. For decay time measurement, $K(\tau)$ for each profile is

$$K(\tau) = (1/\sqrt{2\pi\sigma^2})\exp[-0.5((\tau - \tau_{\text{center}})/\sigma)^2] \quad K > 0 \tag{4.6a}$$

$$K(\tau) = 0 \quad \tau \leq 0 \tag{4.6b}$$

$$R = \sigma/\tau_{\text{center}} \tag{4.6c}$$

where τ_{center} is the center of the distribution and σ is the standard deviation of the width. In effect, one assumes there is a Gaussian weighted distribution of exponentials contributing to the system impulse response. R is the fractional width of different Gaussian distributions. The condition in Eq. (4.6b) accounts for distributions so wide that they would otherwise have significant, but impossible amplitudes at negative decay times. $R = 0.25$ is narrow enough to have a negligible $K(\tau)$ below 0, but $R = 0.5$ would have a significant contribution below zero. $R = 0.01$ is indistinguishable from a pure single exponential decay.

Where quenching is present, the bimolecular quenching rate constant, k_2, is assumed constant across the distributions, although for different classes of sites the rate constant may vary. For decay curves for a single site, Eq. (4.5) is used but τ in the exponential decay part depends on quencher concentration and k_2.

$$D(t) = \int_0^\infty K(\tau)\exp[-t/(\tau/\{1 + k_2\tau[Q]\})]d\tau \tag{4.7}$$

where $[Q]$ is the quencher concentration. For systems with more than one distribution of sites, the right-hand side of Eq. (4.6) is summed over all sites. After generation of $D(t)$ it was normalized to an appropriate amplitude. For double Gaussian simulations, we used peaks centered at $\tau = 5$ and $\tau = 15$ with R of 0.20 or 0.25. Each component was allowed to contribute 50% of the total emission intensity under unquenched conditions.

For computing $D(t)$'s we assumed that data were from a time correlating single photon counting (SPC) with a peak channel count of 10^4, a common experimental value. The decays were fitted down to 25 counts which keeps the Poisson statistics nearly Gaussian while still containing useful information. The 400 to 1 fitting range is much better than a purely analog instrument, although some digital oscilloscopes now appear to approach this. All data fits were performed by nonlinear least squares using a simplex or Marquardt algorithm.[56] Deconvolution[57] was ignored since, while it does increase experimental difficulty, it generally has a minimal effect on the extracted parameters unless the τ's are comparable to, or shorter than, the flash width.

The quality of fits to simulated decays was judged by the usual criteria: the reduced chi-square, χ_r^2 and visual inspection of the residual, r_i, and weighted residuals, r_{wi}, plots.[56, 57] χ_r^2 is given by

$$r_i = D(t_i) - D_{calc}(t_i) \tag{4.8a}$$

$$\chi_r^2 = (1/\nu) \sum (r_i/\sigma_i)^2 \tag{4.8b}$$

$$\nu = N - p - 1 \tag{4.8c}$$

$$r_{wi} = r_i / \sigma_i \tag{4.8d}$$

where D and D_{calc} are the observed and calculated best fits for the model, respectively, r is the residual or difference between the observed and calculated fit, σ is the standard deviation for each point, ν is the number of degrees of freedom, and p is the number of parameters fit in the model. The summation is over all N data points used in the fit. Since r is on the order of σ for a good model, a good fit will give a random distribution of r_{wi}'s around zero with an average amplitude of about 1.

For SPC data, the statistics are Poisson and σ_i is defined uniquely by the data

$$\sigma_i = D(t_i)^{1/2} \tag{4.9}$$

A good fit will give r_i's on the order of σ_i and the summation in Eq. (4.8b) then roughly equals N. For a reasonable N (>50) ν is approximately N and χ_r^2 is about 1. Thus, a χ_r^2 much greater than 1 suggests a faulty model. Extensive tables of χ_r^2's versus ν are available to determine if the models were reasonable.[58] For our discussion, it is sufficient to recognize that a χ_r^2 much above 1 would indicate a poor model.

4.7.2. Uniqueness: a Caveat

As has long been recognized it is extremely difficult to accurately fit experimental decay curves to sums of exponentials, especially for relatively close lifetimes (< factor of 2).[55, 56, 59, 60] That is, one can get good fits but with parameters that are physically meaningless. The same is true of many different types of models. However, the point is so important as to justify repeating. Earlier we gave several examples.[55, 56] We repeat some information on the more extreme one. Consider five functions. The first is the Gaussian distribution of Eq. (4.6) with $\tau = 15$ and a wide $R = 0.25$ and is denoted by D_G(t). The remaining four functions are

$$D_1(t) = 4582.5 \exp(-t/11.092) + 5395.3 \exp(-t/18.379) \tag{4.10a}$$

$$D_2(t) = 2005.7 \exp(-t/7.6144) + 8022.8 \exp(-t/16.774) \tag{4.10b}$$

$$D_3(t) = 7910.7 \exp(-t/13.434) + 1977.9 \exp(-t/21.179) \tag{4.10c}$$

$$D_4(t) = 702.2 \exp(-t/6.977) + 5892 \exp(-t/13.223) + 3406 \exp(-t/19.72) \tag{4.10d}$$

Mathematically these are radically different functions. D_1, D_2, and D_3 are all double exponential decays, but their preexponential factors deviate radically and the lifetimes differ noticeably. The ratio of preexponentials for the fast and slow components vary by a factor of 16! D_1 has comparable amplitudes, while D_2 has a ratio of short to long of 4, and D_3 has a ratio of short to long of 1/4. D_4 is a sum of three exponentials. *All five functions vary from a peak of about 10^4 to 25, and all four functions, if overlaid, are virtually indistinguishable.* To amplify these differences, we assume that the Gaussian distribution, D_G, is the correct decay function and then show the deviations of the other functions from D_G. These results are shown in Figure 4.10. The double exponential D_1 fits the distribution decay essentially perfectly. Even D_2 and D_3 are a very credible fit. D_4 matches D_G so well that the differences are invisible on this scale, and it is not even plotted.

To further appreciate the indistinguishability of these functions, we show the error bounds of σ, assuming that the data came from an SPC instrument. Clearly, Poisson noise always greatly exceeds the differences except for D_3. For 10^4 peak counts in SPC data, differentiation between any of these functions is impossible except perhaps for D_3. Noise hides the differences. While we have selected only a small set of possible exponentials, clearly there is a continuum of possible lifetimes and preexponential factors that would be similarly indistinguishable.

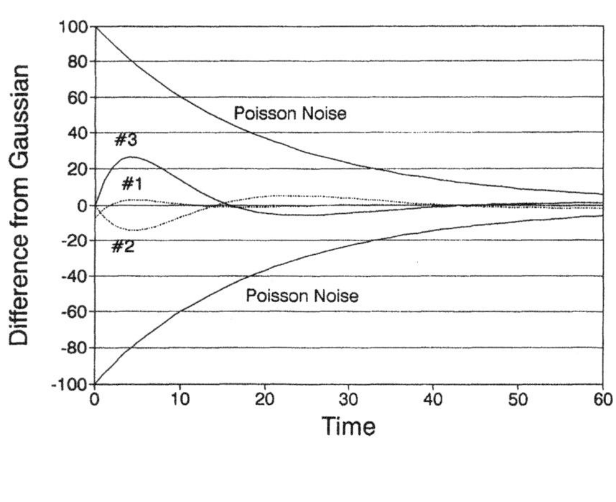

Figure 4.10. Differences between Eqs. (4.10) from the single Gaussian distribution of Eq. (4.5) with R = 0.25. #1 is D_1–D_G, #2 is D_2–D_G, and #3 is D_3–D_G. D_4–D_G is so close to zero that it does not even show up on this plot. The curves labeled "Poisson Noise" represent one standard deviation SPC decay data. All functions, if overlaid, are essentially indistinguishable. (Adapted from Ref. 55.)

In summary, for complex data fitting, it is *caveat emptor*. The uniqueness problem in nonlinear fitting of experimental data is so complex and ambiguous that one should take satisfactory fits of complex models as merely a suggestion that the proposed model may be the underlying reality. To further reinforce our warning we now present some additional examples of systems that have been fit in the literature and which are similarly poorly defined. We also present some of our data fitting and models and again reiterate our warning. A good fit does not prove a model.

4.7.3. Simulation Results

4.7.3.1. Lifetime Quenching Measurements

We first address the question of how wide a Gaussian distribution can be distinguished from a discrete single or a double exponential model by lifetime measurements. We reduced noise-free simulated decays with different R's using either single or double exponential models, and examined the reduced chi-square and the residuals plots. Figure 4.11 shows the reduced chi-squares for single and double exponential fits versus R. In practice, a reduced chi-square of 0.5 or greater for noise-free data would generally produce noticeable systematic deviations in the r_{wi}'s for the same data with Poisson noise on it. Thus, a Gaussian distribution up to about $R = 0.15$ is indistinguishable from a single exponential model and a Gaussian distribution up to about $R = 0.8$ cannot be differentiated experimentally at our noise level from a double exponential. For $R < 0.3$, a two-exponential fit will give two nearly equal preexponential factors at approximately $\tau_{center}(1 \pm R)$. Thus, if the lifetimes are not too dissimilar, nearly equal preexponential factors on a double exponential fit suggests a Gaussian distribution while very different preexponential factors indicates the inadequacy of a single Gaussian distribution model.[55]

While an unquenched single Gaussian distribution cannot be differentiated from a discrete double decay, does quenching lifetime data reveal the existence of the

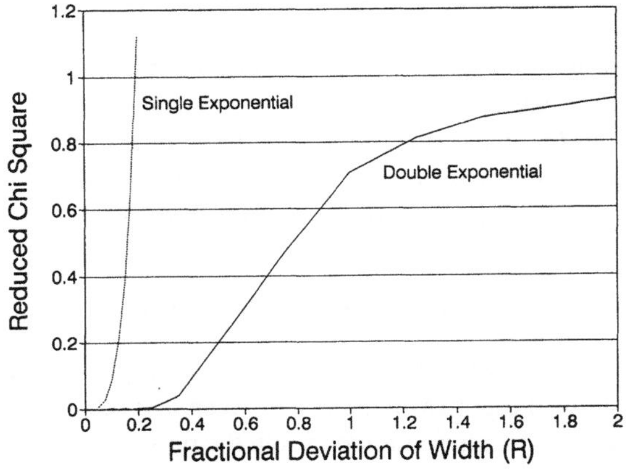

Figure 4.11. Reduced chi-square for fitting a single Gaussian distribution function of decays with either a discrete single or double exponential model as a function of distribution width (R).

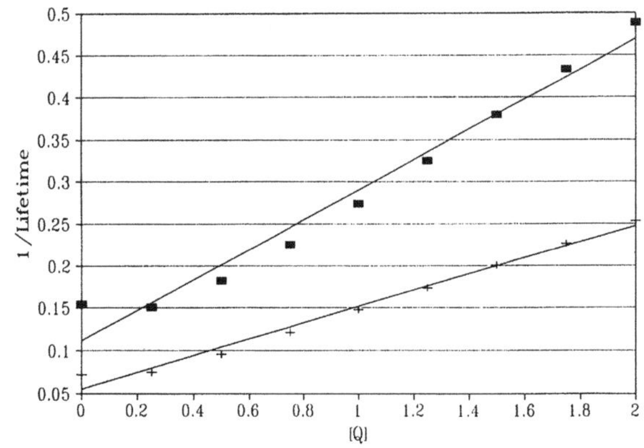

Figure 4.12. Stern–Volmer lifetime plots for quenching of decays given by a single Gaussian distribution (R = 0.4). The decays were fit to two exponential decays, and each decay was then treated as though it was independently quenched. The decays were generated using a τ-independent k_2. (Adapted from Ref. 4.)

underlying Gaussian? To test this, we assumed a Gaussian distribution with the rather wide $R = 0.4$ and generated quenched decays for different $[Q]$'s and $k_2 = 0.1$ and $\tau_{center} = 1$. For each decay, we fit the data to a dual exponential, which gave an essentially perfect experimental fit. We then treated the long and short lifetimes as though they were actually single component lifetimes and made Stern–Volmer lifetime plots for each lifetime. Figure 4.12 shows the data and the best fits. From these two plots we calculate k_2's of 0.1783 and 0.0960 for the short- and long-lived components respectively.

The noise-free Stern–Volmer lifetime plots are clearly curved, which indicates a failure of a two discrete site model. However, this is a difficult nonlinear least-squares fitting problem, and the unquenched apparent lifetimes are within a factor of two of each other. Thus, for real data, it is much more difficult to pick up on the nonlinearities and exclude a discrete two-site model. For distributions with smaller R's, of course, fitting becomes too difficult for reliable model testing at least at 10^4 counts in the peak channel.

Can an unquenched double Gaussian be differentiated from a discrete double or triple exponential decay? We fit a double Gaussian distribution ($\tau_{center} = 5$ and 15; R's = 0.2; 50% intensity from each) to discrete double and triple exponentials. Even with the double exponential fit, the r_i's are well below the SPC noise level and the triple was essentially a perfect fit. Therefore, a discrete double or triple exponential decay would be indistinguishable from the true underlying double Gaussian distribution at 10^4 peak counts.[55]

Of course, the problem in decay time measurements is improved noticeably if the noise level can be reduced. This is possible, but expensive. SPC instruments with peak counts of 500,000 have been developed.[59]

4.7.3.2. *Intensity Quenching Measurements*

We turn now to intensity quenching measurements (I_0/I versus $[Q]$). As we will show, these measurements are even less sensitive for detecting complex models than are lifetime measurements. On the plus side, however, they show a remarkable ability to model complex systems using what amounts to a simple set of lumped parameters that contain the critical system features.

As a test of whether a single Gaussian distribution could be revealed, we examined the Stern–Volmer intensity data for a single Gaussian distribution with a τ-independent k_2 as a function of R. The deviations from linearity were absolutely minimal except for the extremely wide Gaussian with $R = 4$. However, even for $R = 4$, real data are likely to conceal the small nonlinearity.[54] Indeed, for $R = 4$ the truncated Gaussian is so distorted that it is really no longer Gaussian, but highly asymmetric versus τ. There was no discernible nonlinearity for $R = 2$.

The only way that one can obtain greatly hooked Stern–Volmer plots with Gaussian distributions and a τ independent k_2 is to have more than one Gaussian distribution with different k_2's. Figure 4.13 shows intensity Stern–Volmer plots for a double distribution with $\tau_{short} = 5$, $\tau_{long} = 15$, $R_{long} = R_{short} = 0.25$, and $k_{2short} = 0.50$ and $k_{2long} = 0.01$. We have assumed that each site contributes 50% of the unquenched steady state luminescence intensity.

To test whether one can differentiate between a two-site discrete model and a dual distribution function, we calculated intensity Stern–Volmer plots for a two-component model as a function of R. These are also shown in Figure 4.13. What is remarkable is that even for the quite wide $R = 0.25$, there is no experimentally detectable difference between two discrete sites and two continuously variable distribution of sites. Only when one gets to $R = 0.5$ does the data deviate noticeably. However, even though the shape has changed, it is still well fit by a dual discrete site model with different parameters.

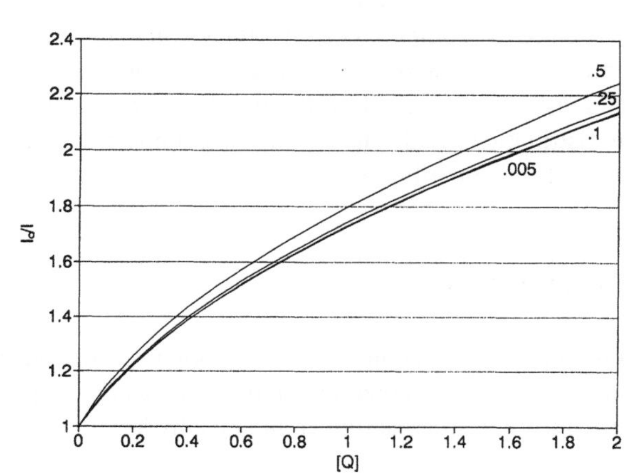

Figure 4.13. Intensity Stern–Volmer plot for a double Gaussian distribution of lifetimes as a function of distribution width R. The two lifetimes are 5 and 15 with bimolecular quenching rate constants of 0.50 and 0.01, respectively. The unquenched intensity contribution is 0.5 for each lifetime. The R for each distribution is indicated next to the curve. (Adapted from Ref. 54.)

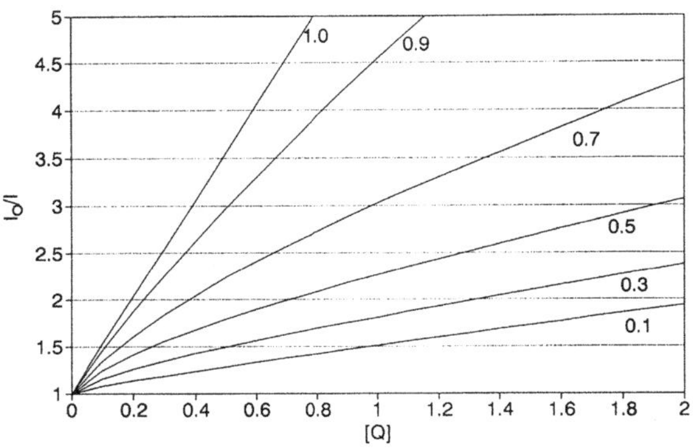

Figure 4.14. Intensity Stern–Volmer plots for a double Gaussian distribution of lifetime as a function of the fraction of contribution of short-lived unquenched intensity. The fractional contribution of the short component is shown by each curve. (Adapted from Ref. 54.)

Another issue is how much of a contribution from two sites is required to produce nonlinear Stern–Volmer plots? Figure 4.14 shows Stern–Volmer plots for another dual distribution data set. $\tau_{short} = 5$, $\tau_{long} = 15$, $R_{long} = R_{short} = 0.25$, and $k_{short} = 1.0$ and $k_{long} = 0.025$. However, the fractional contribution of the short-lived component to the total unquenched steady-state luminescence was varied. Clearly, the curvature is pronounced and experimentally detectable from 0.1 to 0.9; not surprisingly, it is more pronounced for comparable contributions from both sites. This last feature is due to the fact that in the limit of pure fast or slow components, the plots become linear.

The poor ability to distinguish quenching models by intensity data is not very comforting for those trying to model fundamental processes using only intensity data. However, the result is a nice feature for fitting sensor data. The complexity of a multidistribution site system is virtually completely masked in intensity quenching data, and one can replace the complex system with a simple discrete two-site model. The parameters in the discrete two-site model actually represent lumped parameters for the true model. Further, they even closely approximate the fractional contributions from each of the distributions with the quenching parameters being close to those values for the center of each distribution.

4.7.4. Physical System Results

To illustrate the problems encountered in supported sensors we present data of two actual systems that we have studied in detail. We focus on oxygen quenching of metal complex dispersed in a silicone rubber and on a hydrophilic silica.

4.7.4.1. Silicone Supports Two State Quenching Model

The downward curvature of the Stern–Volmer intensity plots necessitates a model more complex than a single species quenched bimolecularly. We have evaluated numerous models.[26] The only one that satisfactorily duplicates the data involves the sensor existing in two distinctly different environments with both being quenched but with different rate constants. This model yields the intensity Stern–Volmer equation:

$$I_0/I = 1/\{f_{01}/(1 + K_{SV_1}[Q]) + f_{02}/(1 + K_{SV_2}[Q])\} \qquad (4.11)$$

where the f_{0i}'s are the fraction of the total emission from each component under unquenched conditions and the K_{SV_i}'s are the associated Stern–Volmer quenching constants for each component. However, other workers have developed an equation that fits equally well based on nonlinear gas solubility.[61] Since the luminescence decays in some of these systems are certainly not single exponentials, this is a clear indication that a multisite model is required. Therefore, it is likely that both heterogeneity and nonlinear gas solubility are simultaneously applicable.

Figure 4.7 shows the best fits to the experimental data using Eq. (4.11). Although the data are fit within experimental error, the two-state model is certainly just an approximation. More complex distributions of sites with different quenching constants could fit the data. The success of the two-state model is not surprising given the well-known ability of two exponentials to accurately mimic complex decay curves (see above). Further, τ data indicate that a more complex model is needed for a full description.

The two-site model of Eq. (4.11) also works well for fitting the quenching data of a large number of $ReL(CO)_3NCR^+$ complexes with different α-diimines and alkyl groups embedded in a silicone rubber.[62] Further, it provides considerable insight into the nature of the binding sites as a function of the structure and size of the α-diimine.

We have also carried out an extensive study of oxygen-sensing properties of $[Ru(Ph_2phen)_3]Cl_2$[63] in a diverse series of polymers, most with a common polydimethylsiloxane (PDMS) component. Systematic variations in the polymer properties have been made in order to delineate the structural features important for satisfactory use of supports for oxygen sensors. Most measurements were made using homo- or copolymers containing a PDMS region, although some measurements were made on small-ring siloxane polymers. In particular, quenching behavior was examined as a function of the type and amount of polar copolymer cross-linkers; these were added to enhance the solubility of polar sensor molecules in an otherwise nonpolar polymer. In addition, hydrophobic silica was added to alter quenching properties. A domain model was used to explain the variations in oxygen quenching properties as a function of additives and cross-linkers. The relative affinity of the different domains for the complex and the efficacy of the domains for oxygen quenching control the overall behavior of the sensing response. Guidelines for design of suitable polymer supports for oxygen sensors were proposed.

Even though the two-site model has shortcomings, it is excellent for fitting intensity quenching curves. Thus, it has excellent predictive and calibration properties, has a chemically sound basis, and (at least for inorganic complex sensors) is preferable to the less accurate power law calibration equation.

Our results demonstrated clearly that the lifetime data are more sensitive to subtleties of the micromechanistic photophysics. In this case we were able to establish inadequacies of the two-component model that were not detected by intensity quenching measurements alone. It is also clear that resolution of the detailed mechanism in these complex polymer systems will require even better lifetime data than we are able to obtain with a conventional flash lamp–based time-correlated photon counting system.

These results show that the complexes in silicone rubber, as well as silica supports (discussed below), show considerable promise as O_2 sensors. Further, they serve to demonstrate the complexity present in persistent microheterogeneous systems.

4.7.4.2. Silica Supports

Fumed hydrophilic silica (Cab-O-Sil) is a new and interesting solid support with considerable promise as a sensor matrix and as a support for an immobilized photocatalyst.[33] This airy powder has the remarkable property of forming, under compression, a dense, very high surface area, optically transparent glass that can be used for a variety of spectroscopic, photophysical, and photochemical studies. The disks (1 g/cm^{-3}) are clear with a slight cloudiness, but are approximately 30 times more dense than the uncompacted powder. The surface area of pressed and unpressed Cab-O-Sil powder are both 200 m^2/g. Virtually instantaneous stabilization of intensity readings in the quenching experiments establishes a very porous structure. The heavy quenching at higher O_2 pressures shows that the entire disk volume is accessible to gas.

Figure 10.9 shows the intensity versus $[O_2]$ quenching curve for three complexes in Cab-O-Sil disks. The solid lines are the best fits to a Freundlich adsorption model (see below). All quenching is dynamic with no static component.

4.7.4.3. Silica Quenching Models

Even in the absence of quencher, the decays are extremely nonexponential, clearly a sign of persistent heterogeneity. Three and four exponentials were routinely required to fit the data. With so many exponentials, as well as the pressure dependence of the fitting parameters, it is unlikely that the τ's represent true components. Thus, the decay probably represents distributions. The nonlinear intensity Stern–Volmer plots also establish a heterogeneous quenching model.

For such a complex system we envisioned several possible quenching mechanisms. The complex decay curves clearly indicate a heterogeneous distribution of lifetime sites. Site quenching could be by gaseous O_2 and/or adsorbed O_2. For adsorbed O_2 the Langmuir and Freundlich adsorption isotherms are possibilities for predicting surface O_2 concentrations.

We have examined numerous models including discrete and continuous τ distributions. Remarkably, the best fit to all the intensity quenching data involved quenching by surface-bound O_2 where the adsorption was described by a Freundlich isotherm:

$$\theta = aP^{1/n} \tag{4.12}$$

coupled with a single Stern–Volmer quenching constant. θ is the fractional surface coverage of the adsorbate, a is a collection of constants, P is the equilibrium gas phase pressure, and n is an empirical parameter related to adsorption intensity. The Langmuir model assumes that all sites have the same binding energy while the Freundlich model acknowledges that there is a distribution of site energies with preferential binding to the most energetic sites. This yields a two-parameter ($aK_{SV}[Q]_{adsM}$ and n) Stern–Volmer intensity quenching equation.

$$I_0/I = 1 + (K_{SV}[Q]_{ads_M})aP^{1/n} = 1 + K_{SV_{effective}}P^{1/n} \tag{4.13}$$

where K_{SV} is a normal Stern–Volmer quenching constant, $K_{SV_{effective}}$ is the effective Stern–Volmer constant in the modified Stern–Volmer Eq. 12, $[Q]_{ads_M}$ is the maximum surface concentration of quencher, and a and n are the Freundlich parameters of Eq. (4.12).

Figure 4.9 shows the best Freundlich quenching model plots. While more complex models could fit the data, this simple two-parameter model fits all the data within experimental error. Since the decays are highly nonexponential, the parameters represent lumped parameters that are some average over the available sites, lifetimes, and K_{SV}'s. A detailed discussion of the different models tried is given elsewhere.

An interesting point is that our original RTV 118 quenching data is due in part to the presence of hydrophobic silica filler in the polymer.(64) We have examined in considerable detail the effect of silica filler in polymer supports.(33) It is noteworthy that the hydrophobic filled polymers give much less hooked results than the hydrophilic silica. Indeed, the hydrophobic silica gives nearly ideal one-site quenching results.(64)

4.8. Conclusions

Advances in understanding the photophysics and photochemistry of transition metal complexes offer opportunities to utilize these materials as luminescence sensors and probes. However, heterogeneity has an enormous effect on luminescence, quenching, and photochemistry. An intimate understanding of the detailed interactions between the complexes and their environment will be necessary before the rational design of new high-performance sensors and probes can be achieved. Such understanding is fairly well established for homogeneous media, but it is still in its infancy in solid supports (surfaces and polymers). We still lack an understanding of the impact of probe charge, size, and shape and their relationship to binding with polymers and other supports as well as sensor characteristics. Also, seemingly monolithic materials can show distinct domains with very different properties such as hydrophobicity and permeability, and this domain structure can play a pivotal role in sensor behavior.

The temperature dependence of luminescent metal complexes can be controlled by molecular design that affects the energy gap between the emitting state and the deactivating *d–d* or by altering the preexponential factor for thermal deactivation. The sometimes large temperature dependencies of lifetime and quantum yields for metal complexes also suggest their use as temperature sensors.

We envision several major areas of significant future activity. For those seeking specific molecular probes, the design of features that give site selective binding while still preserving the optimum luminescence properties is still at a rudimentary level. Adequate pH and CO_2 sensors still remain to be demonstrated. The design of both molecular probes and sensors will be enhanced by an understanding of the details of the interactions of the complexes with the substrate or support. Developments in lifetime measurements (phase and pulsed), time-resolved emission spectra, and mathematical methods of fitting data coupled with the enormous computing power of desk top computers will revolutionize the handling of these complex and fascinating systems.

Our computations clearly show a number of points:

- Fitting luminescence decay data to sums of exponentials, even with rather good statistics, can present very serious problems in data fitting and in uniqueness of the solutions. This difficulty can severely cloud interpretation of data from microheterogeneous systems.
- A single Gaussian distribution of conformations (unless pathologically wide ($R > 1$)) shows little detectable nonlinearity in Stern–Volmer intensity quenching curves.
- However, for reasonably wide single distributions lifetime decays can provide a warning that a single discrete model is inappropriate even when the intensity Stern–Volmer plots give no warning of system complexity.
- Large differences in K_{SV}'s for a dual Gaussian distribution site model are required to produce nonlinear intensity Stern–Volmer plots. When the differences in K_{SV}'s are large enough to produce nonlinear intensity Stern–Volmer plots, wide ranges in the fractional contributions of the sites to the total intensity can yield detectable curvature.
- An advantage of the inability to detect single Gaussian distributions by intensity data is that intensity quenching data (even complex distribution functions of two sites) can be reliably modeled using a discrete two-site model. This has obvious practical implications in sensor design and calibration.
- In spite of the problems of lifetime measurements, they are much more sensitive to microheterogeneity than intensity measurements and must form a significant component in any attempt to unravel details of microheterogeneous sensors.

Acknowledgments

We gratefully acknowledge support by the National Science Foundation (CHE 91-18034).

References

1. R. A. Lieberman, M. T. Wlodarczyk, eds., *Chemical, Biochemical, and Environmental Sensors*, Vol. 1172, The International Society for Optical Engineering, Bellingham, Washington, 1988.
2. E. P. Diamandis and T. K. Christopoulos, Europium chelate labels in time-resolved fluorescence immunoassays and DNA hybridization arrays, *Anal. Chem. 62*, 1149A–1157A (1990).
3. O. Wolfbeis, *Molecular Luminescence Spectroscopy Methods and Applications: Part 2* (S. G. Schulman, ed.), pp. 129–281, John Wiley and Sons, New York (1988).
4. W. W.-S. Lee, K.-Y. Wong, and X.-M. Li, Luminescent dicyanoplatinum(II) complexes as sensors for the optical measurement of oxygen concentrations, *Anal. Chem. 65*, 255–258 (1993).
5. Entire issue: *J. Chem. Ed. 60*(10), 1983.
6. A. Juris, V. Balzani, F. Barigellitti, S. Campagna, P. Belser, and A. vonZelewsky, Ru(II) Polypyridine complexes: Photophysics, photochemistry, electrochemistry, and chemiluminescence, *Coord. Chem. Rev. 84*, 85–277 (1988).
7. J. N. Demas and B. A. DeGraff, Design and applications of highly luminescent transition metal complexes, *Anal. Chem. 63*, 829A–837A (1991).
8. F. A. Cotton and G. Wilkinson, *Advanced Inorganic Chemistry*, Chapter 20, John Wiley and Sons, New York (1980).
9. J. N. Demas and G. A. Crosby, Quantum efficiencies of transition metal complexes. I. *d–d* luminescences, *J. Am. Chem. Soc. 92*, 7262–7270 (1970).
10. J. N. Demas and G. A. Crosby, Quantum efficiencies of transition metal complexes. II. Charge transfer luminescences, *J. Am. Chem. 93*, 2841–2847 (1971).
11. J. V. Caspar, E. M. Kober, B. P. Sullivan, and T. J. Meyer, Application of the energy gap law to the decay of the charge transfer excited state, *J. Am. Chem. Soc. 104*, 630–632 (1982).
12. J. V. Caspar and T. J. Meyer, Application of the energy gap law to nonradiative, excited state decay, *J. Phys. Chem. 87*, 952–957 (1983).
13. L. Sacksteder, A. P. Zipp, E. Brown, J. Streich, J. N. Demas, and B. A. DeGraff, Luminescence studies of pyridine α-diimine rhenium(I) tricarbonyl complexes, *Inorg. Chem. 29*, 4335–4340 (1990).
14. E. Krausz and J. Ferguson, The spectroscopy of the $[Ru(bpy)_3]^{2+}$ system, in: *Progress in Inorganic Chemistry* (S. J. Lippard, ed.), Vol. 37, pp. 293–390, John Wiley and Sons, New York (1989).
15. W. J. Dressick, J. Cline III, J. N. Demas, and B. A. DeGraff, Energy degradation pathways and binding site environment of micelle bound ruthenium(II) photosensitizers, *J. Am. Chem. Soc. 108*, 7567–7574 (1986).
16. J. N. Demas and B. A. DeGraff, On the design of luminescence based temperature sensors, *SPIE 1796*, 71–80 (1992).
17. D. H. W. Carstens, Spectroscopic Studies of Complexes of Rhodium(III) and Iridium(III), Ph.D. Dissertation, University of New Mexico, Albuquerque, 1969.
18. L. R. Allen, P. P. Craft, B. Durham, and J. Walsh, Substitution reactions of Ruthenium (II) complexes containing 2,2-bipyridine and 1,10-phenanthroline, *Inorg. Chem. 26*, 53–56 (1987).
19. R. J. Watts, J. S. Harrington, and J. Van Houten, A stable monodentate 2,2-bipyridine complex of iridium (III): A model for reactive intermediates in ligand displacement reactions of tris 2,2-bipyridine metal complexes, *J. Am. Chem. Soc. 99*, 2179–2187 (1977).

20. P. E. Hoggard and G. B. Porter, Photoanation of the tris(2,2-bipyridine)ruthenium(II) cation by thiocyanate, *J. Am. Chem. Soc. 100*, 1457–1463 (1978).
21. W. J. Dressick, K. I. Raney, J. N. Demas, and B. A. DeGraff, Properties of osmium(II) photosensitizers in aqueous and sodium lauryl sulfate micellar media, *Inorg. Chem. 23*, 875–880 (1984).
22. G. A. Reitz, J. N. Demas, E. Stephens, and B. A. DeGraff, Inter- and intramolecular interactions of surfactant active rhenium(I) photosensitizers, *J. Am. Chem. Soc. 110*, 5051–5059 (1988).
23. L. Sacksteder, M. Lee, J. N. Demas, and B. A. DeGraff, Long lived, highly luminescent rhenium(I) complexes as molecular probes: Intra- and intermolecular excited state interactions, *J. Am. Chem. Soc. 115*, 8230–8238 (1993).
24. A. P. Zipp, L. Sacksteder, J. Streich, A. Cook, J. N. Demas, and B. A. DeGraff, Luminescence of Phenium(I) complexes with highly sterically hindered α-diimine ligands, *Inorg. Chem. 32*, 5629–5632 (1993).
25. J. R. Lakowicz, *Principles of Fluorescence Spectroscopy*, New York, Plenum Press (1983).
26. E. R. Carraway, J. N. Demas, B. A. DeGraff, and J. R. Bacon, Photophysics and photochemistry of oxygen sensors based on luminescent transition-metal complexes, *Anal. Chem. 63*, 337–342 (1991).
27. W. P. Krug and J. N. Demas, Charge transfer luminescence from an α-diimine complex of ruthenium(II), *J. Am. Chem. Soc. 101*, 4394–4396 (1979).
28. G. Orellana, M. C. Moreno-Bondi, E. Segovia, M. D. Marazuela, Fiber-optic sensing of carbon dioxide based on excited-state proton transfer to a luminescent ruthenium(II) complex, *Anal. Chem. 64*, 2210–2215 (1992).
29. R. W. Harrigan, G. D. Hager, and G. A. Crosby, Evidence for multiple-state emission from ruthenium(II) complexes, *Chem. Phys. Lett. 21*, 487–490 (1973).
30. M. E. Lippitsch, J. Pusterhofer, M. J. P. Leiner, and O. S. Wolfbeis, Fibre-optic oxygen sensor with the fluorescence decay time as the information carrier, *Anal. Chim. Acta 205*, 1–6 (1988).
31. O. S. Wolfbeis, L. Weis, M. J. P. Leiner, and W. Ziegler, Fiber-optic fluorosensor for oxygen and carbon dioxide, *Anal. Chem. 60*, 2028–2030 (1988).
32. J. R. Bacon and J. N. Demas, Determination of oxygen concentrations by luminescence quenching of a polymer immobilized transition metal complex, *Anal. Chem. 59*, 2780–2785 (1987).
33. E. R. Carraway, J. N. Demas, and B. A. DeGraff, Photophysics and oxygen quenching of transition-metal complexes on fumed silica, *Langmuir 7*, 2991–2998 (1991).
34. R. B. Rexwinkel, S. C. J. Meskers, H. P. J. M. Dekkers, and J. P. Riehl, Ionic strength dependence of the enantioselective quenching of $Tb(2,6\text{-pyridinedicarboxylate})_3^{2+}$ luminescence by resolved $Ru(1,10\text{-phenanthroline})_3^{2+}$, *J. Phys. Chem. 96*, 5725–5733 (1992).
35. C. V. Kumar, J. K. Barton, and N. J. Turro, Photophysics of ruthenium complexes bound to double helix DNA, *J. Am. Chem. Soc. 107*, 5518–5523 (1985).
36. A. Kirsch-De Mesmaeker, G. Orellano, J. K. Barton, and N. J.Turro, Light-dependent interactions of ruthenium(II) polypyridyl complexes with DNA probed by emission spectroscopy, *Photochem. Photobiol. 52*, 461 (1990).
37. D. H. Metcalf, S. W. Snyder, S. Wu, G. L. Hilmes, J. P. Riehl, J. N. Demas, and F. S. Richardson, Excited-state chiral discrimination observed by time-resolved circular polarized luminescence measurements, *J. Am. Chem. Soc. 11*, 3082–3083 (1989).
38. D. H. Metcalf, S. W. Snyder, J. N. Demas, and F. S. Richardson, Chiral discrimination in electronic energy-transfer processes between dissymmetric metal complexes in solutions, *J. Am. Chem. Soc. 112*, 5681–5895 (1990).
39. R. M. Hartshorn and J. K. Barton, Novel dipyridophenazine complexes of ruthenium(II): Exploring luminescent reporters of DNA, *J. Am. Chem. Soc. 114*, 5919–5925 (1992).
40. J. N. Demas and G. A. Crosby, The measurement of photoluminescence quantum yields, *J. Phys. Chem. 75*, 991–1024 (1971).
41. K. Mandel, T. D. L. Pearson, and J. N. Demas, New luminescent quantum counter systems based on a transition metal complex, *Inorg. Chem. 20*, 786–789 (1981).

43. A. P. Schapp (ed.), *Singlet Molecular Oxygen, Benchmark Papers in Organic Chemistry*, Vol. 5, Dowden, Hutchinson, and Ross, New York (1976).
44. E. C. Blossey, D. C. Neckers, A. L. Thayer, and A. P. Schapp, *J. Am. Chem. Soc. 95*, 5820–5822 (1973).
45. S. L. Buell and J. N. Demas, Heterogeneous preparation of singlet oxygen using an ion exchange bound tris(2,2′-bipyridine)ruthenium(II) photosensitizer, *J. Phys. Chem. 87*, 4675–4681 (1983).
46. J. N. Demas, E. W. Harris, and R. P. McBride, Energy transfer from metal complexes to oxygen, *J. Am. Chem. Soc. 99*, 3547–3551 (1977).
47. W. J. Vining, J. V. Caspar, and T. J. Meyer, Influence of environmental effects on excited state lifetimes. The effect of ion pairing on metal-to-ligand charge transfer excited states, *J. Phys. Chem. 89*, 1095–1099 (1985).
48. M. C. Moreno-Bondi, O. S. Wolfbeis, and M. J. P. Leiner, Oxygen optrode for use in fiber-optic glucose biosensor, *Anal. Chem. 62*, 2377–2380 (1988).
49. J. I. Peterson, R. V. Fitzgerald, and D. K. Buckhold, Fiber-optic probe for vivo measurements of oxygen partial pressures, *Anal. Chem. 56*, 62–67 (1984).
50. E. D. Lee, T. C. Werner, and W. R. Seitz, Luminescence ratio indicators for oxygen, *Anal. Chem. 59*, 279–283 (1987).
51. C. W. Berndt and J. R. Lakowicz, Electroluminescent lamp-based phase fluorimeter and oxygen sensor, *Anal. Biochem. 201*, 319–325 (1992).
52. D. R. James, Y. Liu, N. O. Petersen, A. Siemiarczuk, B. D. Wagner, and W. R. Ware, Recovery of underlying distributions of lifetimes from fluorescence decay data, *Fluorescence Detection, SPIE 743*, 117–122 (1987).
53. E. R. Carraway, J. N. Demas, and B. A. DeGraff, Luminescence quenching mechanism for microheterogeneous systems, *Anal. Chem. 63*, 332–336 (1991).
54. J. N. Demas and B. A. DeGraff, Luminescence sensors: Modeling of microheterogeneous systems and model differentiation *SPIE 1681*, 2–11 (1992).
55. J. N. Demas and B. A. DeGraff, Luminescence based sensors: microheterogeneity and temperature effects, *Sensors and Actuators B, 11*, 35–41 (1992).
56. J. N. Demas and S. E. Demas, *Interfacing and Scientific Computing on Personal Computers*, Chapter 8, Allyn & Bacon, Needham Heights, Massachusetts (1990).
57. J. N. Demas, *Excited State Lifetime Measurements*, Academic Press, New York (1983).
58. P. R. Bevington, *Data Reduction and Error Analysis for the Physical Sciences*, McGraw-Hill, New York (1969).
59. D. R. James, Y-S. Liu, P. DeMayo, and W. R. Ware, Distributions of fluorescence lifetimes: Consequences for the photophysics of molecules adsorbed on surfaces, *Chem. Phys. Lett. 120*, 460–465 (1985).
60. A. Seimiarczuk and W. R. Ware, Temperature dependence of fluorescence lifetime distributions in 1,3-di(1-pyrenyl)propane with maximum entropy method, *J. Phys. Chem. 93*, 7609–7618 (1989).
61. X.-M. Li, F.-C. Ruan, and K.-Y. Wong, Optical characteristics of a ruthenium(II) complex immobilized in a silicone rubber film for oxygen measurement, *Analyst 118*, 289–292 (1993).
62. L. Sacksteder, J. N. Demas, and B. A. DeGraff, Design of oxygen sensors based on quenching of luminescent metal complexes: Effect of ligand size and heterogeneity, *Anal. Chem.* (submitted).
63. W. Xu, R. C. McDonough III, B. Langsdorf, J. N. Demas, and B. A. DeGraff, Oxygen sensors based on luminescent metal complexes: Interactions with the polymer supports, *Anal. Chem.* (submitted).
64. I. Klimant and M. J. P. Leiner, Recent investigations in optical oxygen sensing: The behavior of ruthenium complexes in silicone matrices, in: Abstracts, 1st European Conference on Optical Chemical Sensors and Biosensors, Graz, Austria, April 12–15, 131 (1992).

5

Fluorescence Probes Based on Twisted Intramolecular Charge Transfer (TICT) States and Other Adiabatic Photoreactions

W. Rettig and René Lapouyade

5.1. Introduction

Fluorescence probes combine the advantage of high detection sensitivity with the possibility of built-in specific sensing functions for various microscopic properties in the immediate surrounding of the probe.

If only the sensitivity aspect is used, the probes act as labels, and one of the most well-known applications is their use in fluorescence immunoassays.[1] In this case, antigenes labeled with the fluorescence dye are used, and only those which have combined with the corresponding genes to be detected are retained in the test-tube. Thus the measured fluorescence intensity is directly related to the number of gene–antigene complexes. The sensitivity achievable in this way is comparable to that of radio immunoassays which use antigenes labeled with radioactive species. The reason for this high sensitivity in these two cases as well as in chemiluminescence immunoassays lies in the possibility of detecting single particles, photons in the case of luminescence, α or β particles in the case of radioimmunoassays. The advantage of the fluorescence method lies in the flexibility of the labeling dye to be used which allows variation of properties such as color (or fluorescence spectrum) and fluorescence lifetime. In this way, different functions can be labeled and detected simultaneously, offering, e.g., the possibility for automatic recognition of amino acids in the sequencing of proteins or of DNA-bases in genome sequencing. A further major advantage of fluorescence is that high photon count rates can be achieved such that

W. Rettig • W. Nernst Institute for Physical and Theoretical Chemistry, Humboldt University at Berlin, D-10117 Berlin, Germany. *René Lapouyade* • Photophysique et Photochimie Moléculaire, URA (CNRS) No. 348, Université de Bordeaux I, F-33405 Talence, France.

Topics in Fluorescence Spectroscopy, Volume 4: Probe Design and Chemical Sensing, edited by Joseph R. Lakowicz. Plenum Press, New York, 1994.

analysis times can be strongly reduced, and even the giant task of human genome analysis, which would take several human generation by conventional methods, could be shortened to several years only by using fluorescence correlation spectroscopy for single-molecule detection.[2]

The possibility for a choice of the color (absorption and fluorescence spectra of the dye) is an important point because it allows optimization of the detection conditions and can help to avoid factors negatively influencing the sensitivity. Such factors are, e.g., the fluorescence of other components (e.g., in blood) or of the biological material itself. A general strategy is to use labeling dyes absorbing at wavelengths different from the interfering material. If this is not possible, labels possessing a very large Stokes shift (energy difference between absorption and emission maxima) can be used because fluorescence from the interfering material usually possesses much smaller Stokes shifts, and thus, by detecting only the strongly Stokes shifted fluorescence, emission from the label and from the interfering material can be distinguished.

The labeling can also be done by fluorescence lifetime differences, e.g., introduced by quenching pathways connected with different surroundings. This can be used for lifetime imaging methods (Chapter 1, this volume) or for distinguishing complexes of one and the same fluorescence label with, e.g., different DNA-bases. In this case, the fluorescence label is not only a label but incorporates a function which senses the environment and can therefore be regarded as a sensing fluorescence probe.

Sensing fluorescence probes have been in use for a number of years in biological and medical applications, mainly for determining the polar properties of microdomains[4] or for detecting changes in membrane potentials[5] or in intracellular ion concentration.[6] They are gaining increasing importance in these latter aspects, i.e., fluorescence sensing of ions (development of ion-analytic and pH-sensing fluorescence probes) as well as new aspects such as sensing of the microviscous properties of the surrounding (free volume fluorescence probes).

This review will cover various ways of achieving large Stokes shifts and/or sensing properties in a fluorescence dye. It will give examples for each of these cases and outline the present status and some future possibilities. In a general sense, Stokes shift and sensing property are linked because they derive from a single source, namely, from the fact that an adiabatic photoreaction takes place i.e., a reaction occurring on the excited state hypersurface and leading from the excited precursor (E^*) to the excited product species (P^*) which can fluoresce.[7–9] Product fluorescence is often strongly redshifted leading to the desired "large-Stokes-shift" property. The kinetics of the reaction are mostly sensitive to the environment, leading to the possibility of dual fluorescence (simultaneous E^* and P^* fluorescence) with environment-sensitive ratio of the two bands.

5.2. Adiabatic Photochemical Reaction Mechanisms or How to Produce Large Stokes Shifts

Adiabatic photoreactions in the sense introduced by Förster[7] involve a reaction occurring on either the lowest singlet or triplet excited hypersurface from E^* to P^*. No distinction is made regarding the detailed electronic nature of E^* and P^*. Therefore, reactions where E^* and P^* belong to different symmetry species and hence involve a crossing of these lower-lying excited state surfaces (this is normally connected with the term "nonadiabatic transition"[10]) nevertheless belong to the class of adiabatic photoreactions, as well as of course those where E^* and P^* correlate directly without crossing of excited states. Following Kasha's rule, in the case of an excited-state crossing, the reaction always proceeds on the lower one of the two potential energy surfaces, which constitutes the S_1-hypersurface. The crossing point can be regarded as the top of an energy barrier and is developed into a "conical intersection"[10, 11] allowing for the possibility of following a reaction profile without potential barrier by "surrounding" of the cone.[12, 13]

Figure 5.1 summarizes the kinetic schemes for various adiabatic photoreactions. In Figure 5.1a and b, both precursor E^* and product P^* can fluoresce leading to the simplest case of dual fluorescence and to the possibility of equilibration between E^* and P^* (back reaction k_{PE}). If the ground state P corresponding to P^* is an unstable product (living only for a very short time) then there is no net formation of ground state photoproducts, and after a photochemical cycle, the system is regenerated in its original precursor ground state E, i.e., the system is stable against illumination although a photoreaction has taken place (Figure 5.1a). If P is a stable species and different from the precursor ground state, illumination changes the sample composition, which is the conventional meaning of photochemistry. This case is depicted in Figure 5.1b. It can be detrimental for the use as fluorescence probe because it leads to "photobleaching," i.e., the disappearance of precursor material E. The lifetime of P depends on the size of the ground state barrier $\Delta G^‡$, and depending on this barrier and on the size of the ground state free energy difference ΔG between precursor E and product P, thermal population of P may also be possible before irradiation. This strongly complicates the analysis of fluorescence spectra and lifetimes and necessitates advanced methods like the "global and compartmental analysis."[14]

Figure 5.1c depicts the case where the excited product P^* is extremely short-lived corresponding to strong fluorescence quenching, e.g., because of a very narrow energy gap between P^* and P. This situation is normally called a photochemical funnel.[15] If a permanent ground state photoproduct P' is formed, the reaction is called a "diabatic photoreaction" according to Turro *et al.*[8] because P' formation involves a nonadiabatic (= diabatic) S_1/S_0 transition. This view is focused on the mechanism of transformation of E into P', which is a multistep sequence $E \rightarrow E^* \rightarrow P^* \rightarrow P \rightarrow P'$, and focusing on the initial steps $E \rightarrow E^* \rightarrow P^*$ clearly identifies this part of the reaction as an adiabatic one. Dark P^* states are often the cause of "intramolecular fluorescence quenching"

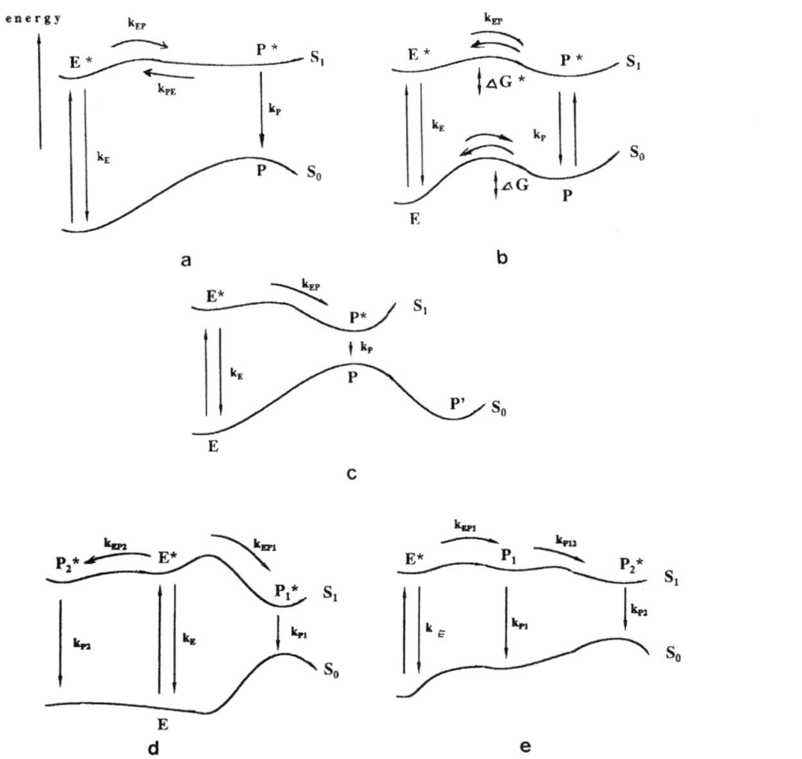

Figure 5.1. Various adiabatic photochemical reaction mechanisms (see text for details). (a) Simple case of dual fluorescence; (b) illumination changes sample (i.e., photochemistry); (c) strong fluorescence quenching (photochemical funnel); (d) competitively coupled product species; (e) consecutively coupled product species.

(because k_{EP} competes with k_{E} thus quenching the fluorescene of the precursor state E^*)[16] and they are also involved in the primary step of the visual process.[9, 15]

As will be shown below, more elaborate fluorescence probes can involve the interplay of several product species. These can either be coupled competitively (Figure 5.1d) or consecutively (Figure 5.1e). An example for the competitive case (Figure 5.1d) are donor–acceptor (D–A) stilbenes where a channel k_{EP1} leads to a nonradiative product $P_1{}^*$ (double-bond twisting leading to trans–cis isomerization, i.e., a permanent ground state product). If the parallel reaction k_{EP2} leading to a fluorescent second product $P_2{}^*$ is much faster, then the "intramolecular fluorescence quenching" by k_{EP1} can strongly be reduced and highly fluorescent dyes can result.[17] Figure 5.1e shows the case of consecutive photoreactions. The example refers to proton-transfer dyes and will be elaborated below (Section 5.2.2).

In the following, some of the most well-known adiabatic photoreaction mechanisms relevant to fluorescent probes are briefly introduced.

5.2.1. Twisting and Charge Transfer: The TICT Mechanism

The discovery of the dual fluorescence of the simple donor–acceptor substituted benzene derivative 4-*N,N*-dimethylaminobenzonitrile (DMABN) by Lippert *et al.* in 1962[18] and the subsequent model compound studies by Grabowski *et al.*[19–23] including rigidized and pretwisted compounds such as the indoline derivative MIN, the tetramethylbenzonitrile TMABN, and ayanolenzquinuclidne CBQ gave birth to the idea of twisted intramolecular charge transfer (TICT) states (Figure 5.2). This was the start for an expanding area of physical chemistry, physical organic chemistry, and organic chemistry connected with mechanistic and kinetic questions of electron transfer, and with a rationalization of the excited state behavior of many dye systems. Applicational aspects are growing in various fields such as tailor-making of fluorescence dyes,[9, 24] sensing of free volume in polymers,[25, 26] fluorescent pH or ion indicators,[27, 28] fluorescent solar collectors,[24] and electron transfer photochemistry for the destruction of harmful chlorinated aromatics.[29, 30] TICT states have been reviewed several times in recent years[9, 19, 31–34] while focusing on different particular aspects.

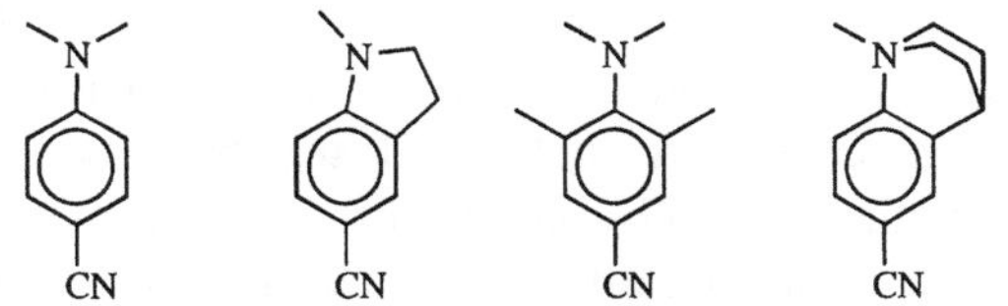

DMABN MIN TMABN CBQ

The TICT model was put forward to account for the observation that the dual fluorescence of DMABN with its "normal" band (*B* band) at around 350 nm and its "anomalous" one (*A* band, around 450 nm in medium polar solvents) depends on the

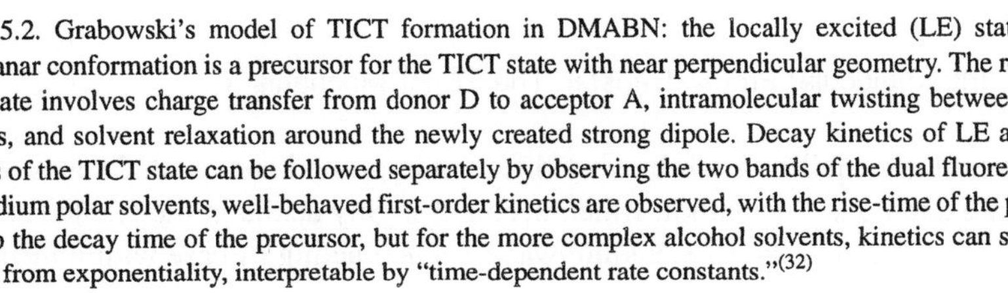

Figure 5.2. Grabowski's model of TICT formation in DMABN: the locally excited (LE) state with near-planar conformation is a precursor for the TICT state with near perpendicular geometry. The reaction coordinate involves charge transfer from donor D to acceptor A, intramolecular twisting between these subunits, and solvent relaxation around the newly created strong dipole. Decay kinetics of LE and rise kinetics of the TICT state can be followed separately by observing the two bands of the dual fluorescence. For medium polar solvents, well-behaved first-order kinetics are observed, with the rise-time of the product equal to the decay time of the precursor, but for the more complex alcohol solvents, kinetics can strongly deviate from exponentiality, interpretable by "time-dependent rate constants."[32]

conformational freedom of the dimethylamino (DMA) group: For compounds like MIN, where the DMA group is more or less fixed to a coplanar conformation with the benzonitrile skeleton, and where the lone pair orbital on the amino nitrogen is nearly parallel to the carbon p-orbitals constituting the benzonitrile π-system, only the B band is observed. For the pretwisted compounds TMABN and CBQ, where the nitrogen lone pair is nearly in-plane with the benzonitrile skeleton and perpendicular to the π-orbital system, on the other hand, only the "anomalous" A band is observed. The conclusion regarding DMABN was therefore that there exists a reaction path in the excited state leading from the near planar conformation (emitter of the B band) to a photochemical product with an energetic minimum at the perpendicular conformation (emitter of the A band). These two emitters were called B^* state and A^* state and were shown to possess a mother-daughter relationship later substantiated by direct kinetic measurements. In many cases, the back-reaction $A^* \rightarrow B^*$ also occurs leading to an excited state equilibrium. The ground state of DMABN is known to possess an energy maximum for the perpendicular conformation (the rotational barrier), therefore emission from the perpendicular excited-state minimum occurs to a repulsive potential and is expected to lead to structureless spectra.

Equations (5.1) and (5.2) can be used to predict possible new TICT systems.[9, 35] Whether or not the energetic minimum of the A^* (TICT) state is lower than that of the precursor B^* state (inequality Eq. (5.1) fulfilled) sensitively depends on the electron donor–acceptor properties of the subsystems which can be quantified by ionization (or oxydation) potential and electron affinity (or reduction potential) of donor D and acceptor A (Eq. (5.2)).

$$E_{B^*} - E_{\mathrm{TICT}} > 0 \tag{5.1}$$

$$E_{\mathrm{TICT}} = IP(D) - EA(A) + C + E_{\mathrm{solv}} \tag{5.2}$$

The B^* state responds much less to changes in donor and acceptor properties than the TICT state, and Eq. (5.1) can often be easily fulfilled by increasing donor and/or acceptor strength. In addition to these two factors which deliver the decisive part of the reaction driving force, polar solvent stabilization E_{solv} and the mutual Coulombic attraction C of the linked donor and acceptor radical anion/cation also help to preferentially stabilize the TICT state with respect to the precursor B^* state.

Some typical structural elements that occur in TICT compounds are collected in Figure 5.3.

5.2.2. Intramolecular Proton Transfer: The ESIPT Mechanism

Acid/base equilibria can be very different in ground and excited states. A well-known example is β-naphthol, which becomes highly acidic in the excited state and transfers a proton to the surrounding water solvent acting as base within the excited

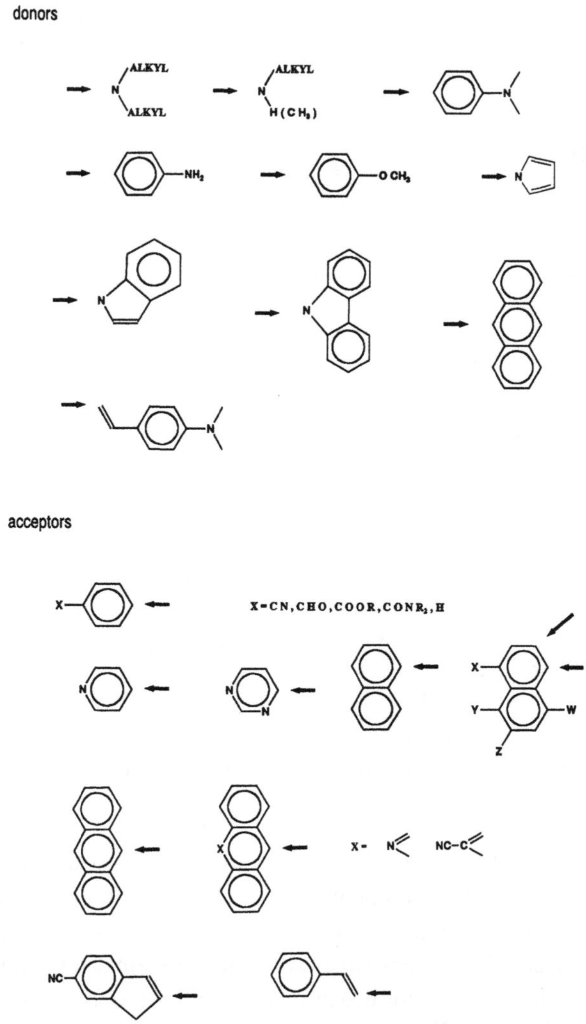

Figure 5.3. Common structural elements active as donors and acceptors in TICT compounds. The relevant linking positions are marked by arrows.

state lifetime. The corresponding product P^* is the excited naphtholate anion with a strongly redshifted fluorescence.[36]

If the molecule carries its own base, excited-state proton transfer can occur intramolecularly (ESIPT = excited-state intramolecular proton transfer[37]) and becomes more or less independent of the surrounding solvent. The ESIPT reaction is extremely fast (subpicosecond kinetics[38]) and occurs also in rigid glasses and at very low temperatures.[26, 39] Very often, only the ESIPT product P^* fluoresces, and this is the source of extremely large Stokes shifts which are fairly independent of medium

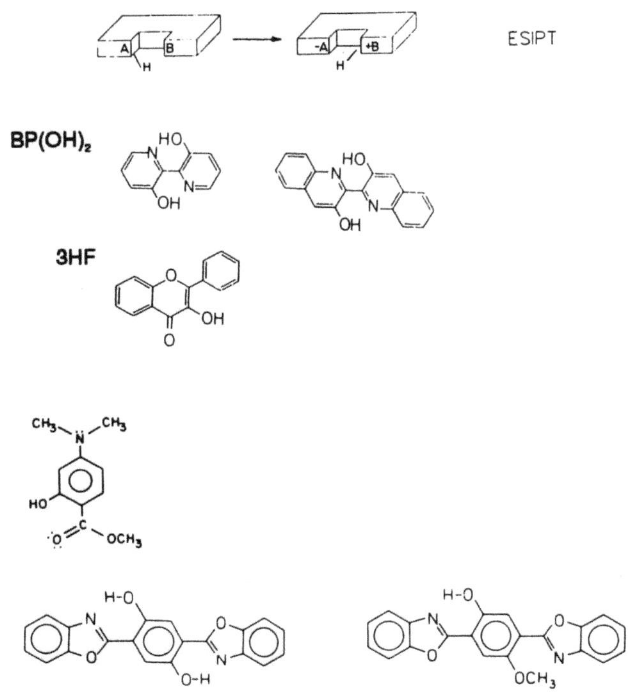

Figure 5.4. Some ESIPT structures. In the *top* part of the figure, the intramolecular proton transfer is sketched schematically, exemplifying the small spatial need necessary for this reaction.

and temperature. Thus, ESIPT dyes are ideal candidates for use as fluorescence labels in order to avoid interference with other fluorescent material present in the sample to be analyzed. A further advantage of the large Stokes shift is the virtually complete disappearance of spectral overlap between absorption and emission. This makes ESIPT dyes a promising class of dyes for use in fluorescent solar concentrators. These dyes could gain a large factor in efficiency if reabsorption of fluorescence could be avoided through the use of such "large-Stokes-shift" dyes instead of the conventional highly brilliant but "small-Stokes-shift" dyes.[24, 40]

The structures of some ESIPT dyes are shown in Figure 5.4. In a few cases like 3HF[41] and $BP(OH)_2$,[42] the fluorescence quantum yield can be quite high, and laser emission from the ESIPT product state can even be observed.[43, 44]

As a rule, however, most of the ESIPT dyes exhibit weak to very weak fluorescence quantum yields, i.e., the quantum yield of the nonradiative processes is near unity. Such dyes, if they show little permanent photodestruction, can be used as ultraviolet (UV)-stabilizers of polymers, such as Tinuvin (a hydroxy-benzotriazole),[45] because they efficiently convert UV radiation, harmful for the polymer, into harmless heat. The mechanisms of these nonradiative decay paths are often linked to energetically low-lying $n\pi^*$-states, but in a few cases a consecutive two-step photochemical mechanism according to Figure 5.1e has also been found, linking ESIPT (the luminescent product) and TICT mechanisms (the quenching channel, identifiable by its dependence on medium viscosity and the electron-donor–acceptor properties of the compound).[24, 46]

5.2.3. Intramolecular Folding: The Excimer/Exciplex Mechanism and Dewar Isomerization (Butterfly Mechanism)

Two other adiabatic photoreactions both involving intramolecular folding can be used for probing the local viscous properties of the surrounding: One is the well-known formation of excited dimers or complexes, the excimer or exciplex mechanism[47, 48] which often exhibits luminescent products. In conjunction with intramolecular bridging (so-called intramolecular excimers or exciplexes, Figure 5.5), it is well established for characterizing viscous oils and other "soft" polymers.[49, 50] It fails, however, for "hard" polymers like polymethylmethacrylate PMMA and other polymer glasses, because these do not yield enough freedom of movement for the folding reaction, which needs a rather large reaction volume.

In this case, a less well-known adiabatic photoreaction can be favorably used, that connecting aromatic compounds with their Dewar isomer.[51] In our case, the excited state reaction leading from anthracene toward the corresponding Dewar isomer (Figure 5.6) can be favorably used.[26, 52] This reaction possesses a driving force toward a nonradiative funnel F situated along the excited-state reaction path toward the Dewar isomer. This funnel is linked with a folding (butterfly motion) of the anthracene molecule, which requires considerably less reaction volume than excimer formation. Therefore, this reaction occurs even in glassy polymers (e.g., PMMA at room temperature and well below) and can be followed by the shortening of the fluorescence decay of the precursor state, the excited anthracene.[26] The decays observed are nonexponential, indicative of a distribution of lifetimes, and are consistent with the idea that the total free volume in a glass corresponds to a distribution of microscopic free volume elements of different size.

Butterfly-type folding has also been discussed in the context of rhodamine dyes adsorbed on surfaces.[53] In this case, too, nonexponential fluorescence decays are observed. A comparison of the consequences of the different reaction mechanisms with respect to free volume sensing is given in Section 5.4.

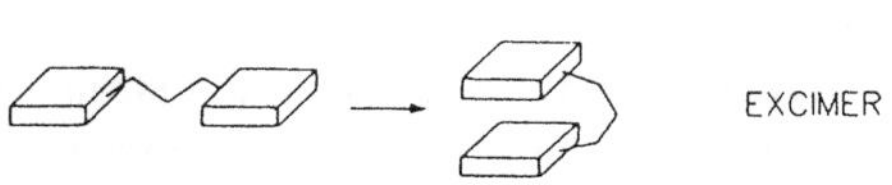

Figure 5.5. The formation of intramolecular excimers usually involves the folding of a flexible chain, and the displacement of the large chromophores into a sandwich-type arrangement necessitates a large reaction volume.

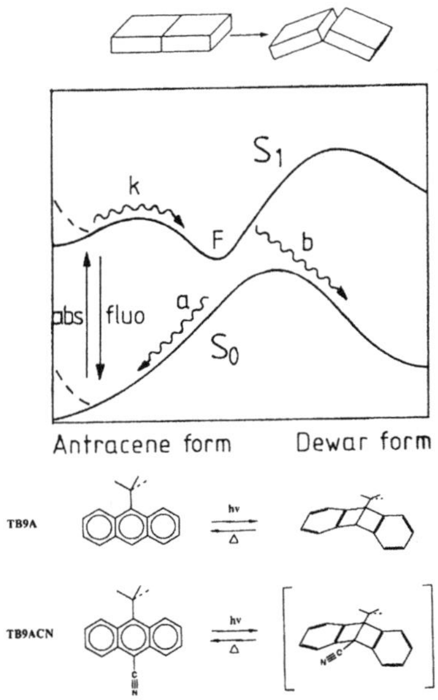

Figure 5.6. Schematic representation of S_0 and S_1 energy profiles for DEWAR formation in TB9A and TB9ACN.[26] The excited state funnel F is very close to the ground state surface and therefore leads to fluorescence quenching (identifiable with rate constant k). Most of the molecules return to the anthracene form via pathway a, while only a few proceed to the Dewar form (pathway b), because F is placed to the left of the ground state barrier. The steric effect of the tert-butyl substituent is indicated by the broken line. Without this "prefolding" of the anthracence form, Dewar formation is not observed. The *top* part of the figure contains a schematic description of the butterfly-type folding process, while the *bottom* part contains examples of actual molecules.

5.3. Examples of Polarity Probes

Two principal ways exist to use a dye as a sensor of local polarity (or of microscopic electric fields): (1) monitoring the polarity-induced shift of the energy levels, e.g., the red shift of the fluorescence; and (2) monitoring changes in fluorescence intensity induced by the polarity- or field-induced modulation of nonradiative rates. As these compete with the fluorescence emission, the fluorescence intensity (and lifetime) is correspondingly modulated. (3) In some cases, the radiative rates are also solvent sensitive; this is usually connected with the formation of luminescent products.

Often all three approaches ((1) to (3)) are simultaneously active, e.g., in the dual fluorescence of the classical TICT compounds like DMABN where both the energy of the long-wavelength emission (factor (1)) as well as the ratio between the two bands (factors (2) and (3) above) varies. The simultaneous presence of two bands allows very accurate ratio measuring, and small changes in this ratio can be detected more accurately as if only changes in absolute quantum yields have to be observed. These probes with dual fluorescence will be called "type-3."

Examples of the classical "type-1" compounds are collected in Figure 5.7a.[4, 54–57] In this case, mostly acceptor variations of naphthyl amines were used. With the eventual knowledge of the TICT mechanism it became apparent that all these compounds belong to the TICT class,[4, 58, 59] and mostly the very highly polar TICT state is emissive and correspondingly strong redshifts are observable. The acceptor end is often used to attach the probe covalently to the biological material. By varying the

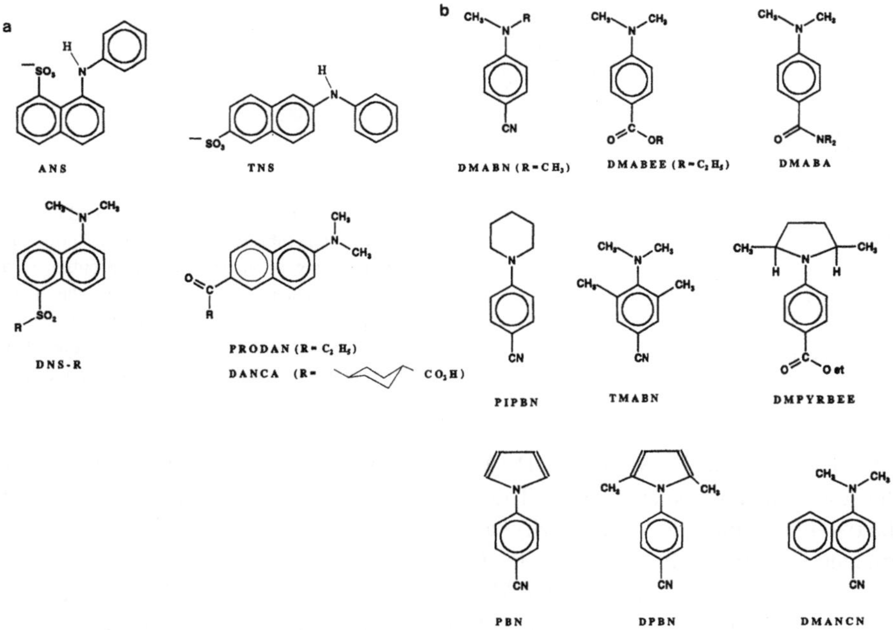

Figure 5.7. (a) Widely used biological probes of the TICT type.[4, 54–57] (b) TICT compounds with dual fluorescence and interesting properties amenable to tailor-making of fluorescence probes. Functionalization and inclusion in larger molecular assemblies is easily possible, e.g., at the positions indicated by *R*. This holds especially for esters and amides which have been used as probes in large polymers[97–101] or in TICT-labeled cyclodextrins.[149]

donor function, the properties can be changed in a predictable manner: If the ionization potential (IP) of the donor is lowered the TICT transition should be favored and more redshifted TICT fluorescence should be observable. These approaches have been used in the synthesis as well as in the theoretical prediction of new possible labels, some of which are collected in Figure 5.7b. Here, DMABN and derivatives, as well as the naphthalene derivative DMANCN,[60] are outstanding due to their dual fluorescence property. DMABN has already been used in probing the interior of cyclodextrin cavities[61–63] and in the determination of the micropolarity inside micelles.[62, 64] Various isomers and derivatives of DMANCN have recently been studied theoretically,[59] and donor substitution in 1-position was found to be more favorable for TICT behavior, in accordance with expectations[9] and with experimental evidence.[59] If the donor becomes sufficiently strong, like in DPBN as compared to PBN, TICT fluorescence is even observable in nonpolar solvents[65] and in the gas phase.[9] Another approach to enhance the kinetic channel leading to the TICT state is pretwisting (i.e., nonplanar ground state conformations) as in TMABN or PIPBN[2, 66, 67] or manipulation of the state correlations i.e., topology of the reactive hypersurface[9, 13, 68] as in the

a)

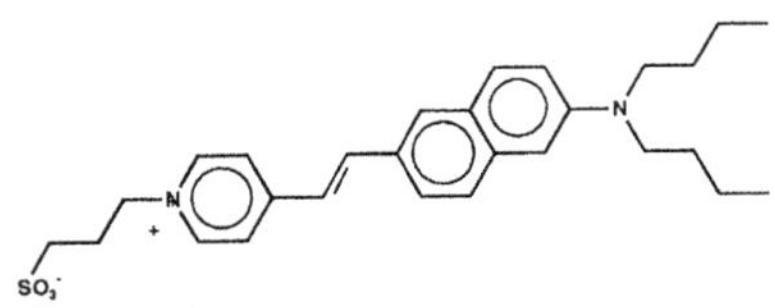

b)

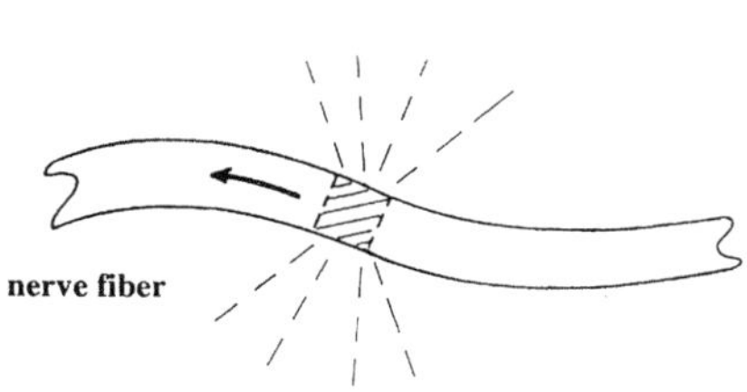

Figure 5.8. (a) Example for a compound used in the study of membrane potentials and (b) principle of functioning of visualizing nerve pulses in a living cell. When a nerve pulse passes, the membrane potential changes, and this induces a change in the fluorescence intensity of the probe. The temporal and spatial profile of these changes can be followed by time-resolved methods.

ester derivatives of DEABN, e.g., DEABEE: both these compounds also show TICT fluorescence in nonpolar solvents. Thus, both the dual fluorescence property (ratio of the two bands and solvent sensitivity) as well as the location of the TICT band can be affected in a predictable manner in the tailor-making of dyes.

The compounds DMABN, DEABN, PIPBN, and DMANCN shown in Figure 5.7b, which exhibit dual fluorescence, belong to the type-3 compounds, because the kinetics of their TICT formation reaction strongly depends on solution polarity.[8, 12, 69] In other dyes, TICT fluorescence is not visible but nevertheless affects the precursor fluorescence through its quenching action which depends on the local electric field (type-2). This is especially helpful when fast temporal changes of such fields have to be monitored. Figure 5.8 contains some dyes, mainly merocyanines and stilbazolium derivatives, which have been used in the development of probes for membrane potentials, e.g., of nerve cells.[5, 70–74] Some studies indicate that the effect of these dyes occurs through a combined mechanism of type-1 (electrochromism) and type-2 (quantum yield changes).[5, 75] The type-2 effects can be viewed in conjunction with the photophysical studies on stilbene, azastilbene, and biphenyl derivatives[17, 70, 76, 77] which indicate nonradiative channels activated through twisting of single and/or double bonds. The stilbene case is described in detail below in Section 5.6.3. The twisting of bonds with character intermediate between single and double, as in the merocyanines and azastilbenes, is describable within the theoretical model of biradicaloid states which contains both single- and double-bond twist and where TICT formation (single-bond twisting) is viewed as a special case in a more general scheme.[15, 32, 33, 78–81]

5.4. Examples of Free Volume Probes

Microviscosity has been a long-standing subject in the characterization of biological materials such as membranes[54] but has also been discussed in connection with bulk polymers and low-viscosity solvents. Free volume effects derive from the discontinuous nature of matter which is composed of molecules of different size and shape. In solvents and also in membranes, the "free space" between these molecules is rapidly fluctuating and can be described by a size-distribution-function which is more or less frozen for rigid polymer materials. The free space provides an additional possibility for movement, e.g., diffusion of molecules, and therefore the usual Stokes–Einstein relation, (Eq. (5.3)), which states that diffusion coefficient D and macroscopic viscosity η are inversely proportional and hence their product should be a constant, is no longer observed under high-viscosity conditions but the product exceeds the value for large viscosities indicating an additional diffusion channel and hence a microviscosity smaller than the macroscopically measurable one.[82] Various microviscosity or free volume theories have been put forward, the most well known being that of Gierer and Wirtz[83] which results in Eq. (5.4),[82] or the so-called Williams–Landel–Ferry (WLF) equation[84] which is capable of quantifying the relative free volume.

$$\mathrm{D} \cdot \eta = \mathrm{A} \tag{5.3}$$

$$\mathrm{D} \cdot \eta = \mathrm{A} + \mathrm{B} \cdot \eta^{\mathrm{x}} \tag{5.4}$$

A free volume fluorescence probe has to allow the measurement of some diffusion process occurring in the excited state. Depending on the necessary reaction volume, the size of the diffusing moieties and the amplitude of the corresponding large amplitude motion, and on the distribution of the free volume of the medium to be investigated, different fractions of the free volume are probed, and therefore different free volume probes can give different answers regarding one and the same material.

Various photochemical reaction mechanisms lend themselves for variation of the above parameters, and the more important ones are summarized above. In general, one can expect for the volume necessary for the reaction to occur that it decreases in the sense Excimer > TICT > Butterfly > ESIPT mechanism. Therefore, for a given size distribution of microscopic free volume voids, the fraction of the total free volume usable for the reaction and thus amenable to probing increases in the same sense with the ESIPT mechanism being the outstanding extreme, because this reaction cannot even be stopped in a rigid matrix at very low temperature.[39]

In a totally independent approach, highly fluorescent but rigid molecules can be used and their reorientational movement probed in solution, mostly by time-resolved fluorescence anisotropy.[85] For rodlike molecules like *p*-oligo-phenyls, this movement corresponds to the reorientation of the long molecular axis the length of which can be easily varied.[86] The use of fluorescence probes in liquid and solid media and on surfaces has been treated recently in several reviews and books.[87]

In the following, we will concentrate on the probes involving a photochemical reaction and give some examples of how some of the mechanisms have been used in developing free volume fluorescence probes.

5.4.1. Excimer Probes

After the initial use of pyrene and other excimer-forming molecules in biological applications,[54] it was soon recognized that excimer formation can be strongly enhanced and the probe concentrations therefore strongly reduced if the reaction partners are chemically linked, e.g., by a methylene chain. Thus, intramolecular excimers were developed, some of which are collected in Figure 5.9.

In a series of papers, Diphant has been used to probe the microviscous properties of various polymer oils, and free volume parameters have been extracted.[49, 88–90] In a comparative study of Excimer and TICT probes, it could be shown that the response of these probes is frozen out at lower temperatures, as can be expected from the large reaction volume necessary, whereas the TICT reaction still shows sizable rates at these high-viscosity conditions.[26] Moreover, this study also showed that the free volume fraction measured by the TICT probes is larger than that measured by Excimer probes.

5.4.2. TICT Probes

The twisting motion in the TICT mechanism is also a large-amplitude motion which is easily stopped by very high-viscosity conditions, and therefore TICT molecules with various rotating moieties can be used in the development of free volume

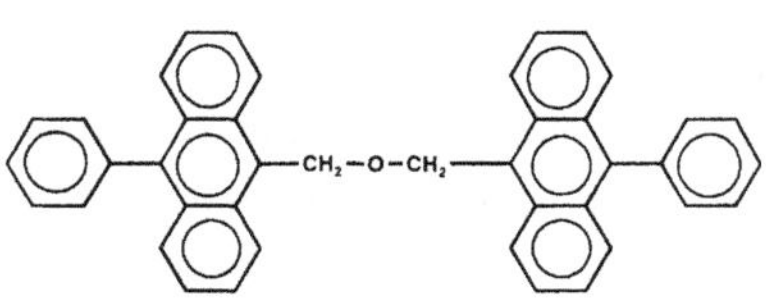

DIPHANT

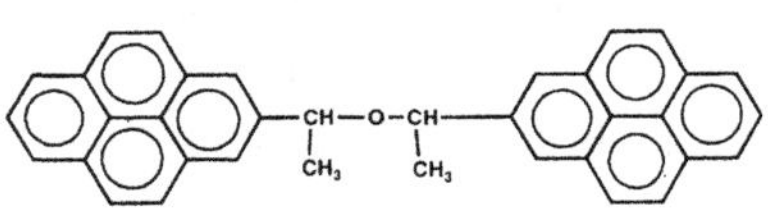

meso-bis[1-(2-pyrenyl)ethyl]ether

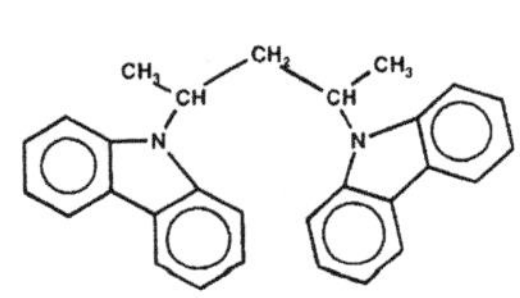

meso-2,4-di(N-carbazolyl)pentane

Figure 5.9. Chemical structure of some intramolecular excimers used as free volume fluorescence probes.

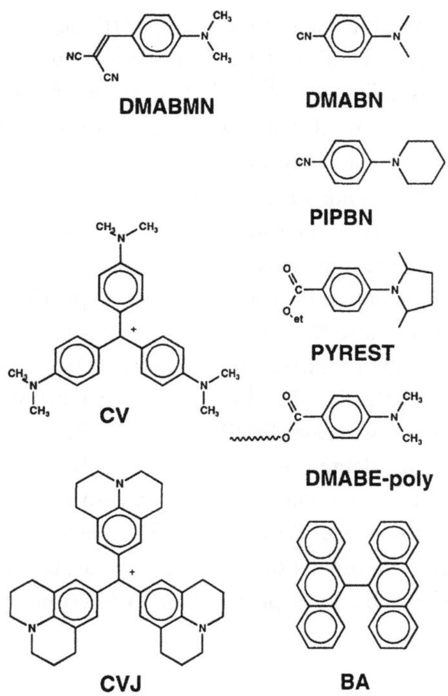

Figure 5.10. TICT probes used in probing of polymers.[25, 26, 91–103]

probes. Figure 5.10 shows some of the compounds used in this connection.[25, 26, 82, 91–94] An overview of this field has been given recently.[80]

The malonitrile DMABMN has become popular due to its applications for monitoring the degree of polymerization during a polymerization reaction.[95] It belongs to the type 2 probes with viscosity-dependent quantum yield of a single fluorescence band, and when a certain degree of polymerization, i.e., microviscosity is attained, fluorescence quantum yield increases steeply. The strong quenching action in low-viscosity conditions is thought to derive mainly from the effect of double-bond twisting at the acceptor end. This can be shown by model compounds where the dimethylamino group is bridged. Recently, a pressure study using this compound has been conducted.[96] This approach allows the viscosity to be varied in a controlled way under isothermal conditions.

The dual fluorescing type 2 compounds DMABN and its ester derivatives have been used to extract detailed free volume parameters, especially for the case where the probes are covalently linked to the polymeric backbone, e.g., by an oxymethylene chain of variable length.[97–101] As the TICT reaction requires a comparatively large reaction volume, the measurable effects are especially strong in solutions of these labeled polymers but they tend to disappear for the pure polymer due to its too-large microviscosity. In this case, the small effects which still persist can be amplified by a technique called red-edge-excitation (REE). This has been shown for DMABN and derivatives in solid PMMA glasses[25, 102, 103]: On excitation into the maximum of the absorption band, only the normal, short-wavelength fluorescence band can be observed. Excitation into the red edge of the absorption, however, induces dual fluorescence, the level of which depends on excitation wavelength. REE can be interpreted in terms of the inhomogeneity of the sample's surrounding. In a rigid polymer where the surrounding cannot adjust in a relaxation process, REE corresponds to site selection, i.e., selective excitation of molecules with a specific surrounding which in this case favors the transition to the TICT state. In this way, various rigid polymers have been compared, and their relative free volumes determined.[25, 103]

DMABN suffers from the fact that dual fluorescence is only observable for polar media. Therefore, the pretwisted ester DMPYRBEE has been developed which shows dual fluorescence also in alkane solvents.[9] This probe allowed measurement of nonpolar polymeric siloxane oils and a comparison with the corresponding measurements using an EXCIMER probe. As expected from the decreased reaction volume necessary for the TICT photoreaction, the latter is usable down to much lower temperatures (higher viscosities) and probes a larger fraction of free volume.[26]

5.4.3. Butterfly Probes

A photoreaction which involves still less reaction volume is that linked with Dewar isomerization (butterfly-type intramolecular folding) of certain anthracenes in the excited state.[26, 52] As outlined in Figure 5.6, anthracene fluorescence is quenched by the adiabatic pathway to the funnel F with subsequent nonradiative return to the ground state hypersurface and formation of either the Dewar product or of starting material. This reaction occurs even in rigid polymers at temperatures far below the glass transition temperature and can be observed either by measuring the fluorescence quantum yield or—better—the fluorescence decay which exhibits a pronounced shortening when the reaction is possible. Unlike the TICT case, these effects are also observed when exciting into the bulk absorption band, i.e., REE is not needed here. The shortening of the fluorescence decays is connected with nonexponentiality, and the latter can be translated into lifetime distributions which in principle give an indication for the distribution of the free volume available for the reaction. An example of such an evaluation in terms of lifetime distributions is given in Figure 5.11.

It can be seen that the most probable lifetime $<\tau>$ becomes shorter and the width of the distribution broader as the temperature is increased. For similar temperatures in liquid solution, the lifetime is unmeasurably short ($\tau < 0.1$ nsec). Therefore, $<\tau>$ and the width of the distribution reflect the effect of the surrounding in slowing down k and not a classical activation barrier effect.

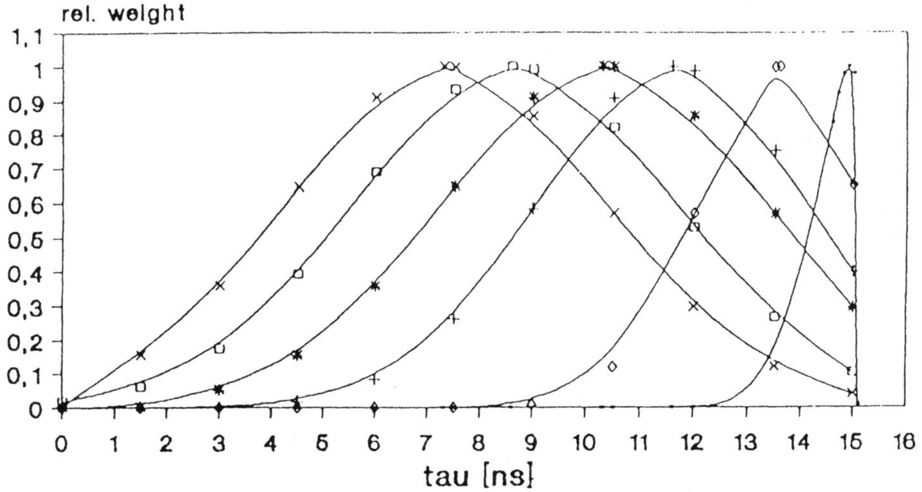

Figure 5.11. Decay time distribution for different temperatures recovered by fitting Gaussian lifetime distributions to the observed nonexponential decays of TB9ACN in PMMA.[26] x, 261 K; □, 243 K; *, 224 K; +, 203 K; ◇, 175 K; •, 146 K.

5.5. How to Construct Proton- and Ion-Sensitive Analytical Probes: Principles and General Scheme of Use

5.5.1. Generating Sensitivity through Introduction of TICT-Pathways

The conventional pH- or ion indicators apply chromophoric systems which change on protonation/ion complexation and therefore exhibit either changes in the absorption and/or the emission spectra. The new generation of fluorescence indicators use the same principle but the changes in the chromophore systems occur as an adiabatic photoreaction in the excited state and are thus decoupled from the absorption behavior: the construction principle in this case involves a dye with a photoreaction attached to a "recognizing element" (the basic site or the ion complexing groups). An important condition is that the photoreaction be sensitive to the status of the recognizing element (complexed or uncomplexed). Various adiabatic photoreactions can be used as outlined above. The most important ones are linked with TICT formation and will be reviewed below. As recognizing elements, amino or hydroxylate groups have been widely used for pH sensing, whereas for ion recognition either macrocycles (e.g., crown ethers) of different size[104] or complexing groups derived from EDTA (ethylene–diamine–tetraacetate) have been applied (e.g., BAPTA, see Figure 5.22 below). An example of how a long-wavelength absorbing dye system like Rhodamine B (Rh B) can be developed into a TICT-molecule[27] and further to a pH- and ion-sensing probe is shown in Figure 5.12.

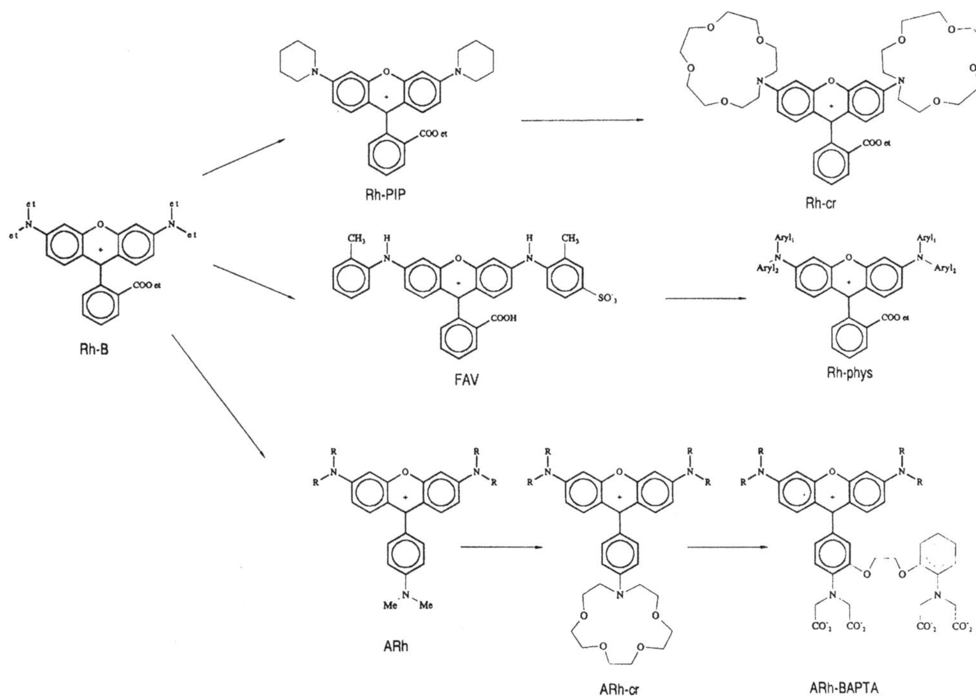

Figure 5.12. Various possibilities of introducing a fast TICT channel into a Rhodamine dye like Rhodamine B (see text).

Rh B itself shows already some TICT activity through twisting of the diethylamino groups which can be strongly enhanced by way of the pretwisting effect, as exemplified, e.g., in the piperidino-Rhodamine Rh-PIP.(105) From this source, the crown-derivative Rh-cr should also show a sizable TICT-formation rate. The other approach is to lower the ionization potential of these amino groups, e.g., by attaching aryl groups. A well-known example is fast acid violet (FAV)(106) which shows very fast and viscosity-dependent fluorescence decay times. Using other aryl-substituents should allow to vary the basicity of the TICT-active nitrogen and therefore should provide a possibility to develop pH-probes for the physiological pH-range (Rh-phys). A second, much more important TICT channel is introduced when the carboxylate group is replaced by an amino function (ARh). This leads very efficiently to a nonemissive TICT state which can be blocked by protonation.(107) The ratio of fluorescence quantum yields in neutral and acidic ethanol thus approaches 300.(108) The amino function could further be developed into an ion-sensing crown (ARh-cr) or 1,2-bis-(2-aminophenoxy)ethane-N,N,N′,N′-tetra-acetic acid (BAPTA)-function (ARh-BAPTA).

5.5.2. General Use as Indicators

Conventional absorptiometric and fluorimetric pH indicators show a shift of band positions in absorption and emission spectra between the protonated and deprotonated forms. This feature allows the spectroscopic measurement of the acid dissociation constant in the ground state, K_a, and also the evaluation of the dissociation constant in the excited state, K_a^* (Eq. (5.5)), from the Förster cycle under the assumption of equivalent entropies of reaction in the two states.[109–112]

$$\mathrm{p}K_a^* = \mathrm{p}K_a - 0.625\frac{\nu_a - \nu_b}{T} \tag{5.5}$$

where ν_a and ν_b are the energies (in cm^{-1}) of the 0–0 transition in the protonated and anionic forms, respectively.

In practice a simple spectrophotometric titration of the probe with an acid (base) solution leads to the pK_a using Eq. (5.6).

$$\mathrm{p}K_a = pH - \log[\varepsilon/(\varepsilon_M - \varepsilon)] \tag{5.6}$$

where ε is the observed extinction coefficient and ε_{M} the extinction coefficient of the unprotonated (protonated) species at a wavelength at which the absorption of the protonated (unprotonated) species is negligible.

For the determination of the dissociation constant in the excited state, several methods have been used: the Förster cycle,[109–111] the fluorescence titration curve,[113] the triplet–triplet absorbance titration curve,[114] but all involve the assumption that the acid-base equilibrium may be established during the lifetime of the excited state, which is by no means a common occurrence. A dynamic analysis using nanosecond or picosecond time-resolved spectroscopy is therefore often needed to obtain the correct pK_a^* values.[115]

When, after a detailed photophysical study, all the characteristics of the ground and excited states prototropic reactions of a molecule are known, in return we have in hand a probe which can quantitatively monitor the pH of its microscopic surrounding. In fact, as a rule, the absorption and the fluorescence spectra of a probe are only measurably dependent on pH over ca. 2 pH units around its pK_a. Since probes with a wide range of pK_a values—currently from 1 to 9—are available, pH values from 0 to 10 are addressable. For multiple reasons fluorescence measurements are preferred to direct absorption spectra.

1. Absorption and extinction coefficients are generally less pH dependent than fluorescence spectra and quantum yields because the radiative rates often compete with intra- and intermolecular relaxation precesses.
2. The sensitivity and specificity of the fluorescence measurement allow to probe one molecule[116, 117] and nanometer space in some picoseconds leading to a real-time and real-space monitoring of molecular species.

3. Fluorescence being an emitted light, its measurement can be done outside the medium, which when it presents a high optical density, preclude the use of the absorption spectra.

Technically, measurements of pH using molecular fluorescent probes are performed by any of the following model methods:

1. When the fluorescence spectra of the probe shifts on protonation, two emission wavelengths with opposite proton-sensitive response are chosen to give a pH-dependent emission intensity ratio. In this ratio method a number of ion-independent factors that affect the signal intensity like photobleaching, variations in probe concentration, and illumination instability are eliminated.
2. If there is no emission shift, but only variation in the fluorescence quantum yield on protonation, one follows the fluorescence intensity at one emission wavelength when the acid and the base forms of the probe are excited. All the advantages of the ratio method described with two emission wavelengths also apply to this fluorescence excitation method.
3. One excitation and one emission wavelengths can be used for following rapid kinetics but unlike the ratiometric methods, this technique does not yield absolute values.

It is important to emphasize that these considerations also apply to indicators with other ionic specificities and provided that their spectra are well resolved from those of the pH indicators, simultaneous measurements of ion concentrations and pH are now commonly practiced.(118)

5.6. pH Indicators

Rather than to draw up an exhaustive list of pH sensing fluorescent probes which can be found in several reviews(112, 119) we will try to define different classes of compounds whose fluorescence is sensitive to protonation and give some illustrative examples.

The first pH indicators studied possessed the acid-base site (phenol, aniline, or carboxylic acid) as an integral part of the fluorophore. Structurally, in the most general sense, pH sensitivity is due to a reconfiguration of the fluorophorets π-electron system that occurs on protonation. Consequently, the acid and the base forms often show absorption shifts and also, when the two forms fluoresce, emission shifts or at least, when only one form emits, a pH-dependent fluorescence intensity. This class of compounds has been reviewed(112) and the best structures have to be designed according to the medium probed and the technique used. After a short consideration of physiological pH indicators we will describe the main photophysical processes sensible to protonation.

5.6.1. Physiological pH Indicators

Recently a variety of fluorescent pH indicators have been described that determine intracellular pH values. The parent chromophores are mainly 7-hydroxycoumarin, pyranine, or fluoresceine. From the latter a series of fluorescent, long-wavelength, benzo[c]-xanthene dyes has been characterized for pH measurement in both excitation and emission ratio applications. The two general classes of these indicators are seminaphtofluoresceins (SNAFLs) and seminaphtorhodafluors (SNARFs) which are substituted at the 10 position with oxygen or nitrogen, respectively (Figure 5.13).[120] These probes may be excited at 488 or 514 nm with an argon ion laser and they have pK_a values between 7.6 and 7.9 which make them able to monitor physiological pH. As they are spectrally well separated from calcium indicators such as Fura and Indo-1[121] they are suitable for simultaneous determination of pH and Ca^{2+} transients.[118] Concerning the electronic structure and prototropism in the ground and excited states, it appears that the oxygen at the 3-position (Figure 5.13) is the site of protonation in the pH range 5–10. Rigidizing the nitrogen in 10-position of the SNARF derivatives with propylene chains, as in C.SNARF-X (Figure 5.13) produces the anticipated increase in quantum yield and shifts the absorption and emission maxima to longer wavelengths, relative to the unbridged derivatives. As in the case of 7-aminocoumarin derivatives,[122] the fluorescence enhancement is likely to be due to the inability of the rigidized nitrogen to form a nonemissive TICT excited state through twisting of the dialkylamino group. Since the fluorescence enhancement is approximately twofold larger for the protonated form (≈ 3×) than for the deprotonated form (≈ 1.5×) it is likely that a TICT state plays a larger role in the fluorescence quenching of the acidic form than in the basic form, indicating that the electron density on the nitrogen in 10-position is more delocalized into the π-electron system of the fluorophore in the acidic form than in the basic form (Figure 5.14).[120]

5.6.2. Fluorescent Probes with an Efficient Intramolecular Fluorescence Quenching Process in the Base Form Possibly Related to the Formation of a Nonemissive TICT State

When highly fluorescent ionic fluorophores, such as acridinium or xanthene dyes, are linked to a flexible electron-donor-substituted aryl group, a low-lying intramolecu-

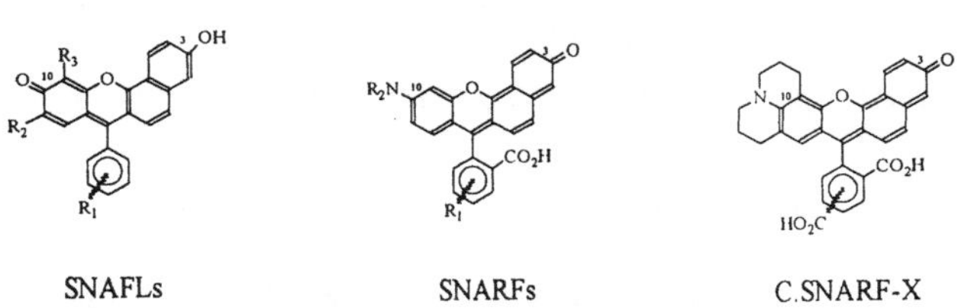

Figure 5.13. Structures of SNAFL and SNARF dyes.

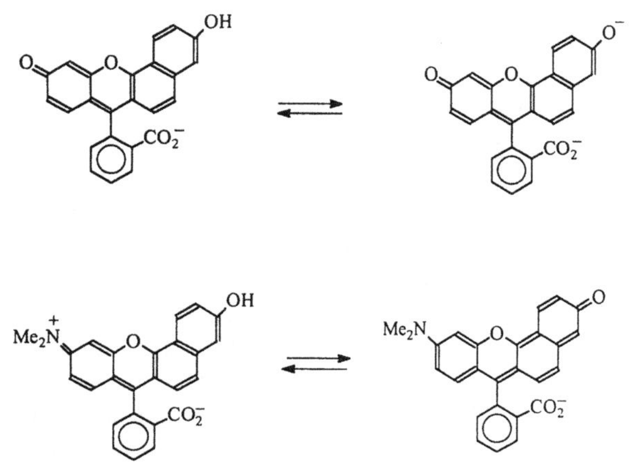

Figure 5.14. Predominant fluorescent species involved in the prototropic equilibrium of SNAFL and SNARF indicators.

lar charge transfer state is produced, resulting in a state inversion and a quenching of the fluorescence. Intramolecular rotational motion leading to a nonemissive TICT-like state with charge localization was invoked to explain this observation. When the ionization potential of the donor group is increased sufficiently, e.g., by protonation, such that the energy of the CT state is raised above that of the first locally excited state, the fluorescence of the fluorophore is restored.[9]

5.6.2.1. 9-Arylacridinium Ions

Absorptive and emissive properties of 9-arylacridinium ions (Figure 5.15) have been studied and shown to be dependent on the electron-donating properties of the aryl group. While for PAc$^+$ and *o*-MPAc$^+$ the first excited state remains largely localized on the acridinium chromophore, interaction of this chromophore with the electron-donor-substituted aryl group in APAc$^+$, DMAPAc$^+$, and *o*-MDMAPAc$^+$ produces a low-lying intramolecular charge transfer state, resulting in a state inversion. This state inversion leads to the appearance of a new long-wavelength absorption in the visible region and to complete quenching of the acridinium fluorescence. Protonation of the amino function cancels this state inversion thereby causing a dramatic chromofluoro-ionophoric effect, i.e., proton-induced change of both color and fluorescence properties (*o*-MDMAPAc$^+$: $pK_a = 3.83$, $\phi_F = 0$; *o*-MDMA-PAc$^+$H$^+$: $\phi_F = 0.7$).[123]

5.6.2.2. Rhodamine Derivatives

Rhodamine derivatives are widely used as laser dyes, and intramolecular rotational motion of a dialkylamino group leading to a nonemissive TICT state has been

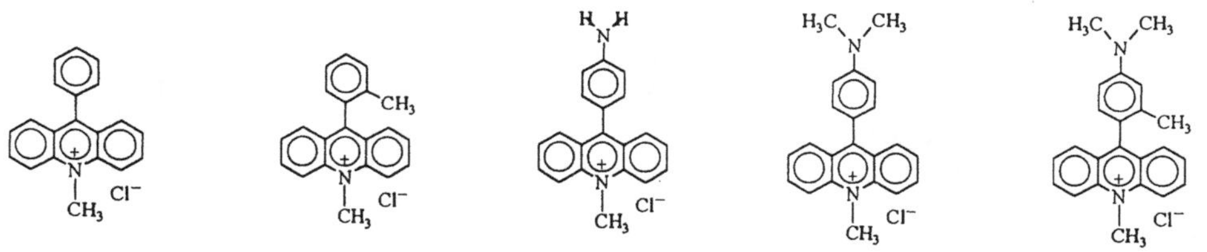

Figure 5.15. Structures of 9-arylacridinium ions.

proposed to explain the quenching effect of polar solvents and high temperature observed with Rh-B (Figure 5.16).[105] Julolidine-type bridging as in Rh-101 suppresses this quenching while decreasing the ionization potential of the dialkylamino groups by aryl substitution as in FAV increases further the non radiative TICT channel. Finally, the carboxy-phenyl group in rhodamines, which acts as electron acceptor, can be replaced by a dialkylamino donor group to yield ARh which shows a very efficient quenching channel due to TICT formation (ϕ_F(EtOH) = 0.005). On protonating ARh, the dialkylamino group on the flexible phenyl ring acts as a weak electron acceptor, similar to the carboxyl group in ordinary Rhodamines. The corresponding TICT energy is raised, and the fast quenching channel is closed, leading to a strong increase (more than two orders of magnitude) of the fluorescence quantum yield and a slight shift from 570 to 584 nm of the fluorescence spectrum.[27, 108] A similar TICT channel is responsible for the strong intramolecular fluorescence quenching in triphenylmethane dyes like malachite green (MG) (Figure 5.16), and studies including bridged derivatives established the importance of the twisting motion.[9, 16]

5.6.3. Donor–Acceptor and Donor–Donor Substituted Stilbenes

Contrary to the Rhodamine and triphenylmethane dyes there is an important class of laser dyes, the donor–acceptor (D–A) stilbene derivatives such as the well-known laser dye DCM, where TICT formation is the source of favorable fluorescence properties.[124] This can occur because these dyes possess a nonradiative decay channel, that of double-bond twisting, the importance of which is diminished by the competing formation of a fluorescent TICT state.[17, 125] When a dimethylamino group acts as the donor (D) the absorption spectra of the D–A derivatives show a long wavelength band with CT character which disappears on protonation resulting in blue-shifted spectra resembling that of stilbene.[126] The fluorescence spectra of the protonated derivatives are the mirror image of the absorption spectra, giving evidence for the absence of any prototropic behavior in the excited state. The fluorescence quantum yield of the protonated D–A stilbenes is reduced by a factor of more than 2

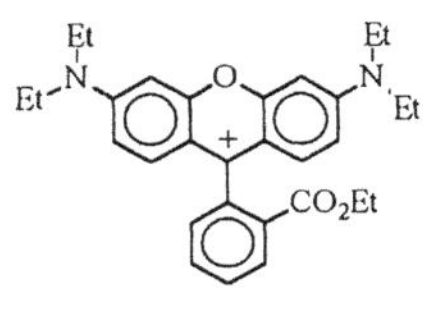

Rh-B

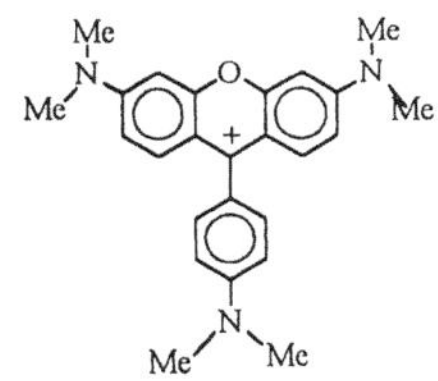

ARh

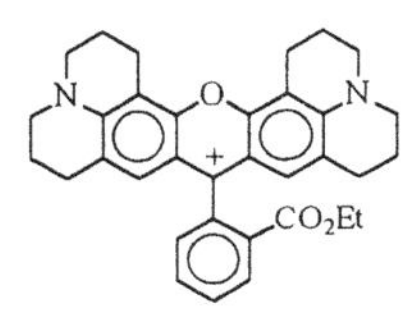

Rh-101

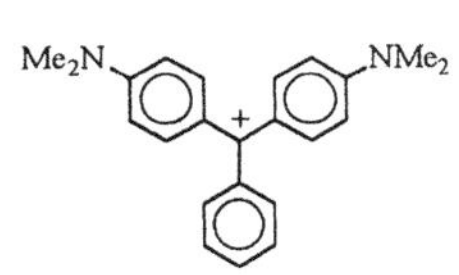

MG

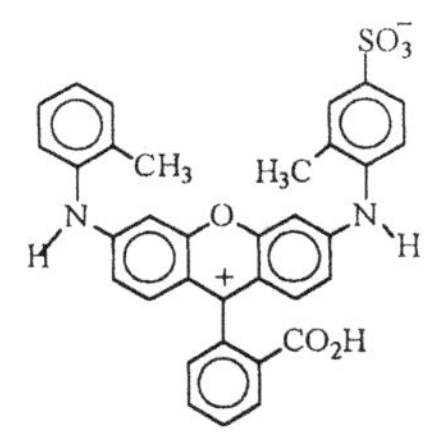

FAV

Figure 5.16. Various structures of triphenylmethane and Rhodamine dyes.

for 4-dimethylamino-4 cyanostilbene (DCS) in MeOH-H_2O (1v/3v) (Figure 5.17) as a result of the unavailability of the TICT state.

The fluorescence quantum yield of donor–donor (D–D) stilbenes, e.g., 4,4-diaminostilbene (DAS) ($pK_a = 3.95$), is also increased relative to that of stilbene. This may be due to delocalization of the nonbonding electrons from the amino group[127] with formation of a possible TICT state.[128] Upon protonation at pH = 2 in water, a blue shift of 44 nm is observed for the first absorption band which then resembles the absorption spectrum of stilbene, but two fluorescence maxima occur at 349 and 461 nm. Both bands possess the same excitation maximum at 292 nm. The 349 nm fluorescence band is likely to derive from the fully protonated DAS while the 461 nm is the contribution from a partially deprotonated species in which only one amino group has lost a hydrogen ion. This deprotonation in the excited state has also been found with other D–D stilbenes, e.g., the crown-derivative DDS with two amino functions (Figure 5.17), but in acidic acetonitrile only the deprotonated species fluroesces with a quantum yield slightly reduced as compared to the free ligand.

In conclusion, protonation of the donor group of D–A stilbenes leads to a short wavelength shift of both absorption and fluorescence spectra and a decrease of the fluorescence quantum yield while doubly protonated D–D stilbenes exhibit a monodeprotonation in the excited state and emit an additional long-wavelength bond.

DCS-Crown DDS-Crown

a)

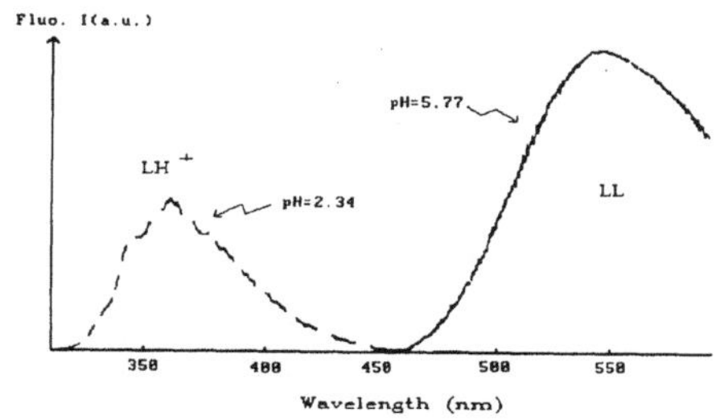

b)

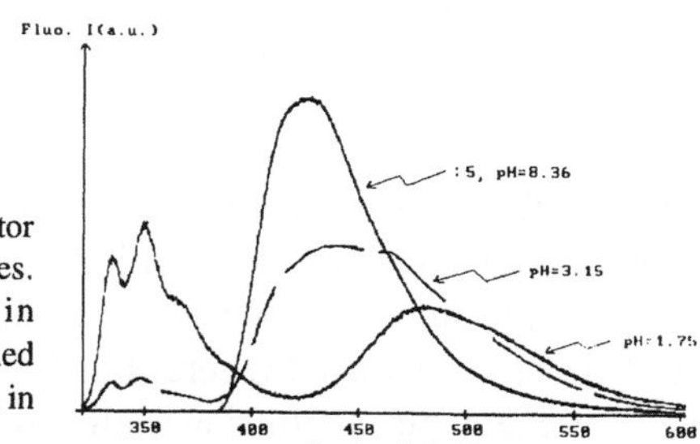

Figure 5.17. Formulas of some donor–acceptor and donor–donor stilbene-Crown derivatives. Emission spectra of (a) DCS-Crown in MeOH/H_2O (lv/4v) (———) and in acidified MeOH/H_2O (lv/4v) (- - -) and (b) DDS-Crown in MeOH/H_2O (lv/4v) at various pH values.

5.6.4. "Fluor–Spacer–Receptor" Systems with a Photoinduced Electron Transfer as a Quenching Process of the Fluorescence

De Silva recently reviewed the "fluor–spacer–receptor" systems which have been designed such that the fluorescence is switched off by an intramolecular photoinduced electron transfer (PET) process which in turn can be suppressed by the entry of a proton (or other ion) into the receptor restoring the emission of the fluor unit (Figure 5.18).[(119)] All the common classes of organic proton receptors (amine, carboxylate, pyridine) and chromophores with the major types of excited states [π–π^*, internal charge transfer (ICT), ligand-to-ligand charge transfer (LLCT) and metal-to-ligand charge transfer (MLCT) states] have been associated through polymethylene spacers, with the CH_2 spacer being the favorite since it allows fast PET rates and is most convenient for synthesis via benzylic functionalities. In the ground state, most of these fluorescent ion signaling systems show only very weak interactions between the fluorescent

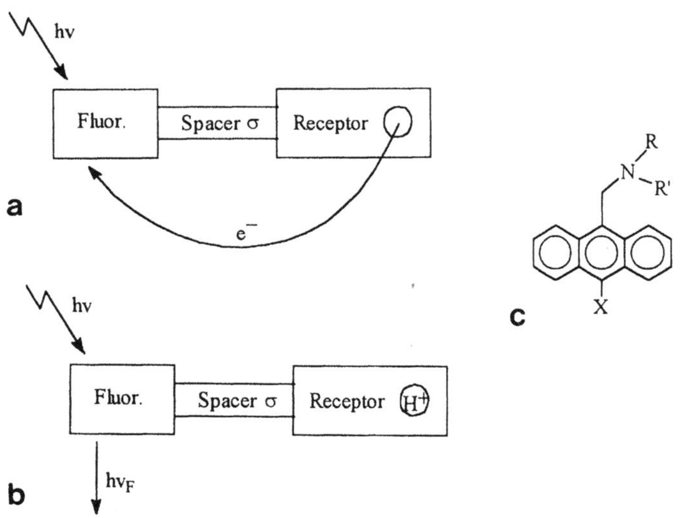

Figure 5.18. Schematic representation of photoinduced processes in a "fluor–spacer–receptor" signalling system (a) when ion free, (b) when ion bound, and (c) one example.

molecular part (fluor) and the ion receptor separated by the all-σ-bonded spacer but a strong long-range interaction develops in the form of an electron transfer from the proton-free receptor to the fluor when the unprotonated molecule is photoexcited. When the thermodynamic driving force for PET (ΔG_{ET}), as calculated from the Weller equation[(129)] is exergonic, a diffusion controlled rate constant can be expected, two or three orders of magnitude larger than the radiative rate constant of the fluor. If protonation of the basic receptor raises its oxidation potential until a sufficiently positive ΔG_{ET}, corresponding to positive energy differences between CT and B^* state [Eq. (5.1)]; PET should be strongly inhibited resulting in a proton-induced enhancement of fluorescence.

These systems possess the simplest possible variation in that only one parameter, i.e., the fluorescence quantum yield is proton-controlled. The insulation of the fluor and receptor modules in the ground state by the alkyl spacer leads to essentially pH-invariant absorption (position, shape, and intensity) and fluorescence spectra (shape and position only). The same pK_a values are obtained from fluorimetry or absorption spectroscopy so that the detection sensitivity of excited state experiments can be used for the measurement of binding constants of the ground state.

A slightly more complex system results from the pH sensors based on $[Ru^{II}(bpy)_3]^{2+}$ with an *N*-methylaniline basic site, electronically separated from the bipyridine ligand by a methylene spacer[(130)] (Figure 5.19). Fluorescence of $Ru^{II}(NMePh)_2$ (Figure 5.19) ($\phi_F = 0.023$) is quenched to a certain extent compared to the model compound RuIIMe2 ($\phi_F = 0.033$) owing to PET. In the pH range 3.8–1.8

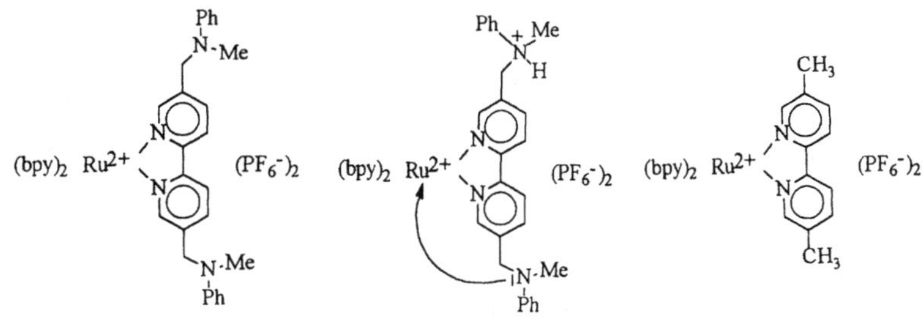

Figure 5.19. Molecular structure of pH sensors based on Di(2,2′-bipyridyl)(5,5′-diaminomethyl-2,2′-bipyridyl)-ruthenium II complexes.

the emission intensity drops while the emission spectral maximum and shape are retained. This is attributed to PET from the free amino group of monoprotonated Ru^{II}(NMePh)(NH^{+}MePh) to the $[Ru^{II}(bpy)_3]^{2+}$ moiety, the reduction potential of which is lowered by the vicinity of the additional positive charge. From pH 1.8 to 0.6, both side-chain amine groups are protonated and the long-wavelength emission shift (from 605 to 653 nm) and the fluorescence quenching ($\phi_F = 0.009$) are ascribed to the positive charges in the vicinity of the $[Ru^{II}(bpy)_3]^{2+}$ luminophore probably leading to a Ru–N bond fission in the excited state via a *d–d*-state, to a five-coordinated intermediate which may hydrate to restore six-coordination. The monoaquo complex would then thermally return to the original trisbipyridyl coordination, since these sensors are photostable. The red shift of the emission is attributed to the interaction of the protonated side chain amine across the methylene spacer with the localized charge acquired by the bipyridyl ligand in the MLCT excited state of the $[Ru^{II}(bpy)_3]^{2+}$ moiety.

With these particular MLCT luminophores, two mechanisms are involved: PET and another one originating from the presence of proximal charges.

5.7. Ion Complexing Probes

5.7.1. Monoaza-15-Crown-5 Stilbenes Forming Emissive TICT States

When protonation of a probe leads to dramatic effects on the electronic absorption and emission as described above, one can expect that cations other than protons might be capable of inducing similar effects if these ions can be made to bind to the basic atom of the probe molecule. A simple but effective method to do so exists in the formal replacement of the amino substituents of chromophores by aza crowns. In the resulting chromoionophores the amine nitrogens possess simultaneously an electron-donor

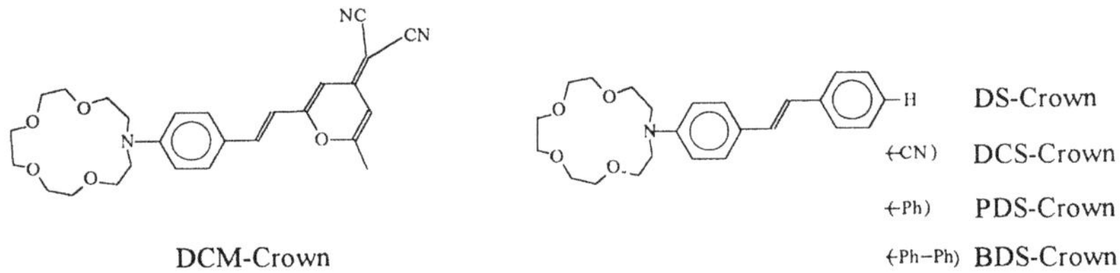

Figure 5.20. Molecular structure of DS-Crowns.

function to complexed cations as well as to the chromophore (Figure 5.20). The majority of the chromo- and fluoroionophores described until now meet this construction scheme.[131] In polar solvents (CH_3CN) and with sufficiently strong D–A interaction the resulting fluoroionophores have almost identical photophysical properties as their "uncrowned" counterparts, with a dimethylamino group instead of the macrocycle.[132] On complexation by cations, the crown-compounds show hypso- and hypochromic shifts of the absorption spectrum and a quenching of the fluorescence whereas the emission spectrum is only slightly blueshifted and the fluorescence lifetime almost unchanged.[133] This fluorescence close to the emission of the uncomplexed probes can be interpreted by breaking of the cation-nitrogen interaction in the excited state before emission occurs as a result of the reduced electron density on the nitrogen when the chromophore is excited. The slight blue shift of the spectrum with respect to that of the free ligand is as expected from the proximity of dipole and ion.[132] This assignment has been reinforced by a study of the DS-crown series (Figure 5.20) where the electron affinity of the acceptor part is decreased along DCM-crown > DCS-crown > DS-crown and the relative stabilization of the precursor state B^* is increased upon phenyl and biphenyl substitution of DS-crown in the series DS-crown < PDS-crown < BDS-crown (Figure 5.20) in order to form a less and less polar excited state which finally leads to the nitrogen–cation interaction remaining unbroken during the fluorescence lifetime. Indeed, from PDS-crown, a dual emission (Figure 5.21) has been observed, respectively, from the complexed ligand involving the nitrogen of the chromophore, as in the ground state, and the cation–ligand separated by the solvent (TICT fluorescence). The former is increased by less polar solvents and when the TICT state formation becomes less favorable by a relative stabilization of the precursor state B^* (Eqs. (5.1) and (5.2)) as in the compounds of Figure 5.20.[132]

5.7.2. Fluorescent Calcium Indicators in Current Use in Molecular Biology

Most of the fluorescent calcium indicators and their cell-permeant acetoxymethyl (AM) esters are variations of the nonfluorescent calcium chelator BAPTA and have been proposed by Tsien.[134–136] Among them Fura-2 and Indo-1 (Figure 5.22) are particularly used for measuring Ca^{2+} in single cells by imaging or flow cytometry.[6]

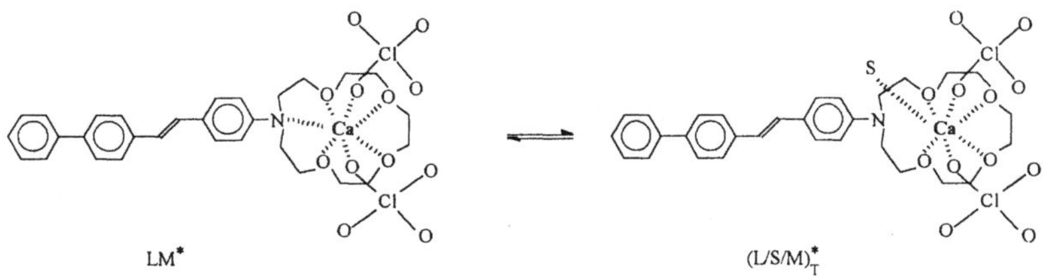

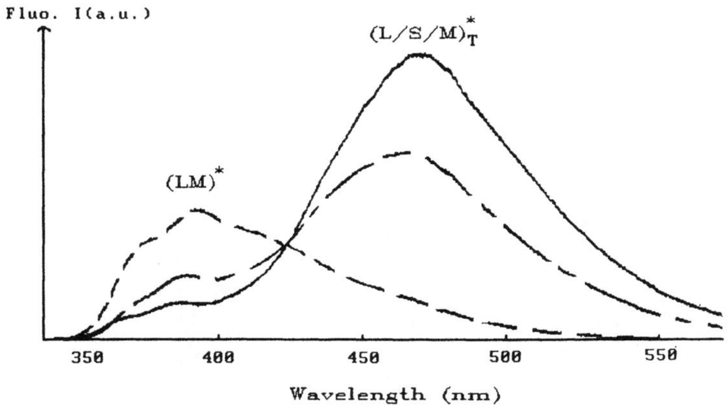

Figure 5.21. Fluorescence spectra of PDS-Crown/Ca^{2+} in (———) in CH_3CN, (- — -) *n*-butyronitrile BuCN, and (- - -) CH_2Cl_2. *Above:* Presumed excited state equilibration between a metal ligand complex where the amino group is engaged in complexation (LM^*) or where a solvent molecule takes over the complexing function ($L/S/M^*_T$).

The absorption shift of Fura-2 that occurs on binding can be determined by scanning the excitation spectrum between 300 and 400 nm while monitoring the emission at ≈ 510 nm. The emission maximum of Indo-1 shifts from 490 nm in Ca^{2+} free media to 405 nm when the dye is saturated with Ca^{2+} allowing the ratiometric measurements of $[Ca^{2+}]$ essentially independent of the extent of dye loading, cell thickness, photobleaching, or dye leakage (Figure 5.22).

Structurally these compounds combine a eight-coordinate tetracarboxylate chelating site with stilbene chromophores. Of the six dyes proposed in Ref. 121 only Indo-1 leads to a short-wavelength fluorescence shift when bound to Ca^{2+}. It is also the probe with the smallest D–A interactions as revealed by the shortest wavelength of fluorescence of the free ligand. As a consequence, the decrease of the electron density of the nitrogen binding site is not sufficient to break completely the interaction with Ca^{2+} in the excited state and the spectrum is considerably blueshifted.

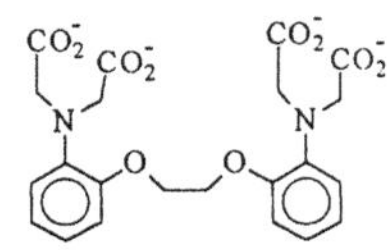

BAPTA

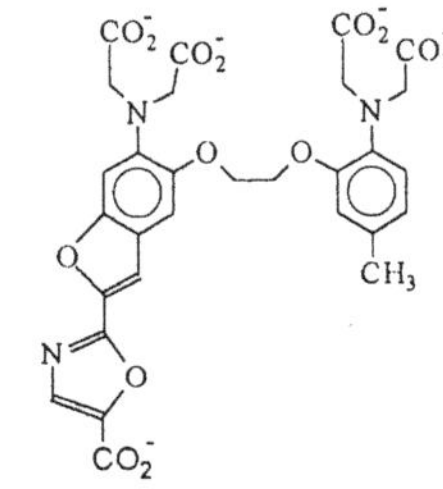

Fura-2

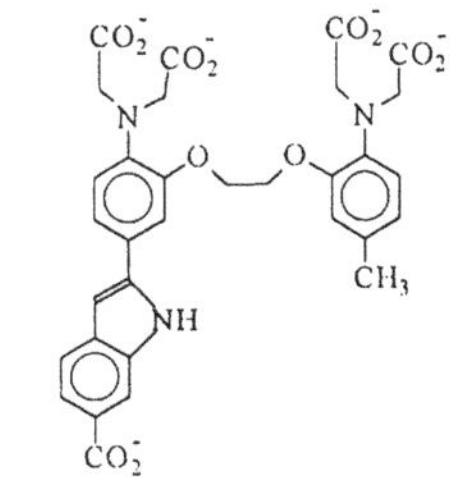

Indo-1

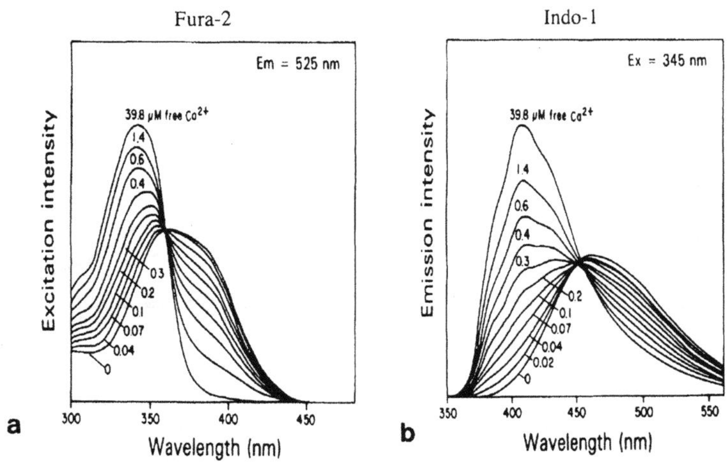

Figure 5.22. Structures and electronic spectra of Fura-2 and Indo-1: (a) excitation spectra of Fura-2; (b) emission spectra of Indo-1, in increasing concentrations of free calcium (from molecular probes).

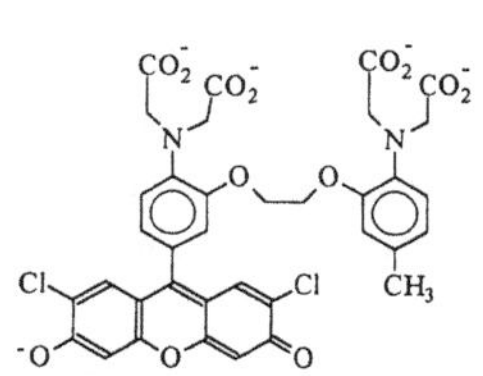

Fluo-3

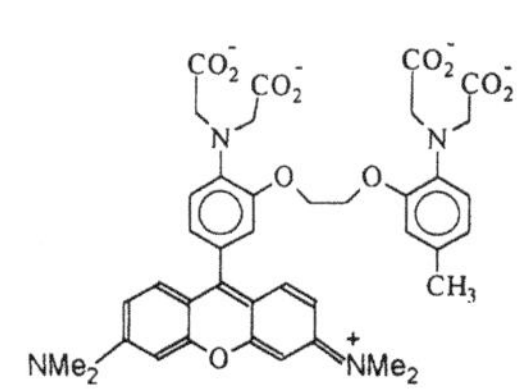

Rhod-2

Figure 5.23. Molecular structure of Fluo-3 and Rhod-2.

Fluo-3 and Rhod-2 (Figure 5.23) were later developed by Tsien for use with long-wavelength excitation sources.[135] Their quantum yields of fluorescence are very low in Ca^{2+}-free medium, presumably because of an intramolecular electron transfer from the ionophore to the fluorophore in a nonemissive TICT state. Binding of Ca^{2+} raises the energy of the TICT state above that of the fluorophore and ϕ_F is increased by up to 80-fold with Fluo-3. In accordance with this mechanism neither Fluo-3 nor Rhod-2 undergoes significant shifts in emission or excitation wavelength on binding to calcium, which precludes ratio imaging.

5.7.3. Other Fluoroionophores with Enhanced Fluorescence in the Presence of Cations

5.7.3.1. Acridinium Azacrowns

Replacing the amino function of the 9-arylacridinium featured in Figure 5.15 by an azacrown ether resulted in systems where ions other than protons had a significant effect on the electronic absorption and emission. Verhoeven *et al.*[123] prepared and studied the chromo- and fluoroionophoric response of $ACPAc^+$ and $ACoMPAc^+$ (Figure 5.24). In analogy with the dimethylamino analogues (Figure 5.15) the protonation raises the energy of the nonemissive TICT above that of the acridinium excited state and switches on the fluorescence from 0 to 100% for $ACoMPAc^+$ and to 24% for $ACPAc^+$. With alkaline or alkaline earth metal-ions, no state inversion could be achieved and the probes remained nonfluorescent. But with the Ag^+ ion (a soft Lewis acid) which binds more covalently with the nitrogen of the azacrown, the fluorescence is again switched on more efficiently with $ACoMPAc^+$ ($\phi_F = 0.072$) than with $ACPAc^+$ ($\phi_F = 0.014$).

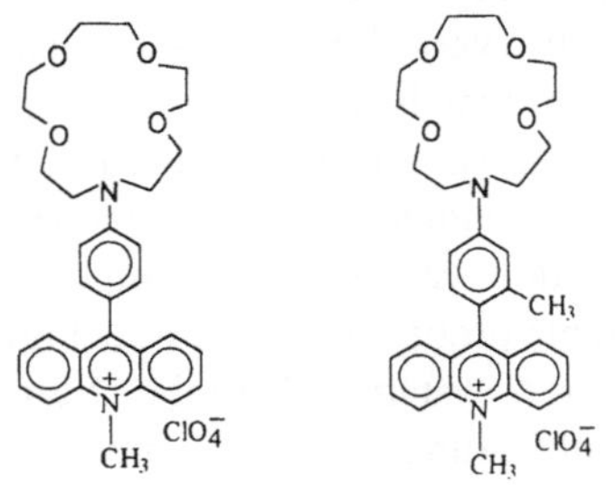

ACPac⁺ ACoMPAc⁺

Figure 5.24. Structure of $ACPac^+$ and $ACoMPAc^+$.

5.7.3.2. *Re(I) Fluoroionophores*

In fac-(bpy)Re(I) $(CO)_3$-A (where bpy is 2,2′-bipyridine and A is an aromatic amine), the d–π(Re)$\rightarrow\pi^*$ (bpy) MLCT fluorescent excited state is strongly quenched via intramolecular aniline-Re charge transfer leading to a nonfluorescent LLCT state. By incorporating the donor amino group belonging to the A moiety into a crown-macrocycle, Schanze and Mac Queen[137] have provided a new luminescent cation sensor whose quantum yield of fluorescence raises from 0.0017 (without cation) to 0.012 (when Ca^{2+} is saturated). Emission quantum yield and lifetime data of the complexed probe are interpreted by a kinetic model in which the rate constant of fluorescence and the dissociation of the cation from the macrocycle in the excited state are in competition.

5.8. Basic Ideas for Future Developments

A large number of sensitive fluorescent probes are now available for the measurement of pH and of the concentration of ions.

Each indicator has to be designed according to the ion probed in its particular environment and the technique used, so that the possibility of preparation of a universal molecular system is unlikely. The indicators have to fulfill some minimum requirements. Obviously, the binding properties of the ionophore have to be matched with the target ion to be monitored in its environment. But also the electronic interaction of the ionophore with the fluorophore in the ground or the excited states has to be significantly altered by the presence of the ion.

Some of the research directions under current consideration contain the following ideas:

1. With conjugated D–A systems which increase their dipole moment during the electronic excitation, the charge transfer (CT) contribution leads to a long-wavelength CT transition in absorption which can be suppressed with an ion on the donor side[131] or increased when it is on the acceptor side.[138] With respect to the fluorescence, most of the known results show only a very slight short-wavelength shift of the spectra when complexed probes are excited, which possess strong D–A interactions.[133] This shift can be increased with less polar suitable probes because then the coordination of the cation is partially or totally maintained in the excited state.[132] For cases where the ratio method with one excitation wavelength is compulsory, e.g., with emission ratio imaging, the design of suitable probe could thus be achieved.

2. Nevertheless, the D–A systems where the D group is part of the ionophore possess the inconvenience of leading, in the presence of cations, to a fluorescence at short wavelength where the free probe has its charge transfer absorption band. To overcome this difficulty D–D conjugated systems with one D substituent part of the ionophoric group have been designed, which in the presence of cations lead to a $D^{\delta+}$–$D^{\delta-}$–M^+ system with a long-wavelength emission (Figure 5.17).[139]

For this type of "integrated" fluoroionophores, the photochemical processes can be accelerated and can lead to very fast and reversible photochemical ion release or ion takeup. One example has been described recently[140] where Ca^{2+} or Li^{+} is ejected in some picoseconds. By this way, the application of biologically useful chelators, which have their binding constant altered by an irreversible photoreaction taking at best some milliseconds,[141] can be extended to ultrashort time scales.

3. Virtually all PET probes devised until now function in the same way in that they disfavor ET on complexation because the complexing site is an integral part of the donor system. A principally different behavior with increased TICT formation or PET on complexation is expected if the complexing site is part of the acceptor system. A simple way of realizing this is to introduce complexing sites at the place of the cyano group in DMABN, e.g., in the corresponding dimethylamino- or azacrown amides.[142] Thus two classes of probes with dual fluorescence can be constructed with opposite complexation response (DMABN-cr, DMABAm-cr).

4. As a rule, TICT states formed starting from ionic dye systems show strongly reduced TICT fluorescence quantum yields due to enhanced nonradiative transitions.[143] It is thus preferable to use nonionic dyes as basis. For example, the cationic Rhodamine chromophore in Rhod-2 (Figure 5.23) could be replaced by the neutral chromophore of the SNARF compounds in Figure 5.13. Then the chance of observing multiple fluorescence bands and large quantum yields is strongly increased. A further possibility is to start with dye systems known to possess unusually large TICT fluorescence quantum yields such as 5,5′-biperylenyl.[144, 145]

5. As shown above with the example of Rhodamine dyes, TICT channels can be introduced into most other dye systems giving a large flexibility in choosing absorption and emission wavelengths to adapt to given conditions of the material to be investigated.

6. As a possible source for new long-wavelength dye systems, biradicaloid dyes possessing energetically close-lying frontier molecular orbitals could be used.[146] An example of such dye systems already in extensive use in imaging are squaric acid derivatives like SQ.[147, 148]

Acknowledgments

Support by the Bundesministerium für Forschung und Technologie (project 05 414 FAB1 and 05 5KT FAB9), and the EC-Large Scale Installations Program (GE 1-0018-D(B)) is gratefully acknowledged.

References

1. T. M. Li and R. F. Parrish, Fluorescence and immunodiagnostic methods, in: *Topics in Fluorescence Spectroscopy*, Vol. 3 (J. R. Lakowicz, ed.), pp. 273–288, Plenum Press, New York (1992).

2. R. Rigler, J. Widengren, and Ü. Mets, Interactions and kinetics of single molecules as observed by fluorescence correlation spectroscopy, in: *Fluorescence Spectroscopy, New Methods and Applications* (O. S. Wolfbeis, ed.), pp. 13–24, Springer-Verlag, Berlin (1993).
3. H. Szmacinski and J. R. Lakowicz, Lifetime-based sensing, in: *Topics in Fluorescence Spectroscopy*, Vol. 4 (J. R. Lakowicz, ed.), pp. 295–334, Plenum Press, New York (1994).
4. D. J. Cowley, Polar pocket with nonpolar lining, *Nature 319*, 14 (1986).
5. P. Fromherz, K. H. Dambacher, H. Ephardt, A. Lambacher, C. O. Müller, R. Neigl, H. Schaden, O. Schenk, and T. Vetter, Fluorescent dyes as probes of voltage transients in neuron membranes. Progress report, *Ber. Bunsenges. 95*, 1333 (1991).
6. A. P. Jackson, M. P. Timmerman, C. R. Bagshaw, and C. C. Ashley, The kinetics of calcium binding to Fura-2 and Indo-1, *FEBS Lett. 216*, 35–39 (1987).
7. T. Förster, *Fluoreszenz Organischer Verbindungen*, Verlag Vandenhoeck & Ruprecht, Göttingen (1951).
8. N. J. Turro, J. McVey, V. Ramamurthy, and P. Lechtken, Adiabatische Photoreaktionen organischer Moleküle, *Angew. Chem. 91*, 597–612 (1979).
9. W. Rettig, Charge separation in excited states of decoupled systems—TICT compounds and implications regarding the development of processes and the development of new laser dyes and the primary processes of vision and photosynthesis, *Angew. Chem. Int. Ed. Engl. 25*, 971 (1986).
10. H. Köppel, W. Domcke, and L. S. Cederbaum, Multimode molecular dynamics beyond the Born-Oppenheimer approximation, *Adv. Chem. Phys. 57*, 59–246 (1984).
11. W. Rettig and G. Wermuth, The kinetics of formation of twisted intramolecular charge transfer (TICT) states in p-substituted dialkylanilines: Consequences of conical intersections along the reaction coordinate, *J. Photochem. 28*, 351 (1985).
12. W. Rettig, Solvent polarity dependent formation dynamics of TICT states. I. Differential solvatokinetics, *Ber. Bunsenges. Phys. Chem. 95*, 259 (1991).
13. W. Rettig, Photoinduced electron transfer in twisted π-systems, in: *Dynamics and mechanisms of Photoinduced Electron Transfer and Related Phenomena* (N. Mataga, T. Okada, and H. Masuhara, eds.), pp. 57–69, Elsevier Science Publishers, Amsterdam (1992).
14. M. M. H. Khalil, N. Boens, M. Van Der Auweraer, M. Ameloot, R. Andriessen, J. Hofkens, and F. C. De Schryver, Compartimental analysis of the fluorescence decay surface of the exciplex formation between 1-methylpyrene and triethylamine, *J. Phys. Chem. 95*, 9375–9381 (1991).
15. J. Michl and V. Bončacič-Koutecký, *Electronic Aspects of Organic Photochemistry*, John Wiley & Sons, New York (1990).
16. M. Vogel and W. Rettig, Efficient intramolecular fluorescence quenching in triphenylmethane-dyes involving excited states with charge separation and twisted conformations, *Ber. Bunsenges. Phys. Chem. 89*, 962 (1985).
17. R. Lapouyade, K. Czeschka, W. Majenz, W. Rettig, E. Gilabert, and C. Rullière, Photophysics of donor-acceptor substituted stilbenes. A time-resolved fluorescence study using selectively bridged dimethylamino cyano model compounds, *J. Phys. Chem. 96*, 9643–9650 (1992).
18. E. Lippert, W. Lüder, and H. Boos, Fluoreszenzspektrum und Franck-Condon-Prinzip in Lösungen Aromatischer Verbindungen, in: *Advances in Molecular Spectroscopy* (A. Mangini, ed.), p. 443, Pergamon Press, Oxford, England (1962).
19. Z. R. Grabowski, K. Rotkiewicz, A. Siemiarczuk, D. J. Cowley, and W. Baumann, Twisted intramolecular charge transfer states (TICT). A new class of excited states with a full charge separation, *Nouv. J. Chim. (New J. Chem.) 3*, 443 (1979).
20. K. Rotkiewicz, K. H. Grellmann, and Z. R. Grabowski, Reinterpretation of the anomalous fluorescence of P-N,N-dimethylamino-benzonitrile, *Chem. Phys. Lett. 19*, 315 (1973).
21. K. Rotkiewicz, Z. R. Grabowsksi, A. Krowczynsksi, and W. Kühnle, Picosecond chemical relaxation of excited p-cyano-N,N-dialkylanilines, *J. Lumin. 12*, 877 (1976).

22. Z. R. Grabowski, K. Rotkiewicz, W. Rubaszewska, and E. Kirkor-Kaminska, Spectroscopy and kinetic of the twisted internal charge-transfer (TICT) excited state formation in p-substituted dialkylanilines, *Acta Phys. Pol. A 54,* 767 (1978).
23. Z. R. Grabowski, K. Rotkiewicz, and A. Siemiarczuk, Dual fluorescence of donor-acceptor molecules and the twisted intramolecular charge transfer (TICT) states, *J. Lumin. 18,* 420 (1979).
24. W. Rettig, Anwendungsaspekte adiabatischer Photoreaktionen, *Nachr. Chem. Tech. Lab. 39,* 298 (1991).
25. K. A. Al-Hassan and W. Rettig, Free volume sensing fluorescent probes, *Chem. Phys. Lett. 126,* 273 (1986).
26. W. Rettig, R. Fritz, and J. Springer, Fluorescence probes based on adiabatic photochemical reactions, in: *Photochemical Processes in Organized Molecular Systems* (K. Honda, ed.), p. 61, Elsevier Science Publishers, Amsterdam (1991).
27. W. Rettig, Kinetic studies on fluorescence probes using synchrotron radiation, in: *Fluorescence Spectroscopy: New Methods and Applications* (O. S. Wolfbeis, ed.), pp. 31–47, Springer-Verlag, Berlin (1993).
28. (a) R. A. Bissell, A. P. De Silva, H. Q. N. Gunaratne, P. L. M. Lynch, G. E. M. Maguire, and K. R. A. S. Sandanayake, Molecular fluorescent signalling with 'fluor-spacer-receptor' systems: Approaches to sensing and switching devices via supramolecular photophysics, *Chem. Soc. Rev.*, 187–195 (1992); (b) R. A. Bissell, A. P. De Silva, H. Q. N. Gunaratne, P. L. M. Lynch, G. E. M. Maguire, C. P. McCoy, and K. R. A. S. Sandanayake, Fluorescent PET (Photoinduced Electron Transfer) Sensors *Top. Curr. Chem. 168,* 223 (1993).
29. J. L. Habib Jiwan and J. P. Soumillion, Electron transfer photochemistry initiated from a twisted intramolecular charge transfer state used as an electron donor and as an acceptor, *J. Photochem. Photobiol. A: Chem. 64,* 145–158 (1992); J. P. Soumillion, Photoinduced electron transfer implying organic anions, in: *Topics in Current Chemistry, Vol., Photoinduced Electron Transfer* (J. Mattay, ed.), Springer Verlag, Berlin, pp. 93–141 (1993).
30. G. Schopf, W. Rettig, and J. Bendig, Quenching of TICT fluorescence by electron donors, *J. Photochem. Photobiol. A: Chem.*, in press.
31. Z. R. Grabowski and J. Dobkowski, Twisted intramolecular charge transfer (TICT) excited states: Energy and molecular structure, *Pure Appl. Chem. 55,* 245 (1983).
32. E. Lippert, W. Rettig, V. Bonačić-Koutecký, F. Heisel, and J. A. Miehé, Photophysics of internal twisting, *Adv. Chem. Phys. 68,* 1 (1987).
33. W. Rettig, Charge transfer in self-decoupling π-systems, in: *Modern Models of Bonding and Delocalization, Serie: Molecular Structure and Energetics,* Vol. 6 (J. F. Liebman and A. Greenberg, eds.), pp. 229–282, VCH Verlagsgesellschaft mbH, Weinheim, Germany (1988).
34. W. Rettig and W. Baumann, Luminescent and nonluminescent twisted intramolecular charge transfer (TICT) states: Dipole moments and application, in: *Photochemistry and Photophysics*, Vol. 6 (J. F. Rabek, ed.), pp. 79–134, CRC Press, Boca Raton, Florida (1992).
35. M. Zander and W. Rettig, Fluorescence studies on solvent-induced intramolecular charge separation in symmetric biaryls, *Chem. Phys. Lett. 110,* 602 (1984).
36. Th. Förster, Die pH-Abhangigkeit der Fluoreszenz von Naphthalinderivaten, *Z. Elektrochem. 54,* 531–53 (1950).
37. J. Waluk, H. Bulska, A. Grabowska, and A. Mordzinski, Excited state proton transfer analysed by state-orbitals correlations, *Nouv. J. Chim.* (*New J. Chem.*) *10,* 413 (1986).
38. F. Laermer, T. Elsaeser, and W. Kaiser, Femtosecond spectroscopy of excited-state proton-transfer in 2-(2′-hydroxyphenyl)benzothiazole, *Chem. Phys. Lett. 148,* 119 (1988).
39. (a) N. P. Ernsting, A. Mordzinski, and B. Dick, Excited-state intramolecular proton transfer in jet-cooled 2,5-Bis-(2-benzothiazole)hydroquinone, *J. Phys. Chem. 91,* 1404 (1987); (b) B. Dick and N. P. Ernsting, Excited-state intramolecular proton transfer in 3-hydroxyflavone isolated in solid argon: Fluorescence and fluorescence-excitation spectra and tautomer fluorescence rise time. *J. Phys. Chem. 91,* 4261 (1987).

40. S. Demmig and H. Langhals, Leichtlösliche, lichtechte Perylen-Fluoreszenzfarbstoffe, *Ber. Bunsenges. Phys. Chem. 21*, 225 (1988).
41. O. A. Salman and H. G. Drickamer, The effect of pressure and freezing on luminescence. I. 3-hydroxyflavone and indole, *J. Chem. Phys. 77*, 3329–3336 (1982).
42. H. Bulska, Intramolecular cooperative double proton transfer in (2,2′-bipyridyl)-3,3′-diol, *Chem. Phys. Lett. 98*, 398 (1983).
43. P. Chou, D. McMorrow, T. J. Aartsma, and M. Kasha, The proton transfer laser. Gain spectrum and amplification of spontaneous emission of 3-hydroxyflavone. *J. Phys. Chem. 88*, 4596 (1984).
44. J. Sepiol, H. Bulska, and A. Grabowska, The dihyodroxy derivative of 2,2′-bipyridyl as a new proton-transfer lasing dye, *Chem. Phys. Lett. 140*, 607 (1987).
45. (a) G. Woessner, G. Goeller, J. Rieker, H. Hoier, J. J. Stezowski, E. Daltrozzo, M. Neureiter, and H. E. A. Kramer, Ultraviolet stabilizers of the 2-(hydroxyphenyl)benzotriazole class—influence of the solvent on the absorption spectra and photochemical deactivation mechanism, *J. Phys. Chem. 89*, 3629–3636 (1985); (b) H. E. A. Kramer, The lightfastness or the non-photochemistry of dyes, *Chimia 40*, 160–169 (1986).
46. D. Gormin and M. Kasha, Triple fluorescence in aminosalicylates. Modulation of normal proton-transfer and twisted intramolecular charge-transfer (TICT) fluorescence by physical and chemical perturbations, *Chem. Phys. Lett. 153*, 574 (1988).
47. H. Beens and A. Weller, Excited molecular π-complexes in solution, in: *Organic Molecular Photophysics*, Vol. 2 (J. B. Birks, ed.), John Wiley & Sons, London (1975).
48. A. Weller, Singlet- and triplet-state exciplexes, in: *The Exiplex* (M. Gordon and W. R. Ware, eds.), p. 23, Academic Press, New York, San Francisco, London, (1975).
49. L. Bokobza, C. Pham-Van-Cang, L. Monnerie, J. Vandendriessche, and F. C. De Schryver, Investigation of the mobility of polybutadienes: 1. Excimer fluorescence technique, *Polymer 30*, 45 (1989).
50. L. Bokobza, C. Pham-Van-Cang, C. Giordano, L. Monnerie, J. Vandendriessche and F. C. De Schryver, Relation between excimer formation in small probes and free-volume theory in polymer melts, *Polymer 28*, 1876 (1987).
51. (a) N. C. Yang, R. V. Carr, E. Li, J. K. McVey, and S. A. Rice, 2,3-Naphtho-2,5-bicyclo[2.2.0]hexadiene, *J. Am. Chem. Soc. 96*, 2297–2298 (1974); (b) R. V. Carr, B. Kim, J. K. McVey, N. C. Yang, W. Gerhartz, and J. Michl, Photochemistry and photophysics of 1,4-dewarnaphthalene, *Chem. Phys. Lett. 39*, 57–60 (1976).
52. B. Jahn and H. Dreeskamp, Die photochemische Valenzisomerisierung 9-alkylsubstituierter Anthrazene zu 9,10-Dewarisomeren, *Ber. Bunsenges. Phys. Chem. 88*, 42–47 (1984).
53. K. Kemnitz, N. Tamai, I. Yamazaki, N. Nakashima, and K. Yoshihara, Site-dependent fluorescence lifetimes of isolated dye molecules adsorbed on organic single crystals and other substrates, *J. Phys. Chem. 91*, 1423–1430 (1987).
54. A. Azzi, The application of fluorescent probes in membrane studies, *Q. Rev. Biophysics 8*, 237 (1975).
55. (a) W. Nowak, P. Adamczak, A. Balter, and A. Sygula, *J. Mol. Struct.* (*Theochem.*) *139*, 13 (1986); (b) J. Catalan, P. Perez, J. Laynez, and F. G. Blanco, Analysis of the solvent effect on the photophysics properties of 6-propianyl-2-(dimethylamino)naphthalene (PRODAN), *J. Fluorescence 1*, 215–233 (1991).
56. R. B. Macgregor and G. Weber, *Nature 319*, 273 (1986).
57. J. R. Lakowicz, *Principles of Fluorescence Spectroscopy*, Plenum Press, New York (1983).
58. E. M. Kosower and D. Huppert, Excited state electron and proton transfers, *Ann. Rev. Phys. Chem. 37*, 127 (1986).
59. W. Nowak and W. Rettig, Theoretical study of singlet excited states of donor–acceptor naphthalene derivatives, *J. Mol. Struct.* (*Theochem.*) *283*, 1–12 (1993).
60. A. A. Ayuk, W. Rettig, and E. Lippert, Temperature and viscosity effects on an excited state equilibrium as revealed from the dual fluorescence of very dilute solutions of 1-dimethylamino-4-cyanonaphthalene. *Ber. Bunsenges. Phys. Chem. 85*, 553 (1981).

61. (a) A. Nag and K. Bhattacharyya, Twisted intramolecular charge transfer emission of dimethylaminobenzonitrile in α-cyclodextrine cavities, *Chem. Phys. Lett. 151*, 474 (1988); (b) K. Bhattacharyya, Dual luminescence of dimethylaminobenzonitrile in γ-cyclodextrin, *J. Chem. Soc. Faraday Trans. 86*, 53 (1988).
62. K. Bhattacharyya and M. Chowdhury, Environmental and magnetic field effects on exciplex and twisted charge transfer emission. *Chem. Rev. 93*, 507–535 (1993).
63. G. S. Cox, P. J. Hauptmann, and N. J. Turro, *Photochem. Photobiol. 39*, 597 (1984).
64. O. S. Khalil and A. J. Sonessa, Critical micelle concentration. Fluorescence of 4-dimethylaminobenzonitrile in surfactant solutions, *Mol. Photochem. 8*, 399 (1977).
65. W. Rettig, F. Marschner, Molecular conformation and biradicaloid charge transfer states in substituted N-phenyl-pyrroles, *New J. Chem. 14*, 819 (1990).
66. K. Rotkiewicz and W. Rubaszewska, Intramolecular charge transfer state and unusual fluorescence from an upper excited singlet of a nonplanar derivative of p-cyano-N,N-dimethylaniline, *J. Lumin. 27*, 221 (1982).
67. W. Rettig and R. Gleiter, Dependence of intramolecular rotation in p-cyano-N,N-dialkylanilines on the twist angle. A fluorescence, UV absorption, and photoelectron spectroscopic study. *J. Phys. Chem. 89*, 4676 (1985).
68. W. Rettig, G. Wermuth, and E. Lippert, Photophysical primary processes in solutions of p-substituted dialkylanilines, *Ber. Bunsenges. Phys. Chem. 83*, 692 (1979).
69. J. Hicks, M. Vandersall, Z. Babarogic, and K. B. Eisenthal, The dynamics of barrier crossings in solution: The effect of a solvent polarity-dependent barrier, *Chem. Phys. Lett. 116*, 18–24 (1985).
70. H. Ephardt and P. Fromherz, Fluorescence and photoisomerization of an amphiphilic aminostilbazolium dye as controlled by the sensitivity of radiationless deactivation to polarity and viscosity, *J. Phys. Chem. 93*, 7717–7725 (1989).
71. H. Ephardt and P. Fromherz, *J. Phys. Chem. 93*, 7717 (1986).
72. P. Fromherz and A. Heilemann, Twisted internal charge transfer in (aminophenyl)pyridinium. *J. Phys. Chem. 96*, 6864–6866 (1992).
73. L. B. Cohen, B. M. Salzberg, H. V. Davila, W. N. Ross, D. Landowne, A. S. Waggoner, and C. H. Wang, Changes in axon fluorescence during activity: Molecular probes of membrane potential, *J. Membrane Biol. 19*, 1 (1974).
74. R. K. Gupta, B. M. Salzber, A. Grinvald, L. B. Cohen, K. Kamino, S. Lesher, M. B. Boyle, A.S. Waggoner, and C. H. Wang, Improvements in optical methods for measuring rapid changes in membrane potential, *J. Membrane Biol. 58*, 123 (1981).
75. L. M. Loew, S. Scully, L. Simpson, and A. S. Waggoner, Evidence for a charge-shift electrochromic mechanism in a probe of membrane potential, *Nature 281*, 497 (1979).
76. A. M. Klock and W. Rettig, The excited-state behavior of donor-acceptor biphenyl and phenylanthracene: A comparative study, *Polish J. Chem. 67*, 1375–1388 (1993).
77. M. Van Damme, J. Hofkens, F. C. De Schryver, T. G. Ryan, W. Rettig, and A. M. Klock, Solvent dynamics and twisted intramolecular charge transfer in 4-cyano-4′-butyloxybiphenyl (4COB), *Tetrahedron 45*, 4693–4706 (1989).
78. V. Bonačič-Koutecký, J. Koutecký, and J. Michl, Neutrale und geladene Diradikale, Zwitterionen, Trichter auf der S1-Hyperpläche und Protonentranslokation; ihre Bedeutung für den Sehvorgang und andere photochemische und photophysikalische Prozesse, *Angew. Chem. 99*, 216 (1987); *Angew. Chem. Int. Ed. Engl. 27*, 170 (1987).
79. V. Bonacic-Koutecky and J. Michl , Charge-transfer-biradical excited states: Relation to anomalous fluorescence. "Negative" S1-T1 splitting in twisted aminoborane, *J. Am. Chem. Soc. 107*, 1765 (1985).
80. W. Rettig and W. Baumann, Luminescent and nonluminescent twisted intramolecular charge transfer (TICT) states: Dipole moments and applications, in: *Photochemistry and Photophysics*, Vol. 6 (J. F. Rabek, ed.), pp. 79–134, CRC Press, Boca Raton, Florida (1992).

81. W. Rettig, B. Strehmel, and W. Majenz, The exited states of stilbene and stilbenoid donor–acceptor dye systems. A theoretical study, *Chem. Phys. 173*, 525–537 (1993).
82. M. Vogel and W. Rettig, Excited state dynamics of triphenylmethane-dyes used for investigation of microviscosity effects, *Ber. Bunsenges. Phys. Chem. 91*, 1241 (1987).
83. (a) A. Spernol and K. Wirtz, Zur Mikroreibung in Flüssigkeiten, *Z. Naturforsch. 8A*, 522–532, 1953; (b) A. Gierer and K. Wirtz, Molekulare Theorie der Mikroreibung, *Z. Naturforsch. 8A*, 532–538 (1953).
84. M. L. Williams, R. F. Landel, and J. D. Ferry, Temperature dependence of relaxation mechanisms in amorphous polymers and other glass forming liquids, *J. Am. Chem. Soc. 77*, 3701–3707 (1955).
85. (a) D. V. O'Connor and D. Phillips, *Time-Correlated Single Photon Counting*, Academic Press, New York (1984); (b) R. F. Steiner, Fluoresence anisotropy: Theory and applications, in: *Topics in Fluorescence Spectroscopy*, Vol. 2 (J. R. Lakowicz, ed.), Plenum, New York (1991), p. 1.
86. S. Canonica, A. A. Schmid, and U. P. Wild, The rotational diffusion of p-terphenyl and p-quaterphenyl in non-polar solvents, *Chem. Phys. Lett. 122*, 529–534 (1985).
87. (a) M. F. Winnick, Photophysics of preassociated pyrenes in aqueous polymer solution and in other organized media, *Chem. Rev. 93*, 587 (1993); (b) V. Ramamurthy (ed.), *Photochemistry in Organized and Constrained Media*, VCH Publishers, New York (1991); (c) J. F. Rabek (ed.), *Progress in Photochemistry and Photophysics*, Vol. 6, CRC Press, Boca Raton, Florida (1992).
88. L. Bokobza, C. Pham-Van-Cang, C. Giordano, L. Monnerie, J. Vandendriessche, and F. C. De Schryver, Studies of the mobility of probes in poly(propylene oxide): 2. Excimer fluoresence technique, *Polymer 29*, 251 (1988).
89. L. Bokobza L, E. Pajot-Augy, L. Monnerie, A. Castallan, and H. Bouas-Laurent, Excimer fluorescence as a probe of mobility in bulk polymers, *Polymer Photochem. 5*, 191 (1984).
90. E. Pajot-Augy, L. Bokobza, L. Monnerie, A. Castallan, and H. Bouas-Laurent, Mobility of macromolecular chains as studied by means of excimer fluoresence, *Macromolecules 17*, 1490 (1984).
91. R. O. Loutfy and K. Yee Law, Spectrochemistry and spectroscopy of intramolecular charge-transfer complexes. N,N-dialkylaxinobenzylidenemalononitriles, *J. Phys. Chem. 84*, 2803 (1980).
92. K. A. Al-Hassan and T. Azumi, The red edge effect as a tool for investigating the origin of the anomalous fluorescence band of 9,9′-bianthryl in rigid polar polymer matrices, *Chem. Phys. Lett. 150*, 344 (1988).
93. D. Anwand, F. W. Müller, B. Strehmel, and K. Schiller, Determination of the molecular mobility and the free volume of thin polymeric films with fluorescence probes, *Makromol. Chem. 192*, 1981–1991 (1991).
94. B. Strehmel, V. Strehmel, H. J. Timpe, and K. Urban, Time-resolved fluoresence measurements for the characterization of epoxy systems, *Eur. Polym. J. 28*, 525–533 (1992).
95. R. O. Loutfy, Fluoresence probes for polymer free-volume, *Pure Appl. Chem. 58*, 1239 (1986).
96. Z. A. Dreger, J. M. Lang, and H. G. Drickamer, High pressure study of flexible fluorescent dye molecules in solid polymeric media. III. *p-N,N*-dimethylaminobenzylidenemalononitrile (DMABNM)—reanalyzed. *Chem. Phys., 169*, 369–371 (1993).
97. S. Tazuke, R. Kun Guo, and R. Hayashe, How does the polymer main chain influence the side-chain mobility? A fluorescence probe study by means of twisted intramolecular charge-transfer phenomena, *Macromolecules 21*, 1046 (1988).
98. S. Tazuke, R. K. Guo, and T. Ikeda, Time-resolved fluoresence spectroscopy of a twisted intramolecular charge transfer compound bonded to polymers, *J. Phys. Chem. 94*, 1408 (1990).
99. R. Hayashi, S. Tazuke, and C. W. Frank, Twisted intramolecular charge-transfer phenomenon as a fluorescence probe of microenvironment. Effect of polymer concentration on local viscosity and microscopic polarity around a polymer chain of poly(methyl methacrylate), *Macromolecules 20*, 983 (1987).
100. S. Tazuke and R. K. Guo, Effects of polymer microenvironment on the thermodynamics of the twisted intramolecular charge transfer fluorescence, *Macromolecules 23*, 719 (1990).

101. R. K. Guo and S. Tazuke, Microheterogeneity of local segment mobility in solid polymethacrylates as studied by a polymer-bound twisted intramolecular charge transfer chromophore, *Macromolecules 22*, 3286 (1989).

102. K. A. Al-Hassan and T. Azumi, The role of free volume in the twisted intramolecular charge transfer (TICT) emission of dimethylaminobenzonitrile and related compounds in rigid polymer matrices. *Chem. Phys. Lett. 146*, 121 (1988).

103. K. A. Al-Hassan and T. Azumi, Fluorescence decay characteristics of 4-dimethylaminobenzontrile in different polymer matrices: A time-resolved study in the nanosecond range, *Chem. Phys. Lett. 163*, 129 (1989).

104. L. F. Lindoy, in: Mixed-donor macrocycles: solution and structural aspects of their complexation with transition and post-transition ions, *Cation Binding by Macrocycles* (Y. Inoue and G. W. Gokel, eds.), p. 599, Marcel Dekker, New York (1990).

105. M. Vogel, W. Rettig, R. Sens, and K. H. Drexhage, Structural relaxation of rhodamine dyes with different N-substitution patterns: A study of fluorescence decay times and quantum yields. *Chem. Phys. Lett. 147*, 452–460 (1988).

106. C. J. Tredwell and A. D. Osborne, Viscosity dependent internal conversion in the rhodamine dye, fast acid violet 2R, *J. Chem. Soc. Faraday Trans. II 76*, 1627 (1980).

107. W. Rettig, Kinetic studies on fluorescence probes using synchrotron radiation, in: *Fluorescence Spectroscopy. New Methods and Applications* (O. S. Wolfbeis, ed.), pp. 31–47, Springer-Verlag, Berlin (1993).

108. P. Plaza, N. Dai Hung, M. M. Martin, Y. H. Meyer, M. Vogel, and W. Rettig, Ultrafast internal charge transfer in a donor-modified rhodamine, *Chem. Phys. 168*, 365–373 (1992).

109. S. G. Schulman and J. D. Winefordner, Influence of pH in fluorescence and phosphorescence spectrometric analysis, *Talanta 17*, 607–616 (1970).

110. Z. R. Grabowski and A. Grabowska, The Förster cycle reconsidered, *Z. Phys. Chem. (Wiesbaden) 101*, 197–208 (1976).

111. E. Van Der Donckt, Acid-base-properties of excited states, *Prog. React. Kinet. 5*, 273–299 (1970).

112. J. F. Ireland and P. A. H. Wyatt, Acid-base-properties of electronically excited states of organic molecules. *Adv. Phys. Org. Chem. 12*, 131 (1976).

113. A. Weller, Fast reactions of excited molecules, *Prog. React. Kinet. 1*, 187–214 (1961).

114. G. Jackson and G. Porter, Acidity constants in the triplet state, *Proc. R. Soc. Lond. Ser. A 260*, 13–30 (1961).

115. H. Shizuka, Excited state proton-transfer reactions and proton-induced quenching of aromatic compounds, *Acc. Chem. Res. 18*, 141–147 (1985).

116. M. Orrit and J. Bernard, Single pentacene molecules detected by fluorescence excitation in a p-terphenyl crystal, *Phys. Rev. Lett. 65*, 2716–2719 (1990).

117. L. Kador, D. E. Horne, and W. E. Moerner, Optical detection and probing of single dopant molecules of pentacene in a p-terphenyl host crystal by means of absorption spectroscopy, *J. Phys. Chem. 94*, 1237–1248 (1990).

118. R. J. Gillies and R. Martinez-Zaguilan, Regulation of intracellular pH in BALB/c 3T3 cells, *J. Biol. Chem. 266*, 1551–1556 (1991).

119. R. A. Bissell, E. Calle, A. P. De Silva, S. A. De Silva, H. Q. N. Gunaratne, J. L. Habib-Jiwan, S. L. A. Peiris, R. A. D. D. Rupasinghe, T. K. S. D. Smarasinghe, K. R. A. S. Sandanayake, and J. P. Soumillion, Luminescence and charge transfer. Part 2. Aminomethyl anthracene derivatives as fluorescent PET (photoinduced electron transfer) sensors for protons, *J. Chem. Soc. Perkin Trans. 2*, 1559–1564 (1992).

120. J. E. Whitaker, R. P. Haughland, and F. G. Prendergast, Spectral and photophysical studies of benzo[c]xanthene dyes: Dual emission pH sensors, *Anal. Biochem. 194*, 330–344 (1991).

121. G. Grynkiewicz, M. Poenie, and R. Y. Tsien, A new generation of Ca^{2+} indicators with greatly improved fluorescence properties. *J. Biol. Chem. 260*, 3440–3450 (1985).

122. W. Rettig and A. Klock, Intramolecular fluorescence quenching in aminocoumarines: Identification of an excited state with full charge separation, *Can. J. Chem. 63*, 1649–1653 (1985).
123. S. A. Jonker, F. Ariese, and J. W. Verhoeven, Cation complexation with functionalized 9-arylacridinium ions: Possible applications in the development of cation-selective optical probes, *Rec. Trav. Chim. Pays-Bas 108*, 109–115 (1987).
124. W. Rettig, W. Majenz, R. Lapouyade, and G. Haucke, Multidimensional photochemistry in flexible dye systems, *J. Photochem. Photobiol. A: Chem. 62*, 415–427 (1992).
125. W. Rettig and W. Majenz, Competing adiabatic photoreaction channels in stilbene derivatives, *Chem. Phys. Lett. 143*, 335–341 (1989).
126. J. F. Letard, R. Lapouyade, and W. Rettig, Structure-photophysics correlations in a series of 4-(dialkylamino) stilbenes: Intramolecular charge transfer in the excited state as related to the twist around the single bonds, *J. Am. Chem. Soc. 115*, 2441–2447 (1993).
127. K. J. Smit and K. P. Ghiggino, Effect of solvent on the photochemistry of trans-4,4′-diaminostilbene-2,2′-disulphonic acid, *J. Photochem. 34*, 23–34 (1986).
128. J. F. Letard, R. Lapouyade, and W. Rettig, Relaxation pathways in photoexcited electron-rich stilbenes (D–D stilbenes) as compared to D–A stilbenes, *Chem. Phys. Lett. 222*, 209–216 (1994).
129. A. Z. Weller, Photoinduced electron transfer in solution: Exciplex and radical ion pair formation free enthalpies and their solvent dependence, *Z. Phys. Chem. Neue Folge, Wiesbaden 133*, 93–98 (1982).
130. R. Grigg and W. D. J. A. Norbert, Luminescent pH sensors based on di(2,2′-bipyridyl) (5,5′-diaminomethyl-2,2′-bipyridyl)ruthenium (II) complexes, *J. Chem. Soc. Chem. Commun.* 1300–1302 (1992).
131. H. G. Lohr and F. Vögtle, Chromo- and fluoroionophores. A new class of dye reagents. *Acc. Chem Res. 18*, 65–72 (1985).
132. J. F. Letard, R. Lapouyade, and W. Rettig, Synthesis and photophysical study of 4-(N-monoaza-15-crown-5)stilbenes forming TICT states and their complexation with cations, *Pure Appl. Chem. 65*, 1705–1712 (1993).
133. J. Bourson and B. Valeur, Ion-responsive fluorescent compounds. 2. Cation-steered intramolecular charge transfer in a crowned merocyanine, *J. Phys. Chem. 93*, 3871–3876 (1989).
134. R. Y. Tsien, New calcium indicators and buffers with high selectivity against magnesium and protons: Design, synthesis and properties of prototype structure, *Biochemistry 19*, 2396–2404 (1980).
135. A. Minta, J. P. Y. Kao, and R. Y. Tsien, Fluorescent indicators for cytosolic calcium based on rhodamine and fluorescein chromophores, *J. Biol. Chem. 264*, 8171–8178 (1989).
136. (a) A. Minta and R. Y. Tsien, Fluorescent indicators for cytosolic sodium, *J. Biol. Chem. 264*, 19449–19457 (1989); (b) A. P. Jackson, M. P. Timmerman, C. R. Bagshaw, and C. C. Ashley, The kinetics of calcium binding to Fura-2 and Indo-1. *FEBS Lett. 216*, 35–39 (1987); (c) W. H. Wier, M. B. Cannell, J. R. Berlin, E. Marban, and W. J. Lederer, Cellular and subcellular heterogeneity of $[Ca^{2+}]_i$ in single heart cells revealed by Fura-2, *Science 235*, 325–328 (1987).
137. D. B. Mac Queen and K. S. Schanze, Cation-controlled photophysics in a Re(I) fluoroionophore, *J. Am. Chem. Soc. 113*, 6108–6110 (1991).
138. J. Bourson, J. Pouget, and B. Valeur, Ion-responsive fluorescent compounds. 4. Effect of cation binding on the photophysical properties of a coumarin linked to monoaza- and diazacrown ethers, *J. Phys. Chem. 97*, 4552–4557 (1993).
139. J. F. Letard and R. Lapouyade, to be published.
140. M. M. Martin, P. Plaza, N. Dai Hung, Y. H. Meyer, J. Bourson, and B. Valeur, Photoejection of cations from complexes with a crown-ether-linked merocyanine evidenced by ultrafast spectroscopy, *Chem. Phys. Lett. 202*, 425–430 (1993).
141. S. R. Adams, J. P. Y. Kao, and R. Y. Tsien, Biologically useful chelators that take up Ca^{2+} upon illumination. *J. Am. Chem. Soc. 111*, 7957–7968 (1989).
142. D. Braun, R. Lapouyade, and W. Rettig (in preparation).

143. W. Rettig, Photoinduced charge separation via twisted intramolecular charge transfer states, in *Topics in Current Chemistry, Vol. 169: Electron Transfer I* (J. Mattay, ed.), pp. 253–299, Springer Verlag, Berlin (1994).
144. W. Rettig, W. Majenz, R. Lapouyade, and M. Vogel, Adiabatic photochemistry with luminescent products, *J. Photochem. Photobiol. A: Chem. 65*, 95–110 (1992).
145. J. Dobkowski, W. Rettig, B. Paeplow, K. H. Koch, K. Müllen, R. Lapouyade, and Z. R. Grabowski, Intramolecular charge transfer states in symmetric biaryls. The unusual case of biperylenyl, *New J. Chem. 18*, 525–533 (1994).
146. R. Zahradnik and J. Fabian, Auf der Suche nach tieffarbigen organischen Verbindungen, *Angew. Chem. 101*, 693 (1989).
147. V. Bonačič-Koutecký, K. Schöffel, and J. Michl, Electronic states of cyclobutadiene heteroanalogues. Critical biradicaloids, *J. Am. Chem. Soc. 111*, 6140 (1989).
148. V. Kamat, S. Das, K. G. Thomas, and M. V. George, Photochemistry of squaraine dyes. 1. Excited singlet, triplet, and redox states of bis[4-(dimethylamino)phenyl]squaraine and bis[4-(dimethylamino)-2-hydroxyphenyl]squaraine. *J. Phys. Chem. 96*, 195–199 (1992).
149. K. Hamasaki, H. Ikeda, A. Nakamura, A. Ueno, F. Toda, I. Suzuki, and T. Osa, Fluorescence sensors of molecular recognition. Modified cyclodextrins capable of exhibiting guest-responsive twisted intramolecular charge transfer fluorescence, *J. Am. Chem. Soc. 115*, 5035–5040 (1993).

6

Red and Near-Infrared Fluorometry

Richard B. Thompson

6.1. Introduction

This is a review of practical aspects and methods for performing fluorescence measurements in the red and near-infrared (near-IR) regions (for our purposes, 600–1000 nanometers) of the spectrum. After summarizing the background and rationale for performing fluorometry in this wavelength region, we shall successively consider aspects of excitation sources; detectors and aspects of optical components; the major classes of compounds fluorescing in the IR; scattering and absorbance in the near-IR, with emphasis on tissue; and finally conclusions and future prospects. While recent results are highlighted, examples have been included for didactic value, rather than for a comprehensive overview. For in-depth treatment of instrumentation, the reader is referred to Birch.[1]

6.2. Background and Rationale

At first glance, performing fluorometry with emission and perhaps excitation in the IR seems counterintuitive. In particular, few detectors as good as common photomultiplier tubes are available in the IR, there is nothing like the broad palette of fluorescent probes in the ultraviolet (UV) and visible with which to work, and water absorbs significantly at certain wavelengths. However, performing fluorometry in the IR has attractive advantages for several applications. First, the relative rarity of fluorescent materials emitting at wavelengths longer than visible makes possible fluorescence measurements in matrices such as sea water, tissue, and serum which exhibit substantial interfering fluorescence at shorter wavelengths. Second, the ν^4 dependence of the intensity of scattered light means that many samples that appear murky in visible light, such as macroscopic tissue samples, are substantially transpar-

---

Richard B. Thompson • Department of Biological Chemistry, University of Maryland School of Medicine, Baltimore, Maryland 21201.

Topics in Fluorescence Spectroscopy, Volume 4: Probe Design and Chemical Sensing, edited by Joseph R. Lakowicz. Plenum Press, New York, 1994.

ent in the IR; this creates the prospect of many different *in vivo* fluorometric measurements. Third, the availability of laser diodes with all their advantages has created the prospect of applying fluorescence measurements to a range of new applications. Fourth, some IR fluorophores are in fact quite good, exhibiting very high detectability due to high absorbance ($> 10^5\ M^{-1}\ cm^{-1}$ at the long wave peak) and quantum yield. Finally, the high transparency of optical fiber in the IR suggests that fiber optic chemical sensors should be designed for operation in the IR as well. Each of these attributes will be discussed in more detail below.

6.3. Excitation Sources

One of the most compelling reasons for doing fluorescence measurements in the IR is the availability of light sources, particularly diode lasers. The wavelength characteristics of several light sources are indicated in Figure 6.1; we shall consider each in varying degrees of detail, with emphasis on their application in frequency-domain time-resolved measurements. Of course, any probe can be excited at shorter wavelengths, but IR fluorophores typically exhibit strong absorbance in a narrow longwave band, and absorb poorly at shorter wavelengths.

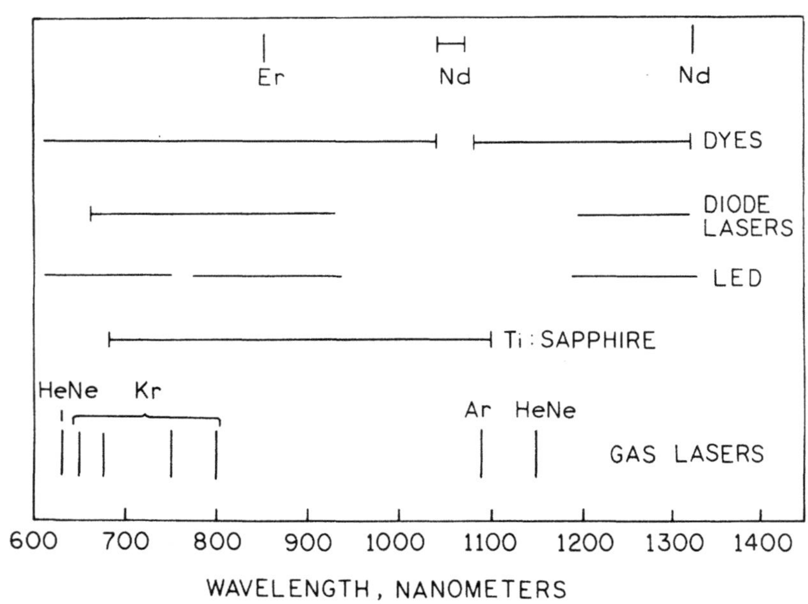

Figure 6.1. Wavelength coverage of commercially available excitation sources for near-infrared fluorometry.

Not depicted in Figure 6.1 are incandescent sources and arc lamps. Quite a large variety of these are evidently available, and most will have some IR output.[2] Note that the high-pressure xenon arc lamps commonly used in fluorometry have less output in the IR than the UV and shortwave visible; how much light is necessary depends on the experiment. Low-pressure gas-filled lamps customarily used for calibration or portable fluorometers may or may not have significant output in the IR; mercury lamps have little output beyond 580 nm, but krypton lamps with subnanosecond flashes and usable output at 752 nm are available commercially.[3] As is well known, lamps of all kinds are difficult to modulate directly for time-resolved fluorometry, but can be modulated using a Pockels cell with some success.[4] Although lamps are typically cheap, their polychromaticity, lack of collimation, extended source size, and heat production all require some accommodation in the instrument design. In view of the rather specialized, application-specific nature of IR fluorometry, the great flexibility of wavelength offered by the lamp seems less important.

6.3.1. Gas and Dye Lasers

Beyond the 633 nm line of HeNe, there are relatively few gas laser lines of interest. The krypton (Kr) lines exhibit only modest power and are inefficient besides: a large-frame ion laser that produces 25 W all-lines in the visible with argon produces only 6 W in the visible and infrared with krypton. Furthermore, such Kr lasers are not available commercially in a modelocked configuration, which is important for time-resolved measurements. While many other IR gas laser lines having low gain are known besides those depicted in Figure 6.1,[5] the likelihood of their being offered commercially is modest in view of the competition from diode lasers. Only rarely do the superior coherence length and monochromaticity of the gas lasers justify their higher cost compared with diode lasers.

Dye lasers[6–8] may be either laser- or flashlamp-pumped, and offer complete coverage of the near-IR spectral range using different dyes. Dye lasers synchronously pumped by modelocked neodymium:yttrium-aluminum-garnet (Nd:YAG) or argon ion lasers have become premier light sources for time-resolved fluorometry; they have been discussed in detail elsewhere in this series.[7] It is worth noting that many of the IR dyes are overtly less efficient and stable than the benchmark Rhodamine 6G. Synch-pumped dye lasers usually have a cavity dumper attachment which permits reduction of the repetition rate with little loss in average power; the pulse energy is correspondingly increased manyfold. Dye lasers pumped by flashlamp, excimer, or metal vapor lasers provide pulsed output with rather high energies. Unfortunately, the pulses from such lasers tend to be too prolonged, variable, and too infrequent for time-resolved applications, and these lasers are often expensive and temperamental as well.

6.3.2. Light-Emitting Diodes

Light-emitting diodes (LEDs) are very cheap, efficient, widely available at many wavelengths, and can be modulated to high frequencies relatively easily for phase fluorometry.[9] Their basic configuration and physics are rather similar to those of a laser diode (see below), and in fact laser diodes can be thought of as a subset of LEDs. Unfortunately, they are not monochromatic, they emit only modest powers (even superluminescent diodes), their output is typically uncollimated and is incoherent, and in some respects they must be handled with the same care as laser diodes. In addition, the color and intensity of their emission change overtly with temperature. For instance, a typical LED such as the Siemens LD-271 has a peak wavelength of 950 ± 20 nm with a bandwidth of more than 50 nm (Figure 6.2), and although it typically emits 12 mW of power at a forward current of 100 mA, the output half-angle is greater than 25 degrees (Figure 6.3). The specified "switching time" of this LED is 1 μsec to go from 10 to 90% intensity; others are as fast as 50 nsec. Even LEDs designed for the purpose are difficult to couple with optical fiber: the Siemens SFH 407 couples only 2% of its radiant output into a 200 μm core, step index optical fiber, and much less into 50 μm gradient index telecommunications fiber. However, the output of LEDs in a steady-state mode in the range where their output is a linear function of current can be more

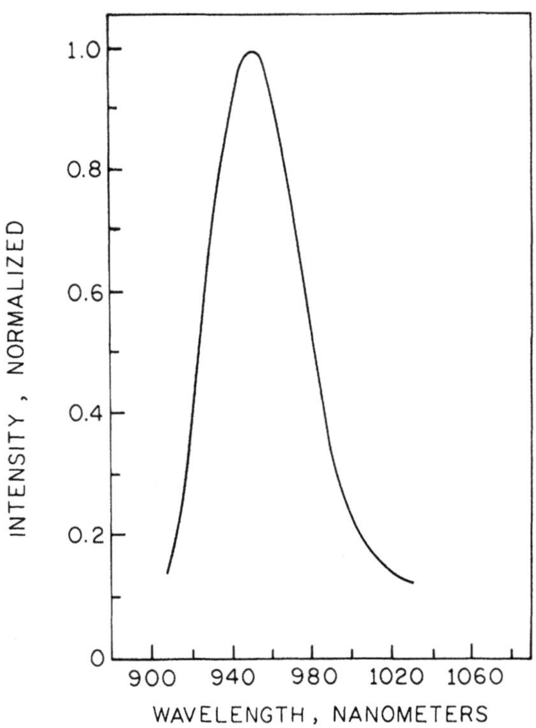

Figure 6.2. Wavelength-dependent output of Siemens LD-271 light-emitting diode. Redrawn from Siemens literature.

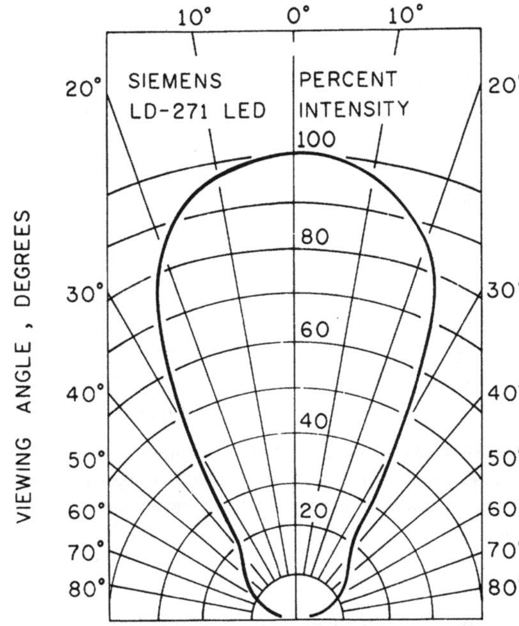

Figure 6.3. Angular distribution of light output from Siemens LD-271 light-emitting diode. Redrawn from Siemens literature.

precisely controlled than any other light source. In particular, Winefordner *et al.*(10) showed that fluorescence intensities could be measured with a precision of better than 0.002% using a photodiode detector. Such results suggest that ratiometric intensity probes such as the seminaphthofluoresceins (SNAFLs) and seminaphthorhodafluors (SNARFs),(11) Fura's, and Indo's,(12) and anisotropy measurements might be made over a much broader dynamic range. Despite these advantages, for most purposes anything an LED can do, a laser diode can do better due to its higher brightness, collimatibility, monochromaticity, and modulatability.

6.3.3. Titanium:Sapphire Lasers

The titanium:sapphire (Ti:sapphire) laser has recently emerged as an important source for IR fluorometry.(13, 14) It is powerful (up to 1.5 W continuous wave power at peak wavelengths), broadly tunable (up to 700–1050 nm, depending on pump strength (Figure 6.4)), solid state, mode-lockable, and can produce pulses of even femtosecond duration with surprising ease. The output can be frequency doubled or tripled using commercial devices with good efficiency, making the Ti:sapphire very versatile indeed. The broad tunability is also very efficient, since by comparison it may take a day to change dyes in a dye laser. Its major drawbacks are that refined, convenient versions remain relatively expensive, and it is best pumped with a relatively expensive

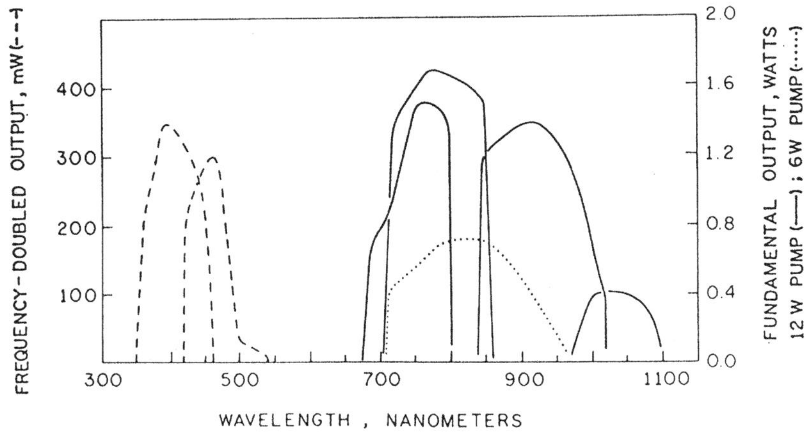

Figure 6.4. Wavelength distribution of output from a picosecond pulsed titanium:sapphire laser. Output at the fundamental with 6-watt visible multiline argon pump power (·····) and <12 watts pump power (—) are shown with the right-hand scale; the frequency-doubled output is shown (- - -), on the left-hand (expanded) scale.

multiwatt argon laser. For time-resolved fluorometry, there are other issues as well. For both frequency and time domain measurements[7, 15] it is usually necessary to lower the repetition rate of the laser from about 80 MHz by 20-fold or more. On synchronously pumped dye lasers, this is customarily done with a cavity dumper, which permits energy to build up in a particular pulse in the cavity before releasing it at longer intervals; effectively, a pulse train at 4 MHz has nearly the same average power as the fundamental 80 MHz pulse train. However, the nature of the semipassive modelocking used in most current modelocked Ti:sapphire lasers requires that a pulse picker be used, such that the average power of the beam is dramatically decreased. A second issue is the high precision synchronization required ($\leq$ 1 Hz) in frequency domain fluorometry between the frequency of the pulse repetition and the cross-correlation frequency;[16] Gratton and others have shown that this can be done with the Ti:sapphire (E. Gratton and P. Tso, personal communication), and at least one manufacturer has introduced a product for this purpose. A worthwhile feature for any Ti:sapphire laser is a nitrogen cavity purge. Losses due to water vapor-OH stretching overtones or oxygen absorbance can be significant at several wavelengths. Purging with prepurified grade dry nitrogen effectively increases power at these wavelengths, at modest cost. The Ti:sapphire laser promises to be a very important tool for research in IR fluorometry.

The broad tunability of the Ti:sapphire laser accounts for its ability to produce extremely short pulses when modelocked: several companies produce models with specified femtosecond pulsewidths. While shorter pulses are, to a first approximation, better, there are some additional tradeoffs to be borne in mind. First, with very few exceptions all common time-resolved fluorescence experiments may be carried out with picosecond pulse excitation, so the additional expense of femtosecond pulses would seem unjustified. Second, the specified efficiency of frequency doubling femtosecond pulses compared with picosecond pulses in commercial lasers is not as good as one would expect based on pulse duration. In particular, one manufacturer specifies 150 mW output of frequency-doubled picosecond pulses (< 2 psec) at 395 nm with a large frame pump, whereas the output is only 200 mW with the same pump power and 10-fold shorter pulses under otherwise similar conditions. The major reason for this is that the doubler (and the rest of the optics) are designed to prevent broadening the femtosecond pulses, and thus the doubling crystal interaction length is kept somewhat shorter than for the laser emitting picosecond pulses, which would otherwise be expected to double with much less efficiency. A third concern is that even simple optical components such as dielectric mirrors can cause pulse broadening; mirrors designed to minimize this have recently become commercially available.[18]

There are other concerns with the Ti:sapphire laser as well. The preferred pump laser is a multiline argon ion laser. In this case, more power is better not only because it provides more output from the Ti:sapphire, but also because it expands the tuning envelope significantly (Figure 6.4). Care should be taken when using pump lasers from different manufacturers than that of the Ti:sapphire to match as closely as possible pump beam size, divergence, and mode quality; some manufacturers do not guarantee output performance of their Ti:sapphires when pumped by anyone else's argon laser. Similarly, some manufacturers offer argon lasers with active beam stabilization. While this is not necessary *per se* for Ti:sapphire pumping, some users will find it convenient not to have to wait up to 2 h for the argon laser to reach a thermal steady state before beginning their experiment.

Modelocking Ti:sapphire lasers is done differently than for argon or Nd:YAG lasers. The passive, Kerr-lens modelocking approach used with some current Ti:sapphire lasers is more akin to classic modelocking techniques using saturable absorbers[6, 7] than the active mode locking technique (gain modulation with an electro-optic modulator) commonly used with argon and Nd:YAG lasers.[7] In particular, the gain of the cavity is not actively modulated; rather, in Kerr lens modelocking changes in the optical properties of the Ti:sapphire crystal itself as the beam passes through it defocus the beam for nonselected modes; modelocking is initiated by dithering the cavity length or gain at the outset. Other Ti:sapphire lasers are modelocked using regenerative schemes, which are like standard active AM modelocking except that the master oscillator frequency source is not an external synthesizer, but rather the laser cavity itself. A photodiode monitors the laser output and fluctuations in cavity length are fed back into the intracavity modulator as changes in modulation frequency.

From the standpoint of time domain (e.g., time-correlated single photon counting) experiments the method of modelocking is not too crucial as long as the pulse jitter is modest (some picoseconds), and the pulse intensity doesn't vary too much; if the time-to-amplitude converter is being started instead of stopped by the excitation pulse, it may be immaterial. From the standpoint of the frequency domain, however, the method of modelocking is an issue since the cross-correlation method[19] demands that a frequency offset of $\pm$ 1 Hz between the fundamental modulation frequency (i.e., the pulse repetition rate)[16, 20] and the signal applied to the last dynode of the PMT. Maintaining the offset is more important and usually more demanding than accurately producing a particular modulation frequency; if the modulation frequency varies 100 Hz (1 part per million), the effect on measured phase angles and modulations is negligible, but a 50 part per billion variation in the offset is unacceptable. Some workers[17] have used the Ti:sapphire itself as the master oscillator and divided down the 80–100 MHz repetition rate to 10 MHz to use as timing signal for a frequency synthesizer. However, the broad tuning range of the Ti:sapphire means that the cavity mode spacing is not the same at all wavelengths, and thus the characteristics of the laser cavity as a master oscillator in this respect are not wavelength-independent. Recently, synchronization devices that overcome this effect by active cavity-length adjustment have been introduced, but they have not yet been tested in phase fluorometry.

6.3.4. Diode Lasers

As we have stated above, one of the reasons for pursuing applications of IR fluorometry in the first place is the availability and virtues of diode lasers. Laser diodes are inexpensive ($15 apiece for the mass-produced lasers used in CD players, thousands of dollars for high-power or distributed feedback modelocked lasers used for high-speed telecommunications), durable (250,000 mean time between failures at room temperature, much less at elevated temperature), small and light (laser pens), rugged (laser gunsights withstand 1000 × g impacts), and efficient (tens of percent from wall plug to laser light). Laser diodes represent a relatively mature technology with many manufacturers worldwide, and a large ancillary technology base of experience, hardware, and test equipment. While it is beyond our scope to consider diode lasers in any great detail,[1, 6, 21] we will examine several features germane to IR fluorometry, especially time-resolved measurements in the frequency domain.

What are the characteristic features of laser diodes which are most important to us? Perhaps it is best to consider them in comparison to lasers that are more familiar to us, such as HeNe's or argon ion lasers. For instance, from an electrical point of view the diode laser is much different. The laser diode requires a few tens of milliamps current at a few volts potential to operate, compared with thousands of volts to initiate and hundreds of volts to maintain the discharge in a gas laser. The diode laser is much more efficient, exhibiting tens of percent conversion of the electrical energy to optical power; the corresponding figure for gas lasers is tenths of a percent. Thus diode lasers are easily battery-powered. Note that the diode laser is more fragile electrically: it can be destroyed by transient current surges or static discharges. The gas laser exhibits a relatively small (1 mm or so), well collimated (a few mrad), monochromatic (a few GHz without intracavity devices), and coherent beam. By comparison, the emission

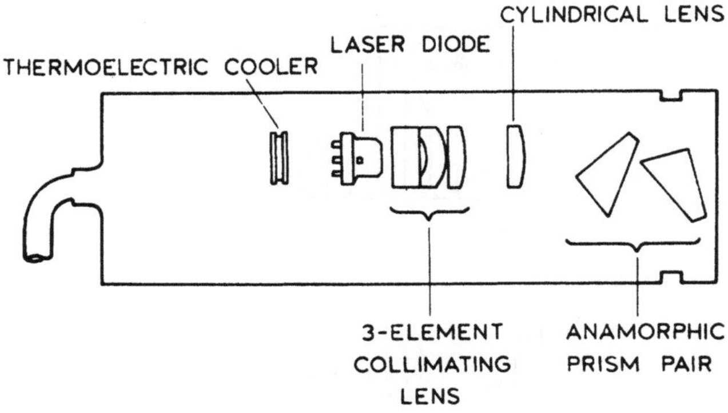

Figure 6.5. Diode laser head. The laser diode output passes successively through a three-element collimating lens, cylindrical lens, and anamorphic prism pair to become a fully collimated, circular, anastigmatic beam. Redrawn from Melles Griot product literature.

from a bare laser diode doesn't even seem like a laser beam: due to the cavity size and shape, it diverges dramatically and astigmatically, and is overtly less coherent. As a result, often the most expensive parts of a laser diode device for a particular application are the optics needed to collimate and circularize its beam into something useful. For example, in its current original equipment manufacturer lasers the cost of the laser is much overshadowed by the three-element collimating lens, cylindrical lens, and anamorphic prism pair (Figure 6.5) Melles Griot incorporates to get good beam quality. Simpler optical designs are available, with correspondingly lower performance and cost. Moreover, the output intensity, threshold current, mode structure, and wavelength of the emission all vary with temperature (Figures 6.6 and 6.7), such that temperature control at the least is always necessary. Laser diodes are customarily packaged with a

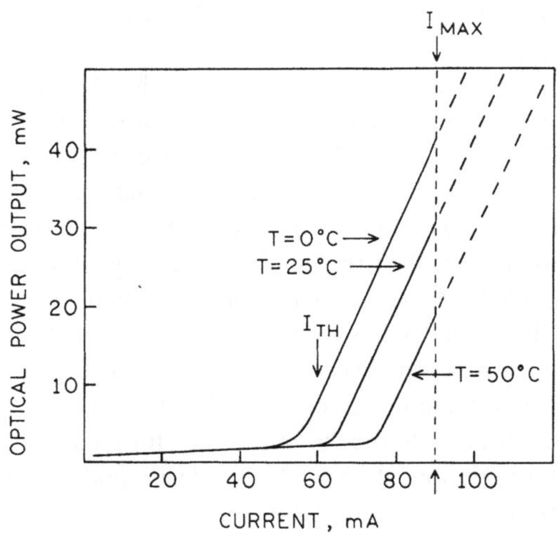

Figure 6.6. Laser diode intensity as a function of temperature and current. Threshold for this diode at 25 C is 60 mA and the maximum current is 90 mA. Redrawn from Melles Griot product literature.

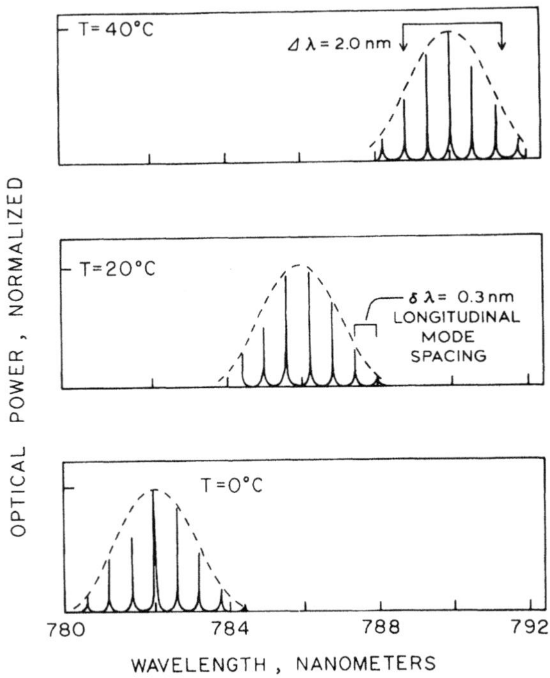

Figure 6.7. Temperature dependence of a hypothetical laser diode. Several modes are shown oscillating within the 2-nm bandwidth envelope of this laser; the mode spacing and position change somewhat with temperature as well. Note that some longer-wavelength lasers exhibit significant increases in the bandwidth with temperature.

separate photodiode mounted on the back of the laser diode to monitor the optical output. Thus while the laser diode itself is cheap, the additional cost of optics, a power supply and thermostat for the thermoelectric cooler, and circuitry to monitor the laser diode output using the built-in photodiode (perhaps with feedback to the power supply) can be manyfold greater than that of the diode. Nevertheless, laser diode systems are typically less expensive than "low end" air-cooled argon or helium–cadmium (HeCd) lasers.

To date, few time-resolved fluorescence measurements in either the time or frequency domain have been performed using laser diodes. This is doubtless due to the lack of useful fluorescent probes in this wavelength regime. Nevertheless, in the time domain Ishibashi[(22)] showed that very reproducible subnanosecond pulses could be readily obtained from laser diodes and used time-correlated single photon counting to measure the lifetime of a cyanine dye. Lakowicz *et al.*[(23)] showed that the harmonic content of a pulsed picosecond laser diode could be used to perform phase fluorometry. They found, however, that jitter in the time delay between the application of the current pulse to the diode and the onset of lasing began to degrade the results at high frequency. Minimizing this jitter is the object of a good deal of research inasmuch as it is a main contributor to the "bit error rate" for gigabyte per second digital telecommunications through optical fiber.[(24)] Thus laser diodes seem to be good sources for time domain fluorometry for all but the most demanding applications.

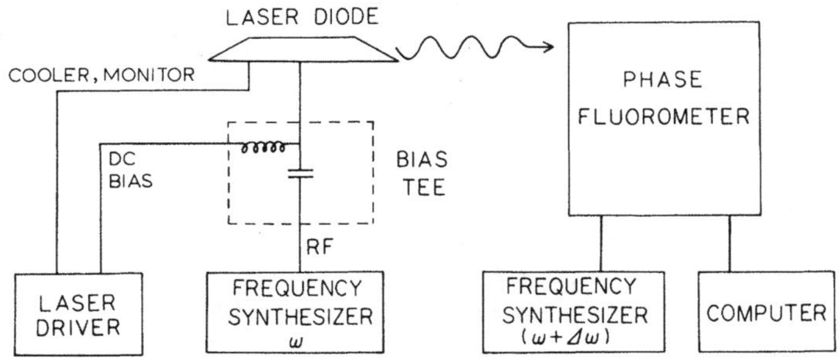

Figure 6.8. Incorporation of directly modulated laser diode into phase fluorometer. Reproduced from Ref. 25 with permission.

Continuous-wave laser diodes are also well suited for frequency-domain fluorescence lifetime measurements. We found that it was relatively straightforward to directly amplitude modulate the output of a laser diode and use it to perform phase fluorometry.[25] The apparatus for doing this is depicted in Figure 6.8 and is now available commercially from Melles Griot and others; the bias-tee permits an alternating current (AC) signal to be superimposed on the direct current (DC) which biases the laser diode above threshold; the rest of the apparatus is just a standard phase fluorometer. The main caveat associated with this approach is that as the laser diode is run at higher current, it requires more radio frequency power to modulate the laser output to a comparable extent. The phenomenon is depicted in Figure 6.9 for a Sharp LT024 modulated at 10 MHz: at just barely over threshold (bias = 50 mA, laser output = 7 mW), only 1 mW RF power is required to achieve 40% modulation; but at higher power (DC = 70 mA, laser power = 19.3 mW), more than 10-fold higher radio

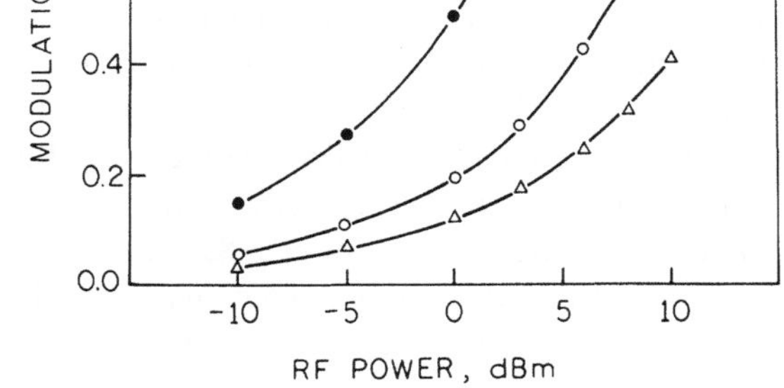

Figure 6.9. Effective modulation of laser output as a function of radio frequency power applied to a Sharp LT024 laser diode at 30 MHz. Data are shown for DC bias of 50 mA (●, 7 mW laser output), 60 mA (○, 14 mW), and 70 mA (△, 19 mW). Reproduced from Ref. 25 with permission.

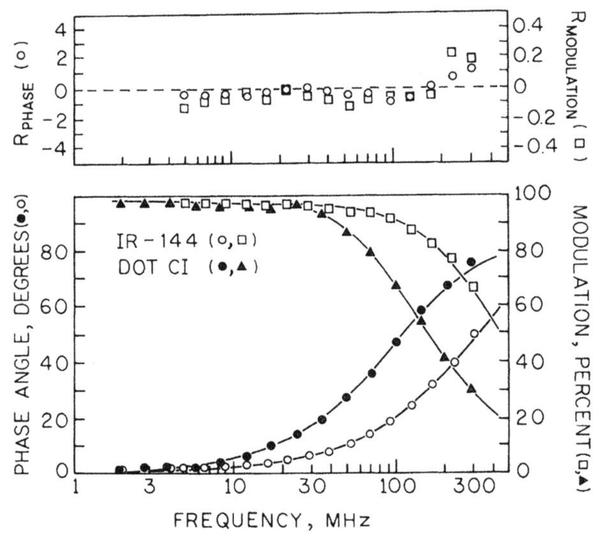

Figure 6.10 Frequency-dependent phase shifts and demodulations (bottom panel) for IR-144 (open symbols) and DOTCI (solid symbols), together with the best single exponential fit to each data set. The residuals for the IR-144 data set are depicted in the *upper panel*. Reproduced from Ref. 25 with permission.

frequency (RF) power is needed to achieve the same degree of modulation. Moreover, as the RF power is increased, some distortion may appear in the optical output. Despite these caveats, excellent results (Figure 6.10) are obtainable up to the approximately 400 MHz frequency limit of the photomultiplier tubes used in that study. It seems likely that the approach is probably usable to at least 1 GHz or better; verification of this awaits the introduction of microchannel plate photomultiplier tubes (MCP-PMTs) with suitable IR photocathodes (see below).

6.3.5. External Modulation

Lacking a modelocked laser or a directly modulated diode laser, it is necessary to externally modulate a CW laser or lamp in order to perform phase fluorometry. The commonly used Pockels cell modulators built with potassium dihydrogen phosphate (KDP) are satisfactory beyond 1000 nm, as are the classical Debye–Sears acousto-optic modulators(19) and TeO_2 acousto-optic modulators.(26) Recently, traveling wave singlemode waveguide modulators fabricated in $LiNbO_3$ have been incorporated into a variety of integrated optic devices (Figure 6.11).(27) These devices are commercially available for wavelengths as short as 830 nm and frequencies in the GHz regime. They have modest power consumption and may provide better modulation of a CW laser diode than directly modulating the diode, especially for peak output. While $LiNbO_3$

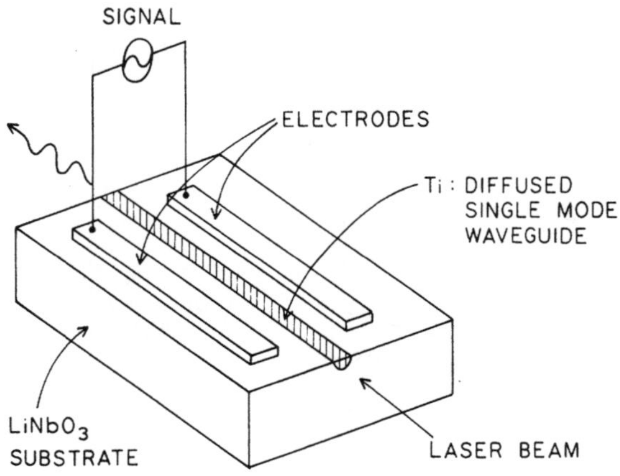

Figure 6.11. Traveling wave single-mode $LiNbO_3$ electro-optic modulator.

does not attenuate too badly at shorter wavelengths, the requirement that the device waveguide be singlemode and thus have a small cross-sectional area means that only modest powers (10 mW or so) can be accommodated at visible wavelengths without photorefraction changing the waveguide parameters. However, they are likely to be a very good choice for Ti:sapphire lasers in the IR that are not modelocked, or diode lasers that will otherwise chirp (shift wavelength) as they are modulated, particularly well above threshold.

6.4. Detectors and Optics

From the standpoint of detectors and optics, some recent developments have made IR fluorometry much easier than in the past. We shall discuss detectors first, emphasizing frequency domain applications, then the other optical components. In the IR spectral region under discussion, the detectors of interest all work by variations of the photoelectric effect, whereby the incident photon creates either a small current or a small voltage. As one moves into the mid-IR, the photon energies begin to approach kT at room temperature, and it becomes necessary to cool the detector to reduce noise. At far IR wavelengths, the detection mechanisms are more thermal in nature, and one is measuring a temperature change. From our point of view, there are basically three important detectors: the photomultiplier tube, the charge-coupled device, and the photodiode.

6.4.1. Photomultiplier Tubes

The principles and operation of PMTs have been described at length elsewhere,[28, 29] so we shall concern ourselves primarily with spectral sensitivity and frequency domain

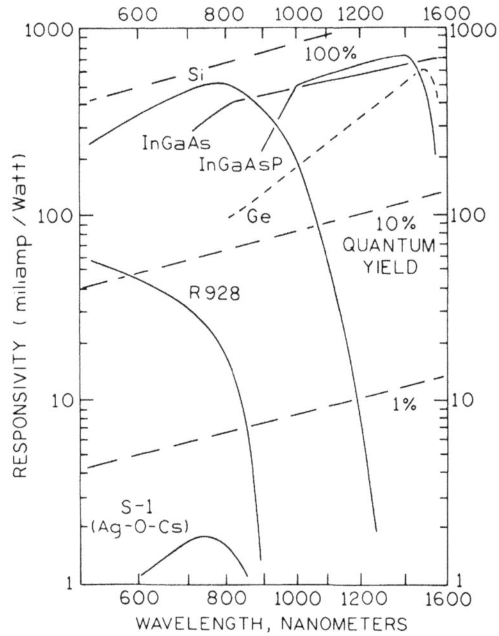

Figure 6.12. Photocathode response for various materials as a function of wavelength.

measurements. Of course, PMTs are essentially constructed of two components, the photocathode and the electron multiplier, which may in turn be a dynode chain or a set of microchannel plates.[7] The business end is the photocathode, which emits electrons into the electron multiplier when struck by a photon. Its propensity for doing so is a steep function of the wavelength of the photon; representative photocathode responses are depicted are depicted in Figure 6.12. For many years the best IR-sensitive photocathode material was silver oxide/cesium, also known as S-1. Its very modest quantum yield may explain why infrared fluorometry was little studied until recently, due to a lack of sensitive detectors. The multialkali photocathode of the Hamamatsu R928 PMT is an improvement on older "Bialkali" photocathodes, with 10-fold better response than S-1 at its peak, and usable response past 900 nm. The R2658 from Hamamatsu has an indium–gallium–arsenide (cesium) InGaAs (Cs) photocathode with improved performance out to 1010 nm, and 100-fold lower dark current than the S-1; its gain is somewhat more modest than that of the R928, a nine-stage tube. Dark current can be further reduced about 100-fold by cooling the PMT to –20 °C or so, and thus is particularly important at long wavelengths where responsivity is poor. It should be recalled that the electron multiplier portion of the PMT is basically a superb amplifier, with at least 60 dB gain and extremely low noise. For most studies to 900 nm or so, the R928 is still hard to beat as a detector.

The electrons in a PMT take some time to traverse the dynode chain to the anode (tens of nanoseconds), and moreover do not arrive all at the same time; the specified rise time caused by the "transit time spread" of the R928 is of the order of a few nanoseconds. This phenomenon ultimately limits the time response of the PMT. In phase fluorometry this means that the frequency response of the R928 drops dramatically above 300 MHz, depending somewhat on the wiring of the dynode chain. While this tube is not widely used for time-correlated single photon counting, the most common time domain method, we might anticipate that one could measure lifetimes of the order of 100 psec with reasonable accuracy using either the time or the frequency domain. To look at much faster processes the usual dynode chain must be replaced with a microchannel plate electron multiplier, i.e., a microchannel plate photomultiplier tube.[(7)] The fastest available MCP-PMT's have frequency response beyond 10 GHz in the frequency domain, and are capable of resolving lifetimes in the high femtosecond regime, which previously could only be observed with streak cameras[(30)] or cumbersome upconversion techniques. These devices are discussed elsewhere in this volume; for our purposes it is only important to point out that recently MCP-PMTs with multialkali photocathodes sensitive to light of 800 nm are now available (Hamamatsu R3809U). Note that while the circuitry for performing cross-correlation with a standard PMT is straightforward,[(31)] for the higher frequency range achievable with MCP-PMT's, the cross-correlation can be internal[(32)] or performed by external mixing.[(33)] For GHz frequencies a significant amount of complex and relatively costly microwave circuitry is required, and the components tend not to be as broadband.

6.4.2. Photodiodes and Avalanche Photodiodes

Other detectors important in the IR regime include photodiodes, avalanche photodiodes, and charge-coupled devices, all of which are discussed elsewhere in some detail.[(1, 34)] Of course silicon (Si) photodiodes provide excellent responsivity through most of the wavelength range of interest, with InGaAs and germanium (Ge) detectors covering the rest. It should be noted that even though these semiconductor detectors have relatively high quantum yields, they are also somewhat noisy at ambient temperatures (particularly Ge) and their outputs require substantial amplification to provide a usable signal. For instance, a picowatt of light impinging on the active area of a photodiode might generate nearly a picoamp of current, comparable to the noise output of the diode (over all frequencies); one might anticipate amplifying such a signal one million-fold (60 dB) to get a measurable signal. A PMT provides 60 dB of gain in its dynode chain, with very little noise; it is for this reason that PMTs are usually better detectors than PIN diodes in the visible wavelengths. To minimize the noise contributed by the photodiode itself (expressed as noise-equivalent-power (NEP), the optical power that produces a signal strength equal to the noise), three expedients are common. First, the size of the photodiode active area is minimized as much as possible, since the noise is proportional to it; areas of less than 1 mm^2 are common to achieve subpicowatt noise levels. Such small areas are also necessary for the highest frequency response: suitably designed commercial PIN diodes respond up to 60 GHz.[35] Note that a small target requires good focusing and precise positioning of the optical input beam. Second, the semiconductor detector is often packaged together with a small amplifier in a "can," which protects the photodiode signal from electrical noise until it can be amplified. Third, and perhaps most important, photodiodes are often operated with very narrowband signals and synchronous detection (using a lock-in amplifier) that discriminates very well against the bulk of the noise, which is broadband and predominantly at low (<200 Hz) frequencies. For applications not requiring great sensitivity but high frequency response, photodiodes remain the detector of choice.

Avalanche photodiodes (APDs) are a subset of photodiodes designed to be operated in a "Geiger mode" where the p-n junction is biased excessively such that electron-hole pairs created by impinging photons create additional pairs in an avalanche effect. Significant gain (>20 dB) is created by this effect, such that some current devices may be operated as photon counters. APDs also have a very high frequency response: beyond 5 GHz. Since no PMT has responsivity and high frequency response in the 1300 and 1500 nm windows for fiber optic telecommunications, APDs are the detectors of choice for that application. These APDs are made from InGaAs and are the result of intensive development. Berndt[36] has demonstrated the use of APDs in phase fluorometry with cross-correlation detection; note that APDs in this role have better frequency response, but are less sensitive and no less expensive than a conventional PMT.

6.4.3. Infrared Optics

Finally, there are some additional practical aspects to performing fluorometry in the infrared. First, IR light is difficult, but not always impossible to see. Even weak emissions at 700 nm can be seen in a darkened room, but a 790-nm laser beam is difficult to see even under ideal conditions. Safety is thus more an issue with IR laser beams than visible ones, and protective eyewear is more necessary. For routine use IR indicator cards (Quantex, 2 Research Ct., Rockville, Maryland 20850; and Eastman Kodak, 343 State St., Rochester, New York 14652-3512) and an IR viewer (FJW Optical Systems, 629 S. Vermont St., Palatine, Illinois 60067-9904; and Electrophysics Corp., 373E Rt. 46 West, Fairfield, New Jersey 07004) are virtually indispensable. The transmittivity, refractive indices, and especially reflectivity of certain optics can differ significantly in the IR. The enhanced aluminum mirrors commonly used in some fluorometers have overtly poorer reflectivity beyond the visible, as do typical broadband visible dielectric coatings (Figure 6.13). However, broadband mirrors designed for use with laser diodes, along with gold-coated mirrors exhibit much better performance in the infrared. Because of the difference in refractive indices with wavelength (dispersion) of optical glasses as well as fused silica, optical trains aligned in the visible may not be aligned as well in the infrared, particularly for imaging. Note that because

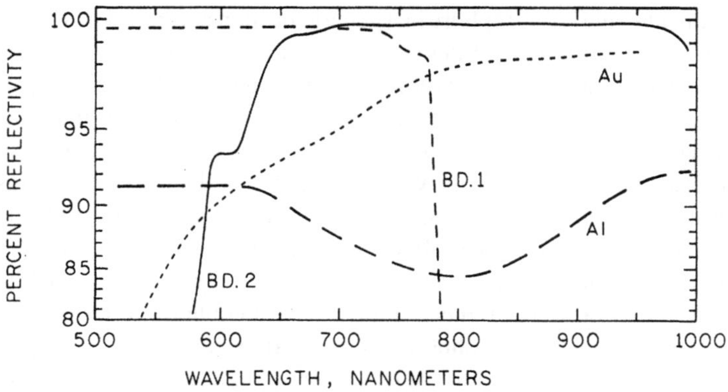

Figure 6.13. Mirror reflectivity as a function of wavelength for two multilayer dielectric reflectors optimized for visible (BD.1, - - -) and near-IR use (BD.2, ___), as well as protected aluminum (Al, __ __) and gold coatings (Au, ····). Redrawn from Newport Corporation product literature.

of the lower resolution and scattering of the long-wavelength light, optical components need not be so tightly specified in terms of figure, mounting, or scratch and dig. Where molded glass or plastic optical components might be unacceptable in the visible, in the infrared they might be perfectly adequate. Most fluorometers have quartz or synthetic fused silica optical components because of their broadband transmission and low photoluminescence, particularly with UV excitation. An IR fluorometer might be made with glass or plastic components with no loss in performance, but at much lower cost.

6.5. Infrared Fluorophores

Compared to the UV and visible regions of the spectrum, only a small fraction of known fluorescent organic and organometallic molecules emit in the IR. In the UV and visible regions, many molecules having a vast array of different characteristics useful in answering research questions, or in more practical applications are known. For instance, the structure and fluorescence properties of diphenylhexatriene make it useful for monitoring fluorescence anisotropy,[37] the pH-dependent fluorescence of hydroxy-pyrenetrisulfonate makes it useful in pH sensors,[38] and the solvent-sensitive fluorescence of the anilinonaphthalenesulfonates has made them useful in many biophysical studies.[39, 41] By comparison, the palette in the IR is much less versatile. While this review cannot discuss all known IR fluorophores, we can briefly examine the properties of the most important groups. First, we will discuss the generic properties of infrared fluorophores, then examine four classes of IR fluorophores in more depth.

If we consider an IR fluorophore in a generic fashion compared to a shorter-wavelength fluorophore, some generalities emerge. Fluorophores tend to be relatively rigid, electron-rich molecules such as aromatic compounds so that (simplistically) the molecule can resonantly be excited by a photon, and then retain that energy long enough to reemit it as a photon. For this to happen at UV or visible wavelengths requires a relatively small molecule with perhaps two or three aromatic rings. For the absorption and emission to occur at longer wavelengths, the quantum-mechanical particle must reside in a larger box: effectively, there must usually be a rigid, extended set of conjugated bonds or fused aromatic rings. Roughly speaking, such molecules are relatively large for non-polymers, often poorly soluble even in organic solvents, and typically easily oxidized or photooxidized. Furthermore, to a first approximation the emissive rate is proportional to the oscillator strength and thus the wavelength of emission.(42) For typical UV and visible fluorophores, fluorescence lifetimes typically range from 1 to 10 nsec, with several examples known which exhibit lifetimes of 20 nsec, or longer. By comparison, relatively few infrared fluorophores are known with lifetimes longer than 2 nsec, even when quenchers have been scrupulously removed.(25) The above are of course generalizations, and ultimately exceptions are likely to be found.

6.5.1. Cyanines

The family of dyes termed the cyanines (after their blue colors) are by far the most highly developed and flexible IR dyes, and are the most widely used. Cyanines were first synthesized in the 1800s in Germany by simple procedures, and it was shortly thereafter discovered that dyes such as cryptocyanine could act as sensitizers for photographic film.(43–46) This fueled a substantial development effort by companies such as Eastman Kodak and led to the rich chemistry now available. The basic structure of cyanines is depicted in Figure 6.14, together with a synthetic route taken by Waggoner and his colleagues: two equivalents of a nitrogenous base such as a trimethylindolenine derivative are condensed with a conjugated alkene-*bis*-aldehyde equivalent.(47) Several hundred cyanines are now known to the art, differing in the structures of their nitrogenous bases and the length of the alkenyl chain connecting them. The choice of base and chain length determines the spectroscopic properties of the fluorophore to a large degree: the homologous series of reactive *bis*-trimethylindole cyanines with three, five, and seven carbons in their alkyl chains emit at 590, 669, and 777 nm, respectively (Figure 6.15).(47–49) Thus chemists were able early on to design cyanines for particular wavelength ranges, and cyanines are known with up to seven double bonds in the alkyl chain, and emission to 1260 nm in the IR.(8)

Cyanines have been widely used as laser dyes, and as saturable absorbers in modelocked and Q-switched laser systems.(8, 50) The propensity of most cyanines to photooxidize which makes them useful in photographic film and as saturable absorbers makes them less than desirable as fluorophores in other applications. The use of

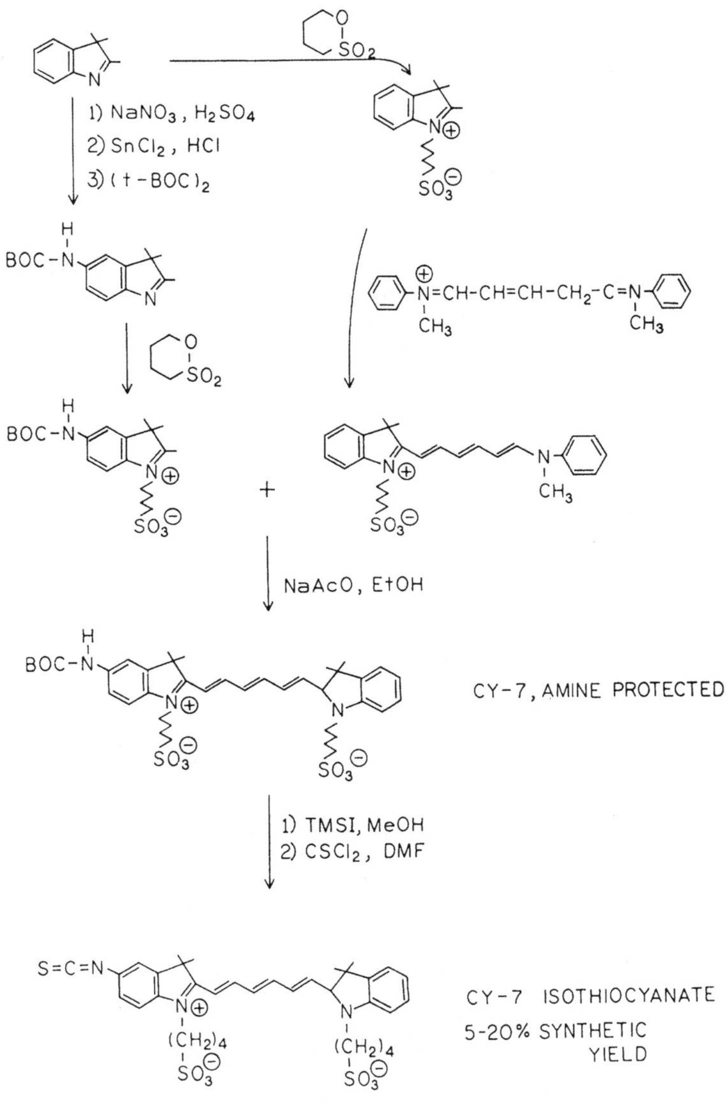

Figure 6.14. Synthesis of isothiocyanatocyanine dye for protein derivatization; redrawn from Ref. 47.

cyanines as laser dyes probably is due to their availability, high extinction, and adequate quantum yields, rather than their other characteristics. In particular, their Stokes' shifts are usually quite small, resulting in a limited dye laser tuning range. Furthermore, some cyanines are known to form non-fluorescent "J-aggregates,"(51) and undergo excited state reactions such as photoisomerizations. Cyanines often

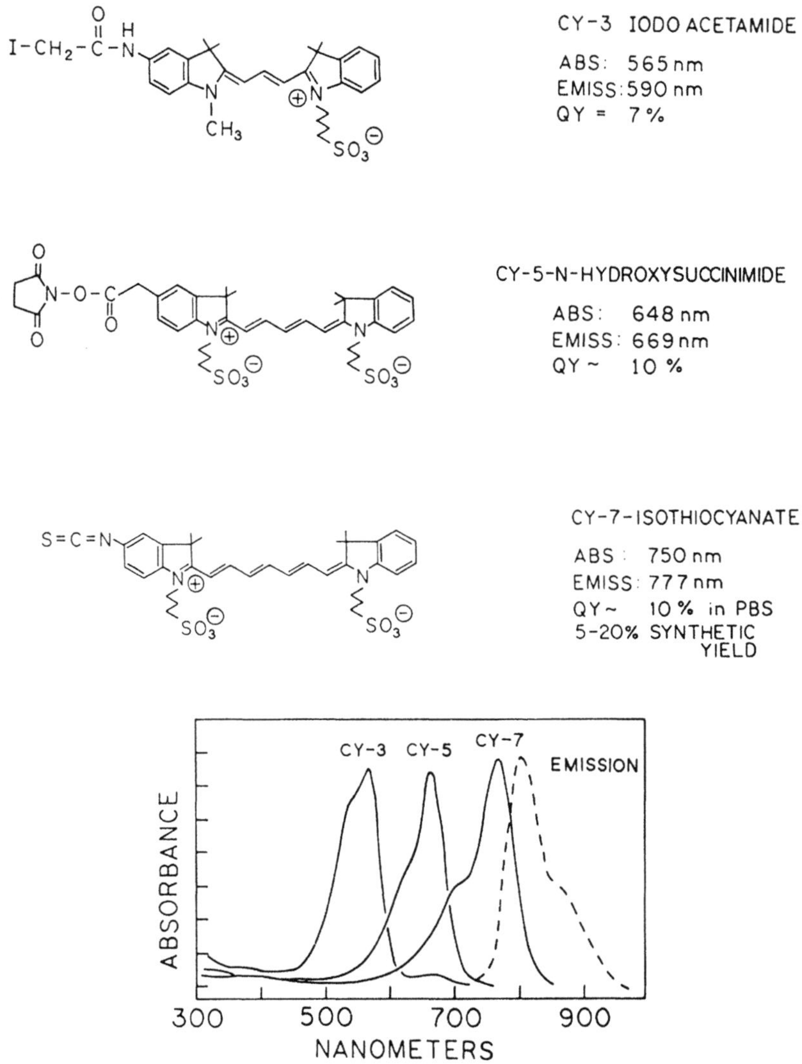

Figure 6.15. Waggoner reactive cyanines and their absorbance spectra.

exhibit greater fluorescence in nonpolar solvents, but this may be due in part to their propensity to aggregate in aqueous solutions. Most cyanines do not have excited states of polarity differing much from that of the ground state, and thus exhibit very modest solvent-dependent shifts in emission wavelength. Some more recently developed laser dyes[52] have larger Stokes' shift and probably improved stability, their emission may also be more solvent sensitive. Recently, Terpetschnig and Lakowicz[53] described the synthesis of squarylium cyanines, which seem to exhibit improved photostability.

Relatively few IR cyanine dyes have been synthesized for biochemical/biophysical applications. We shall mention only in passing the *bis*-alkylindocarbocyanine family of dyes used in fluorescence recovery after photobleaching (FRAP) microscopy studies, as well as potential-sensitive dyes, since the derivatives commonly used absorb and emit in the visible.[(54)] The former fluorophores are optimized for photobleaching studies, and thus are designed not to be photostable. The recently developed Toto™ and Yoyo™ indocarbocyanine derivatives[(54)] are cyanine nucleic acid stains which offer improved performance in comparison to the cationic intercalators such as ethidium bromide and the acridines, but the cyanines are relatively new and there is little data on them as yet. Probably the most interesting IR cyanine dyes from our point of view are the covalent derivatives synthesized by Waggoner *et al.*[(47–49)] and commercially available from Research Organics (4353 East 49th St., Cleveland, Ohio 44125-1083); see Figure 6.15. These reactive fluorophores have emission peaks at 550, 650, or 750 nm for CY-3, CY-5 and CY-7, respectively, behaving spectrally like the corresponding unsubstituted *bis*-(trimethylindolenine)alk(*n*-)enes. Waggoner has also described the synthesis of iodoacetamidyl- and isothiocyanato-derivatives as well as *N*-hydroxysuccinimidyl cyanines, but at present these seem not to be commercially available. Again, these molecules are relatively new and much more experience is necessary before they can be compared with fluorophores such as fluorescein or the rhodamines for applications such as microscopy, cell sorting, or fluorescence immunoassay. Recently, a fluorescence-based DNA sequencer has appeared (LiCor, Inc., Lincoln, Nebraska) which utilizes a new covalent IR cyanine dye and a laser diode for excitation. Finally, it should be noted that one cyanine dye, Indocyanine Green, is currently approved for *in vivo* use in humans as a probe of liver function and blood volume determination;[(55)] thus the dye may serve as a "springboard" for through-skin *in vivo* fluorescence experiments.

The styryl series of laser dyes are structurally similar to the cyanines, except that one of the heterocyclic bases has been replaced by a styryl moiety. They are more efficient laser dyes than the corresponding cyanines because they exhibit larger Stokes' shifts, but otherwise share most of the characteristics of the cyanines. There seems to be little development of styryl dyes in the infrared underway as biophysical probes.

In summary, the cyanines are currently the most popular IR dyes due to their well-known, flexible chemistry, but they have significant drawbacks which may be less of a problem with another variety of dye.

6.5.2. Oxazines

The second major category of IR fluorescent dyes are the oxazines, as exemplified by the laser dye Oxazine 1 and Nile Blue (Figure 6.16). Probably the major advantage of the oxazines is their photochemical stability, which is typically much better than that of the cyanines or rhodamines.[(52)] They are synthesized by condensation of the appropriate *p*-nitrosoaniline derivatives with the corresponding phenol.[(45)] These dyes

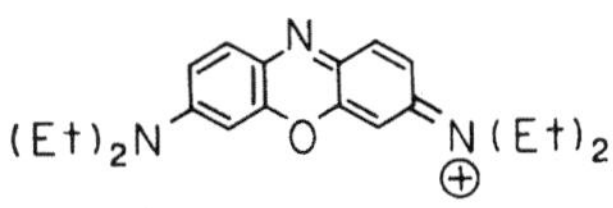

OXAZINE 1
EXC=643 nm, EM=658 nm

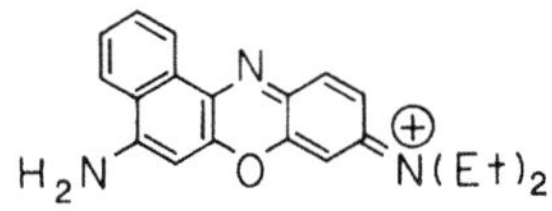

NILE BLUE
EXC=638 nm, EM=660nm

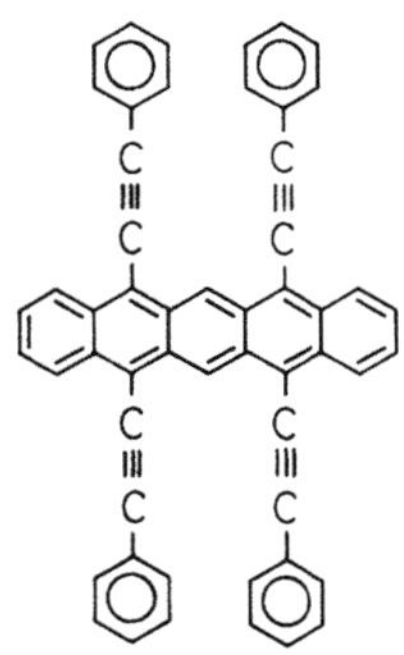

5,7,12,14-TETRAKIS (PHENYLETHYNYL) PENTACENE
EXC = ? EM=740 nm

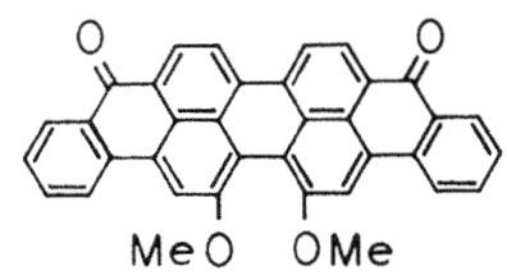

16,17-DIMETHOXY VIOLANTHRONE
EXC=660nm EM=713 nm

Figure 6.16. Oxazine and polynuclear aromatic hydrocarbon infrared fluorophores.

have relatively modest Stokes' shifts and are not known to absorb and emit very far to the red; Oxazine 1 emission peaks at about 660 nanometers. The phenoxazine Nile Blue is often contaminated by the corresponding oxazone Nile Red, which is readily formed by weak acid hydrolysis of the former;[57] it can be seen in the structure of Nile Blue that the unsubstituted amine is sterically hindered by the ring hydrogen, but the corresponding oxygen in Nile Red is less so. Nile Red exhibits solvent sensitivity in its emission, but it absorbs and emits at much shorter wavelengths than Nile Blue.[54] The oxazines are relatively compact compared to cyanines with similar spectral properties (compare Oxazine 1 with diethylthiadithiacarbocyanine iodide (DTDCI)), and may be easier to solubilize. While there has been little development of oxazine probes apart from Nile Blue as a lipid and membrane probe, they may provide fertile ground for future development.

6.5.3. Polynuclear Aromatic Hydrocarbons

The fused-ring or polynuclear aromatic hydrocarbons (PAHs) represent a relatively diverse group of fluorophores; some examples are depicted in Figure 6.16, and others are compiled.[50, 58, 59] These fluorophores are relatively undeveloped compared to the cyanines, in that many fewer syntheses are known, especially of substituted derivatives. Some ester and ether derivatives of the violanthrones have been synthesized for use in IR-emitting peroxyoxalate chemiluminescence formulations,[58, 59] but they seem unremarkable. Unlike the cyanines, which have strong market forces driving synthetic development, PAHs are of low commercial interest and have spawned little synthetic development. In particular, many of the PAHs are synthesized by multistep processes from nontrivial starting materials such as anthraquinones, so modifying syntheses to make a variety of derivatives may be difficult; compare the cyanines or oxazines. While some PAHs such as perylene, pyrene, and coronene are very useful fluorophores in the UV-visible range of the spectrum, they have been little use in the IR. Mainly this is because IR-emitting PAHs must be relatively large, rigid molecules which can be difficult to solubilize. For instance, 5,7,12,14-tetrakis(phenylethynyl)pentacene (TPEP) is soluble only to the extent of 10^{-4} molar in organic solvents, even in trichlorobenzene. Some of these fluorophores have good quantum yields, with 16,17-dimethoxyviolanthrone yielding 28% and 1.9, 5.10-di(perinaphthalene) anthracene (DPNA) yielding 51%, both in organic solvents. For many of these compounds it is difficult to measure their quantum yields in aqueous solvents because they are so insoluble. Note that sulfonating PAHs such as rubrene to impart water solubility sometimes dramatically alters their fluorescence.[60] Finally, some PAHs such as benzo[a]pyrene are well-known carcinogens, suggesting that structurally analogous PAHs with epoxidizable "bay" moieties should be handled with appropriate precautions.

6.5.4. Phthalocyanines

The phthalocyanines, naphthalocyanines, and certain of their metal derivatives (Figure 6.17) are infrared fluorophores.[61–64] As a class, they are exceptionally stable compounds, with copper (Cu) phthalocyanine (not a fluorophore) remaining intact above 300 °C in air. First used for textile dyeing in the last century and still widely used, there is a rich chemistry of phthalocyanines. Most derivatives can be made by prolonged heating of a phthalimide or phthalic acid derivative with a metal in powder or salt form at elevated temperature. Several derivatives absorb in the near-IR, and either fluoresce or phosphoresce. The electronic transitions of phthalocyanines are complex and have been extensively studied, at least in part because the symmetry of the molecule makes theoretical calculations of its spectroscopic behavior more tractable. Unsubstituted phthalocyanines and naphthalocyanines are, as a class, very insoluble in solvents other than, for instance, nitrobenzene. Sulfonated phthalocyanines are water soluble and exhibit spectra comparable to the parent derivative. Photoluminescent phthalocyanines (Pcs) include SiPc, ZnPc, and Pc itself. These compounds have been little used for practical infrared fluorometry to date; however, Diatron Corporation in California has recently introduced a new class of Pc derivatives with reputedly much improved properties in this regard.[65]

6.5.5. Other Infrared Fluorophores

Other types of compounds are also fluorescent in the IR and near-IR; in most of these cases only one or a few examples are known, and/or the compound is proprietary or incompletely characterized. Among these are certain transition metal:polypyridyl complexes. Some containing osmium emit from their metal:ligand charge transfer bands at wavelengths somewhat longer than the better known ruthenium tris(bipyridyl) derivatives[66–69] (Figure 6.17). Solar energy and immunoassay research provided the impetus for development of the ruthenium complexes; the osmium complexes may be sufficiently analogous to be comparably useful. Some anthraquinone dyes such as Calcocid Quinizol Blue BP (American Cyanamide)[58] have been described as fluorescent in the IR, but the emissions of these materials are mostly incompletely characterized. One recently developed series of probes that shows promise are the Bodipy™ derivatives produced by Molecular Probes, Eugene, Oregon[54, 70] (Figure 6.17). While the longest-wavelength derivatives emit just below 700 nm, there seems

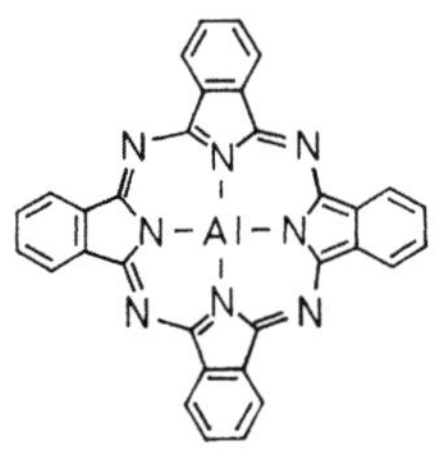

Al PHTHALOCYANINE
EXC=662nm EM=680nm

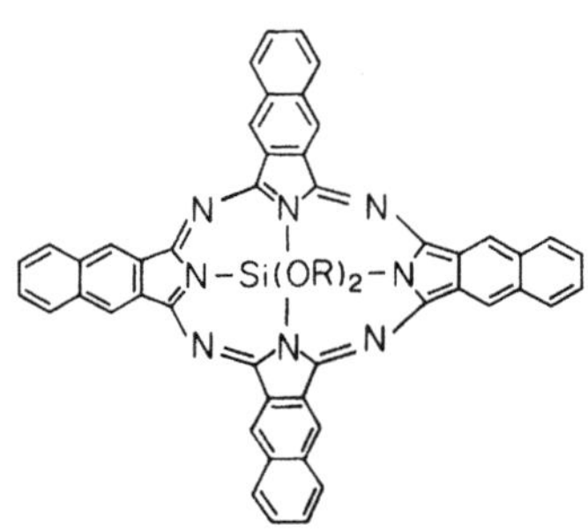

bis (triALKYLSILOXY) SILICON NAPHTHALOCYANINE
EXC = 772 nm , EM = 780 nm

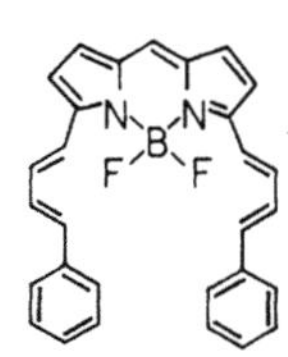

BODIPY™ 665/676
EXC=665nm , EM=676 nm

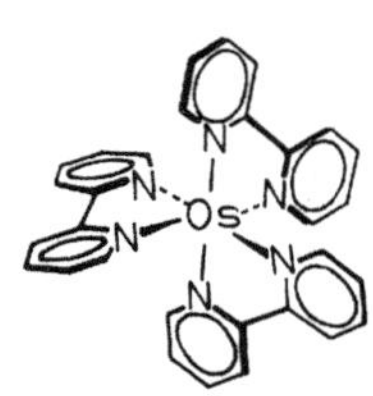

TRIS (BIPYRIDYL) OSMIUM
EXC = 650nm , EM = 700nm

Figure 6.17. Phthalocyanine, naphthalocyanine, Bodipy™, and Tris(bipyridyl) osmium infrared fluorophores.

to be the possibility of extending this to longer wavelengths. Two advantages of these probes are the availability of different derivatives, including haloacetyl, succinimidyl, and carboxylate derivatives, and improved photostability. It may be that in the future Bodipy™ derivatives may offer the flexibility of the cyanines with fewer of the drawbacks.

6.6. Scattering, Absorbance, and Interfering Fluorescence

An important reason for performing fluorometry in the IR is the greatly reduced influence of scattering and, in some instances, of optical absorbance in this wavelength regime compared with the visible or UV. In the case of human or animal tissues, scattering and absorbance are sufficiently reduced that one can contemplate performing not only fluorometry but also so-called "photon migration" studies.[71] While photon migration studies experimentally are similar to time-resolved fluorometry experiments, they lie somewhat outside our scope; however, optical attenuation measurements performed with such studies in mind are quite germane to our discussion. Similarly, detailed treatment of scattering phenomena is beyond our scope; we shall make a brief and empirical overview.

The importance of IR fluorometry in this respect lies in our ability to perform fluorometry on certain samples as if they were transparent, rather than in a front face mode at visible or ultraviolet wavelengths. Thus a laser beam at 415 nm might be expected to only penetrate a few microns into the skin of a subject due to strong scattering and absorbance by hemoglobin (see below), whereas an 830-nm light beam would penetrate some centimeters due to 16-fold lower scattering and much reduced absorbance as well. We can imagine interrogating implanted fluorophores through the skin for a variety of diagnostic purposes, such as blood glucose monitoring.[72] Other materials are likely to be more transparent in the near-IR for the same reasons, such as plastics, glass-fiber composites, paper, cloth, turbid water, paint, and some building materials. This in turn suggests other applications beyond the biomedical, such as quality control of materials and structures, security, and process control.

Relatively few data exist on the attenuation of various bodily tissues due to scattering and absorbance; the data in Figure 6.18 are a compilation.[73] Attenuation is expressed as an effective attenuation coefficient μ_{eff} expressed in cm^{-1}, which is the sum of contributions of attenuation due to scattering and absorbance:

$$\mu_{\mathrm{eff}} = \mu_{\mathrm{scat}} + \mu_{\mathrm{abs}} \tag{6.1}$$

The fraction of light (expressed as percent) transmitted a particular distance into the medium is an exponential function of the effective attenuation coefficient, in a manner analogous to Beer's law:

$$\mathrm{T}(\lambda) = \mathrm{e}^{-\mu x} \tag{6.2}$$

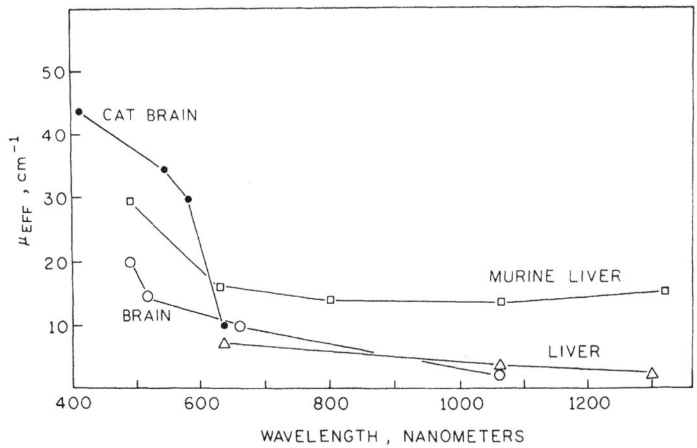

Figure 6.18. Attenuation of tissue as a function of wavelength. $\mu_{EFF} = \mu_{SCAT} + \mu_{ABS}$; $T = e^{-\mu x}$.

where T(λ) is the fraction of light transmitted through the tissue at wavelength λ, μ is the effective attenuation coefficient at that wavelength, and x is the depth into the tissue. While absorbance is largely isotropic, scattering is anisotropic in that it depends on several factors, including the angle of incidence and polarization of the light entering the medium; the size, shape, and refractive indices of the particles causing the scattering, and the geometry of the light collection optics. At present it is difficult to accurately assess the importance of all these factors in determining the scattering behavior of biological tissues, so scattering (and thus attenuation) have been measured empirically on samples of various tissues; results of such measurements are depicted in Figure 6.18. Unfortunately, these measurements have been performed with a few laser lines instead of broadly tunable sources; thus interpolation between the (sparse) points in Figure 6.18 should be done with caution, if at all.

In mammalian tissue at visible wavelengths, the principal absorbing species is of course hemoglobin, which is due to its strong absorbance and high concentration (approx. 8 m*M* in whole blood). Both deoxy- and oxy-hemoglobin exhibit significant absorbance at wavelengths longer than 700 nm; absorbance spectra are depicted in Figure 6.19.[74] Thus for some "average" tissue with blood comprising 10% of the total volume and hemoglobin at normal levels, we can expect the absorbance of one cm. of this average tissue due to hemoglobin to be 0.6 at 805 nm, the isobestic point between oxy- and deoxyhemoglobin. Little seems to be known about other IR absorbers; species such as porphyrins and their metabolites and some other heme and metalloproteins are typically present only at low levels.

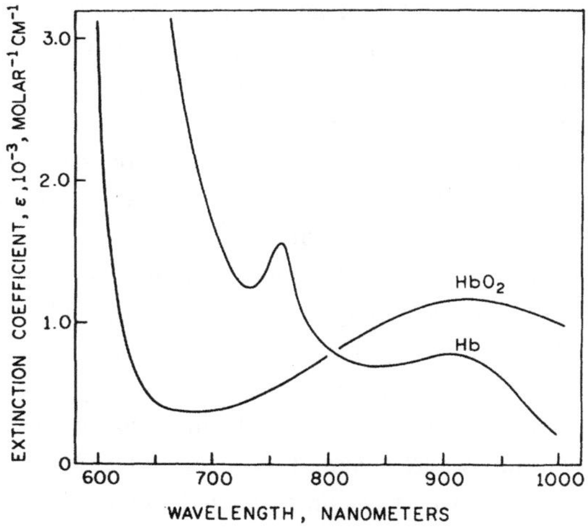

Figure 6.19. Absorbance of hemoglobin in the near-infrared. Redrawn from Ref. 74.

It is important to note that even some very simple molecules have significant absorbance in the IR, even though they appear colorless in the visible. Water absorbs measurably at 977, 942, 906, 823, and 724 nm, the band at 942 nm being easily measurable in a spectrophotometer (Figure 6.20).[75] The absorbance is modest and an intensity measurement in a cuvette should require only a slight correction; however,

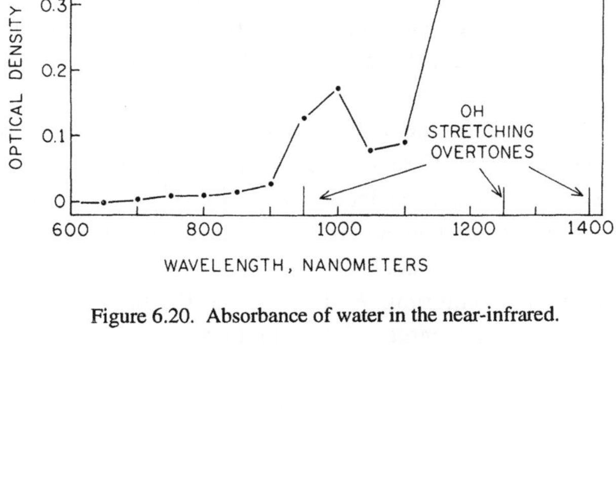

Figure 6.20. Absorbance of water in the near-infrared.

intensity measurements through kilometers of atmosphere or optical fiber might well need to take such effects into account. Indeed, the minima of lowest attenuation in silica or glass fibers at 1.3 and 1.5 μ exist due to the absence of -OH stretching overtones in those frequency bands. The absorbance of sea water increases in the IR to a degree indicating that other solutes are in part responsible, probably including chlorophyll. Other natural waters remain uncharacterized in this respect.

The other major reason for performing fluorescence experiments in the IR (particularly *in vivo* experiments) is that interfering fluorescence is likely to be very modest. Among natural materials only metalloporphyrins such as chlorophyll and a small subset of PAHs fluoresce significantly beyond the visible. Thus compared with the UV and visible wavelength regimes, serum seems hardly to fluoresce at all in the infrared.[76] Nevertheless, there has been little detailed characterization of the infrared fluorescence of any tissue we are aware of. As expected, chlorophyll fluorescence in sea water and other natural waters is significant and widely measured, even remotely from aircraft. This freedom from interfering fluorescence has fueled the development of Fourier transform Raman spectroscopy in the IR; while the Raman scattering is dramatically reduced at longer wavelengths, the decline in interfering fluorescence more than offsets this for a net improvement in signal to noise. Raman scattering is of course a much weaker phenomenon than fluorescence, so we might presume that IR fluorescence will be correspondingly successful. It is the prospect of performing DNA sequencing,[77] fluorescent *in situ* hybridization (FISH) of single copy genes, nonradioactive gel electrophoresis and blotting techniques, microscopy, cell sorting, and immunoassay at unprecedented levels of sensitivity that has fueled the development of IR fluorometry.

6.7. Conclusions

We have summarized certain aspects of IR fluorometry, with an emphasis on the practical. What can we hope to see in the future? First and foremost, we look for the introduction of many new IR dyes. Only a limited amount of work has been done in this respect so far, and the functional equivalent of rhodamine isothiocyanate excitable 200 nm farther to the red remains to be invented.[47–49, 65, 78] Because of the simplicity of LEDs and their easy modulatability, we claim that fluorescence instruments, including time-resolved instruments, will be easier and cheaper to produce in the IR.[10, 22, 25, 72] The lack of interference and potential for high precision suggests that new applications for fluorescence techniques will be found, particularly in the biomedical area. We note that many techniques once found only in the research laboratory are now widely applied, including flow cytometry, magnetic resonance imaging, and blood gas monitoring. We believe that IR fluorometry will in fact be broadly applicable, and will eventually be the preferred way to do fluorometry.

Acknowledgments

The author is grateful for support from the U.S. Office of Naval Research and the University of Maryland Designated Research Initiative Fund.

References

1. D. J. S. Birch and G. Hungerford, Instrumentation for red/near infrared fluorescence, in: *Topics in Fluorescence Spectroscopy, Vol. 4: Probe Design and Chemical Sensing* (J. R. Lakowicz, ed.), Plenum Press, New York (in press).
2. A. J. LaRocca, in: *Infrared and Electro-Optical Systems Handbook: Sources of Radiation*, Vol. 1 (G. J. Zissis, ed.), pp. 51–138, SPIE Optical Engineering Press, Bellingham, Washington (1993).
3. D. J. S. Birch, G. Hungerford, and R. E. Imhof, *Rev. Sci Instrum. 62*, 2405–2408 (1991).
4. ISS, Inc., 2604 N. Mattis Ave., Champaign, Illinois 61821.
5. R. J. Pressley (ed.), *Handbook of Lasers with Selected Data on Optical Technology*, CRC Press, Cleveland, Ohio (1971).
6. O. Svelto, *Principles of Lasers*, 2nd Ed., Plenum Press, New York (1982).
7. E. W. Small, in *Topics in Fluorescence Spectroscopy, Vol. 1: Techniques* (J. R. Lakowicz, ed.), pp. 97–182, Plenum Press, New York (1991).
8. F. P. Schäfer (ed.), *Dye Lasers*, 3rd Ed., Springer-Verlag, New York (1990).
9. *Siemens Optoelectronic Data Book 1987–8*, Siemens, 19000 Homestead Road, Cupertino, California 95014.
10. B. W. Smith, B. T. Jones, and J. D. Winefordner, *Appl. Spectrosc. 42*, 1469–1472 (1988).
11. J. E. Whitaker, R. P. Haugland, and F. G. Prendergast, *Anal. Biochem. 194*, 330 (1991).
12. R. Y. Tsien, *Ann. Rev. Neurosci. 12*, 227 (1989).
13. P. F. Moulton, *J. Opt. Soc. Am. B3*, 125 (1986).
14. P. Lacovara, L. Esterowitz, and R. Allen, *Opt. Lett. 10*, 273–275 (1984).
15. J. R. Lakowicz and I. Gryczynski, in: *Topics in Fluorescence Spectroscopy, Vol. 1: Techniques* (J. R. Lakowicz, ed.), pp. 293–336, Plenum Press, New York (1991).
16. E. Gratton, D. M. Jameson, N. Rosato, and G. Weber, *Rev. Sci Instrum. 55*, 486–494 (1984).
17. E. Gratton and P. Tšo, (submitted for publication).
18. VF. 25 Dielectric Mirrors from Newport Corporation, 1791 Degre Ave, Irvine, California 92714.
19. R. D. Spencer and G. Weber, *Ann. NY Acad. Sci. 158*, 361–376 (1969).
20 K. Berndt, H. Dürr, and D. Palme, *Opt. Commun. 42*, 419–422 (1985).
21. *Melles Griot Optics Guide 5*, pp. 20-2 to 20-16, Melles Griot, Inc., Irvine, California (1990).
22. T. Imasaka, A. Yoshitake, K. Hirata, Y. Kawabata, and N. Ishibashi, *Anal. Chem. 57*, 949–952 (1985).
23. K. W. Berndt, I. Gryczynski, and J. R. Lakowicz, *Rev. Sci Instrum. 61*, 1816–1820 (1990).
24. P. Spano, A. d'Ottavi, A. Mecozzi, B. Daino, and S. Piazzolla, *IEEE J. Quantum Electron. QE-25*, 1440–1449 (1989).
25. R. B. Thompson, J. K. Frisoli, and J. R. Lakowicz, *Anal. Chem. 64*, 2075–2078 (1992).
26. S. Musikant, *Optical Materials: An Introduction to Selection and Application*, Marcel Dekker, New York (1985).
27. R. C. Alferness, in: *Guided-Wave Optoelectronics*, 2nd Ed. (T. Tamin, ed.), pp. 145–210, Springer-Verlag, New York (1990).
28. R. W. Engstrom, *Photomultiplier Handbook*, RCA Solid State Division, Lancaster, Pennsylvania 17604 (1980).
29. J. H. Moore, C. C. Davis, and M. A. Coplan, *Building Scientific Apparatus*, Addison-Wesley, Reading, Massachusetts (1983).

30. T. M. Nordlund, in: *Topics in Fluorescence Spectroscopy, Vol. 1: Techniques* (J. R. Lakowicz, ed.), pp. 183–260, Plenum Press, New York (1991).
31. E. Gratton and M. Limkeman, *Biophys. J. 44*, 315–324 (1983).
32. K. W. Berndt and J. R. Lakowicz, *Rev. Sci. Instrum. 61*, 2557–2565 (1990).
33. G. Laczko, I. Gryczynski, Z. Gryczynski, W. Wiczk, H. Malak, and J. R. Lakowicz, *Rev. Sci Instrum. 61*, 2331–2337 (1990).
34. P. M. Epperson, J. V. Sweedler, R. B. Bilhorn, G. R. Sims, and M. B. Denton, in: *Instrumentation in Analytical Chemistry, 1988–1991* (L. Voress, ed.), pp. 49–56, American Chemical Society, Washington, D.C. (1992).
35. Model 100X 60 GHz photodetector, New Focus, 1275 Reamwood Ave., Sunnyvale, California 94089-2256.
36. K. Berndt, *Opt. Commun. 56*, 30–35 (1985).
37. M. Shinitsky and Y. Barenholz, *J. Biol. Chem. 249*, 2652 (1974).
38. O. S. Wolfbeis, L. J. Weiss, M. J. P. Leiner, and W. E. Ziegler, *Anal. Chem. 60*, 2028–2030 (1988).
39. E. Daniel and G. Weber, *Biochemistry 5*, 1893–1900 (1966).
40. J. Slavik, *Biochim Biophys. Acta 694*, 1–25 (1982).
41. J. R. Lakowicz, R. B. Thompson, and H. Cherek, *Biochim. Biophys. Acta 734*, 295–308 (1983).
42. F. Perrin, *J. Phys. Radium 7*, 390 (1962).
43. F. M. Hamer, *The Cyanine Dyes and Related Compounds*, Wiley-Interscience, New York, 1964.
44. D. R. Sturmer, in: *Kirk-Othmer Encyclopedia of Chemical Technology*, 3rd Ed. (D. Eckroth, ed.), Vol. 7, pp. 335–358 (1978).
45. D. R. Waring and G. Hallas (eds.), *The Chemistry and Application of Dyes*, Plenum Press, New York (1990).
46. J. Fabian and H. Hartmann, *Light Absorption of Organic Colorants*, pp. 162–197, Springer-Verlag, New York (1980).
47. R. B. Mujumdar, L. A. Ernst, S. R. Mujumdar, and A. S. Waggoner, *Cytometry 10*, 11–19 (1989).
48. L. A. Ernst, R. K. Gupta, R. B. Mujumdar, and A. S. Waggoner, *Cytometry 10*, 3–10 (1989).
49. P. L. Southwick, L. A. Ernst, E. W. Tauriello, S. R. Parker, R. B. Mujumdar, S. R. Mujumdar, H. A. Clever, and A. S. Waggoner, *Cytometry 11*, 418–430 (1990).
50. M. Maeda, *Laser Dyes: Properties of Organic Compounds for Dye Lasers*, Academic Press, New York (1984).
51. A. H. Herz, *Photogr. Sci. Eng. 18*, 323 (1974).
52. K. H. Drexhage, in *Dye Lasers*, 3rd Ed. (F. P. Schäfer, ed.), pp. 155–199, Springer Verlag, New York (1990).
53. E. Terpetschnig and J. R. Lakowicz, *Dyes and Pigments 21*, 227–234 (1993).
54. R. P. Haugland, *Handbook of Fluorescent Probes and Research Chemicals*, 5th Ed., Molecular Probes, Eugene, Oregon (1992).
55. U.S. Pharmacopeia, 20th Ed., p. 399, U.S. Pharmacopeial Convention, Easton, Pennsylvania (1979).
56. F. J. Green, *Sigma-Aldrich Handbook of Stains, Dyes and Indicators*, Aldrich Chemical Co., Milwaukee, Wisconsin (1990).
57. P. Greenspan and S. D. Fowler, *Kodak Lab. Chem. Bull. 56*(3), 1–4 (1985).
58. M. M. Rauhut, D. R. Maulding, W. Bergmark, B. G. Roberts, R. A. Clarke, and R. Coleman, *Exploratory Development of Chemiluminescent Materials which Emit Radiation in the Infrared Region: Final Technical Report* to U.S. Army Munitions Command Contract DAAA 21-67-C-0503, Picatinny Arsenal, Dover, New Jersey 07801; October 1968. AD 843278 Unclassified; Distribution limited to DOD and contractors.
59. M. M. Rauhut, B. G. Roberts, D. R. Maulding, W. Bergmark, and R. Coleman, *J. Org. Chem. 40*, 330–335 (1975).
60. A. G. Mohan, S. S. Tseng, M. M. Rauhut, F. J. Arthen, R. G. Dulina, V. M. Kamhi, D. E. McKay, R. J. Manfre, J. D. Ofeldt, and L. S. Vizcarra, *Aqueous Peroxyoxalate Chemiluminescence: Final Report to ONR*, National Technical Information Service AD121396, Alexandria, Virginia (1982).

61. F. H. Moser and A. L. Thomas, *The Phthalocyanines*, Vols. 1 and 2, CRC Press, Boca Raton, Florida (1982).
62. C. C. Leznoff and A. B. P. Lever (eds.), *Phthalocyanines: Properties and Applications*, VCH Publishers, New York (1989).
63. B. D. Berezin, *Coordination Compounds of Porphyrins and Phthalocyanines*, John Wiley & Sons, New York (1981).
64. J. G. Fabian, H. Nakazumi, and M. Matsuoka, *Chem. Rev. 92*, 1197–1226 (1992).
65. La Jolla Blue™, Diatron Corp., 7340A Trade St., San Diego, California 92121.
66. J. N. Demas and B. A. DeGraff, Design and applications of highly luminescent transition metal complexes, in: *Topics in Fluorescence Spectroscopy, Vol. 4: Probe Design and Chemical Sensing* (J. R. Lakowicz, ed.), Plenum Press, New York (in press).
67. R. B. Thompson and L. M. Vallarino, in: *Proceedings of the Conference on Time-Resolved Laser Spectroscopy in Biochemistry* (J. R. Lakowicz, ed.), *Proc. SPIE 909*, 426–437 (1988).
68. L. P. Pan, B. Durham, J. Wolinska, and F. Millett, *Biochemistry 27*, 7180–7184 (1988).
69. C. V. Kumar, J. K. Barton, and N. J. Turro, *J. Am. Chem. Soc. 107*, 5518–5523 (1985).
70. D. Basting, F. P. Schäfer, and B. Steyer, *Appl. Phys. 3*, 81 (1974).
71. B. Chance (ed.), *Proceedings of the Conference on Time-Resolved Spectroscopy and Imaging of Tissues, Proc. SPIE 1431*, (1991).
72. J. R. Lakowicz, *Laser Focus World 28*(5), 127 (1992).
73. W. F. Cheong, S. A. Prahl, and A. J. Welch, *IEEE J. Quant. Elect. QE-26*, 2166–2185 (1990).
74. B. L. Horecker, *J. Biol. Chem. 148*, 173–183 (1943).
75. D. Kryskowski and G. H. Suits, in: *Infrared and Electro-Optical Systems Handbook, Vol. 1: Sources of Radiation* (G. J. Zissis, ed.), p. 259, SPIE Optical Engineering Pub., Bellingham, Washington (1993).
76. O. S. Wolfbeis, in: *Molecular Luminescence Spectroscopy, Part I: Methods and Applications* (S. G. Schulman, ed.), p. 305, Wiley-Interscience, New York (1985).
77. Model 4000 DNA Sequencer, LiCor, Inc., 4421 Superior St., Lincoln, Nebraska 68504.
78. S. Theodoropulos, U.S. Patent No. 4,714,763 (1987).

7

Near-Infrared Fluorescence Probes

Guillermo A. Casay, Dana B. Shealy, and Gabor Patonay

7.1. Introduction

Optical fiber detectors (OFD) are devices that measure electromagnetic radiation transmitted through optical fibers to produce a quantitative signal in response to the chemical or biochemical recognition of a specific analyte. Ideally, an OFD should produce a specific and accurate measurement, continuously and reversibly, of the presence of a particular molecular species in a given sample medium. Additionally, OFD should provide maximum sensitivity and minimal interferences from superfluous ions or molecules to obtain low detection limits. Other attractive features include the miniaturization of the fiber's tip to accommodate single-cell analysis and portable instrumentation to allow *in situ* analysis.

OFDs can be divided into two subclasses: (1) optical fiber chemical detectors (OFCD) which detect the presence of chemical species in samples, and (2) optical fiber biomolecular detectors (OFBD) which detect biomolecules in samples. Each subclass can be divided further into probes and sensors, and bioprobes and biosensors, respectively. As a result of the rapid expansion of optical research, these terms have not been clearly defined and to date, the terms "probe" and "sensor" have been used synonymously in the literature. As the number of publications increases, the terminology should be clarified. Although both probes and sensors serve to detect chemicals in samples, they are not identical. The same situation exists with "bioprobes" and "biosensors." Simply, probes and bioprobes are irreversible to the analyte's presence, whereas sensors and biosensors monitor compounds reversibly and continuously.

Progress in the development of OFD has been parallel to advances in the production of optical fibers. Fiber production and reliability have increased and fiber costs have decreased; therefore, routine use of OFD has become more practical. The increased availability of OFD has produced a surge of publications on its applications including quantitative metal detection, pH determination, immunoassays, and the

---

Guillermo A. Casay, Dana B. Shealy, and Gabor Patonay • Department of Chemistry, Georgia State University, Atlanta, Georgia 30303.

Topics in Fluorescence Spectroscopy, Volume 4: Probe Design and Chemical Sensing, edited by Joseph R. Lakowicz. Plenum Press, New York, 1994.

measurement of hydrophobic interactions.[1–10] A more extensive review of optical fibers has been reported by Wolfbeis.[11, 12]

The intent of this chapter is not to provide an exhaustive review of chemical- and biosensors and probes, but rather to offer a brief overview of existing optical techniques and an indepth analysis of near-infrared (NIR) fluorogenic probes and sensors for the detection of metal ions, solution pH, and biomolecules and to present some of the latest results.

7.1.1. Background

OFCDs require the use of molecular probes such as fluorescent dyes or indicators that can be induced during analysis to produce a detectable spectral change. Several elements must be considered when choosing a molecular probe. Ideally, molecular probes would possess large molar absorptivities and large fluorescence quantum yields to allow a maximum optical signal. Probes should also possess a suitable functionality such as isothiocyanate to allow covalent attachment to a polymeric support. Additionally, the probes should be resistant to photobleaching and should remain stable under photolytic conditions, thus allowing the use of intense excitation sources such as lasers.

If fluorescent dyes are employed, they should have large Stokes' shifts which would reduce the need for high quality optical filters that may or may not adequately filter out scattered radiation. When small concentrations of dye are utilized, the scattered light band is much greater than the fluorescence band and obvious band overlapping is observed. Many times, to achieve adequate masking of the scattered light, much of the fluorescence signal must be sacrificed in the filtering process. Alternatively, if the scattered light is not adequately reduced, a large background signal is collected.

Many conventional OFCD exploit pH indicators as molecular probes.[13–19] The pH indicators allow the detection of a chemical species by measuring the generation or uptake of hydrogen ions in chemical reactions. Some indicators such as *p*-nitrophenol,[13] Congo Red,[14] bromophenol blue,[15] and others[16–19] have been used successfully. Unfortunately, not all chemicals can be detected in this fashion and alternate methods of detection are necessary. Additionally, the use of absorbance spectroscopy by itself may not allow low detection limits since many molecules can interfere with absorbance measurements.

The use of fluorescence spectroscopy in OFCD greatly improves the detection limits of the analysis. The fluorescence signal at low concentrations is directly proportional to the intensity of the incident light source and follows a general equation as

$$I_f = QI_0[1 - \exp(-\varepsilon bc)] \tag{7.1}$$

where I_f is the fluorescence signal, I_0 is the incident power of the light source, Q is the quantum yield of the dye, ε is the molar absorptivity, b is the path length, and c is the concentration. Deviations from linearity are frequently observed at high concentrations.

Fluorescence exhibits signals much greater than absorbance or reflectance with respect to the background.[20] Since the method is affected by the background fluorescence; larger background signals decrease the detection limits and also reduces the sensitivity. In other words, fluorescence is measured against background, while absorption is a differential measurement requiring a reference signal. Additionally, fluorescence spectroscopy introduces more selectivity into the method because much fewer molecules fluoresce than absorb light.[20]

In the literature, fluorescence spectroscopy in OFD has been limited to the use of ultraviolet (UV) or visible dyes as molecular probes.[1] The most common fluorescent dye used in OFD is fluorescein and its derivatives.[21–23] Fluorescein possesses a good fluorescence quantum yield and is commercially available with an isothiocyanate functionality for linking to the polymeric support.[24–26] Additionally, selective laser excitation can be performed because the absorbance maximum of fluorescein coincides with the 499-nm laser line emitted from an argon laser. Unfortunately, argon lasers are costly and bulky, thus limiting the practicality of their use. Similar difficulties exist with other popular commercial dyes.

All OFDs reported in the literature suffer from spectral interferences, long response times, and narrow dynamic responses. Many of these obstacles exist as a result of limitations due to the properties of UV/visible fluorescent dyes. These dyes typically absorb and fluoresce between 300 and 650 nm, a region susceptible to extensive interference, especially from biomolecules (Figure 7.1). The fluorescence of sample impurities combined with the inner effect of the matrix and polymer support greatly increase the signal interference of the analysis.

This factor is particularly significant in OFBD since biological samples or isolates are used. In addition to background interference, fluorescence quenching has been demonstrated in a variety of biomolecules such as thiamine (vitamin B_1),[27] nicotinamide,[28] nucleosides/nucleotides,[29] and pyruvate.[30] To circumvent the obvious limitations associated with the use of UV or visible fluorophores in OFD, the potential

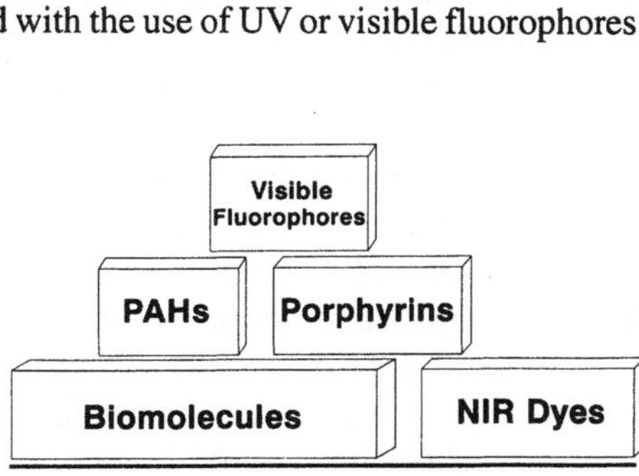

Figure 7.1. Typical absorbance and fluorescence properties of organic compounds.

use of near-infrared fluorogenic dyes as molecular probes with alternative methods of detection is under investigation.

NIR fluorogenic dyes possess spectral properties from 700 to 1100 nm in the NIR of the electromagnetic spectrum. This region is characteristically low in interference; thus interference from impurities is negligible when compared to the noise generated from the detector. In other words, the method becomes detector limited as opposed to background limited. NIR dyes have large molar absorptivities and acceptable quantum yields. They also have reasonably large Stoke's shifts which minimize the band overlap of scattered light. Their spectral properties make them suitable for selective excitation with inexpensive, commercially available laser diodes. Using laser fluorimetry in the NIR region, Ishibahi demonstrated that the fluorescent signal intensity was at least two orders of magnitude better than the signal obtained from a standard Xenon lamp as an excitation source.[31]

7.1.2. Characteristics of the NIR Region

The NIR spectral region is described as the portion of the electromagnetic spectrum from 700 to 3000 nm. The anharmonic stretching of hydrogen together with most other fundamental hydrogen modes determine the major properties of the NIR region. It is characterized by first and second overtones, a combination of high frequency stretching bands of the hydrogen bond. Overtones and similar combinations occur due to the anharmonic nature of the molecular vibration and most of these bands originate with the hydrogen stretching vibrations of CH, OH, and NH bonds because of the light mass of the hydrogen atom.[32] The hydrogen stretching affects different functional groups by causing a band broadening and a shift to the lower frequency. The first overtones are usually OH- stretching vibrations; the second overtones represent CH- stretches. The stretching vibrations of some of the overtones may vary from 1100 nm for a single bond, to approximately 2500 nm for a double bond. Third vibrational overtones, as well as some second overtones, depending upon the band shift, are usually encountered between 700 and 1100 nm.[32] Consequently, the region of 700–1100 nm is relatively free from interference. To promote electronic transitions only molecules with an extensive conjugated system are able to absorb NIR energy.[33] Nevertheless, these molecules exhibit absorbance bands in the NIR region which are several orders of magnitude smaller than their UV bands. Molecules with NIR spectral properties can be natural dyes such as porphyrins (λ_{max} = 619 nm) or synthetic dyes which include squarylium, croconium, phthalocyanine, and carbocyanine dyes. Phthalocyanines (λ_{max} = 690 nm) serve as precursors for naphthalocyanine isomers with absorbance maxima ranging from 720 to 800 nm by the annelation of a phenyl ring which induces the observed bathochromic shift. To date, the phthalocyanine and carbocyanine dyes have been the most widely applied NIR dyes.[34] We are currently investigating naphthalocyanine and cyanine dyes as possible fiber optic probes.[33]

	Length of Chain	λ_{Abs} nm	λ_{Flu} nm
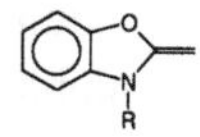	1	490	515
	2	585	620
	3	680	740
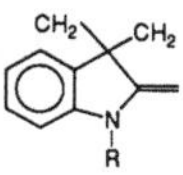	1	550	590
	2	650	690
	3	750	790
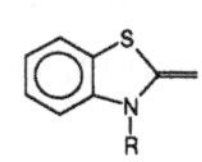	1	560	590
	2	660	690
	3	765	810

Figure 7.2. Absorption and fluorescence properties of carbocyanines dyes as a function of the polymethine chain length. (Reprinted with permission from Ref. 33.)

Although these dye families exhibit large molar absorptivities and quantum yields, the utilization of these dyes has been limited.[31, 35]

The naphthalocyanine (NC) dyes are very stable to degradation by light, heat, and oxidation, but have limited solubility in protic solvents.[34] Naphthalocyanine dyes are planar, rigid molecules which undergo an excitation process resulting in the population of the first singlet excited state.[36] The absorption spectra of 2,3-NC are consistent with $\pi \rightarrow \pi^*$ transitions in the energy region above 769 nm. A typical spectra for these dyes exhibit a stronger band between 322 and 400 nm and a much stronger band between 285 and 333 nm.[37] The significant energy difference between the first two bands reduces the degree of overlapping absorbance,[38] thus producing a narrow absorption band. Although their relatively small Stoke's shifts and limited aqueous solubility may limit their utility in certain applications,[34] the narrow absorption band centered close to the output wavelength of medium-powered laser diodes ($\lambda = 780$ nm) and their sensitivity to the presence of many analytes will permit their implementation into fiber optical technology.

Cyanine dyes are smaller molecules, usually less photostable than NC dyes, but generally have higher aqueous solubility. Cyanine dyes can be described by the structural formula X-$(CR)_n$-Y, where n is a positive, odd integer and X and Y are usually nitrogen atoms.[33] The absorbance maxima of cyanine dyes increase as the polymethine chain increases; for each CH-CH group, an increase of approximately 110 nm is observed[33] as shown in Figure 7.2. Cyanine dyes undergo similar electronic transitions as NC dyes resulting in sharp Soren absorption bands. A large number of publications describe the synthesis of NIR absorbing dyes.[39–43]

7.2. NIR Optical Probe Instrumentation

Commercially available instrumentation for absorbance and fluorescence measurements in the NIR region is scarce. Conventional absorbance spectrophotometers generally extend their spectral range up to around 1100 nm which only encompasses a portion of the NIR region. An even narrower wavelength range is available in most fluorescence instrumentation, extending only to typically about 950 nm.[44] The lack of adequate commercially available instrumentation for NIR measurement requires user fabrication of appropriate instruments. This situation is continuously changing, e.g., NIR instruments like a DNA sequencer (LI-COR, Inc., Lincoln, Nebraska) are now commercially available.

The design and operation of an NIR probe is similar to that of conventional UV-visible OFCD reported in the literature.[21] These probes consist of a light source, a bifurcated fiber, an NIR dye, a polymer matrix, a detector, and other optical components.

An NIR optical fiber detection system has been developed in our laboratory for the determination of metals and solution pH. Additionally, this system has been utilized for biosensor applications. A schematic diagram of the setup is shown in Figure 7.3. The replacement and alignment of the laser diodes were simple due to easy access of different laser diodes with output wavelengths between 670 and 780 nm which have been used as light sources throughout our investigations. Multiline or tunable lasers were used for wavelengths not available with laser diodes. The astigmatism of the diode output beam was controlled by the use of microscope objective lenses of various apertures, depending on the fiber used. A beam splitter diverged the beam to a power meter to control the laser stability. The main beam passed through a chopper connected to a lock-amplifier equipped with a coupler. The laser diode was stabilized by utilizing a feedback circuit controlled by the built-in p- and n-types (PIN) photodetector of the laser diode.

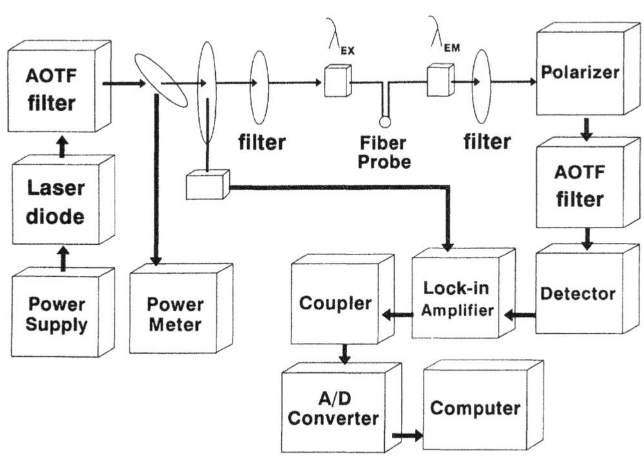

Figure 7.3. Block diagram of a near-IR optical fiber detector. A/D = analog-to-digital; AOTF = acoustic optical tunable filter.

The optical fibers were positioned on an x-y-z translational stage to allow simple and accurate beam focusing. The optical fiber tip where the reaction phase is located was secured perpendicularly to the optical table on a rod's side arm. The side arm could be lowered and raised for a better access to the vials containing the analyte. The detector housing was equipped with an 820 nm bandpass filter (10 nm bandwidth (BW), optical density (OD) > 5) to isolate the fluorescence signal. The collection of fluorescent signal at about 40 nm away from the incident light source ($\lambda_{max} = 780$ nm) is suitable for OFD. For NIR dyes, the fluorescence maximum is generally centered at wavelengths exceeding 800 nm. Thus, using small bandpass filters, less than 3% of the scattered light was received and more than 20% of the fluorescence signal of the dye was collected. Typically, a silicon photodiode (Eg&G UV-100 BG) was employed to detect the fluorescent signal. The optical fiber was also interfaced to the detector of a SLM Aminco 8000C spectrofluorometer.[44] The optical components were similar to the instrument reported by Patonay *et al.*[45]

7.2.1. Light Sources

7.2.1.1. Traditional Sources

Fluorescence excitation is usually accomplished by coupling a light source with appropriate filters or gratings to produce monochromatic radiation near the absorbance maximum of the dye. Traditional light sources of electromagnetic radiation include tungsten-halogen, mercury, and xenon lamps which emit polychromatic light ranging from 340 to 1000 nm. The wavelength of light is chosen by a monochromator or bandpass filters. These light sources distribute their total output energy among the various wavelengths, therefore decreasing the intensity of the source at a given band. Additionally, conventional light sources are bulky and often require special care which renders them less mobile for possible on-site analyses.

7.2.1.2. NIR Lasers

Lasers are attractive light sources for spectral analyses because they provide large output power and narrow spectral linewidth at a reasonable cost. As the reaction phase becomes smaller, fewer dye molecules are available for interaction with the analyte, thus requiring maximum excitation efficiency. Lasers deliver intense radiation in a small diameter spot. The concentration of the monochromatic laser energy in a small diameter increases the excitation efficiency of the dye, thus enhancing the observed fluorescence signal. This is particularly advantageous when minimal quantities of dye are being used as in microscale fiber optic techniques. In addition, their narrow spectral linewidth centered in the NIR region increases the specificity of excitation, thus reducing background signal.

The high output power of the laser is essential for reaching the low levels of detection demanded by the optical probes. Of the high-powered lasers capable of producing NIR radiation, the titanium:sapphire (Ti:sapphire) lasers have been the most widely used due to their tunability. The tuning of this laser allows different output wavelengths to be obtained from a single source. The output wavelength varies from 700 to 1000 nm with a medium radiant power of about 2 W. However, these lasers are large, costly, and require a large input laser energy. They are also difficult to maneuver and maintain.(35) Sometimes these lasers are not particularly suitable for remote applications and compact investigations. Another large laser which operates above 700 nm is the krypton-ion laser which is inefficient, relatively expensive, and consumes large amounts of power.

The most attractive laser source for the NIR region and probably one justification for the expansion of NIR dye applications is the semiconductor laser diode.(46) Semiconductor laser diodes have all the properties of conventional lasers (i.e., monochromatic light, high output energy, and coherent beam). Low-, medium-, and high-powered laser diodes are commercially available with maximum output wavelengths ranging from 680 to 1800 nm. High-powered laser diodes are usually multistripe; low- and medium-powered diodes are single-stripe.(47) They are relatively inexpensive (around $200), compact, and portable. Additionally, they are reliable and durable,(48) and require low input energy for operation, thus can be operated using alkaline batteries. The typical lifetime of a laser diode is about 200,000 h(49) and their typical size is about 5×10 mm with a laser beam diameter of less than 1 mm. The utility of laser diodes in NIR applications has been demonstrated in a low-cost laser diode intracavity absorption spectrophotometer reported by Unger and Patonay.(47) This design afforded lower detection limits than conventional spectrophotometers for the NIR region. In addition, the output wavelength of the diode can be tuned for specific applications.

The gallium aluminum arsenide (GaAlAs) and indium phosphide (InP) laser diodes are among the most used NIR excitation sources (Figure 7.4). These diodes are

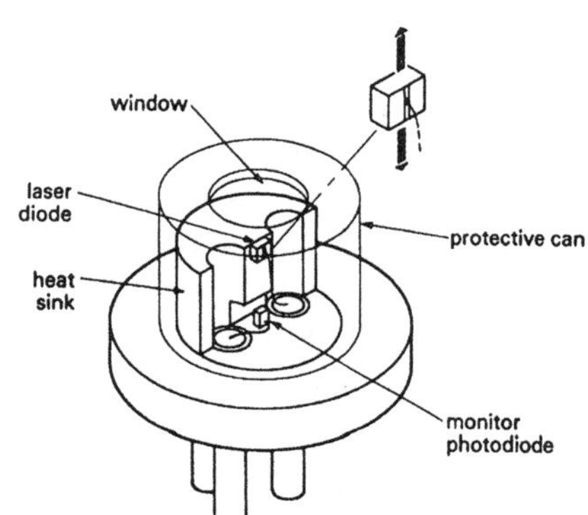

Figure 7.4. Typical laser diode packaging. The heat sink serves as one of the electrical contacts. (Reprinted with permission from Melles Griot.)

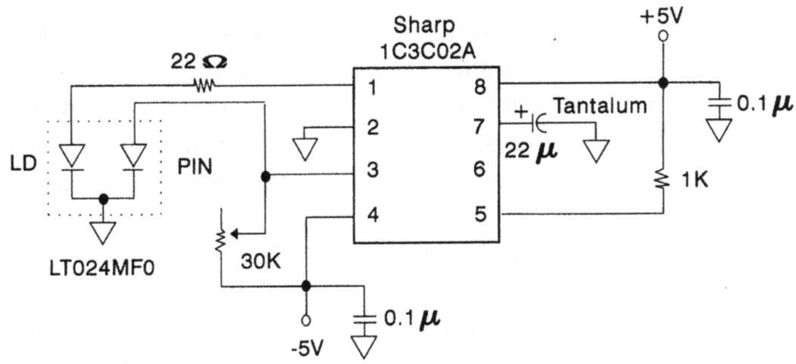

Figure 7.5. Laser diode driver utilizing a Sharp integrated circuit.

commercially available with wavelengths of 750 and 780 nm with 3-, 5-, and 30-mW output powers. A diode driver utilizing a Sharp laser circuit is shown in Figure 7.5. Recent advances in laser diode technology have allowed the production of diodes with wavelength outputs at 740 nm (LDX Optronics, Orlando, Florida) with medium output power. Longer and shorter wavelengths such as 830 and 680 nm are also available. Laser diodes with radiant power output as high as 1 W are also available.[(50)]

Additional laser diode technologies recently reported include a continuous-wave rhodamine 700 dye laser with a maximum wavelength output at 758 nm, powered by two laser diodes each operated with two standard AA batteries (RDT&E division of the US Naval Command, Control, and Ocean Surveillance Center, San Diego, California), and a tunable laser diode with output laser wavelengths of 650, 780, 850, and 1320 nm (New Focus, Mountain View, California).

7.2.2. Detection

Some of the typical parameters or properties utilized for NIR detection are potentiometry,[(5)] absorbance,[(52–54)] refractometry,[(18, 19)] or fluorescence spectroscopy.[(55)] Of these, has proven to be the most valuable detection method in fiber optic applications.[(2, 56)] In standard spectroscopic techniques, the detection limits of a method are greatly determined by the instrument and by the chemical method used for the analysis. However, in OFCD research the detection limits are governed by a series of other variables including the dye, the matrix, and the instrument. By optimizing these variables, low detection limits can be obtained with this technique.

At room temperature, normally 99% of the molecules are in the ground state. On radiant excitation, the absorbed energy promotes electrons to a higher energy state of the chromophore. In chromophores with acceptable quantum yields, the fluorescence emission can be measured as the electrons return to the ground state.

The detection of the analyte by OFCD using fluorescence may be accomplished by using a variety of detectors. The most popular detectors are the photomultiplier tube (PMT) and the solid-state photodiode (SSPD). The choice of detector ultimately depends on the nature of the problem to be solved. Certain detector properties are requisite for optical fiber detection such as low dark noise (the signal generated under ambient conditions) and a high quantum efficiency (the ratio of basic signal elements, photoelectrons, to the total number of incident photons). A high quantum efficiency and low dark current may allow detection limits down to the attomolar level by increasing the detected signal.

Traditionally, the PMT has been the most preferred detector for the UV-visible region. Several companies provide PMTs for detection in the NIR region with costs ranging from \$600 to \$800. More sensitive and sophisticated PMTs such as the microchannel plate photomultiplier (MPP) are also available. The detection limits of the PMT are usually determined by the dark current. The PMT must operate at low temperatures (sometimes as low as –20 °C) to reduce the dark current to an acceptable level. Depending on the cooling unit, it may take 1–2 h to sufficiently cool the PMT low enough so that acceptable dark levels are obtained. In PMT, quantum efficiency (QE) is monitored by the amount of voltage applied to the detector. Typically, 1000–1400 V are applied to PMTs to obtain the maximum QE and as the voltage is increased, the QE of the PMT increases to its operating maximum which is always less than 100%. If the voltage is increased past the QE maximum, the detector undergoes a breakdown phenomena. In addition, the QE decreases as the wavelength increases. A typical QE for a Hamamatsu R928 PMT is less than 2.5% at 800 nm compared to 25% at 300 nm. An MPP with a red multialkali photocathode has a QE of about 5% at 800 nm. Most PMT possess low QE in the NIR region reducing their utility in the NIR region.

Photodiode detectors are more practical for the NIR region. These detectors are more rugged and portable with superior detection capabilities in the NIR region as shown in Figure 7.6. Unlike single channel PMTs which require monochromators or gratings to isolate single wavelengths for detection, multichannel photodiodes are available that are capable of simultaneously detecting all wavelengths resulting in high-speed data acquisition. In principle, a photodiode response is governed by Eq. (7.2)

$$I = I_{\text{photo}} + I_{\text{dk}}\,(e^{qV_0/kT} - 1) \tag{7.2}$$

where I_{photo} is the photocurrent, I_{dk} is the dark current, q is the charge of the electron, V_0 is the voltage across the diode junction, k is the Boltzman constant, and T is the temperature in degrees Kelvin. (At 22 °C, the value of q/kT is 40.)

Typical photodiode detectors consist of a p layer which is made of an electron deficient material; an n layer which is electron abundant; and a depletion region, the p-n junction, located between the two layers. At equilibrium, when no light or current is applied to the system, the p-n junction is in electrostatic equilibrium and the alignment of electrons and electron holes on the two sides of the junction region creates a contact potential voltage. As incident light strikes the surface of the diode, the

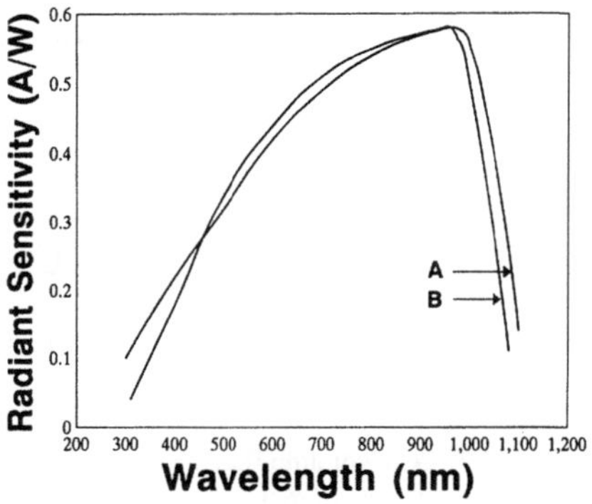

Figure 7.6. Detector efficiency for photodiodes. Spectra for (A) a silicon photodiode and (B) a PIN silicon photodiode are shown. The peak wavelength response of both detectors is at 960 nm.

equilibrium is disrupted. Photons are absorbed and excite electrons into the conduction band. This enables the freed electrons to migrate to the opposite end of the depletion region. The electron holes created in the *n*-region migrate through diffusion to the end of the junction; the free electrons produced in the *p*-region migrate in a similar way. This creates a net charge in the depletion region. When a voltage, or bias, is applied to the system, it produces a current flow which is proportional to the incident radiation. Two modes of operation can be used with these diodes. A photovoltaic mode operates by the migration of the electron-hole pairs to opposite sides of the junction producing a voltage or current if the system is connected to a circuit. The photoconductive mode occurs when a reversed bias is applied across the junction. The current produced by the bias and free carriers is proportional to the light intensity. A photodiode can operate by a reversed bias up to a maximum voltage, the breakdown voltage. If the system is operated above the breakdown voltage which is around 200 V, the photodiode will be damaged. The absorbance of light produces electron-hole pairs which greatly increase the conductance. The absorption of light can be improved by placing a coating on the front or both surfaces of the diode to minimize loss by reflection. This mode of operation may be considered in a certain sense analogous to the operation of a PMT.

Of the photodiode detectors, single-photon avalanche diodes (SPAD) are gaining popularity due to their low bias voltage, size, ease of operation, and simplicity.[55, 57] They are usually made of silicon for UV-visible detection and germanium for NIR detection. These specially made diodes operate at voltage levels above the breakdown voltage. At this voltage, in response to photoexcitation, a surge of electrons moves across the depletion region to recombine with the electron holes. The surge results in a large signal, thus greatly enhancing the detector sensitivity.

The SPAD detector is similar in design to other photodiodes except the electric field of its junction is better separated than in other photodiodes. This design better endures the flow of the avalanche current triggered by a photogenerated carrier and provides better performance than normal SSPD. SPAD have detection limits several orders of magnitude lower than conventional PMTs.[35] In addition, they have been shown to have higher quantum efficiency (QE) values than PMTs in the NIR region.[57]

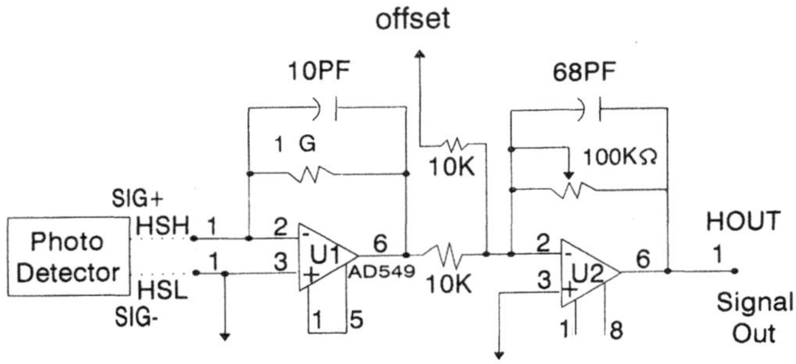

Figure 7.7. Detector amplifier circuit used in our studies. HSH = signal high; HOUT = signal out; HSL = signal low; PF = picofarad.

SPAD have been shown to possess a high QE of 7.5% at 633 nm with a minimal dark signal when operated at 3 V above the breakdown voltage at low temperatures.[58] Recent experiments for single-molecule detection using an SPAD detector have reported a QE of about 40% at 830 nm.[55]

An NIR detection system has been fabricated using photodiode detectors for fiber optic analyses. The complete circuit design is shown in Figure 7.7. Fluorescence radiation was collimated by a planoconvex lens (25.4 mm focal length, 25.4 mm diameter) through three interference filters [two 820 nm (longpass filters) and one 830 nm (bandpass filter), all 25 mm diameter] and focused onto the photodiode with an additional planoconvex lens (31.7 mm focal length, 25.4 mm diameter). The signal from the photodiode was measured as voltage using a current-to-voltage circuit with a built-in amplifier. The measured signal was recorded on a 386-33 MHz personal computer. An analog-to-digital DAS-8 and a 12-bit converter board, interfaced to data acquisition software, were used to analyze and record the data. The detector system operates at low voltage and is very sensitive to NIR detection in the picomolar range. With signal enhancement, the detection could be improved one- to twofold.

7.2.3. Miscellaneous Components

Several other optical components are necessary for the development of OFCD. A laser power drift that may occur can be stabilized with a feedback circuit affording negligible laser power fluctuations during experimentation. To ensure any detected change in the signal is solely due to the interaction of the chromophore with the analyte to be detected, a beam splitter can be used to monitor the laser output. An amplifier discriminator with a signal coupler may be necessary for low-intensity signals. In addition, microscope objective lenses, mirrors, and optical density filters, and other optical components are necessary to direct the laser beam. Furthermore, preamplifiers, amplifier discriminators, photon-counters, couplers, and monochromators can be used to enhance detection. When using laser fluorimetry in the NIR region, SPAD photon-counters are at least two orders of magnitude better than conventional PMTs even if lock-in amplifiers are used.[31]

Acoustic optical tunable filters (AOTF) which are capable of separating a multilinear laser into a single monochromatic light source can be used to select an excitation wavelength.[59] AOTF allow the use of output wavelengths that are not easily available with laser diodes.[60] AOTF may be particularly useful in special applications that require the continuous monitoring of several wavelengths because more than one excitation line may be generated simultaneously. In the same manner, an AOTF filter situated in front of the detector will allow several wavelengths to be detected[61] without changing bandpass filters.

7.2.4. Optical Probe

7.2.4.1. Optical Fiber

Optical fibers which were developed primarily for the communication industry have been successfully implemented into other disciplines such as chemistry, physics, and biomedicine. Mass production of high-quality fibers by a variety of manufacturers has rendered them relatively inexpensive. The use of optical fibers is rapidly increasing because of their size, cost, and easy mode of operation.[62] Table 7.1 shows properties of some commercially available optical fibers. Originally, optical fibers were made of glass but more recently they have become available as organic polymers or metal halides. They are flexible, stable, and resistant to many chemicals. Furthermore, optical fibers can improve safety by allowing the remote analysis of potentially harmful chemicals. A detailed review of the development of optical fibers has been previously published.[63]

Size. Fiber size is very crucial in OFCD development. Small fibers allow miniaturization of the probe so it may be positioned within a single cell to measure various physical parameters. The core size, the size of the path through which light traverses

Table 7.1. Commercially Available Optical Fibers[a]

Fiber	Fiber diameter (μm)	Spectral region (nm)	Typical bandwidth (nm)	Numerical aperture
PMMA/F-PMMA	125–3000	390–1200	20	0.5
Polyimide	50–800	300–2500	20	0.5
Soft glass	12–100	400–2000	20	0.5–0.87
Silica/silicone elastomer	50–2000	240–2000	20	0.22–0.26
Silica/F-silica	25–1000	180–2300	25	0.11–0.24
Doped silica	4–10	Open	1000–2000	0.2

[a]From *General Fiber Optics, Inc, Reference Guide*, Cedar Grove, New Jersey (1990).

the fiber, is between 4 and 3000 μm in diameter, depending on the type of fiber. Plastic fibers made of poly(methyl methacrylate) (PMMA) and fluorinated poly(methyl methacrylate) have core diameters as low as 125 μ and are quite flexible, making them well suited for sensor applications.

Stability. Plastic fibers accommodate a greater load of stress due to bending and vibrations than silica fibers. Plastic fibers can withstand temperatures ranging from –35° to 80°C. For operating at temperatures higher than 80°C, silica fibers must be used. The outer polycarbonate jacket of the plastic fibers makes them rugged and resistant to environmental damage. The PMMA polymer is insoluble in water and other polar solvents; however, PMMA will dissolve in more nonpolar solvents.

References. Because the detected fluorescence signal is a direct response of the dye–analyte complex formed, no reference measurement is required. Also no calibration of the probe is required, although the response function of the probe may be needed.

Disadvantages. Optical fibers transmit broad wavelengths of light as dictated by their original design for communication applications where maximum light transmission was desired. Ideally, optical fibers in OFCD applications would permit only narrow-wavelength bandwidth transmission through a single fiber to improve method selectivity and reduce interfering signals. Furthermore, this could reduce the need for interference or bandpass filters in the OFCD assembly.

7.2.4.2. Fiber Attachments

Although different approaches have resulted in a variety of fiber attachment techniques,[4, 21] three basic designs are being pursued by several research groups; the single fiber strand,[64–66] the single-fiber evanescent,[62, 67] and the bifurcated design[26, 51, 54, 68, 69] as shown in Figure 7.8.

In the single fiber strand technique, a single fiber is used as the source and receiving fiber. The reaction phase is located at the end of the fiber and the separation

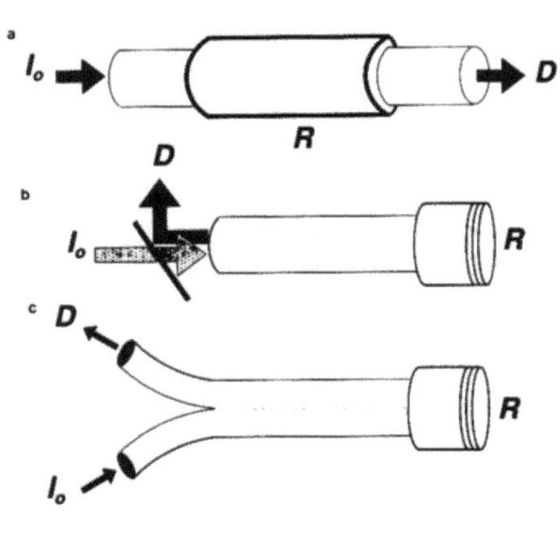

Figure 7.8. Different type of optical fiber probes. I_0 is the incident light source, R is the dye chamber and D is the detector location. (a) The dye is entrapped around a single optical fiber. (b) The same optical fiber is used to route the incident light and the fluorescent signal. (Additional optical components are used to isolate the sample's signal.) (c) One optical fiber is used to carry the incident light and the second fiber is use to carry the sample signal.

of the fluorescence emission is accomplished by the use of optics. In the single fiber evanescent technique, a single fiber is used and the reaction phase is located around the optical fiber. One end of the optical fiber is used for the entrance of the incident light source and the other end is used to route the signal to the detector.

We are interested in the development of an OFCD using bifurcated fibers.[70] In principle, they operate by first transmitting radiation from the light source through an optical fiber. The radiation exits at the distal end of the fiber where the reaction phase is located and where dye molecules susceptible to the presence of an analyte have been immobilized in a polymer matrix. The dye absorbs some of the incident radiation and, consequently, fluoresces. The fluorescence is collected by a second fiber connected in the same reaction phase, and the intensity exits at the other end and is measured by a detector.

The attachment of the two fibers is very crucial for the success of the design. The two optical fibers were attached with PMMA and the final reaction phase located at the tip of the connecting fibers was less than 2.5 mm in diameter. The optical fiber should be constructed such that the maximum possible area for excitation illumination is available and the access of scattered light back to the receiving fiber is restricted. Optimization of the illumination volume is also critical to the success of these attachments because the magnitude of the fluorescence signal is proportional to the interaction with the analyte. If the area receiving the excitation light is too small, a poor emission signal will be observed; however, too large of a section could create an unacceptable reflection.

The ability of an optical fiber to accept light is represented by its numerical aperture (NA). Large NA values indicate that the fiber can accept light from very wide angles, and the position of the light source (including fluorescence of the dye) does not necessarily have to be at the center of the core. Hence high NA values would permit the fiber to accept the fluorescent signal from a wider illuminated region which would afford a larger detected signal. A diagram of the fiber as it receives the fluorescent signal is shown in Figure 7.9.

As the analyte or any other species that may be present changes in concentration, the refractive index (RI) of the sample medium also changes. A large change in the RI may interfere with the fluorescent signal because the numerical aperture of the fiber is proportional to the RI as shown in the relation below:

$$n_o \sin \phi = (n_1^2 - n_2^2)^{1/2} \tag{7.3}$$

where n_o is the RI of the solution, ϕ is the numerical aperture (collection angle), n_1 is the RI of the fiber's core, and n_2 is the RI of the fiber cladding.

For example, the RI of water at 25 °C changes from 1.33262 to 1.3340 when 0.5 *M* K^+ is present. For less concentrated metal solutions, the change in RI is less significant. Potassium solutions less than 0.1 *M* produced a change of less than 0.1% in the NA of the fiber.[65] This small change in the NA will produce only negligible changes in the detected fluorescence intensity.

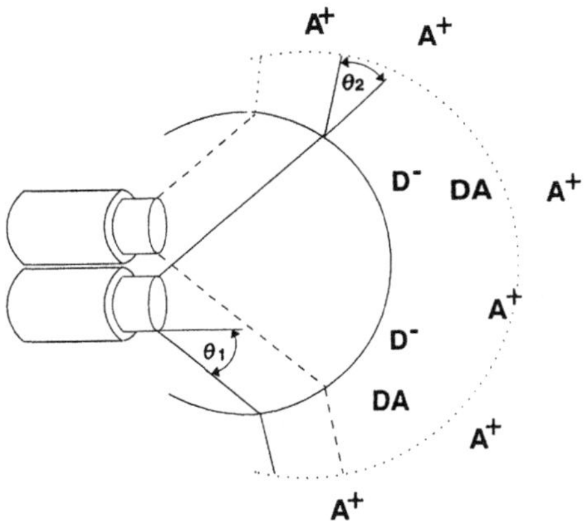

Figure 7.9. The region illuminated by the incident light source, solid line, is calculated by the incident angle θ_1. The signal collected by the optical fiber is represented by area shown with the dashed lines, where D represents the entrapped dye in the polymer matrix (- - -) and A represents the analyte.

7.2.4.3. Matrix

7.2.4.3a. Polymer Matrices. The use of immobilized dyes in a polymer matrix system has become popular in optical fiber applications.(25) In essence, the dyes are entrapped in a polymer matrix and the analyte diffuses through the matrix to interact with the dye. However, the immobilization of the molecular probe dye (MPD) to the distal end of the fiber is one of the most challenging tasks in optical fiber development. In order to collect the signal generated by the dye-analyte complex efficiently, the matrix should not interfere with or quench the fluorescent signal or influence the complexation ability of the dye. Additionally, the matrix must be properly secured to the fiber and leaching of the dye must be prevented. Further consideration should be given to the permeability of the matrix and the effect of the matrix on the response time.

The amount of quenching by the polymer can be calculated by the Stern–Volmer equation,

$$\frac{F}{F_o} = \frac{1}{1 + k_Q\,\tau_o[\mathrm{P}]} \tag{7.4}$$

where F_o is the fluorescence intensity in the absence of polymer, F is the fluorescence intensity in the presence of polymer, P is the concentration of the polymer, k_Q is the diffusion-controlled rate constant, and τ_O is the fluorescence lifetime of the dye. The concentration of the polymer inversely affects the fluorescence intensity of the dye by quenching. Therefore, to effectively reduce quenching of the dye by the polymer, the minimum concentration of polymer must be used.

Many different polymer matrices have been reported in the literature. The use of polyvinyl alcohol (PVA) as a matrix has been reported in the construction of a biosensor specific to NADH.(71) A detailed summary of the polymerization of PVA and the attachment of fluorescein to this polymer was reported by Seitz.(22) Plasticized polyvinyl chloride (PVC) has been used for the detection of lead.(72) PVC was also used as a matrix by Wolfbeis[73] and others[66, 74–76] for potassium determination using the inner filter effect. A polyacrylamide-methylenebis (acrylamide) copolymer (PAMAC) impregnated with fluoresceinamine for solution pH determination was reported by Kopelman.[24] The PAMAC was further photopolymerized at the tip of a silanized submicron fiber tip. Ethylene-vinyl acetate polymers have been used by Luo and Walt[77]; however, the dye was released from the matrix to obtain a response, thus limiting the lifetime of the probe. Rigid rod-type polymers such as poly (phenylquinoline) and poly (biphenylquinoline) have been used by Carey[78] for the determination of nitric acid. The polymer solutions were prepared and stored in an argon atmosphere to prevent crosslinking and degradation.

Poly(2-hydroxyethyl methylmethacrylate) (PHEMA) has been used as a matrix for the detection of metal ions.[79] A near-IR dye (2,3-naphthalocyanine-tetrasulphonic acid) was immobilized in a polymer matrix which was attached to the reaction phase of two optical fibers. A mixture of the matrix and the dye was prepared by mixing PHEMA and dye in a 60/40 ratio. The optimum ratio of polymer and dye were not fully investigated. The dye/polymer mixture was applied to the tip of the probe in 10- to 15-μL aliquots forming a thin coating on the probe after solvent evaporation as shown in Figure 7.9.

Anion-exchange membranes are alternative matrices for the physiological pH range.[80] Seitz and Zhujun[81] reported the use of an anion-exchange resin, amberlite CG-400, as the matrix for the detection of metal ions. Also the use of thin films of Nafion impregnated with a dye has been reported by Bright[82] for humidity measurements and by Patonay *et al.*[44] for pH determination. An optical sensor for sodium hydroxide measurements was reported by Burgess using bromothymol blue entrapped in Nafion.[52]

An alternative immobilization technique may be to covalently bind the dye to the polymer matrix or, ideally, directly to the support. Covalent binding would essentially eliminate the ability of the dye to leach from the matrix. Several NIR dyes which were suitably functionalized in our laboratory have shown good covalent binding efficiencies. Dyes containing reactive isothiocyanate groups have the capability of forming very stable thiourea bonds with amino groups. Dyes functionalized with nucleophilic amino groups can be reacted with electrophiles to form stable carbamates or amides. Covalent bonding has been demonstrated to be more stable and specific than noncovalent associations. Detailed investigations of Nafion polymer films structures have been published by Zen *et al.*[83] Similarly, electropolymerization of cobalt^{II} tetrakis(*p*-hydroxyphenyl)porphyrin directly onto an indium (tin) oxide glass slide has been reported.[51] This technique maybe useful in development of modified glass optical fibers.

A suitable matrix for dye immobilization would permit the dye to respond reversibly and sensitively to the concentration of an analyte or several analytes. The ideal matrix for fiber applications is still in the early stages of development.

7.2.4.3b. Matrix Thickness. The response of the probe can be controlled by the thickness of the membrane.[84] A thicker membrane allows the production of a larger signal due to the presence of more dye, but the response signal is slower. Also, if the polymer matrix or support is larger than 1–5 mm, the tip of the probe may be too large for certain applications. Additionally, thick membranes are difficult to successfully secure on the fiber tip. With thinner membranes, an equilibrium state is achieved more rapidly, and consequently a more rapid optical response is obtained. However, thinner membranes limit the amount of dye available in the matrix for complexation with analytes; thus, lower intensity fluorescence signals are generated. Thinner matrices have been utilized in a variety of applications. Small-diameter optical fibers have been reported for continuous response. A probe with a tip volume of less than 10 μm has been reported by Bobbitt *et al.*[5] The tip was modified with an octadecyl moiety to maximize the surface activation. A submicron chemical sensor has also been reported.[24] The tip of this sensor penetrated and measured pH samples in hole sizes varying from 0.02 to 10 μm.

The signal generated by the complex is governed by several physical phenomena associated with the matrix thickness. As soon as the probe is placed in contact with the analyte, external mass transfer controls the movement of the analyte toward the surface of the optical probe.[84] The osmotic pressure and Gibbs free energy dictate the permeation of the analyte into the matrix. Once the analyte has penetrated the matrix, internal mass transfer resistance controls the movement of the analyte in the matrix. Eventually, the probe reaches a steady state of equilibrium with molecules continuously moving in and out of the matrix.

The thickness of the membrane can be calculated by Fick's law of diffusion:

$$L = (\tau D_c)^{1/2} \tag{7.5}$$

where L is the thickness of the membrane, τ is the time for the response to reach a 95% steady-state signal, and D is the diffusion of the analytes which range from 10^{-8} (large molecules) to 10^{-10} cm^2/sec (typical values for ions).[3] Table 7.2 shows the response time due to membrane thickness.

Table 7.2. Response Times as a Function of Membrane Thickness[a]

External conditions	Thickness (mm)	$t_{1/2}$ (min)
Internal mesh		
pH 6.8–7.4	0.080	3.1
	0.307	5.4
0–30 mg%	0.080	2.4
	0.307	7.8
Internal mesh		
pH 7.1–7.4	0.129	2.5
	0.307	8.5, 9.8, 10.1
0–100 mg%	0.129	5.5
	0.307	5.4, 8.5, 13.3

[a]Ref. 110.

Typically, NIR optical fiber response times ($t_{95\%}$) for potassium solutions less than 5×10^{-4} M were less than 5.0 sec; however, longer response times were observed as the concentration increased. For concentrations above a $1 \times 10^{-3} M$ K^+, a typical $t_{95\%}$ was 1.25 min. The longer fluorescence response of the optical fiber detector at high metal concentrations is attributed to the thickness of the polymer matrix and the diffusion coefficient of potassium (10^{-10} cm^2/sec).

Because the matrix has permeable characteristics, the ability of the molecules to diffuse will depend on the pore size of the permeable polymer and the available free space in the pores. Obviously, larger molecules would not be able to penetrate the matrix and complex with the NIR dye. As the amount of the dye-analyte complexes increase, the amount of free space into which other analytes can diffuse decreases. Therefore, a relation of the mobility with free space is defined as

$$m = C_1 \, P(V_L) = A \exp\left(-\frac{V_{kr}}{V_f}\right) \tag{7.6}$$

where C_1 and A are constants, P is the probability constant, V_L is the hole free volume inside the polymer, V_f is the free volume of the polymer, and V_{kr} is the critical free volume proportional to the van der Waals volume (V_o). The critical volume is also defined as

$$V_{kr} = \beta \, V_o \tag{7.7}$$

where β is a dye-specific constant.[85]

The diffusion of ions through the polymer matrix has been investigated by Muller.[85] The mobility of the different analytes were found to parallel the diffusion constants of oxygen and ammonia and Table 7.3 describes ion diffusion in several polymers. Accordingly, the development of a polymer matrix for a specific application

Table 7.3. Relative Fluorescence Quantum Yield of Crystal Violet, ϕ_r, Oxygen Permeability, P_O, and Ammonia Diffusion Coefficient, D_c, at Room Temperature in Various Polymers[a]

Polymer	ϕ_r	P_O (×10^{15}) cm^2(s.Pa)	D_c (×10^{10}) cm^2/(sec)
Poly(vinyl alcohol) (2% acetate)	1	0.026	1.8
Poly(vinyl alcohol) (12% acetate)	0.73	0.026	6.6
Poly(methyl methacrylate)	0.71	29.1	24.1
Poly(vinyl butyral)	0.66	52.7	24.1
Poly(vinyl acetate)	0.50	81.0	48.2
Poly(*N*-vinyl-2-pyrrolidone)	0.14	1.5	216.0
Polystyrene	0.07	167.1	87.8
Methylcellulose	0.11	99.8	680.0
Cellulose diacetate	0.18	148.8	220.0
Cellulose triacetate	0.16	278.4	132.0

[a]Ref. 85.

Table 7.4. Stability Constants of Several Ionophores[a]

Ionophore	Na^+	K^+	Rb^+	Ca^{2+}	Sr^{2+}	Ba^{2+}
Monensin	5.8	5.0	—	—	—	—
Enniatin B	2.4	2.9	2.7	2.95	2.65	2.9
Valinimycin	0.7	4.9	5.3	2.70	2.2	3.3
Nonactin	2.4	3.7	3.6	—	—	1.7

[a]Ref. 87.

can be chosen from its ability to diffuse an analyte. All of these factors favor the use of NIR dyes, since the NIR dyes are typically large molecules.

7.2.4.3c. Selectivity. In principle optical fibers can be made selective to a particular analyte. The selectivity will be determined primarily by the functional group in the dye and by the nature of the matrix entrapping the dye.

Many dyes complex with their primary analyte due to attractive forces such as ionic charges. These charges are susceptible to nonspecific complexation with interfering analytes with characteristics similar to those of the primary analyte. Both the primary and interfering analytes compete for complexation with the same site. The NIR dyes may be synthesized with specific functional groups that will bond more specifically to an analyte. For instance, an isothiocyanate group forms very stable thiourea chemical bonds with proteins or an amino-modified DNA oligomer. The introduction of more specific and reversible functionalities on the dye molecule should minimize the interference of extraneous molecules or ions.

Another way to increase OFCD selectivity is by varying the matrix. Selectivity can be improved by choosing different polymer matrices or by designing a permeable polymer with active functional groups. However, if the matrix alters the degree of complexation with the dye, its utility is diminished.

Alternatively, the use of ionophores may enhance the selectivity of the matrix. Several ionophores have been used for the detection of metals. Simon reported a neutral ionophore with a Nile Blue derivative and an azo compound for the detection of potassium.[86] Table 7.4 shows the selectivity coefficients of several ionophores.

The construction of highly selective OFMP may also be accomplished by the introduction of crown moieties in the matrix. The size of the crown cavity determines which guest ions may diffuse through the matrix. Only certain metal ions are able to form complexes with given crown moieties as shown in Table 7.5. Different crown cavities have been shown to exhibit high degrees of selectivity for certain ions.[87] Valinomycin is a highly selective neutral carrier for potassium. A dibenzo-18-crown-6 or valinomycin reported by Suzuki showed selectivity for potassium even in the presence of other ions.[54] A silicon–rubber membrane electrode based on valinomycin has been reported by Simon.[88] The membrane was stable over a period of 65 h and maintained the ions properly. A carboxylic polyether antibiotic has been reported by Shirai for the determination of Mg^{II} and Ca^{II}, along with other metals.[69]

Table 7.5. Cavity Ring Size of Several Crown Ethers and Ionic Radius Diameter (Å)[a]

Crown	Cavity (Å)	Ion size	Ionic diameter (Å)
(14)-Crown-4	1.2–1.5	Li^+	1.36
(15)-Crown-5	1.7–2.2	Na^+	1.94
		Mg^+	1.56
		Ca^{2+}	2.12
(18)-Crown-6	2.6–3.2	K^+	2.66
		Ca^{2+}	1.98
		Cs^+	3.30
		Sr^{2+}	2.54
(21)-Crown-7	3.4–4.3	Rb^+	2.94
		Ba^{2+}	2.86

[a]Ref. 87.

7.2.4.4. NIR Molecular Probe Dye

NIR and visible molecular probe dyes utilize the same principle: changes in the microenvironment of the probe molecule induced by an analyte can be directly measured by spectral changes. The measured changes are dependent upon the concentration of the analyte and the extent of interaction between the dye and the analyte; however, the spectral properties of the dye determine the magnitude of the signal.

Few groups of chromophores exhibit absorbance or fluorescence activity due to electronic transitions in the NIR region. Furthermore, very few dyes are commercially available that could be used in probe applications. The limited commercial availability of NIR-absorbing fluorophores, especially functionalized dyes, is probably the major obstacle in the further development of NIR OFCD. However, the synthesis of several NIR dyes has been reported[39–43] and has been applied successfully in NIR OFCD.[17, 44, 79]

In order to prepare successful NIR molecular probe dyes, NIR dyes must meet the following criteria: adequate response to analytes, high lipophilicity and/or reactive functional groups, absorbance maxima compatible with available laser diodes, high fluorescence quantum yield, molar absorptivity, and high photostability.

7.2.4.4a. Response to Analytes. The main requisite for an NIR molecule to serve in a MPD is the ability to change an analytical characteristic in the presence of a particular analyte. Usually a functional group on the dye either associates or covalently attaches to an analyte inducing a change in the dye spectra. Ultimately, the chemical structure of the dye or the functional group will determine the site of complexation or chemical change.

Some biomolecules can induce spectral changes on NIR dyes. The fluorescence intensity of indocyanine green (ICG) increases in intensity when bound to proteins.[35] An increase in the fluorescence intensity signal of a cyanine dye was reported by

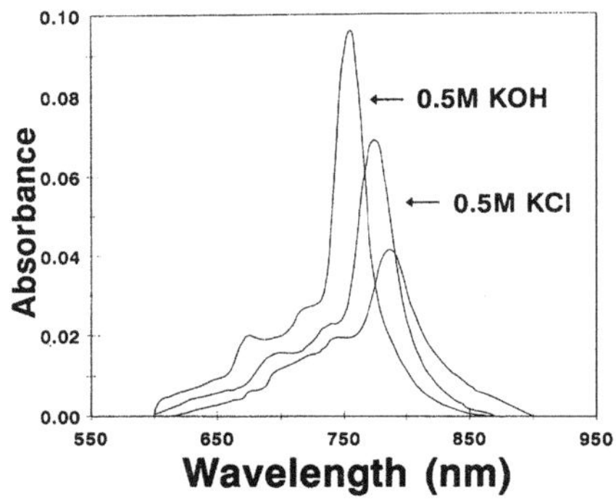

Figure 7.10. Spectral response of a near-IR dye in methanol (λ_{max} = 783 nm) in the presence of KOH (λ_{max} = 755 nm) and KCl (λ_{max} = 771 nm).

Ishibashi.[35] The increase in the fluorescence intensity was due to the separation of the dye dimers by the protein.

Different metal ions have been known to induce spectral changes in NIR dyes depending on the electron donating or electron withdrawing capacity of the metal ions. Since fluorescence is directly proportional to the excitation of the electrons, the disturbance of the electron cloud may affect the fluorescence intensity as the metal ion forms a complex with the NIR dye. To illustrate this phenomena fluorescence spectral changes of an NIR dye in methanol in the presence of varying potassium concentrations are shown in Figure 7.10.

Additionally, NIR dyes sensitive to pH change have been reported by Zen and Patonay[44] and Patonay *et al.*[17] The absorbance and fluorescence of these dyes changed by the action of the hydrogen ions in the chromophore allowing measurements at a particular wavelength.

7.2.4.4b. Lipophilicity and/or Reactive Functional Groups. Successful attachment of dye molecules onto the optical fiber is necessary to avoid leaching of the dye. This can be achieved by immobilization of the dye to the polymer matrix or by covalent binding directly to the optical fiber. In the former approach, NIR probe dyes should be highly lipophilic to enable easy immobilization into polymer matrices. Zen *et al.*[83] demonstrated that cyanine dyes were immobilized in the hydrophobic zone of Nafion films. The dye is stable under these conditions and continuous monitoring of pH can be accomplished for long periods of time. Alternatively, reactive functional groups on the dye can be used to covalently attach the dye molecule to the polymer matrix or, ideally, to the end of the optical fiber. The latter requirement is difficult to fulfill and much effort has been expended in the development of alternative immobilization techniques.

7.2.4.4c. Absorbance Maxima Considerations. For a molecule to absorb and fluoresce in the NIR region it must possess an extensive conjugated system. The fluorescence and absorbance maxima of these NIR dyes are dependent on the chemical structure of the chromophore. The spectroscopic properties of the dyes can be modified for a desired excitation wavelength by expanding the conjugation system or by adding a functional group to the system. In this case, the molecular probe dye can be matched to a particular experiment by controlling the absorbance maximum of the dye, especially to the output wavelength of commercially available lasers.

The availability of low-cost laser diodes has expanded the applications of NIR dyes.[34] NIR dyes have been used in analytical applications such as fiber probes[39, 79] and the detection of caustic brine.[89] However, most of the NIR dye applications have been in electrophotography for the manufacturing of office products such as laser printers and facsimiles. The wavelength of the lasers is not necessarily matched to the absorbance maxima of the dyes. Therefore, an understanding of the spectroscopic properties of NIR dyes and their ability to be chemically tuned is necessary to further expand the use of these dyes.

7.2.4.4d. Fluorescence Quantum Yield and Molar Absorptivity. Large fluorescence quantum yields and molar absorptivities are important characteristics of NIR dyes. The combination of these two characteristics permits lower detection limits or smaller analyte concentrations or thinner immobilization matrices (with fewer dye molecules) to be utilized. The quantum yield of the dyes can be enhanced by increasing the rigidity and the planar orientation of the dyes. Typical quantum yields of cyanine dyes can range from 10^{-4} to 0.73.[33, 90] Some carbocyanine dyes like the 3,3′-diethyl-2,2′-(4,5;4′,5′-dibenzo) thiatricarbocyanine iodide (DDTC)[31] and ICG[35] have molar absorptivities in excess of 150,000 M^{-1} cm^{-1} allowing detection limits in the picomolar range.

7.2.4.4e. Photostability. Many optical fiber applications require the dye to be under laser excitation for long periods of time. The amount of time that a molecule can spend in the excitation state, the bleaching lifetime (τ_b), is determine by the saturated electronic transition

$$\tau_b = \frac{1}{(k_{sat}\,\phi_d)} \tag{7.8}$$

where k_{sat} is the saturated absorption rate and ϕ_d is the photodestruction efficiency.

The large conjugation system required for NIR dyes may render the dyes unstable when continuous excitation is applied. But the best signal-to-noise ratios are obtained when the residence time approaches the average bleaching time of the dye. Therefore, the dye should be able to withstand continuous-wave laser excitations.

In some cases, due to the high output power of lasers which are often used as a light source, the dye may be susceptible to photobleaching. In addition, some NIR dyes may undergo spectral changes when exposed to sunlight over several hours and may also have a limited shelflife, even when kept in the dark. These are some of the present limitations that can be overcome with further development of new NIR dyes. One way to increase stability of NIR is the incorporation of the polymethine chain into cyclic structures.[33]

7.3. Optical Fiber Measurements

7.3.1. Metal Ion Determination

Steps have been taken by the World Health Organization (WHO) and the U.S. Environmental Protection Agency (EPA) to reduce the amount of toxic metal ions in the environment. For example, large concentrations of lead have been shown to be lethal to humans. The maximum amount of lead tolerated in drinking water according to the WHO and the EPA, is 0.05 mg/l[91] and 0.5 mg/l,[92] respectively. For this reason, innovative techniques to measure low concentrations of metal ions are emerging.

Most of the probes for the detection of metal ions use visible dyes. The development of OFCD for the detection of potassium has been reported by several authors. Suzuki[54] reported a sensor based on the absorbance changes ($\lambda_{max} = 513$ nm) of a dye entrapped in a crown-6 or valinomycin matrix. The solution pH had to be strictly maintained at 9.0. Interference of other ions in the signal of interest and poor detection limits (10^{-8} *M*) limited the use of this sensor. Another probe for potassium detection was reported by Ishibashi.[66] The sensor was based on dodecyl-acridine orange dye attached to polyvinyl chloride. The intensity of fluorescence ($\lambda_{em} = 593$ nm) was monitored in the presence of the ion. However, the dye leached from the matrix, thus limiting the probe lifetime. An innovative sensor was reported by Wolfbeis.[73] This sensor exploited the inner filter effect. Two dyes were entrapped in a polymer; one dye was a light absorber and the second was a fluorescer. The absorber dye filtered the light from the fluorescer dye under normal conditions. In the presence of metal ions, the absorber dye undergoes an absorbance shift, therefore, removing the filter effect from the fluorescer dye permitting an increase in fluorescence intensity. The sensor showed response to concentration of ions of about 10^{-7} *M*. The response time of the sensor was slow, preventing its use in rapid on-line measurements.

A fiber optic sensor for the determination of sodium was reported by Burgess.[52] A bifurcated fiber with a reference fiber 5 mm apart from the tip was used to observe the changes of bromothymol blue ($\lambda_{max} = 620$ nm) attached to Nafion in the presence of sodium ions. As the tip was saturated, the probe was renewed with fresh reagent. However, the epoxy holding the fibers was prone to damage from high sodium concentrations of around 2.5 *M* and the sensitivity of analysis was low.

The sensor reported by Shirai[69] used a natural carboxylic polyether antibiotic ($\lambda_{em} = 481$ nm) for the detection of magnesium and calcium. Detection limits of 10^{-5} and 10^{-4} *M*, respectively, were reported; but, interference from other metals was difficult to overcome. Ishibashi[69] used a bulkier hexadecyl-acridine orange dye ($\lambda_{em} = 525$ nm) plasticized in a PVC membrane for the fluorescent detection of ammonium ions. Signal interference due to superfluous ions and poor detection limits of 10^{-5} *M* restricted the use of the probe.

Several papers have been devoted to the detection of toxic metal ions. Some of the probes developed for toxic metals include a selective neutral ionophore for lead reported by Simon.[72] The probe reported absorbance changes at 665 nm with detection limits of about 10^{-12} M. However, maintenance of the solution pH was critical during the experiment, or it had to be monitored as a reference. In addition, any presence of sodium hydroxide in the site would limit the response of the sensor. An optical fiber for the determination of Al^{III}, Mg^{II}, Zn^{II}, and Cd^{II} was reported by Seitz.[81] The fluorescent signal of quinolin-8-ol-5-sulfonate (λ_{em} = 363 nm) was used to monitor the ion concentration. This sensor was subject to variation in the fluorescence intensity due to interfering ions and saturation of the dye by complexation with other ligands.

An NIR optical fiber for the detection of metal ions has been developed.[79] In a controlled environment, the optical probe with immobilized NIR dye was immersed in vials containing different metal ions varying in concentrations from 10^{-9} to 10^{-2} M. The probe response was obtained by the diffusion of the metal ions through the poly(2-hydroxyethyl methacrylate) (PHEMA) polymer matrix where the dye was covalently attached. On complexation of the metal with the dye, the intensity of the fluorescence signal increased.

The probe was selective for potassium determination; no significant fluorescence signal was observed with other metals. However, Na^+ and Li^+ induced a hypsochromic shift of the absorbance maximum in the NIR dye in free solution. Although the NIR dye offered some selectivity for potassium in analysis, the selectivity was enhanced further by the polymer matrix. Furthermore, the sensitivity of the probe to KOH and KCI was almost identical. This may indicate that the sensor response was not influenced by a pH change or the presence of anions. The fluorescence response of the immobilized NIR dye with KOH and KCI is shown in Figure 7.11.

To confirm that displacement of the complexed K^+ by other metals did not occur, the dye/K^+ complex was exposed to other metal ions. Although some variation in the fluorescence signal was observed, it was attributed to fluctuations in the fluorescence quantum yield of the dye due to changes in the ionic strength of the solution. There was clear evidence, however, to indicate ion selectivity.

Different theories have been applied to the mechanism of association and or quenching effects. The treatment of the association of molecules is pertinent to the system under investigation and theory of association should be derived with the

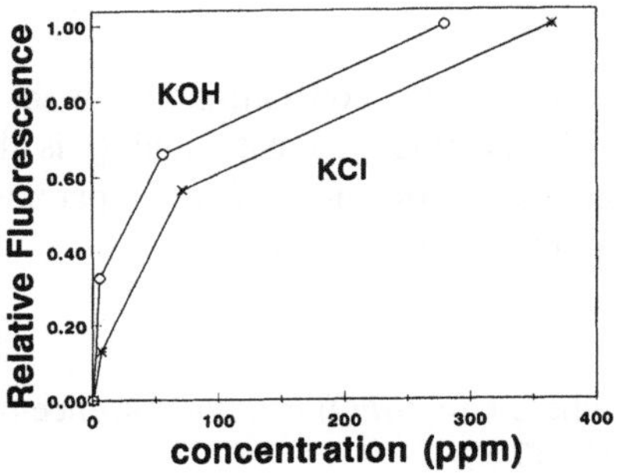

Figure 7.11. Near-IR optical probe response in the presence of KOH (o) and KCl (*) at different molar concentrations.

knowledge of the different constituents taking part in the reaction process. Some mathematical calculations on the fluorescent signal applied to fiber optics have been derived by Walt.[93]

The treatment of the fluorescence data can be represented theoretically by the way the probe was constructed. The number of parameters in the equations is dependent on and directly proportional to the number of components affecting the fluorescence signal. The theory described here is intended for a simple probe where the number of parameters involved in the equation have been kept to a minimum (i.e., the interaction of the NIR dye with the metal ions). In addition, it is assumed that no other components in the probe influence the properties of dye or metal.

The interaction of the dye with a metal ion, assuming a 1:1 stoichiometric relation, can be represented as

$$D^- + M^+ \rightleftharpoons DM \tag{7.9}$$

where D^- represents the uncomplexed NIR dye, M^+ is the free metal ions in solution, and DM is the dye-metal complex.

The equilibrium constant, K_{op}, is written as

$$K_{op} = \frac{[DM]}{[D^-]\,[M^-]} \tag{7.10}$$

where D^- is the dye immobilized in the matrix. (The total concentration of dye (D_T) (= $(D^-) + (DM)$) is the sum of the complexed (DM) and uncomplexed forms).

Substituting (D^-) in Eq. (7.10), the equilibrium constant can be rewritten as

$$K_{op} = \frac{[DM]}{([D_T] - [DM])\, a_{M^+}} \tag{7.11}$$

solving for (DM) and rearranging we obtain

$$[DM] = \frac{K_{op} D_T a_{M^+}}{(1 + K_{op} a_{M^+})} \tag{7.12}$$

The intensity of fluorescence is defined as

$$I_f = c\,d\,k\,I_l\,Q \tag{7.13}$$

where c is the concentration, d is the cuvette's path length, k is ln molar absorptivity, I_l is the laser intensity, and Q is the quantum yield. However, because only the concentration varies, all other parameters are considered constant during the analysis and the equation is written as

$$I_f = k\,[DM] \tag{7.14}$$

where k (= cdI_lQ) is a fluorescence proportionality constant. Substituting (DM) from Eq. (7.8)

$$I_f = \frac{k\,K_{op}\,D_T\,a_{M^+}}{(1 + K_{op}\,a_{M^+})} \tag{7.15}$$

which can be rearranged to the following equation

$$\frac{1}{I_f} = \frac{1}{kD_T} + \left(\frac{1}{kD_T K_{op}}\right)\frac{1}{a_{M^+}} \tag{7.16}$$

A plot of $1/I_f$ versus $1/a_{M^+}$ produces a straight line and the equilibrium constant, K_{op}, can be calculated from the slope. Even though some of the NIR dyes have more than one functional group for interaction with the analytes, the fluorescence intensity is still a function (albeit, more complex) of the concentration of the analyte; therefore, a calibration curve could be constructed.

The potential applications of NIR OFCD determination of metal ions are numerous. The detection of metal contaminants can be accomplished in real-time by using a portable fiber optical metal sensor (OFMD). Metal probe applications developed in the laboratory can be directly transferred to portable environmental applications with minimal effort. The response time of the NIR probe is comparable to its visible counterparts and is much faster than the traditional methods of metal analysis such as atomic absorption spectroscopy, polarography, and ion chromatography. With the use of OFMD results can be monitored on-site resulting in a significant reduction in labor cost and analysis time.

Possible environmental applications of NIR probes include the measurement of trace amounts of metal pollutants in surface and ground waters and water saturated soils and sediments and the detection of lead, chromium, and other heavy metal ions for efforts in pollution control. Also OFCD could assist in the determination of caustic soda and chlorine contents in wastewaters.

7.3.2. Solution pH Determination

The determination of hydrogen ion concentration (pH) is an important part of many analytical procedures. In solution, the color change of acid/base indicators is utilized to determined the solution pH. The color change which parallels absorbance and fluorescence maximum changes can be used in optical fiber research by following the increase or decrease of the intensity at a specific wavelength. Therefore, the pH sensitive dye must undergo a reversible change from the ionized to the nonionized form with a concomitant change in the degree of conjugation of the molecule which in turn will fluoresce into spectral changes.

The use of fluorescent dyes as pH indicators has been well documented in the literature.[14, 21, 23, 26, 94] Numerous indicators with spectral characteristics in the visible region have been tried as potential molecular probes.[13, 15, 16] But only few dyes have been reported for pH measurements in the NIR region.[17, 44]

One of the important aims in the application of optical fibers for solution pH is for physiological use. Peterson reported a probe that could measure pH over the physiological range of 7.0–7.4.[64] The probe had to be calibrated in buffers and could be used for tissue or blood analysis. The signal was determined by a ratio between the signal intensity of two visible wavelengths. Also, a fluorescent sensor for the pH range of 6.5–8.5 was reported by Seitz[80] using a trisodium salt derivative as the indicator.

Several NIR pH-sensitive dyes have been developed for the determination of solution pH in our laboratories.[17,46] A pH-sensitive dye 2,6-*bis*(2′-(1″-ethyl-3″,3″-dimethyl indolin-2″-ylidene)ethyl-idene)cyclohexanone showed two distinct absorption bands. Under basic conditions the dye existed in its ketone form and showed a broad absorbance band with a maximum at 531 nm which decreased with decreasing pH. The absorption spectra underwent a strong bathochromic shift in the presence of acids with a concomitant change in the fluorescence spectra. Around pH 3, the dye solution turned from a deep pink color to a pale green color with the appearance of a narrow absorbance band centered at 709 nm. The dependence of the relative fluorescence intensity of the NIR dye is shown in Figure 7.12. The change in the absorption spectra was attributed to the protonation of the oxygen atom which induced a shift to the cationic enol form of the dye.[17]

Another NIR pH-sensitive dye is a bis-carboxylic acid derivative of the 2-(4′-chloro-7-(2″-(1″-ethyl-3″,3″-dimethylindolenine))-3′,5′-(1‴,3‴-propanediyl-1′, 3′,5′-heptatrien-1′-yl)-1-ethyl-3,3-dimethyl indolenine bromide. The fluorescence signal of the dye in 86%, 75%, 60% and 50% (v/v of nafion) increased with the amount of dye present, following Beer's law. There was an obvious decrease in energy levels of the Q-band and an increase in the energy level of the B-band with an increase in pH.[44] The changes of the dye around pH 11–12.4 were reproducible. The shift in the useful pH range of the probe from the range expected from the pK_a value of the dye was due to the extremely acidic environment in the Nafion membrane. If a lower pH range is desired, the OFCD should be built using a different polymer support that will allow pH ranges comparable to the lower values observed in solution.[46]

Variations in absorption and fluorescence spectra of indicator dyes are usually associated with an alteration in the degree of conjugation on ionization of the

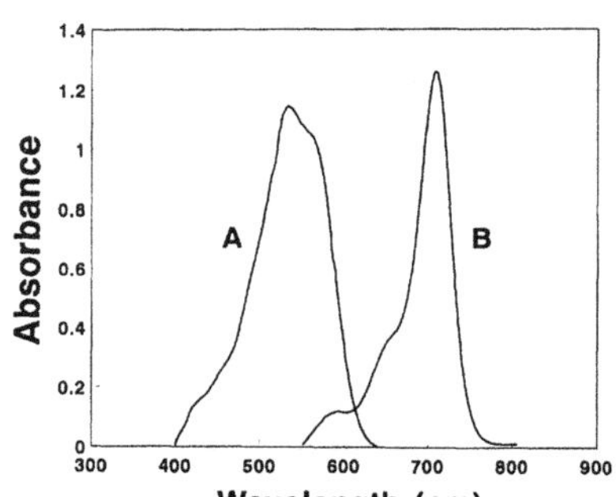

Figure 7.12. Typical absorbance spectra of a near-IR pH-sensitive dye in methanol. The absorbance maxima at pH 6.5 (A) is 535 nm and at pH 2 (B) is 709 nm.

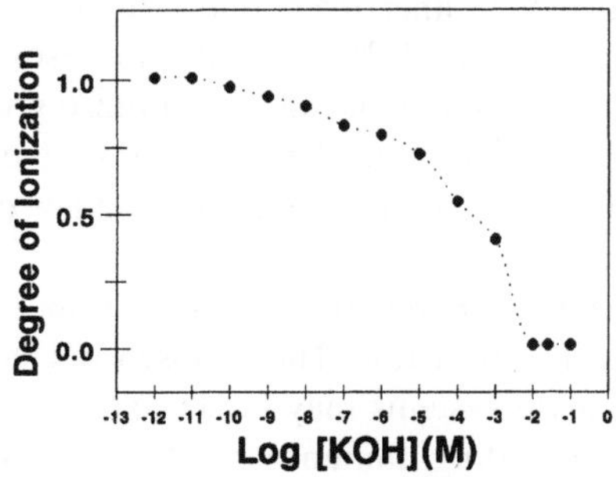

Figure 7.13. Degree of ionization for an NIR dye.

functional groups present in the indicator molecule. The degree of ionization of the NIR dye in the presence of hydrogen ions can be calculated by following the intensity changes at an specific wavelength. The wavelength is usually determined by the absorbance maximum of the dye without the presence of the analyte and followed throughout the investigation and can be represented by Eq. (7.17).

$$D_I = \frac{A - A_0}{A_1 - A_0} \tag{7.17}$$

where A is the absorbance of NIR-dye during analysis, A_0 is the absorbance of NIR-dye in the absence of ions, and A_1 is the absorbance of fully ionized NIR-dye. A plot of the degree of ionization against for a NIR dye is shown in Figure 7.13.

The protonation of the dye assuming a 1:1 ratio and that no other ions form part in the equation can be represented as

$$H^+ + D^- \rightleftharpoons HD \tag{7.18}$$

The optical fiber's response can be determined following the same steps as in the case for the metal ions. Hence we get

$$\frac{1}{I_f} = \frac{1}{kD_T} + \left(\frac{1}{kD_T K_{op}}\right)\frac{1}{10^{-pH}} \tag{7.19}$$

A plot of $1/I_f$ versus pH produces a straight line and the equilibrium constant, K_{op}, can be calculated from the slope.

7.3.3. Biosensors

Numerous enzyme-based biosensors have been described. Wang and Arnold reported the use of a dual-enzyme approach to induce spectral changes in NAD+ in the presence of glutamate.[95] Other techniques employing NAD^+ as an indicator molecule have also been reported.[68, 71, 96] A biosensor for measuring synaptic glutamate and extracellular ammonia using immobilized glutamate oxidase has been described.[97] Trettnak *et al.*[98] developed a biosensor for the determination of cholesterol via digestion by cholesterol oxidase. A detailed review of biocatalytic sensors has been reported elsewhere.[99]

There have been many recent reports of the use of fluorescence immunoassays in biosensors for analytical determination. These biosensors are also called immunosensors. Nonfluorescent analyte systems may be detected via immunometric, antibody sandwich, or competitive binding immunoassay techniques.[100, 101] The appropriate immunoassay for a fluorescent immunosensor depends on the chemical characteristics of the antigen-antibody system under investigation.[102] The simplest technique, the immunometric assay, involves incubation of antigen with a fluorescently-labeled antibody. The resulting antigen-antibody complex produces a fluorescent signal that is directly proportional to bound antigen. The antibody sandwich technique involves incubation of an antigen with a solid-support-bound capture antibody followed by the binding of a fluorescently labeled sandwich antibody. The resulting antibody sandwich produces a fluorescent signal directly proportional to bound antigen. The sensitivity of this technique is increased with increasing amounts of immobilized antibody. The competitive binding technique involves competition between fluorophore-labeled and unlabeled antigen for the limited number of antibody binding sites. The resulting fluorescent signal is inversely proportional to bound unlabeled antigen. The sensitivity of this technique increases with decreasing amounts of immobilized reagent.

Several fluorescent immunosensors have been reported; however, due to instrument availability and the lack of commercial NIR labels, most of the reported techniques utilize UV-visible dyes for detection. Bright *et al.*[103] reported a regenerable immunosensor for detecting human serum albumin (HSA) using an immunometric assay. The fluorescent signal of the visible label, dansyl chloride, attached to the immobilized antibody was generally quenched dynamically by the solvent. Upon complexation, HSA shielded the label from quenching, thus greatly enhancing the observed fluorescent signal. This method resulted in significant nonspecific binding to the antibody thus increasing the background fluorescence. Additionally, the method lacked adequate sensitivity with detection limits around 1 ppm.

Ogasawara[5] fabricated an immunometric biosensor for the measurement of riboflavin binding protein. Fluorogenic riboflavin was noncovalently attached to an optical fiber. On complexation with the protein, the fluorescent signal of riboflavin was quenched. Leaching of the riboflavin was observed during the analysis. Several immunosensors for the determination of benzo(a)pyrene (BaP) have been reported by Vo-Dinh *et al.*[6–9] Antibodies induced against BaP-immunoglobulin G (IgG) were immobilized on the fiber tip. The inherent fluorescence of BaP was measured after binding to the antibodies. Detection limits as low as 200 p*M* were obtained. A biosensor reported by Vo-Dinh *et al.* utilized immobilized anti-IgG.[104] The IgG immunosensor is based on the principle of competitive binding between IgG and fluorescein isothiocyanate (FITC)-labeled IgG for the anti-IgG immobilized on the distal sensing tip of a quartz optical fiber.

An immunosensor based on a competitive fluorescence energy-transfer immunoassay was reported by Anderson[105] for the measurement of phenytoin. Texas red-labeled antibody was incubated with a phenytoin derivative. On displacement of the derivative by the antigen, the change in the fluorescence signal was recorded. Detection limits approached 5 μM with response times ranging from 5 to 30 min.

Tatsu *et al.*[106] reported a novel immunosensor using immobilized liposomes doped with carboxyfluorescein and dinitrophenyl (DNP) hapten on the tip of an optical fiber. On complement-mediated immunolysis by anti-DNP-antibody, the fluorescent signal of the liberated carboxyfluorescein was measured.

The detection of flu viruses via a fluorescent sandwich immunoassay was reported by Bucher.[10] However, the method sensitivity was too low for direct detection of the virus. A novel sandwich immunoassay was described by Ogert[107] for the detection of Botulinum Toxin A. Antibodies specific for *Clostridium botulinum* were covalently attached to the surface of a tapered fiber. After the capture of the antigen, a sandwich was formed with a rhodamine-labeled anti-toxin IgG, and the evanescent wave was measured. The assay was highly specific with detection limits near 5 ppb.

The applications of visible fluorescent immunosensors described are all susceptible to large interferences from biomolecules such as bilirubin and porphyrins. A more comprehensive review of immunosensors has been published by Robinson.[108] The spectral interferences associated with visible fluorophores have prompted the design of an NIR optical immunosensor techniques with high sensitivity and low interference.

An NIR biosensor coupled with an NIR fluorescent sandwich immunoassay has been developed.[109] The capture antibody was immobilized on the distal end of an optical fiber sensor. The probe was incubated in the corresponding antigen with consecutive incubation in an NIR-labeled sandwich antibody. The resulting NIR-labeled "antibody sandwich" was excited with the NIR beam of a laser diode, and a fluorescent signal that was directly proportional to the bound antigen was emitted. The sensitivity of the technique increased with increasing amounts of immobilized receptor. There are several factors involved in the preparation of the sandwich type biosensor. A schematic preparation of the sandwich optical fiber is shown in Figure 7.14.

7.3.3.1. NIR Immunosensor Surface Activation Time

The surface activation time of the polymer required for maximum site activity for binding antibodies was determined by evaluating the fluorescence intensity of a series of probes incubated in inorganic acids at different times. Knowledge of the surface activation time is necessary to obtain maximum activation, thus allowing maximum antibody immobilization on the probe. Consequently, lower detection limits may be achieved.

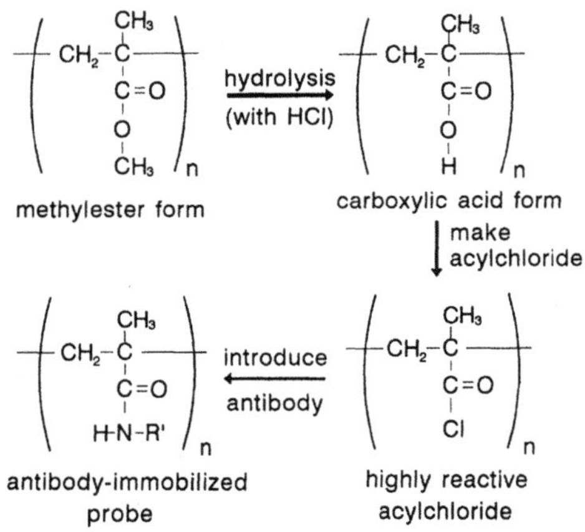

Figure 7.14. Typical chemistry used for the preparation of a sandwich-type bioprobe.

Different probes were activated by incubating using acyl chloride technology for 0.5, 1, 5, 10, 15, 20, and 50 min, respectively. Anti-human IgG was then immobilized on each of the probes. To evaluate the binding activity of these activated probes each probe was placed in a commercially available dye, IR-144 (2.5×10^{-5} *M*), or other NIR dye synthesized in our laboratories. The fluorescence response for each dye was monitored for approximately 720 sec. This way the number of activated sites were proportional to the fluorescence signal, since only the activated sites could immobilized the dye. The same procedure was repeated using three acyl chlorides (PCl_3, PCl_5, and $SOCl_2$) separately to determine the acid which will provide the most sites. The probes made with PCl_3 incubated for 30–60 sec produced the maximum signal response. Times longer than 5 min caused the probe to disintegrate and activation times less than 30 sec produced a small fluorescence signal due to inadequate activation. The optimized conditions were used in all subsequent experiments.

After the probes were incubated with PCl_3, the degree of activation was determined using a competing agent, ethyl acetate, which possessed similar functional groups as the fluorescent dyes but did not fluoresce. The probe was immersed in a vial containing water in order to establish a baseline followed by immersion in a 10^{-4} *M* ethyl acetate solution. The probe was placed in a 10^{-4} *M* NIR dye solution. The observed fluorescent signal remained constant since the competing agent completely saturated the sites in the probe. By bypassing the competing agent complexation, the fluorescent signal increased considerably.

7.3.3.2. Antibody-Immobilized Incubation Period

In order to determine the minimum incubation time required for maximum antibody immobilization on the probe's reaction phase, the activated probes were

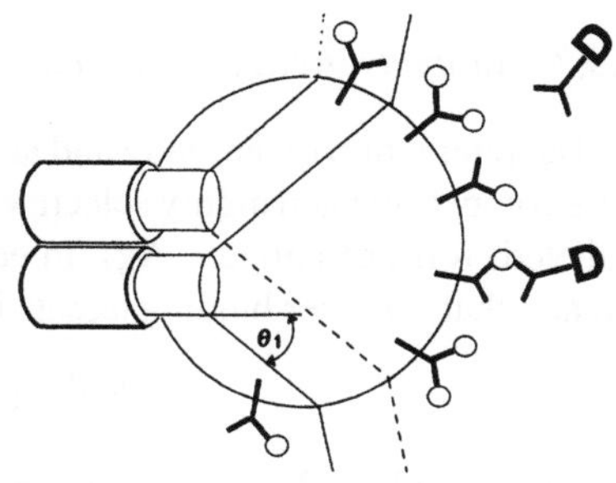

Figure 7.15. Schematic of a NIR optical sandwich-type bioprobe. D, NIR, dye; >-, antibody; o, antigen.

incubated for variable amounts of time with antibody. Following incubation with the antibody, the probe was placed in a 10^{-4} *M* solution of NIR dye. The fluorescence of the NIR dye complexation with the antibody was directly proportional to the amount of antibody immobilized on the probe. The probe was monitored from 24 to 50 h. Incubation periods over 50 h were not attempted since the fluorescence signal reached a maximum steady state at 30 h. The optimal antibody immobilization time was determined to be 45–48 h. All of the successive probes were incubated for approximately 48 h.

The NIR labeling of antibodies was performed using techniques similar to that of visible labeling techniques. Amino groups on the antibody reacted with the isothiocyanate group on the NIR dye molecules to form thiourea linkages. During our studies NIR dyes with -SCN functional groups have been used extensively. A detailed synthesis of these NIR dyes has been described elsewhere.[39–43] A final sandwich probe is shown in Figure 7.15.

The detection limits of the antibody-sandwich bioprobe was monitored by immersing the probe into vials containing variable antigen concentrations ranging from 10 to 100 ng/ml as shown in Figure 7.16.

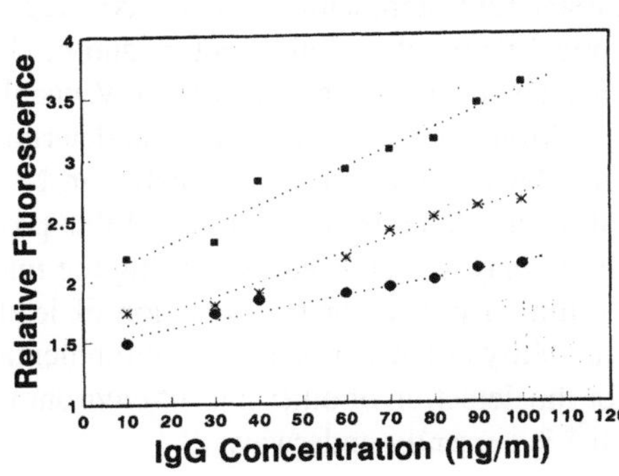

Figure 7.16. Near-IR optical probe response with changes in immunoglobulin G (IgG) concentration. Experimental data of 50 ppm (circles), 100 ppm (stars), and 200 ppm (squares) concentrations of dye is shown.

7.3.3.3. *Antibody–Antigen Interactions*

The interaction of antibody and antigen can be represented in a simple form based on the competitive and highly selective antibody (Ab) and an appropriately labeled (or unlabeled) form of antigen (Ag). In equilibrium the binding process assuming a 1:1 ratio and that no other binding agents is competing for the same site can written as

$$[\mathrm{Ab}] + [\mathrm{Ag}] \rightleftharpoons [\mathrm{Ab{\cdot}Ag}] \tag{7.20}$$

where (Ab) and (Ag) represent the antibody and the antigen concentrations, respectively. Because the equation describing equilibrium of the antibody-antigen complexation is identical to the metal-dye complexation of the chemical sensors described previously, K_{op} can be determined similarly from the slope of a plot of $1/I_f$ versus 1/(Ag).

Direct labeling of NIR fluorophores to antibodies for immunochemical techniques in conjunction with fiber optic sensing is capable of improving the selectivity, sensitivity, and simplicity of the method while increasing its reliability and accuracy. Furthermore, the results indicate that this technique can be useful as an alternative method for investigation and detection of specific disease-related antigens in biological samples. A present limitation of this method is the need for relatively large concentration of immobilized antibody but this should improve in the near future. At large antigen concentrations, the antibody is the limiting factor resulting in nonlinearity. In addition, the retention of antibody "activity"—the ability of the immobilized antibody to recognize its specific antigen—influences both the method sensitivity and selectivity.

7.4. Conclusion

Exploring the fluorescence in the NIR region of the spectrum for potential optical fiber applications is advantageous due to the inherently low biomolecular interference in this region. In addition, the availability of inexpensive optical fibers coupled with laser diodes with output wavelength in the NIR region provides a new opportunity to expand OFD applications into the NIR region. It is therefore beneficial to develop NIR optical fibers chemical detectors and biodetectors for solving tasks that, in the past, have been solved using standard UV/visible fluorescence techniques.

Molecular probe dyes for the determination of potassium, lithium, and sodium have been identified. Additionally, an NIR probe selective for potassium has been fabricated. The detection limits of this probe are in the ppm range. Lower detection limits may be achieved by varying the matrix which allows the entrapment of ions. Preliminary data for the detection of lead and cadmium demonstrate the potential capability of these probes for environmental applications. The development of OFMP for the detection of other ions of environmental interest such as Be^{2+}, Hg^{2+}, As^{3+}, and Ni^{2+} is currently underway.

The data shown here demonstrate the use of an NIR dye immobilized in a polymer matrix which is sensitive to changes in metal ion concentration and solution pH. Additionally, the utility of NIR dyes in immunosensors has been shown. Although the use of NIR dyes in fiber optic techniques has only recently been demonstrated, the range of potential applications is abundant. As functionalized NIR dyes become commercially available or new, large-scale syntheses are reported, the use of NIR dyes in bioanalytical applications will most certainly expand.

Acknowledgments

Some of the research described here was supported in part by a grant from the National Institutes of Health (Grant 1 R01 AI 28903-01 A2) and in part by a grant from the National Science Foundation (CHE-890456).

References

1. J. Janata, Chemical sensors, *Anal. Chem. 64*, 196R–219R (1992).
2. O. S. Wolfbeis, Chemical sensors—survey and trends, *Fresenius J. Anal. Chem. 337*, 522–527 (1993).
3. J. Janata, Ion optodes, *Anal. Chem. 64*, 921A–927A (1992).
4. W. R. Seitz, Chemical sensors based on fiber optics, *Anal. Chem. 56*, 16A–34A (1984).
5. F. K. Ogasawara, Y. Wang, and D. R. Bobbitt, Dynamically modified, biospecific optical fiber sensor for riboflavin binding protein based on hydrophobically associated 3-octylriboflavin, *Anal. Chem. 64*, 1637–1642 (1992).
6. T. Vo-Dinh, G. D. Griffin, K. R. Ambrose, and M. J. Spaniak, Fiberoptics immunofluorescence spectroscopy for chemical and biomedical monitoring, in: *Polyaromatic Hydrocarbons: A Decade of Progress* (M. Cooke and A. J. Dennis, eds.), pp. 885–890, Battelle Press, Columbus, Ohio (1988).
7. T. Vo-Dinh, G. D. Griffin, and K. R. Ambrose, A portable fiberoptic monitor for fluorimetric bioassays, *Appl. Spectrosc. 40*, 696–700 (1986).
8. R. M. Santella, T. Vo-Dinh, B. J. Tromberg, M. J. Spaniak, and J. P. Alarie, Development of antibody-based fiber-optic sensors for detection of a benzo[a]pyrene metabolite, *Anal. Chem. 60*, 1901–1908 (1988).
9. T. Vo-Dinh, J. P. Alarie, R. W. Johnson, M. J. Sepaniak, and R. M. Santella, Evaluation of the fiber-optic antibody-based fluoroimmunosensor for DNS abducts in human placenta samples, *Clin. Chem. 37*, 532–535 (1991).
10. D. J. Bucher, A. Mikhail, S. Popple, P. Graves, G. Meiklejohn, D. S. Hodes, K. Johansson, and P. E. Halonen, Rapid detection of Type A Influenza viruses with monoclonal antibodies to the M protein (M1) by enzyme-linked immunosorbent assay and time-resolved fluoroimmunoassay, *J. Clin. Microbiol. 29*, 2484–2488 (1991).
11. O. S. Wolfbeis, *Fiber Optic Chemical Sensors and Biosensors*, Vol. 1, CRC Press, Boca Raton, Florida (1991).
12. O. S. Wolfbeis, *Fiber Optic Chemical Sensors and Biosensors*, Vol. 2, CRC Press, Boca Raton, Florida (1991).
13. M. A. Arnold and T. J. Ostler, Fiber optic ammonia gas sensing probe, *Anal. Chem. 58*, 1137–1140 (1988).

14. T. P. Jones and M. D. Porter, Optical pH sensor based on the chemical modification of a porous film, *Anal. Chem. 60*, 404–406 (1988).
15. Y. V. Kostov, Immobilized bromophenol blue as sensing element in optical pH sensor, *Sensors and Actuat. 8*, 99–101 (1992).
16. Z. M. Shakhsher and W. R. Seitz, Optical detection of cationic surfactants based on ion pairing with an environment-sensitive fluorophore, *Anal. Chem. 62*, 1758–1762 (1990).
17. G. Patonay, G. A. Casay, M. Lipowska, and L. Strekowski, Spectroscopic studies of a near-infrared absorbing pH sensitive aminodienone-carbocyanine dye system, *Talanta 40*, 935–942 (1993).
18. M. C. Moreno, M. Jimenez, C. P. Conde, and C. Camara, Analytical performance of an optical pH sensor for acid-base titration, *Anal Chim Acta 230*, 35–40 (1990).
19. R. Narayanaswamy and F. Sevilla III, Reflectometric study of the acid-base equilibria of indicators immobilized on a styrene/divinylbenzene copolymer, *Anal. Chim. Acta 189*, 365–369 (1986).
20. F. V. Bright, Bioanalytical applications of fluorescence spectroscopy, *Anal. Chem. 60*, 1031A–1039A (1988).
21. M. A. Arnold, Fiber-optic chemical sensors, Anal. Chem. 64, 1015A–1025A (1992).
22. D. C. Sundberg, Z. Zhujun, Y. Zhang, M. Wangbai, R. Russell, Z. M. Shakhsher, C. L. Grant, and W. R. Seitz, Poly(vinyl alcohol) as a substrate for indicator immobilization for fiber-optic chemical sensors, *Anal. Chem. 61*, 202–205 (1989).
23. Y. Kawabata, T. Kamichika, T. Imasaka, and N. Ishibashi, Fiber-optic sensor for carbon dioxide with a pH indicator dispersed in a poly(ethylene glycol) membrane, *Anal Chim Acta, 219*, 223–229 (1989).
24. W. Tan, Z.-Y. Shi, and R. Kopelman, Development of submicron chemical fiber optic sensors, *Anal. Chem. 64*, 2985–2990 (1992).
25. C. Munkholm, D. R.Walt, F. P. Milanovich, and S. M. Klainer, Polymer modification of fiber optic chemical sensors as a method of enhancing fluorescence signal for pH measurement, *Anal. Chem. 58*, 1427–1430 (1986).
26. L. A. Saari and W. R. Seitz, pH sensor based on immobilized fluoresceinamine, *Anal. Chem. 54*, 821–823 (1982).
27. E. Koller, M. Kriechbaum, and O. S. Wolfbeis, 1-Aminopyrene-3,6,8-trisulfonate: A fluorescent probe for thiamine and pyridinium ion, *Spectroscopy 3*, 37–43 (1988).
28. R. Huttenrauch, S. Fricke, and K. Matthey, On the relation between water structure and fluorescence quenching, *Pharmazie 37*, 225–230 (1982).
29. Y. Kubota, Y. Motoda, Y. Shigemune, and Y. Fujisaki, Fluorescence quenching of 10-methylacridinium chloride by nucleotides, *Photochem. Photobiol. 37*, 1099 (1979).
30. T. Ando and H. Miyata, Pyruvate as a fluorescence quencher: A new spectroscopic assay for pyruvate reactions, *Anal. Biochem. 129*, 170–175 (1983).
31. T. Imasaka, A. Yoshitake, and N. Ishibashi, Semiconductor laser fluorimetry in the near-IR region, *Anal. Chem. 56*, 1077–1079 (1984).
32. W. Kaye, Near-infrared spectroscopy, I. Spectral identification and analytical applications, *Spectrochim. Acta 6*, 257–287 (1954).
33. G. Patonay and M. D. Antoine, Near-infrared fluorogenic labels: New approach to an old problem, *Anal. Chem. 63*, 321A–327A (1991).
34. J. Fabian, H. Nakazumi, and M. Matsuoka, Near-infrared absorbing dyes, *Chem. Rev. 92*, 1197–1226 (1993).
35. K. Sauda, T. Imasaka, and N. Ishibashi, Determination of protein in human serum by high-performance liquid chromatography with semiconductor laser fluorometric detection, *Anal. Chem. 58*, 2649–2653 (1986).
36. M. Matsuoka, *Infrared Absorbing Dyes*, Plenum Press, New York, (1990).
37. A. M. Schaffer and M. Gouterman, Porphyrins XXV. Extended Huckel calculations on location and spectral effects of free base protons, *Theoret. Chim. Acta 25*, 62–82 (1972).

38. L. Lee, N. H. Sabelli, and P. R. LeBreton, Theoretical characterization of phthalocyanine, tetraazaporphyrin, tetrabenzoporphyrin, and porphyrin electronic spectra, *J. Phys. Chem. 86*, 3926–3931 (1982).
39. T. A. Shatskaya, M. G. Gal'pern, V. R. Skvarchenko, and E. A. Luk'yanets, Phthalocyanines and allied compounds. XXX.* Tetra-6,7-bibenzobarreleno-2,3-naphthalocyanines, *Zhurnal Obshchei Khimii 57*(10), 2364–2368 (1987).
40. A. Vogler and H. Kunkely, Template synthesis and optical spectra of zinc-2,3-napthalocyanine, *Inorgan. Chim. Acta 44*, L209–L210 (1980).
41. L. Strekowski, M. Lipowska, and G. Patonay, Facile derivatizations of heptamethine cyanine dyes. *Synth. Comm. 22*(17), 2593–2598 (1992).
42. L. Strekowski, M. Lipowska and G. Patonay, Substitution reactions of a nucleofugal group in heptamethine cyanine dyes. Synthesis of an isothiocyanato derivative for labeling of proteins with a near-infrared chromophore, *J. Org. Chem. 57*, 4578–4580 (1992).
43. R. B. Mujumdar, L. A. Ernst, S. R. Mujumdar, and A. S. Waggoner, Cyanine dye labeling reagents containing isothiocyanate groups, *Cytometry, 10*, 11–19 (1989).
44. J.-M. Zen and G. Patonay, Near-infrared fluorescence probe for pH determination, *Anal. Chem. 63*, 2934–2938 (1991).
45. J. Hicks, D. Andrews-Wilberforce, and G. Patonay, A near-infrared laser diode excitation source for an SLM 8000C Fluorometer, *Anal. Instrum. 19*, 29–47 (1990).
46. G. Patonay, J.-M. Zen, and T. Czuppon, Near-infrared laser diodes in monitoring applications, *SPIE's International Symposium OE/Environmental and Process Monitoring Technologies, 1637*, 142–150 (1992).
47. E. Unger and G. Patonay, Near-infrared laser diode intracavity absorption spectrometry, *Anal. Chem. 61*, 1425–1427 (1989).
48. T. Okazaki, T. Imasaka, and N. Ishibashi, Optical-fiber sensor based on the second-harmonic emission of a near-infrared semiconductor laser as light source, *Anal Chim Acta 209*, 327–331 (1988).
49. Sharp User Manual, Sharp Instruments, Japan, (1986).
50. Spectra Diode Catalog, Spectra Diode Laboratory, San Jose, California (1993).
51. T. L. Blair, J. R. Allen, S. Daunert, and L. G. Bachas, Potentiometric and fiber optic sensors for pH based on an electropolymerized cobalt porphyrin, *Anal. Chem. 65*, 2155–2158 (1993).
52. R. J. Berman, G. D. Christian, and L. W. Burgess, Measurement of sodium hydroxide concentration with a renewable reagent-based fiber-optic sensor, *Anal. Chem. 62*, 2066–2071 (1990).
53. T. Rosatzin, P. Holy, K. Seiler, B. Rusterholz, and W. Simon, Immobilization of components in polymer membrane-based calcium-selective bulk optodes, *Anal. Chem. 64*, 2029–2035 (1992).
54. K. Suzuki, H. Ohzora, K. Tohda, K. Miyazaki, K. Watanabe, H. Inoue, and T. Shirai, Fibre-optic potassium ion sensors based on a neutral ionophore and a novel lipophilic anionic dye, *Anal Chim Acta 237*, 155–164 (1990).
55. S. A. Soper, Q. L. Mattingly, and P. Vegunta, Photon burts detection of single near-infrared fluorescent molecules, *Anal. Chem. 65*, 740–747 (1993).
56. P. A. Johnson, T. E. Barber, B. W. Smith, and J. D. Winefordner, Ultralow detection limits for an organic dye determined by fluorescence spectroscopy with laser diode excitation, *Anal. Chem. 61*, 861–863 (1989).
57. L.-Q. Li and L. M. Davis, Single photon avalanche diode for single molecule detection, *Rev. Sci. Instrum. 64*, 1524–1529 (1993).
58. R. G. W. Brown, K. D. Ridley, and J. G. Rarity, Characterization of silicon avalanche photodiodes for photon correlation measurements. 1: Passive quenching, *Appl. Opt. 25*, 4122–4126 (1986).
59. P. J. Treado, I. W. Levin, and E. N. Lewis, Near-infrared acousto-optic filtered spectroscopic microscopy: A solid-state approach to chemical, *Imaging. Appl. Spect. 46*, 553–559 (1991).

60. C. D. Tran and R. J. Furlan, Spectrofluorometer Based on Acoustic-Optic Tunable Filters for rapid scanning and multicomponent sample Analyses, *Anal. Chem. 65*, 1675–1681 (1993).
61. C. D. Tran and R. J. Furlan, Acoustic-optic tunable filter as a polychromator and its application in multidimensional fluorescence spectrometry, *Anal. Chem. 64*, 2775–2782 (1992).
62. R. E. Dessy, Waveguide as chemical sensors, *Anal. Chem. 61*, 1079A–1094A (1989).
63. J. O. W. Norris, Current status and prospects for the use of optical fibres in chemical analysis, *Analyst 114*, 1359–1372 (1989).
64. J. I. Peterson, S. R. Goldstein, R. V. Fitzgerald, and D. K. Buckhold, Fiber optic pH probe for physiological use, *Anal. Chem. 52*, 864–869 (1980).
65. R. B. Thompson, Fluorescence-based fiber-optic sensors, in: *Topics in Fluorescence Spectroscopy*, Volume 2 (J. R. Lakowicz, ed.), pp. 345–365, Plenum Press, New York (1991).
66. Y. Kawabata, R. Tahara, T. Kamichika, T. Imasaka, and N. Ishibashi, Fiber-optic potassium ion sensor using alkyl-acridine orange in plasticized poly(vinyl chloride) membrane, *Anal. Chem. 62*, 1528–1531 (1990).
67. S. Muto, A. Fukasawa, M. Kamimura, F. Shinmura, and H. Ito, Fiber humidity sensor using fluorescent dye-doped plastics, *Jpn. J. Appl. Phys. 28*, L1065–L1066 (1989).
68. J. Wangsa and M. A. Arnold, Fiber-optic biosensor based on the fluorometric detection of reduced nicotinamide adeneine dinucleotide, *Anal. Chem. 60*, 1080–1082 (1988).
69. K. Suziki, K. Tohda, Y. Tanda, H. Ohzora, S. Nishihama, H. Inoue, and T. Shirai, Fiber-optic magnesium and calcium ion sensor based on a neutral carboxylic polyether antibiotic, *Anal. Chem. 61*, 382–384 (1989).
70. G. A. Casay, T. Czuppon, N. Narayanan, J. Lipowski, and G. Patonay, Near-infrared fluorophores and semiconductor lasers in analytical applications, *J. Proc. and Qual. Contr. 5*, 35–44 (1993).
71. S. M. Gautier, I. J. Blum, and P. R. Coulet, Bioluminescence-based fibre optic sensor with entrapped co-reactant: An approach for designing a self-contained biosensor, *Anal. Chim. Acta 243*, 149–156 (1991).
72. M. Lerchi, E. Bakker, B. Rusterholz, and W. Simon, Lead-selective bulk optodes based on neutral ionophores with subnanomolar detection limits, *Anal. Chem. 64*, 1534–1540 (1992).
73. H. He, H. Li, G. Mohr, B. Kovacs, T. Werner, and O. S.Wolfbeis, Novel type of ion-selective fluorosensor based on the inner filter effect: An optrode for potassium, *Anal. Chem. 65*, 123–127 (1993).
74. Y. Kawabata, T. Kamichika, T. Imasaka, and N. Ishibashi, Ion-selective optrode using hexadecyl-acridine orange attached on poly(vinyl chloride) membrane, *Anal. Chem. 62*, 2054–2055 (1990).
75. H. He, G. Uray, and O. S. Wolfbeis, Enantioselective optodes, *Anal Chim Acta 246*, 251–257 (1991).
76. W. E. Morf, K. Seiler, B. Rusterholz, and W. Simon, Design of a calcium-selective optode membrane based on neutral ionophores, *Anal. Chem. 62*, 738–742 (1990).
77. S. Luo and D. R. Walt, Fiber-optic sensors based on reagent delivery with controlled-release polymers, *Anal. Chem. 61*, 174–177 (1989).
78. W. P. Carey and B. S. Jorgensen, Optical sensors for high acidities based on fluorescent polymers, *Appl. Spect. 45*, 834–838 (1991).
79. G. A. Casay, T. Czuppon, J. Lipowski, and G. Patonay, Near-infrared fluorescence probes, *International Symposium OE/Biomedical Optics*, Los Angeles, CA, January 16–22 (1993).
80. Z. Zhujun and W. R. Seitz, A fluorescence sensor for quantifying pH in the range from 6.5 to 8.5, *Anal.. Chim. Acta 160*, 47–55 (1984).
81. Z. Zhujun and W. R. Seitz, A fluorescent sensor for aluminum(III), magnesium(II), zinc(II) and cadmium(II) based on electrostatically immobilized quinolin-8-ol sulfonate, *Anal. Chim. Acta 171*, 251–258 (1985).
82. K. S. Litwiler, P. M. Kluczynski, and F. V. Bright, Determination of the transduction mechanism for optical sensors based on rhodamine 6G impregnated perfluorosulfonate films using steady-state and frequency-domain fluorescence, *Anal. Chem. 63*, 797–802 (1991).

83. J-M. Zen, M. Lipowska, and G. Patonay, Near-infrared probe for in situ chraracterization of Nafion thin films, *J. Appl. Polym. Sci. 46*, 1167–1176 (1992).
84. L. D. Bowers, Applications of immobilized biocatalysts in chemical analysis, *Anal. Chem. 58*, 513A–530A (1986).
85. D. Anwand, F. W. Muller, B. Strehmel, and K. Schiller, Determination of the molecular mobility and the free volume of thin polymeric films with fluorescence probes, *Makromol. Chem. 192*, 1981–1991 (1991).
86. E. Bakker, M. Lerchi, T. Rosatzin, B. Rusterholz, and W. Simon, Synthesis and characterization of neutral hydrogen ion-selective chromoionophores for use in bulk optodes, *Anal Chim Acta 278*, 211–225 (1993).
87. B. Dietrich, Coordination chemistry of alkali and alkaline-earth cations with macrocyclic ligands, *J. Chem. Educ. 62*, 954–964 (1993).
88. J. Pick, K. Toth, E. Pungor, M. Vasak, and W. Simon, A potassium-selective silicone-rubber membrane electrode based on neutral carrier, *Anal Chim Acta 64*, 477–480 (1973).
89. M. K. Phelan, C. H. Barlow, J. J. Kelly, T. M. Jinguji, and J. B. Callis, Measurement of caustic and caustic brine solutions by spectroscopic detection of the hydroxide ion in the near-infrared region, 700–1150 nm, *Anal. Chem. 61*, 1419–1424 (1989).
90. R. C. Benson and H. A. Kues, *J. Chem. Eng. Data 22*, 379 (1977).
91. World Health Organization, *Guidelines for Drinking-Water Quality*, Vol. 1, Geneva, (1984).
92. U.S. Environmental Protection Agency and Office of Water and Hazardous Materials, *Quality Criteria for Water*, Washington, D.C. (1976).
93. P. Yuan and D. R.Walt, Calculation for fluorescence modulation by absorbing species and its applications to measurements using optical fibers, *Anal. Chem. 59*, 2391–2394 (1987).
94. L. Brown, P. J. Halling, G. A. Johnston, and C. J. Suckling, Water insoluble indicators for the measurement of pH in water immiscible solvents, *Tetrahedron Lett. 31*, 5799–5802 (1990).
95. A-J. Wang and M. A. Arnold, Dual-enzyme fiber-optic biosensor for glutamate based on nicotinamide adenine dinucleotide luminescence, *Anal. Chem. 64*, 1051–1055 (1992).
96. C. Schelp, T. Scheper, F. Buckmann, and K. F. Reardon, Two fibre-optic sensors with confined enzymes and coenzymes: Development and applications, *Anal Chim Acta 255*, 223–229 (1991).
97. S. Kar and M. A. Arnold, Fiber-optic ammonia sensor for measuring synaptic glutamate and extracellular ammonia, *Anal. Chem. 64*, 2438–2443 (1992).
98. W. Trettnak and O. S. Wolfbeis, A fiberoptic cholesterol biosensor with an oxygen optrode as the transducer, *Anal. Biochem. 184* (1), 124–127 (1990).
99. C. R. Lowe, B. F. Hin, D. C. Cullen, S. E. Evans, L. D. Stephens, and P. Maynard, Biosensors, *J. Chromatogr. 510*, 347–354 (1990).
100. R. Ekins, F. Chu and E. Biggart, Development of microspots multi-analyte ratiometric immunoassay using dual fluorescent-labelled antibodies, *Anal Chim Acta 227*, 73–96 (1989).
101. R. M. Nakamura, Y. Kasahara, and G. A. Rechnitz, *Immunochemical Assavs and Biosensors Technology for the 1990s*, American Society for Microbiology, Washington, D.C. (1992).
102. T. VoDinh, G. D. Griffin, and M. J. Sepaniak, Fiber optic immunosensors, in: *Fiber Optic Chemical Sensors and Biosensors*, Vol. 2 (O. S. Wolfbeis, ed.), pp. 217–257, CRC Press, Inc, Boston (1993).
103. F. V. Bright, T. A. Betts, and K. S. Litwiler, Regenerable fiber-optic-based immunosensor, *Anal. Chem. 62*, 1065–1069 (1990).
104. B. J. Tromberg, M. J. Sepaniak, T. VoDinh, and G. D. Griffin, Fiber-optic chemical sensors for competitive binding fluoroimmunoassay, *Anal. Chem. 59*, 1226–1230 (1987).
105. F. P. Anderson and W. G. Miller, Fiber optic immunochemical sensor for continuous, reversible measurement of phenytoin, *Clin. Chem. 34*(7), 1417–1421 (1988).
106. Y. Tatsu, S. Yamamura, and S. Yoshikawa, Fluorescent fibre-optic immunosensing system based on complement lysis of liposome containing carboxyfluorescein, *Biosens. Bioelectron. 7*(10), 741–745 (1992).

107. R. A. Ogert, J. E. Brown, B. R. Singh, L. C. Shriver-Lake, and F. S. Ligler, Detection of Clostridium botulinum toxin A using a fiber optic-based biosensor. *Anal. Biochem.* *205*(2), 306–312 (1992).
108. G. A. Robinson, Immunosensors, *Biosens. Bioelectron.* *6*(3), 183–191 (1991).
109. M. Daneshvar, G. A. Casay, M. Lipowska, G. Patonay, and L. Strekowski, Investigation of a near-infrared fiber-optic immunosensor, *SPIE—The International Society for Optical Engineering; Chemical, Biochemical, and Environmental Fiber Sensors V 2063*, 128–138 (1993).

8

Fluorescence Spectroscopy in Turbid Media and Tissues

Dieter Oelkrug

8.1. Introduction

Most organic and inorganic materials on Earth are strongly light scattering and many of them are fluorescent. Examples are found in the fields of biology, mineralogy, technology, or chemistry (plant and animal tissue, minerals, pigments, papers and printing products, polymer blends, optically brightened products, microdisperse adsorption systems, stationary chromatographic phases, and so on). Quantitative fluorimetry of such systems starts with the absorption of light that is strongly modified by the physical process of multiple scattering and by possible chemical interactions of the absorbers with the scattering environment. The absorption in scattering media is usually investigated with the methods of regular and diffuse reflectance or transmittance spectroscopy,[1,2] the photometric laws of which are well understood under large area of irradiation and to some extent also under spot irradiation.[3–5] The photophysical and photochemical processes following absorption can be investigated nowadays very sensitively by fluorimetric methods,[6–11] but compared to transparent systems it is still a significant problem to quantify the spectral, temporal, and spatial distribution of the emitted light fluxes correctly. The following chapter presents some of the recent theoretical and experimental developments in fluorescence spectroscopy of turbid media, mixing review and original results and concentrating especially on the limit of dominant scattering, i.e., low absorptivity and high probability of multiple scattering. This limit is the most interesting one since fluorimetry is a powerful method especially in trace analysis and not in highly concentrated systems, and the case of low scattering can easily be extrapolated from the more general results of the high scattering limit.

Dieter Oelkrug • Institute of Physical and Theoretical Chemistry, University of Tübingen, DW-7400 Tübingen, Germany.

Topics in Fluorescence Spectroscopy, Volume 4: Probe Design and Chemical Sensing, edited by Joseph R. Lakowicz. Plenum Press, New York, 1994.

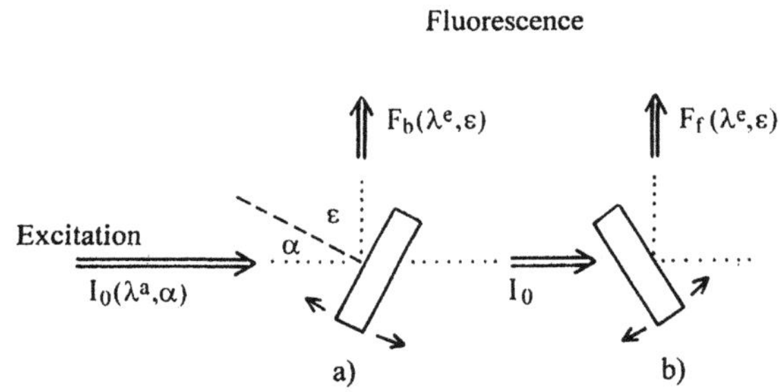

Figure 8.1. Scheme of fluorescence measurement in backward direction (sample a) and forward direction (sample b). For details see text.

8.2. Basic Photometric Quantities

All samples in this chapter are assumed to consist of multiple scattering and fluorescent planar layers of arbitrary thickness d. One surface of the layer is irradiated monochromatically at λ^a with parallel light of intensity I_0 (per unit area perpendicular to the direction of propagation) under the angle α_0 relative to the normal of the surface

Table 8.1. Definitions of Some Photometric Quantities

d, z	Layer thickness, depth coordinate
λ^a, λ^e	Wavelengths of fluorescence excitation (a) and emission (e)
$\alpha, \mu = \cos\alpha$	Angle of irradiation (α_0) and propagation of light
ε	Angle of detection
$\mu_0 I_0 \equiv 1$	Intensity of collimated irradiation, $z = 0$
$\pi I_{0,\text{diff}} \equiv 1$	Intensity of diffuse irradiation, $z = 0$
$\mu_0 I_d = T_c$	Collimated transmittance, $z = d$
$\pi I_{d,\text{diff}} = T_d$	Diffuse transmittance, $z = d$
$\pi J_{0,\text{diff}} = R_0$	Diffuse reflectance, $z = 0$
$\pi J_{0,\text{diff}} = R_\infty$	Diffuse reflectance of a semi-infinite layer
$\pi J^e_{0,\text{diff}} = F_b$	Backward fluorescence intensity at λ_e, $z = 0$
$\pi J^e_{0,\text{diff}} = F_f$	Forward fluorescence intensity at λ_e, $z = d$
σ^a, σ^e	Scattering coefficient of the sample / cm^{-1}
κ^a, κ^e	Absorption coefficient of the sample / cm^{-1}
κ^a_f	Absorption coefficient of the fluorophore / cm^{-1}
$q^e = \kappa^a_f \cdot \varphi^e_f$	Emission coefficient of the fluorophore / cm^{-1}
φ^e_f	Spectral fluorescence quantum yield
$K^{a,e}, S^{a,e}$	Kubelka–Munk absorption and scattering coefficients / cm^{-1}

(see Figure 8.1), or the layer is irradiated diffusely with intensity $I_{0,\mathrm{diff}}$ (per solid angle and unit area). Part of the light is reflected specularly at the surface. The main fraction of light penetrates into the sample, will be absorbed with the absorption coefficient κ^a, and scattered from its original direction with the scattering coefficient σ^a. Both coefficients are defined per unit length of propagating light in arbitrary direction inside the sample, i.e., the coefficients can be tensors. The absorbed light is emitted at λ^e by fluorescence with the emission coefficient q^e. The radiation densities inside the sample are not directly accessible (with the exception of very voluminous low-scattering samples where the detector inside the sample does not disturb the light fluxes too much), but they can be calculated from the experimental quantities and photometric parameters that are summarized in Table 8.1.

8.3. Experimental Methods

8.3.1. Conventional Fluorimeters

The samples are excited and observed in most commercial fluorimeters under fixed geometry where the two light beams form an angle of 90° or less. This geometry can be used also for multiple scattering samples (see Figure 8.1) but it should be borne in mind that most of the nonabsorbed incident radiation is diffusely reflected or transmitted in all direction of the corresponding hemisphere (R_0 and T_d, respectively), i.e., under the same geometrical conditions as the emitted fluorescences F_b and F_f. Therefore, special attention has to be paid to the straylight level of the monochromators, especially when low fluorescence light levels are detected. It is recommended to use double monochromators of good quality or—if only single monochromators are available—an additional series of well-selected band edge and bandpass filters. Part of the incident light is also reflected specularly. This part can be suppressed by setting $\alpha \neq \varepsilon$, as shown in Figure 8.1.

Some features of the spectra obtained with conventional equipment are illustrated with unsaturated hydrocarbons as fluorophores that have been adsorbed from gas or liquid phase on highly porous metal oxide powders as scattering substrates.

Figure 8.2 presents the fluorescence of pyrene on silica gel. The loading is low so that pyrene is predominantly adsorbed as nonaggregated monomers (M_1). The backward fluorescence spectrum F_b of this sample is very comparable to the spectrum in polar solvents and not distorted by reabsorption. However, the forward spectrum F_f is almost completely suppressed in the region of overlap with the *o–o*-transition and hot sidebands of the weak first absorption band S_1. The absorption coefficients of the sample vary widely from $\kappa^a \approx 0.1$ cm^{-1} (S_1-band, $\lambda^a \approx 350$–370 nm) to $\kappa^a \approx 25$ cm^{-1} (S_2-band, $\lambda^a \approx 290$–340 nm), and in a first approximation the excitation spectrum of F_b reflects this variation correctly (Figure 8.2, *left*). The F_f-excitation spectrum, however, has only little in common with the real absorption spectrum of the sample.

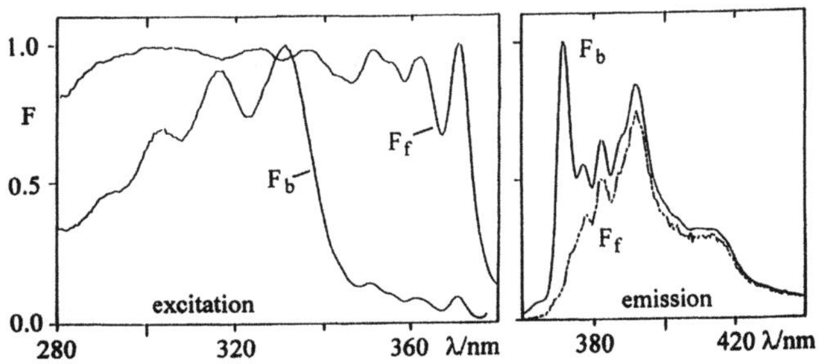

Figure 8.2. Fluorescence emission and excitation spectra of a strongly scattering sample as detected in forward (F_f) and backward (F_b) direction. (Pyrene adsorbed on silica gel, surface loading $c_{pyr} = 5.5 \cdot 10^{-7}$ mol g^{-1}, spec. surface area $a_{silica} = 500$ m^2 g^{-1}).

The weak S_1-band dominates and the strong S_2-band is completely misrepresented. All these facts will be discussed more detailed in Section 8.6 on quantitative fluorimetry.

Figure 8.3 presents a sample similar to Figure 8.2 but with five times higher surface loading. Pyrene is now partly adsorbed as aggregates (M_n) that emit the well-known unstructured excimer spectrum at $\lambda^e_{max} = 475$ nm. Figure 8.4 (top) shows the excitation spectra of pyrene monomers and aggregates (M_1 and M_n) on silica gel with a layer thickness of $d = 1$ cm. The M_n-spectrum is strongly redshifted against M_1 as in the crystal and this could be interpreted with a very high aggregation number n of the surface aggregates. In reality, the strange shape of the M_n-spectrum is only a consequence of inner filtering by the M_1-majority. This becomes evident from the spectra of Figure 8.4 (bottom) that have been taken from the same sample but with a very low layer thickness, i.e., with only a few granula of the adsorbent[12] (for details see Section 8.6.2). The structure of the weak S_1-band is now hidden under the background but the strong S_2-band shows clearly that the "real" spectrum of M_n is very

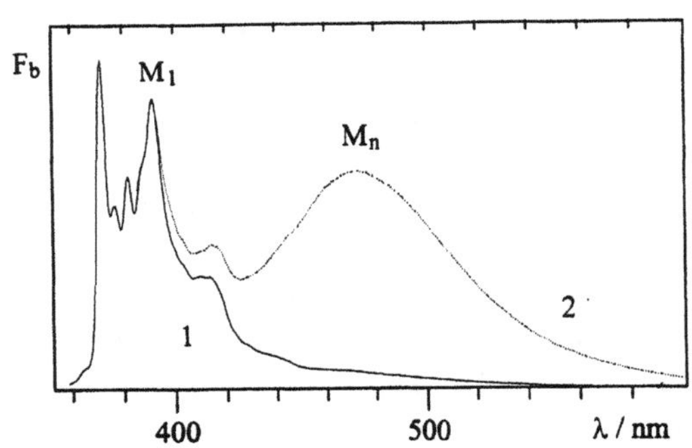

Figure 8.3. Concentration effect on the fluorescence spectra of pyrene on silica gel (1) c = 0.5 μmol g^{-1}, (2) c = 3 μmol g^{-1}. M_1 = monomer, M_n = aggregate fluorescence. λ^a = 340 nm (close to the absorption maximum of M_n).

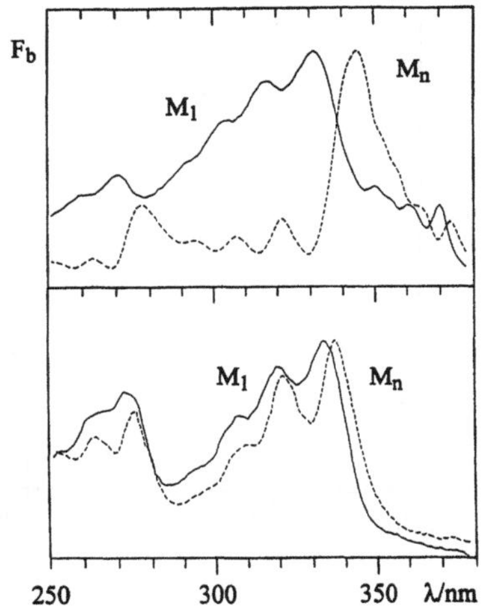

Figure 8.4. Fluorescence excitation spectra of pyrene (sample two of Figure 8.3) measured at λ^e = 392 nm (region of M_1-emission) and λ^e = 500 nm (region of M_n-emission); *upper panel*: thick layer (d = 1 cm), *lower panel*: thin layer (single grains).

similar to that of M_1 with a redshift of 4–5 nm, i.e., much less than in the crystal so that the aggregation number on the surface must be very small.

8.3.2. Diode Array Spectrometers

In cases where remote sensing or free variability of α and ε are desired, this can be achieved with wave guides that are optically adjusted to the entrance or exit port of the sample compartment. The waveguide technique can also be successfully combined with a sensitized diode array spectrometer that allows fast recording of whole spectra (at somewhat lower sensitivity compared to conventional recording with cooled photomultipliers) and time-resolved measurements of not too fast decaying fluorescence processes. Figures 8.5 and 8.6 show examples of fluorescent tissues the spectra of which have been recorded *in vivo* with a gated diode array in combination with a single grating as dispersive element and monochromatized irradiation from a Xenon (Xe)-arc lamp. The spectrum of the palm is obtainable only in the reflectance mode. Fluorescence originates mainly from nicotinamide-adenine dinucleotide (NADH), but one has to be cautious in assigning the spectrum because of possible distortions by additional long-wavelength absorbers, as can be seen in the diffuse reflectance absorption spectrum of the palm also presented in Figure 8.5. The spectrum of the leaf from *Ficus elastica* was recorded in four different ways in order to demonstrate the distortions of a two-component fluorescence system (chlorophyll a and b), where the fluorophores are additionally not homogeneously distributed over the sample depth. Details of the necessary spectral correction will be given in Section 8.6, below.

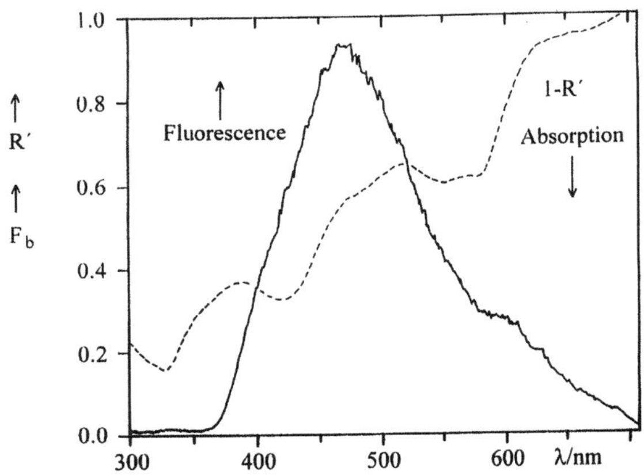

Figure 8.5. Palm of the hand. Fluorescence spectrum (F_b-mode, full line) and diffuse reflectance absorption spectrum (dashed line, normalized to λ = 700 nm, reference: filter paper) measured with a sensitized diode array spectrometer.

8.3.3. Time-Resolved Measurements

In principle, the same techniques are applicable as in transparent media, with the restriction that very fast processes are distorted by the time-of-flight dispersion of the incident and emitted photons (see Section 8.4). Straylight can also produce severe problems in the short time range if high-intensity laser pulses are used as excitation sources. Compared to transparent samples the intensity of straylight, traces of the fundamental line, or of the second harmonics generated at the substrate interface, will be scattered with an extremely high probability into the detector and will easily produce overload artifacts. Using gated diode array detectors the overload effect can be reduced. Figure 8.7 shows as an example the delayed fluorescence spectrum of terthienyl on silica gel that is by several orders of magnitude less intense than the prompt fluorescence. The two emission processes can be clearly separated with the gating technique after delay times of $\Delta t > 50$ nsec because the delayed part can be manipulated with temperature and completely frozen out at $T < 200$K. This effect is very often found in adsorbed species.[13–15] It is explained by a diffusion controlled annihilation process between two adsorbates in the excited triplet state $T_1 + T_1 \rightarrow S_1$

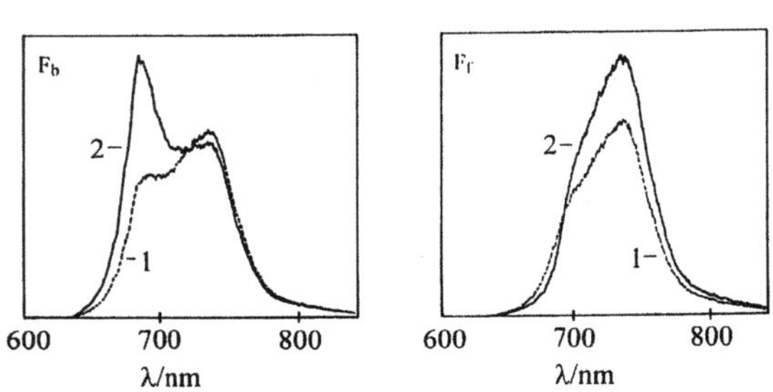

Figure 8.6. Fluorescence spectra of ficus elastica measured in backward direction (F_b, left) and forward direction (F_f, right). Excitation at λ^a = 520 nm. Curve 1: from top surface, curve 2: from undersurface.

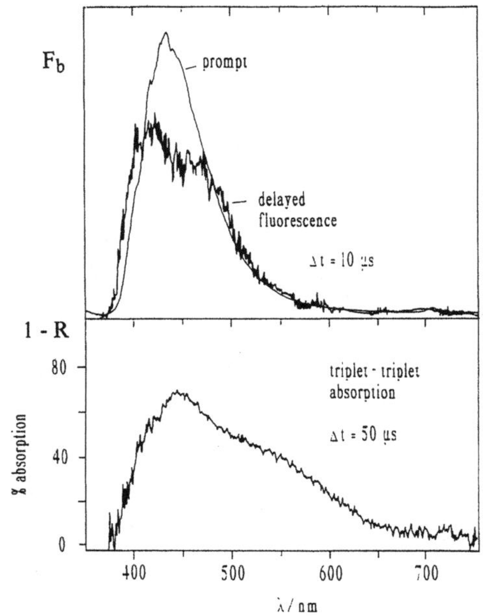

Figure 8.7. Delayed fluorescence and diffuse reflectance transient absorption spectroscopy on scattering substrates. Example: terthienyl on silica gel excited with λ^a = 354 nm (neodymium/yttrium-aluminum-garnet) (Nd/YAG) laser pulse of 10 nsec, 20 mJ), recorded with a gated diode array spectrometer.

$+ S_0 \rightarrow 2S_0 + F_{del}$. The importance of this reaction increases with the triplet formation yield which is very high in terthienyl, with the mobility in the adsorbed state which is generally low for large organic molecules and almost zero at $T < 200$K, and with the triplet lifetime which is at room temperature generally much higher than in solution but not high enough for terthienyl ($\tau = 80\,\mu$sec), so that in total the delayed fluorescence of Figure 8.7 is only a side-reaction. In contrast, quite dominant delayed fluorescence was found in adsorbed acridines,[(13–15)] only because their triplet lifetimes are by several orders of magnitude higher. On low-intensity irradiation it is worth mentioning that, as a rule, the fluorescence in heterogeneous systems is decaying nonexponentially. Examples are chlorophyll fluorescence in living plants[(16)] and numerous aromatic hydrocarbons or dyes at solid/gas interfaces.[(8, 9, 17–21)] Figure 8.8 shows exemplary decay curves of 9,10-diphenylanthracene on alumina, recorded under different experimental conditions. Despite the low reactivity of the fluorophore, its high fluorescence quantum yield, and its low tendency to form aggregates or excimers, the decay curves are clearly nonexponential in almost all cases. A series of explanations have been proposed for such a kind of results:

1. Dispersion of the radiative rate constant by local variations of the refractive index at the solid/gas interface. This could explain the tailing of the decay curves even at very low loadings, with lifetime components that are two to three times as long as the intrinsic radiative lifetimes in solution.[(8)] This could also explain the disappearance

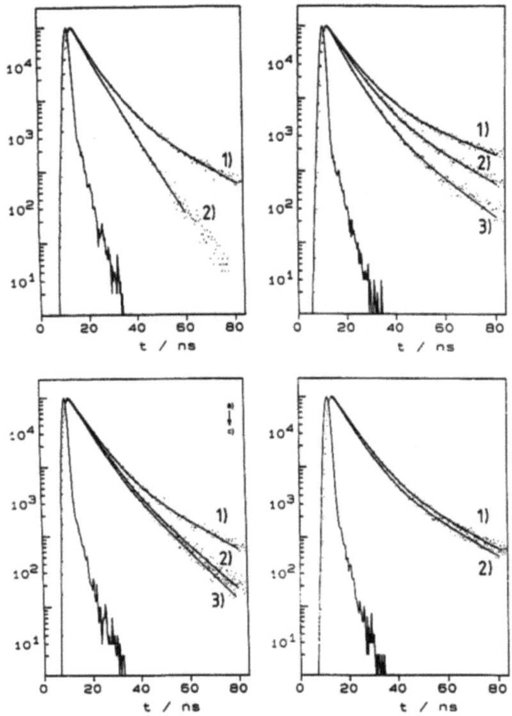

Figure 8.8. Examples of nonexponential fluorescence decay curves; 9,10-diphenylanthracene on alumina for chromatographic purposes (Uhl,Oelkrug, unpublished results) (unnumbered curves: time profiles of the excitation pulse, λ^a = 360 nm). *Upper left*: effect of environment (1) high vacuum, (2) liquid *n*-hexane, $c = 3\ \mu$mol g^{-1}, $\lambda^e = 440$ nm. *Upper right*: effect of fluorescence wavelength (1) $\lambda^e = 500$ nm, (2) 440 nm, (3) 406 nm; $c = 3\ \mu$mol g^{-1}. *Lower left*: effect of surface loading (1) 3 μmol g^{-1}, (2) 0.13 μmol g^{-1}, (3) 0.02 μmol g^{-1}; $\lambda^e = 440$. *Lower right*: effect of sample thickness (1) $d \rightarrow \infty$, (2) $d \rightarrow 0$; $c = 3\ \mu$mol g^{-1}, $\lambda^e = 440$ nm.

of the long tail by adding a dielectric environment with higher refractive index (Figure 8.8, *upper left*).

2. Slow reorganization dynamics in the adsorbate/adsorbent complex after excitation of the Franck–Condon state. This could explain the spectral redshift of fluorescence with time (Figure 8.8, *upper right*, and Figure 8.9). In liquid solution, the excited state equilibrates on the picosecond timescale, but is has been shown[(23)] that this process can slow down on surfaces to 10–50 nsec.

3. Aggregation of the fluorophores on surfaces or in small compartments. The aggregate spectra and decay times usually differ from the monomers and the degree of aggregation increases with fluorophore concentration. The spectra of pyrene on silica (Figure 8.3) are an illustrative example for this behavior. The interaction between 9,10-diphenylanthracenes is expected to be much weaker because of the bulky substituents, and correspondingly smaller are the spectral changes. Nevertheless the concentration dependence of the decay curves (Figure 8.8, *lower left*) may be explained by aggregation. Then also the spectral changes of Figure 8.9 may be due to aggregates.

4. Reabsorption and reemission. In scattering media the effect of reabsorption plays a very important role (see Section 6.3). The consecutive reemission will slow down the decay process (see Section 6.4). The effect is not very big, but it can be

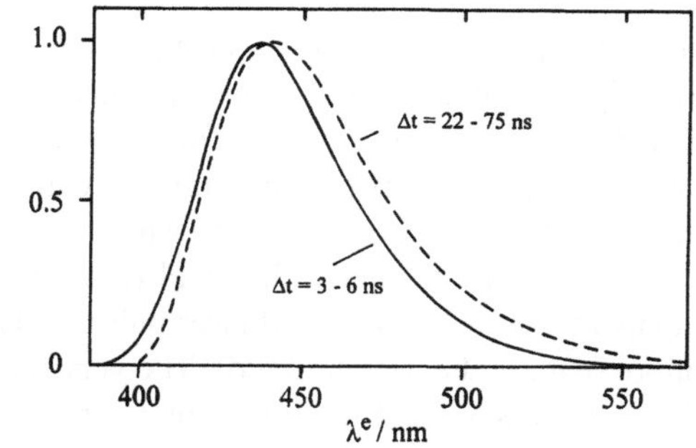

Figure 8.9. Time-resolved fluorescence spectra of 9,10-diphenylanthracene, recorded with time-correlated single photon counting, λ^a = 360 nm. Parameters: gate width and delay time relative to the intensity maximum of the excitation pulse.

avoided by investigating very thin layers, where reabsorption becomes negligible (Figure 8.8, *lower right*).

5. Energy transfer. Energy acceptors can be the substrate[23] as well as suitable adsorbates.[24, 25] In both cases the decay curves become nonexponential and, especially in the initial part, much steeper than the spontaneous decay.

6. Time-of-flight (TOF) dispersion. This effect creates significant tailing in the decay curves of fast fluorescence processes. For details see Section 8.4.4.

7. Cage effect. Electromagnetic radiation will be caged in small compartments of μm-size according to interference. Quantumelectrodynamic calculations propose an increase of the radiation density up to a factor of 10^6, and correspondingly the release of radiation will be damped.

Some methods of quantitative analysis of nonexponential fluorescence decay curves will be shortly described in Section 6.6.

8.3.4. Locally Resolved Measurements

Fluorescence microscopy gains increasing interest in biomedicine or in microphotometric color measurement of printing products with fluorescent pigments and optical brighteners. The local resolution limit of conventional microscopy is in the region of 200–300 nm. This limit increases in scattering media depending on the layer thickness and scattering coefficient of the preparation. Under focused laser irradiation the focal diameter increases approximately with $\sigma^{1/2}$ and linearly with the sample depth. Thus also the resolution of confocal microscopy is reduced.

In total, the fluorescence intensity will increase by scattering, but localization will be less certain because the fluorescence of a given volume element may have its origin from (1) excitation by unscattered primary radiation, (2) excitation by scattered primary radiation, or (3) scattering or reflection of fluorescence that has been originally created in other regions (ghosts). Some of these unwanted effects can be reduced by appropriate illumination and detection. Images that have been produced under large area illumination will contain all three types of fluorescence radiation. Under scanning-spot irradiation with a focused laser beam and two-dimensional diode array detection, the influence of false fluorescence light can be reduced. Computer manipulation of all individually stored images will separate object fluorescence from background fluorescence and will allow to localize ghost reflexes.

Even more elegantly, the local resolution is improved by irradiation with very intense focused femtosecond laser pulses outside the absorption range of the fluorophore (e.g., in the near-infrared). The very intense focus of the laser beam—and only this—will excite the fluorophore by nonresonant two-photon absorption. Artifacts by scattered primary radiation are ruled out and the local resolution is comparable to a confocal microscope. In addition, the damage of the sample by laser light absorption is reduced to a minimum.

In order to quantify the influence of scattering it is desirable to measure the lateral diffusion of primary light and fluorescence directly. This can be done also in a microscope under focused monochromatic irradiation, but instead of scanning as usual the sample or the light source now the detector is moved with very small aperture over the irradiated region. In principle, results like those of Figure 8.10 are obtained. The reflected primary radiation is diffusing substantially into regions outside the area of the incident spot. But the amount of this kind of diffusion is modest compared to backward and forward fluorescence. Fortunately, this huge effect is not the rule in microscopy, because the results of Figure 8.10 have been obtained for relatively thick layers of $d = 200\ \mu m$, whereas microscopic preparations are usually much thinner, which will reduce the extent of light diffusion.

8.3.5. Diffuse Reflectance Spectra of Fluorescent Samples

The intensities of diffuse reflectance and fluorescence are both distributed over the solid angle according to Lambert's cosine-law. An ultraviolet-visible (UV/VIS)-

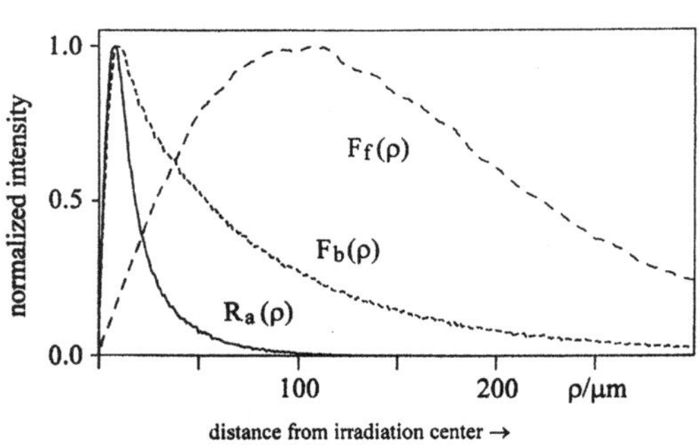

Figure 8.10. Radial distribution of fluorescence S · F(S) and reflectance S · R(S) under focused laser irradiation. Sample thickness = $200\ \mu m$, $\sigma = 300\ cm^{-1}$, $\kappa^a = 150\ cm^{-1}$, $\kappa^e = 0$, e^{-2}-half-width of the laser spot = $12.5\ \mu m$.

spectrometer with diffuse reflectance attachment and conventional beam geometry (light source $\rightarrow$ monochromator $\rightarrow$ sample $\rightarrow$ detector) will therefore measure the sum of reflectance R^a at λ^a and the integrated fluorescence spectrum. The apparent reflectance signal is therefore $R' = \gamma^a R^a + \int \gamma(\lambda^e) F_b(\lambda^e) d\lambda^e$, where $\gamma(\lambda)$ is the response factor of the spectrometer (reflectivity of the gratings, quantum efficiency of the photomultiplier, effective reflectivity of the integrating sphere, and so on). Ignoring for the moment this correction (i.e., $\gamma^a = \gamma^e$), the apparent reflectance of a semi-infinite sample will be $R' = R^a + \phi_F(1 - R^a)$, where ϕ_F is the integral fluorescence quantum yield. In the extreme case of $\phi_F = 1$ the unsatisfactory result $R'(\lambda^a) = 1$ is obtained, independent of the real absorptivity of the sample. The response factor will change this result into $R'(\lambda^a) \lesseqgtr 1$, depending on the wavelength of fluorescence. Spectrometers with inverted beam geometry (light source $\rightarrow$ sample $\rightarrow$ monochromator $\rightarrow$ detector) as it is the case in diode-array spectrometers, will avoid the distortion of R^a in the short-wavelength range but will increase the reflectance in the region of fluorescence emission. This is often a tolerable situation, with the exception of multicomponent systems and photoactive samples that will react under white-light illumination much faster than transparent samples because all light is absorbed in only a small volume adjacent to the layer surface.

The most recommendable geometry of measuring reflectance correctly and gently is that of fluorimeters with two monochromators (light source $\rightarrow$ monochromator(I) $\rightarrow$ sample $\rightarrow$ monochromator(II) $\rightarrow$ detector). Both monochromators have to be started at the same λ and scanned synchronously over the desired wavelength range. The distortion of the spectra is negligible, even in the range of spectral overlap between absorption and fluorescence (apart from very sharp fluorescence bands) and in addition the straylight level, usually being crucial in diffuse reflectance spectroscopy, will be reduced by about two orders of magnitude.

Figure 8.11 shows the diffuse reflectance spectra of a sample with $\phi_F = 0.8$, once scanned in conventional way with an integrating sphere, and once with a fluorimeter. The conventional spectrum is inverted because the coating of the integrating sphere is less reflective in the range of absorption than in the range of emission and will overemphasive fluorescence. The absorption with two monochromators is given correctly, and the spectra show also that photochemical reactions during irradiation can be followed quantitatively by diffuse reflectance spectroscopy.

8.4. Model Calculations

The propagation of light in multiple scattering media is quantified usually on the level of radiative transfer or particle diffusion. Scattering, absorption, and emission are considered as independent statistical processes, and the consequences of wave character are either ignored, like polarization, or added as an additional parameter, like the phase function $P(\mu', \mu)$ that describes the angular distribution of scattered

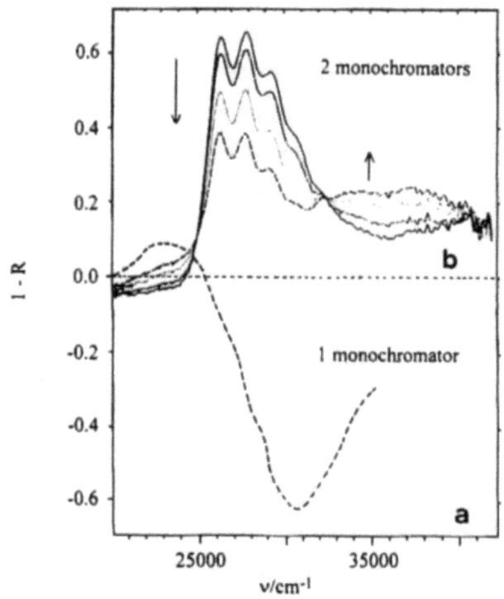

Figure 8.11. Diffuse reflectance absorption spectra of a strongly fluorescent sample (1,6-diphenylhexatriene adsorbed on porous alumina); (a) conventional measurement with monochromatic irradiation and detection via an integrating sphere; (b) measurement in a fluorimeter with two monochromators. Reaction spectra during trans → cis photoisomerization are also given (adapted from Ref. 26).

light in a single scattering process (μ' is the direction of propagation before scattering and μ after scattering; for isotropice scattering one obtains $P(\mu', \mu) = \text{const} = 1$). With the optical constants that have been defined in Section 8.2 the intensity of collimated irradiation inside the sample, $I_c(z)$, and the collimated transmission T_c are given as

$$I_c(z) = e^{-\zeta/\mu_0}, \qquad T_c = e^{-\tau/\mu_0} \tag{8.1}$$

where $\zeta = (\kappa + \sigma)z$ and $\tau = (\kappa + \sigma)d$ are the dimensionless optical depth and optical thickness of the sample, respectively.

In principle, Eq. (8.1) can be applied only for small scattering centers with small scattering cross sections. For extended centers (e.g., a cell, a glass ball) with high scattering cross section, Eq. (8.1) has to be replaced by

$$T_c = (1 - \sigma/n)^{nd} \cdot e^{-\kappa d} \tag{8.1a}$$

where n is the number of scattering centers per unit length and σ/n is the ratio of the scattering cross section to the geometrical cross section of one center. This ratio approaches closer to unity than to zero for large particles so that, for given T_c, the σ-values of Eq. (8.1a) must be much smaller than those of Eq. (8.1).

Under an extended area of irradiation, T_c is only a function of z. Under spot irradiation the lateral coordinates also have to be treated as variables. For simplicity we use cylindrical Gaussian profiles and $\mu_0 = 1$. Then

$$I_c(\vec{r}) = A_0 \cdot e^{-(\zeta + 2\rho^2/\rho_0^2)} \tag{8.2}$$

where A_0 is the intensity at the center of irradiation, $\rho = (x^2 + y^2)^{1/2}$ is the distance from the center perpendicular to the surface normal, ρ_0 is the e^{-2}-halfwidth of the Gaussian profile, and $(\vec{r})$ is the local coordinate in the sample. The diffuse parts of radiation are calculated with the model of radiative transfer.[28] The differential equation for the primary radiation at λ^a is written as

$$\frac{dI^a}{ds}(\vec{r},\vec{\mu}) = -(\kappa^a + \sigma^a)I^a(\vec{r},\vec{\mu}) + \frac{\sigma^a}{2}\int_{-1}^{+1} P(\mu',\mu)I^a(\vec{r},\vec{\mu})d\mu'$$

$$+ (4\pi)^{-1}\sigma^a P(\mu^o,\mu)I^a_c(\vec{r}) \tag{8.3}$$

where ds is the change of the local coordinate in arbitrary direction inside the sample. According to Eq. (8.3) the intensity of the primary radiation decreases by scattering into other directions and by absorption (first term on the right side), and increases by scattering from other directions, where one discriminates "inner sources" originating from light that has already been scattered in former scattering processes (second term) and "outer sources" originating from the incident radiation (third term). The differential equation for the fluorescence radiation at λ^e is written as

$$\frac{dI^e}{ds}(\vec{r},\vec{\mu}) = -(\kappa^e + \sigma^e)I^e(\vec{r},\vec{\mu}) + \frac{\sigma^e}{2}\int_{-1}^{+1} P(\mu',\mu)I^e(\vec{r},\vec{\mu})d\mu'$$

$$+\frac{q^e}{2}\int_{-1}^{+1} I^a(\vec{r},\vec{\mu}')d\mu' + (4\pi)^{-1}q^e I^a_c(\vec{r}) \tag{8.4}$$

The first two terms on the right side have the same sense as in Eq. (8.3), i.e., they describe the (re)absorption and scattering of fluorescence. The third and fourth terms describe the generation of fluorescence by absorption of light at λ^a, assuming the emission process to be isotropic. In principle, Eq. (8.4) can be expanded by a fifth term that considers reemission of the reabsorbed part of fluorescence.

8.4.1. Solution of the Equations of Transfer

In order to describe the fluorescence radiation profile of scattering samples in total, Eqs. (8.3) and (8.4) have to be coupled. This system of differential equations is not soluble exactly, and even if simple boundary conditions are introduced the solution is possible only by numerical approximation. The most flexible procedure to overcome all analytical difficulties is to use a Monte Carlo simulation. However, this method is little elegant, gives noisy results, and allows resimulation only according to the method of trial and error which can be very time consuming, even in the age of fast computers. Therefore different steps of simplifications have been introduced that allow closed analytical approximations of sufficient accuracy for most practical purposes. In a first step, the irradiation density function and the phase function are approximated analytically, so that the differential equations become soluble with diffusion theory of higher order. In a second step, the irradiation density function is inserted as laterally constant, but μ_0 is maintained as a variable parameter. Then conventional diffusion theory can be applied as analytical tool. In a third step, the scattering integrals are substituted by two isotropic antiparallel light fluxes, and the irradiation is also assumed to be isotropic. Then a four-constant four-flux model is obtained, that is an expansion of the Kubelka–Munk (K–M)[29] model and can be solved in closed analytical form.

8.4.2. Spot Irradiation

8.4.2.1. Monte Carlo Simulation

Each incident photon is treated as independent particle whose walk is randomized according to Eq. (8.3). The mean directional change in a scattering process is weighted by the anisotropy parameter g

$$g = \langle \cos\Theta \rangle = \langle \mu, \mu' \rangle = \frac{1}{2} \int_{-1}^{-1} (\mu, \mu') P(\mu, \mu') d(\mu, \mu') \tag{8.5}$$

which becomes unity for perfect forward scattering and zero for isotropic scattering. After being absorbed, the photon enters Eq. (8.4) where it has a certain probability to be emitted as fluorescence, then to be scattered, reabsorbed, or reradiated from the front and back side of the sample, respectively. As a result of the simulation one obtains spatial and temporal radiation density profiles of fluorescence and primary radiation inside the sample and, depending on the experimental boundary conditions of, e.g., spot or pulse irradiation, the lateral fluorescence intensity distribution on the surface, and fluorescence decay curves that have been modified by the time-of-flight (TOF) dispersion of the multiple scattered photons. Figure 8.10 shows a simulation under focused laser irradiation with optical parameters that are typical for thin layer chromatograms or for colored paper printing products. The sample is irradiated with a Gaussian laser profile of e^{-2}, half-width $\rho_0 = 12.5\ \mu$m. The intensities of reflected and emitted radiation are multiplied by the distance ρ from the center of irradiation and by 2π to give the radial photon densities. The area under the curves are equal to the integral reflectance R_0, the integral forward fluorescence F_f, and the integral backward fluorescence F_b. It is evident that the radial distribution of $R_0(\rho)$ is wider than the original laser profile, and that $F_b(\rho)$ and especially $F_f(\rho)$ are much wider because fluorescence is generated inside the sample, i.e., mainly in the region of scattered primary radiation. In addition, fluorescence is only little absorbed so that it has a higher probability to diffuse away from the origin than the incident primary radiation. These results are of some importance in microscopic fluorescence imaging where light diffusion becomes the limiting factor of resolution.

8.4.2.2. *Integration with the P_L-Method*

The intensities $I(\vec{r},\vec{\mu})$ and $I_0(\rho)$ of Eqs. (8.3) and (8.4) are expanded into spherical harmonics

$$I(\vec{r},\vec{\mu}) = \sum_{n=0}^{\infty} \sum_{m=-n}^{m=+n} \left(\frac{2n+1}{4\pi}\right) \cdot I_n^m(\vec{r}) \cdot Y_n^m(\vec{\mu}) \tag{8.6}$$

yielding a series of integrals of increasing complexity. For practical purposes only the first members $n \leq L$ of the series have to be considered (P_L-method). Brun[4] has shown that for "spot" irradiation, i.e., for aperiodic or periodic cylindrical irradiation profiles $I_0(\rho, z = 0)$, the integration has to be carried out at least up to $L = 5$ in order to obtain radial distribution functions that are not significantly different from the Monte–Carlo simulation. The phase function was included in these calculations as

$$P(\mu,\mu') = 1 + 3g(\mu \cdot \mu') \tag{8.7}$$

but for anisotropies only up to $|g| \leq 1/3$. For higher g-values the function

$$P(\mu,\mu') = (1-g^2)\left(1+g^2-2g(\mu\cdot\mu')\right)^{-3/2} \tag{8.8}$$

was inserted.[29] Figure 8.12 shows the influence of the anisotropy on the radial distribution of the diffuse reflectance. All curves are calculated with the same effective scattering coefficient

$$\sigma_{\mathrm{eff}} = (1-g)\sigma \tag{8.9}$$

The curves are broadened in the near-field range for high g-values which are found very often in biological tissues (see lower curve that has been calculated for a typical tissue anisotropy of $g = 0.95$). In all cases, the fluorescence diffuses much wider than the primary radiation but as far as we know no exact values are available for high anisotropy parameters.

8.4.3. Extended Area of Irradiation

8.4.3.1. *Collimated Irradiation, P_1-Approximation*

The unconvenient equations of the former sections can be substituted by the much simpler P_1-approximation that is equivalent to classical diffusion theory. After integra-

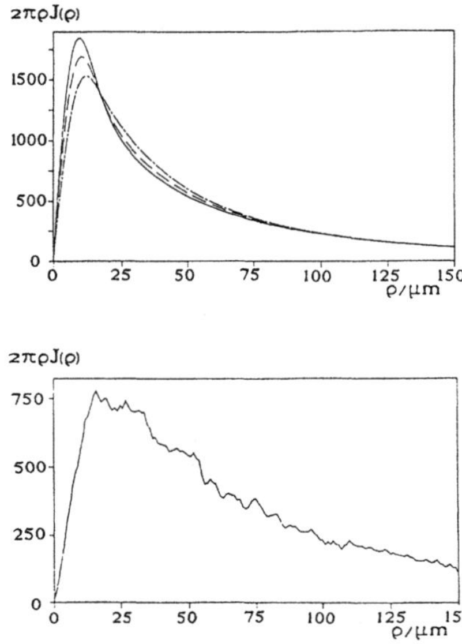

Figure 8.12. Radial distribution of the reflectance as a function of the anisotropy parameter g. The *upper curves* are calculated using Eq. (8.7) (——, $g = 0$; - - -, $g = 1/6$; — · — ·, $g = 1/3$). The *lower curve* is a Monte Carlo simulation ($g = 0.95$) using Eq. (8.8) ($\sigma_{eff} = 500\ cm^{-1}$, $\kappa = 0.67\ cm^{-1}$, e^{-2} -hw = 12.5 μm).

tion, the optical quantities shown in Table 8.1 and the radiation densities inside the sample can be expressed in closed analytical form as functions of σ, κ, g, d, and μ_0. Here only some results for isotropic scattering ($g = 0$) are presented. More detailed formula are given elsewhere.[4] The diffuse reflectance under normal incidence, $\mu_0 = 1$, is

$$R_0 = \frac{3\varpi_0}{1-\eta^2} \cdot \frac{(3+2\eta)(1-\eta)\cdot e^{+\eta\tau} - (3-2\eta)(1+\eta)\cdot e^{-\eta\tau} + 2\eta \cdot e^{-\tau}}{(3+2\eta)^2 \cdot e^{+\eta\tau} - (3-2\eta)^2 \cdot e^{-\eta\tau}} \tag{8.10}$$

where

$$\varpi_0 = \sigma/(\sigma+\kappa), \quad \eta = (3-3\varpi_0)^{1/2}, \quad \tau = (\sigma+\kappa)d$$

Other limiting cases are $\varpi_0 \rightarrow 1$

$$R_0 = \frac{3\tau - (3\mu_0 - 2)(1 - e^{\tau/\mu_0})}{4 + 3\tau} \tag{8.11}$$

and $d \rightarrow \infty$

$$R_\infty = \frac{3\varpi_0}{(3+2\eta)(1+\mu_0\eta)} \tag{8.12}$$

The corresponding equations for transmission are for $\mu_0 = 1$

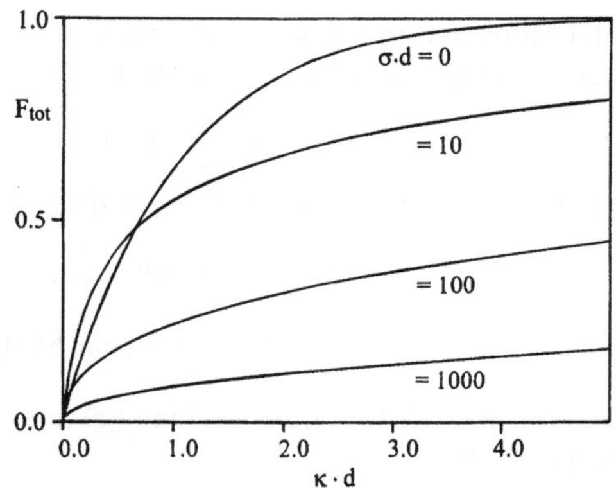

Figure 8.13. Fluorescence intensity under normal incidence ($\mu_0 = 1$) of transparent and scattering samples as a function of the absorption and scattering coefficient. Note the increase of fluorescence when going from transparent to weakly scattering samples.

$$T_{\text{tot}} = e^{-\tau} + \frac{3\varpi_0}{1-\eta^2} \cdot \frac{(3-2\eta)(1-\eta)\cdot e^{-(1+\eta)\tau} - (3+2\eta)(1+\eta)\cdot e^{-(1-\eta)\tau} + 10\eta}{(3+2\eta)^2 \cdot e^{+\eta d} - (3-2\eta)^2 \cdot e^{-\eta d}} \quad (8.13)$$

where the first and second terms are the collimated and diffuse parts of transmission, respectively, and for $\varpi_0 \rightarrow 1$

$$T_{\text{tot}} = \frac{4 + (3\mu_0 - 2)(1 - e^{-\tau/\mu_0})}{4 + 3\tau} \quad (8.14)$$

The total emitted fluorescence radiation without any reabsorption is then

$$F_{\text{tot}} = (1 - R_0^a - T_{\text{tot}}^a) \int_0^\infty q^e dv^e \quad (8.15)$$

Figure 8.13 displays the relation between F_{tot} and the absorption coefficient κ^a under normal incidence $\mu_0 = 1$, with the scattering coefficient σ^a as parameter. For low σ^a - and κ^a-values the fluorescence intensity is higher than in transparent media because the mean pathlength of light in the cuvette increases by scattering, however, for high σ^a the situation is inverted.

8.4.3.2. Diffuse Irradiation, Kubelka–Munk Approximation

On diffuse irradiation, Eqs. (8.10) through (8.15) become much simpler since all terms with the factor $(3\mu_0 - 2)$ vanish, $\int_0^1 (3\mu_0 - 2)\mu_0 d\mu_0 = 0$. Helpwise, collimated irradiation under $\mu_0 = 2/3$ ($\alpha_0 = 48.2°$) has the same effect, but only for weak absorption. With increasing absorption the light fluxes inside the sample deviate more and more from the condition of diffuse irradiation. It has been often shown that the two-flux model derived first by Schuster[30] and then by Kubelka and Munk[28] has formally the same analytical solutions as the P_1-approximation under diffuse irradiation. Kubelka and Munk use formal absorption and scattering coefficients, K and S, that are defined per unit length parallel to the normal of the layered sample. With the abbreviations

$$a = (K + S)/S \quad \text{and} \quad b = (a^2 - 1)^{1/2}$$

the following expressions are obtained for R and T:

$$R_0 = \sinh(bSd) / (a \cdot \sinh(bSd) + b \cdot \cosh(bSd)) \tag{8.16}$$

$$T_d = b / (a \cdot \sinh(bSd) + b \cdot \cosh(bSd)) \tag{8.17}$$

and for the radiation flux densities $I(z)$ and $J(z)$ in positive and negative z-direction, respectively

$$I(z) = I_0 \left(\frac{R_0 - a}{b} \sinh(bSz) + \cosh(bSz) \right) \tag{8.18}$$

$$J(z) = I_0 \left(\frac{aR_0 - 1}{b} \sinh(bSz) + R_0 \cosh(bSz) \right) \tag{8.19}$$

For semi-infinite layers the radiation fluxes simplify to single exponentials

$$I_\infty(z) = I_0 \cdot \exp(-bSz) \tag{8.20}$$

$$J_\infty(z) = R_\infty \cdot I_\infty(z) \tag{8.21}$$

and

$$R_\infty = a - b \tag{8.22}$$

For weakly absorbing, or weakly scattering, or very thin layers, the equations simplify to

$$R_0 \approx \frac{Sd}{1 + aSd}, \quad T_d \approx \frac{1}{1 + aSd} \tag{8.23}$$

$$I(z) \approx I_0 \frac{1 + aS(d - z)}{1 + aSd}, \quad J(z) \approx I_0 \frac{S \cdot (d - z)}{1 + aSd} \tag{8.24}$$

The light fluxes are now linear functions of the depth coordinate z as it is predicted also by Fick's first law for steady-state diffusion without sink. For weak absorption, the equations for T_d and R_0 of the Kubelka–Munk formalism are also directly equivalent to the results of the diffusion approximation. Comparing Eqs. (8.22) and (8.23) with Eqs. (8.11), (8.12), and (8.14) under diffuse irradiation or under $\mu_0 = 2/3$, the Kubelka–Munk coefficients can be expressed by[31–34]

$$S = \frac{3\sigma - \kappa}{4}, \quad K = 2\kappa \tag{8.25}$$

These equations are approximately valid also for anisotropic scattering when σ is substituted according to Eq. (8.9) by $\sigma_{\text{eff}} = (1 - g)\sigma$. The validity, however, decreases with increasing absorptivity of the sample.

The integral true fluorescence intensity is obtained again by Eq. (8.15). Also the partial intensities emerging from the front and back surface are accessible with the Kubelka–Munk formalism in closed analytical form. In order to solve the system of coupled differential equations, the source function

$$dI^e = q^e(I^a(z) + J^a(z))dz \tag{8.26}$$

is inserted in Eq. (8.4) and divided equally into fluxes in $+z$ and $-z$ direction. The fluorescence fluxes are treated in analogy to the primary radiation. After a lengthy procedure of integration[35] the backward fluorescence (in direction of reflectance) and forward fluorescence (in direction of transmittance) at λ^e are obtained as

$$F_b = \frac{K^a \cdot q^e}{2(K^a v^a - K^e v^e)} \cdot \left[(v^a + v^e)(R_0^e - R_0^a) + (v^a - v^e)(1 - R_0^a \cdot R_0^e + T_d^a \cdot T_d^e) \right] \tag{8.27}$$

$$F_f = \frac{K^a \cdot q^e}{2(K^a v^a - K^e v^e)} \cdot \left[(v^a + v^e)(T_d^e - T_d^a) - (v^a - v^e)(R_0^a \cdot T_d^e + R_0^e \cdot T_d^a) \right] \tag{8.28}$$

where $v = K + 2S$. These intensities are normalized to incident primary radiation. They consider reabsorption at λ^e but not reemission. The simpler equation for semi-infinite layers is given in Refs. 36 and 37, and also the equation including reemission is available for this case.[38] In contrast to the primary radiation flux density, $I^a(z) + J^a(z)$, that is monotonically falling with the sample depth coordinate z, the fluorescence flux density $I^e(z) + J^e(z)$ passes through a maximum arising from fluorescence that has been originally created in the deep interior of the sample and therefore has a high probability of being multiply scattered forward and backward before leaving the sample. This part of fluorescence has also a high probability of being reabsorbed (Section 8.6.2).The different depth profiles of primary and fluorescence radiation are illustrated in Figure 8.14.

8.4.4. Time-Resolved Analysis

The equations of transfer do not incorporate time explicitly, but all local variations can be transformed into temporal variations via $dt = dr / c$, where c is the mean velocity of light in the scattering medium. As a consequence of scattering, the incident as well as the emitted photons show TOF dispersion in spatially extended samples. On diffuse

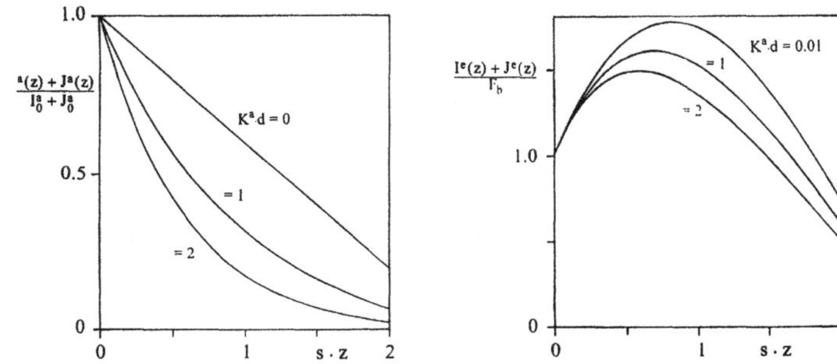

Figure 8.14. Radiation densities in a sample of scattering thickness S·d = 2, with the absorption thickness $K^a{\cdot}d$ as parameter. *Left*: primary radiation; *right*: fluorescence radiation. The radiation densities are normalized to $z = 0$.

irradiation, only the backscattered part of radiation contributes on the average to TOF-dispersion. Thus it is sufficiently accurate in a first approximation to calculate the temporal profiles of incident and emitted radiation by using σ_{eff} or, even simpler, the Kubelka–Munk approximation.

Within this approximation, at least half of the photons that contribute to reflectance of a semi-infinite layer have been scattered only once. On δ-irradiation this part of photons is reflected from the illuminated surface with an exponential time profile $J_0^a(t) = 2^{-1}N_0{\cdot}S^a{\cdot}c{\cdot}\exp(-(S^a + K^a)c{\cdot}t)$, where N_0 is the number of incident photons at $t = 0$. Considering a weakly absorbing and not too strongly scattering sample ($K^a \approx 1$ cm^{-1}, $S^a \approx 10^2$ cm^{-1}), the decay time of the single-scattered photons is in the order of $\tau_1^a \approx 500$ fsec. The second half of photons is multiply scattered and decays

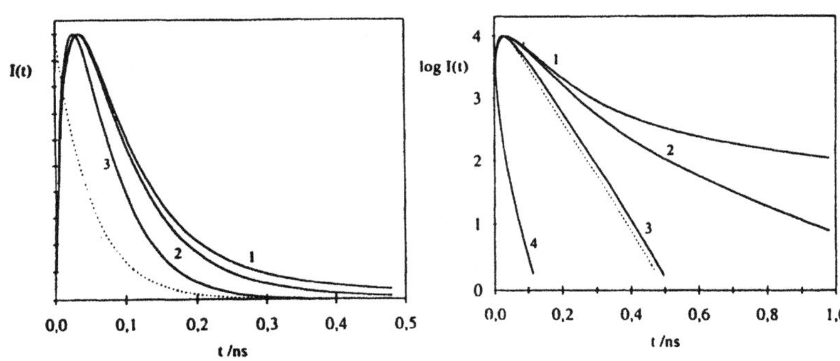

Figure 8.15. Distortion of fluorescence decay curves by time-of-flight dispersion in a scattering sample with τ = 50 ps, d = 1 cm, K^a = 1 cm^{-1}, $S^a = S^e$ = 100 cm^{-1}. Linear (*left*) and semilogarithmic (*right*) presentation. Curve 1: no fluorescence reabsorption; Curve 2: K^e = 0.1 cm^{-1}, Curve 3: K^e = 1 cm^{-1}. The dashed line represents the true fluorescence decay curve, and curve 4 shows the TOF dispersion of the δ-excitation pulse.

nonexponentially. The decay profile has been calculated with numerical methods.[39] It extends to $\tau_\infty^a \approx$ 10–20 psec for the last promille of photons.

The fluorescence is generated inside the sample by photons that are already temporally dispersed. After excitation, the fluorescence is scattered backward and forward and diffuses finally to the sample surface. Kinetically, the diffusion process has to be considered as a consecutive reaction step that creates a maximum in the temporal fluorescence intensity profile.

From a practical point of view the consequences of TOF dispersion are important only for short intrinsic fluorescence decay times of $\tau_0^e < 1$ nsec. Figure 8.15 shows an example with $\tau_0^e = 50$ psec and realistic optical constants of the substrate. The intensity maximum in $F_b(t)$ is formed at $\Delta t \approx 30$ psec after δ-excitation. After this maximum, the fluorescence decays with an effective lifetime of $\tau_{\text{eff}}^e \approx 100$ psec that increases after long times to $\tau_\infty^e > 500$ psec. The long-lived tail disappears as soon as there is some fluorescence reabsorption, and for $K^e = K^a$ there is practically no difference to the intrinsic decay curve (curve 3 in Figure 8.15).

8.5. Determination of Scattering and Absorption Coefficients

Besides the fluorescence coefficient, the optical absorption and scattering coefficients of the sample are the most important parameters in quantative fluorescence spectroscopy of turbid media. In principle two or, if the anisotropy parameter has to be determined, three independent measurements are sufficient to separate the coefficients that appear in all equations as sums or proportions. However, for better accuracy, one of the geometrical parameters (sample thickness, angle of incidence, distance from the irradiated spot) as well as the wavelength of irradiation should be varied over a wide range, and then the data should be fitted with the help of the corresponding model equation.

The following combinations of measurements have already been successfully applied or could be applied with the appropriate equipment

1. Measurement of transmittance T_d and reflectance R_0 of one sample with known layer thickness d. Diffuse irradiation and detection. Evaluation with Eqs. (8.16) and (8.17).
2. Measurement of R_0 on a series of samples with variable d. Diffuse irradiation and detection. Helpwise collimated irradiation under $\mu_0 = 2/3$. Evaluation with Eqs. (8.16) and (8.22).
3. Measurement of collimated transmission T_c according to Eq. (8.1), in addition to points (1) and/or (2). Determination of the anisotropy factor.
4. Measurement of T_d, R_0, R_∞ under variable angle of incidence μ_0 and diffuse or angular resolved detection (detector should have two angular degrees of freedom). Evaluation with Eqs. (8.10)–(8.14).
5. Measurement of lateral resolved reflectance $R_0(\rho)$, $T_d(\rho)$ or $R_\infty(\rho)$ under spot-irradiation. Evaluation with Monte Carlo simulation or P_L-method.

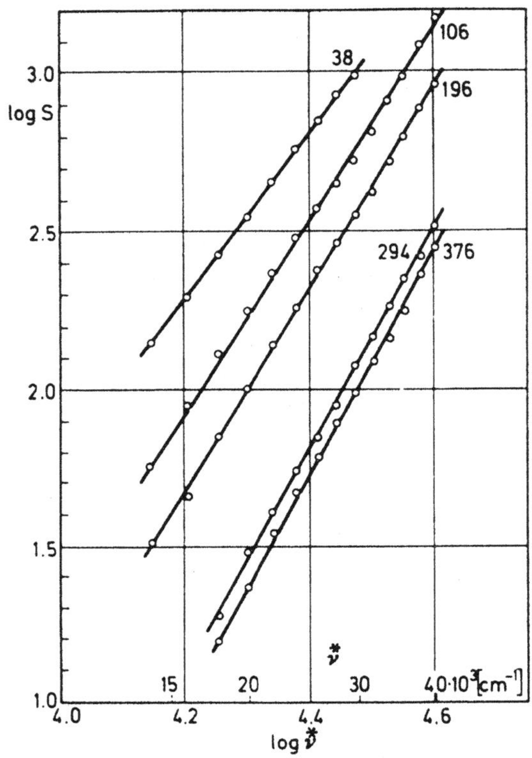

Figure 8.16. KM-scattering coefficients of aerosils as a function of the wavenumber; packing density = 0.107; spec. surface areas in $m^2 \cdot g^{-1}$ are given as parameters. The slopes of the double-logarithmic straight lines are (from top to bottom): 2.6, 3.0, 3.2, 3.5, 3.6. (From Ref. 40.)

6. Measurement of time-resolved $R_0(t)$ or $R_\infty(t)$ under femtosecond-pulse irradiation. Evaluation with Monte Carlo simulation, P_L-method, or numerical KM approximation.

Methods (1) and (2) are conventionally mostly used. In the first systematic study on powdered materials the wavelength λ of irradiation, the packing density π, and the grain diameter δ have been independently varied as parameters.[40] The results are in principle not very different from single scattering. For very small particles ($\delta << \lambda$), S changes with λ^{-n}, where $n = 3.6–2.6$ (see Figure 8.16), i.e., not very far away from $n_{max} = 4$ in Rayleigh's law. S increases linearly with π and, for given π, overlinearly with δ (see Figure 8.17). For large particle diameters ($\delta >> \lambda$), S becomes independent of λ, and decreases approximately with δ^{-1} (for given π). Typical UV/VIS values of S are in the order of $S = 10^1–10^3$ cm^{-1}. The error in the absolute values is relatively high (± 10%), especially because of possible systematic errors in the correction of internal reflection at the sample surfaces, including quartz or polymer plates that often cannot be avoided in order to stabilize the very thin powder layers (typically $d \approx 20–200$ μm) that are necessary to obtain significant results.

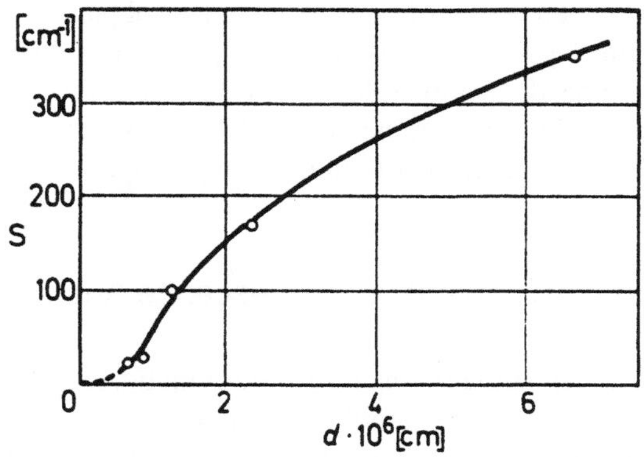

Figure 8.17. KM-scattering coefficients of aerosils as a function of the particle size, $\lambda = 500$ nm. (From Ref. 40.)

Compared to powders, the internal reflection is much more important in systems where the scattering centers are embedded in an isotropic medium whose refractive index n_m is higher than unity. Typical examples are paints or tissues, where the isotropic medium is either a polymer or water, and the scattering centers are fine pigment particles or cell boundaries, respectively. In these cases internal Fresnel reflection has to be considered at the sample surfaces that amounts to $R_{Fre} \approx 0.75$ for $n_m = 1.5$ and diffuse internal irradiation. Fresnel reflection reduces the effective scattering coefficient σ_{eff}, and hence increases the absorption and fluorescence of the sample. It has been shown[(41)] that the increase in radiation density and in absorption or fluorescence, respectively, is localized close to the illuminated surface of the sample. Fresnel reflection has in principle the same effect on σ_{eff} as increasing forward scattering in the volume according to Eq. (8.9). The g-factors of forward scattering have been determined in soft tissues by including the results of collimated transmission T_c according to method (3). The method requires very careful matching of the index of refraction n_m with that of the environment, since otherwise any roughness of the sample surface will deflect the collimated beam and reduce T_c. The reported anisotropy factors of tissues[(42)] are typically $g > 0.9$, and of blood[(42)] $g > 0.99$, indicating an extremely high forward component of scattering that needs future theoretical explanation, (see Eq. (8.1a)).

Method (5) has recently been elaborated by Hubner and Brun[(3)] and offers a quite different approach to the determination of σ, κ, and g. The method requires only one sample of finite or semi-infinite optical thickness. The sample can be absorbing up to $\kappa \approx \sigma$ in contrast to methods (1) or (2) that are limited for practical reason to weakly absorbing samples ($\kappa << \sigma$). Details of the method will be given elsewhere.[(4, 43)] In short, $R(\rho)$-curves as in Figure 8.10 are measured under focused laser irradiation and fitted with the P_5-method to theoretical radial distribution functions. Up to now, a series of inorganic powdered pigments and cellulose samples loaded with organic dyes have been investigated at $\lambda = 633$ nm. The σ_{eff} values range from 380 cm^{-1} for powdered cellulose to 4550 cm^{-1} for rutile. The anisotropy parameter for CaF_2 powder is $g = 0.3$.

In commercial filter paper for chemical purposes, the scattering coefficient parallel to the layer, $\sigma_{\parallel} = 220\ \text{cm}^{-1}$, is significantly different from the scattering coefficient perpendicular to the layer, $\sigma_{\perp} = 360\ \text{cm}^{-1}$. The absorption coefficients of dyed cellulose have been determined in the range of $1 \leq \kappa \leq 200\ \text{cm}^{-1}$. The values are proportional to the analytical concentration of the dye.

8.6. Quantitative Fluorescence Analysis

Quantitative analysis starts with Eq. (8.15) which gives the true total fluorescence flux F_{tot} of the sample relative to the flux of incident radiation. However, the true fluorescence is experimentally only rarely accessible, and questions of analytical interest are among others: how much of F_{tot} is emerging from the sample, how is the emerging part distributed between front and back surface, how are the parts related to the concentration of the fluorophore, how can multicomponent systems be analyzed, how is the fluorescence disturbed by interactions between fluorophore and substrate, how the fluorescence is decaying with time.

8.6.1. Forward and Backward Fluorescence

In this section the ideal case of vanishing reabsorption, $K^e = 0$, is discussed, where $F_b + F_f = F_{tot}$. A large area of the sample should be irradiated close to $\mu_0 = 2/3$, what is a very convenient geometry in most spectrometers, or diffusely via an integrating sphere, what is less convenient but guarantees homogeneous density of irradiation. Under these conditions Eqs. (8.27) and (8.28) are sufficiently accurate for quantitative evaluation.

In layers with very small optical pathlengths ($Sd < 1$) the forward and backward fluorescences are equally intense like in nonscattering media. However, with increasing layer thickness, F_f grows slower than F_b, passes through a maximum, and decreases

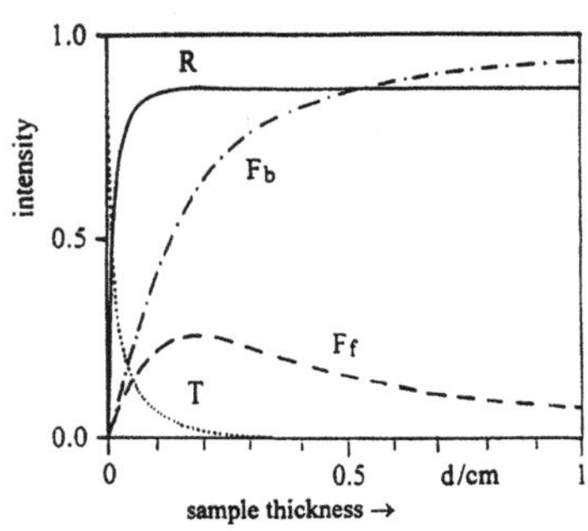

Figure 8.18. Intensities of forward and backward fluorescence of a scattering layer as a function of the layer thickness ($K^a = 1\ \text{cm}^{-1}$, $S^a = S^e = 100\ \text{cm}^{-1}$). The corresponding curves of diffuse reflectance and diffuse transmittance at λ^a are also given.

then to zero for $Sd \rightarrow \infty$. Under this limit all light is emitted as $F_{b,\infty}$. As it can be seen from Figure 8.18 this limit is reached only very slowly (in the example not even at $d = 1$ cm), in strong contrast to the fast saturation of the reflectance (at $d \approx 0.2$ cm) and the almost as fast disappearance of the transmittance (at $d \approx 0.4$ cm). Thus a scattering layer can be treated in many cases as "semi-infinite" in terms of reflectance or transmittance, but not in terms of fluorescence.

Figure 8.19 displays results of a sample with constant layer thickness but variable absorption coefficient K^a of the fluorophore. The abscissa is scaled logarithmically in order to take into account the large K-range in electronic spectroscopy or in analytical chemistry. The layer thickness and scattering coefficient are adjusted to the sample of Figure 8.2. In the low absorption region both F_f and F_b increase monotonically with K^a, but then F_f bends, passes through a broad maximum, and finally decreases despite further increasing absorptivity. The explanation is given by the spatial distribution of fluorescence that is created the closer to the illuminated surface the higher the absorbance. The increasing overall intensity is thus finally overcompensated in F_f by a longer pathlength, i.e., higher probability of backscattering. The calculated results are consistent with the experimental F_f excitation spectrum of Figure 8.2. The weakly absorbing L_b-band ($K^a \approx 0.2$ cm^{-1}, $\lambda^a \approx 350$–370 nm) falls into the region of positive slope, the much stronger absorbing L_a-band ($K^a \approx 50$ cm^{-1}, $\lambda^a \approx 290$–340 nm) into the region of the plateau with extension into the region of negative slope. Thus, L_a is strongly suppressed in the excitation spectrum, and the highest vibrational maxima manifest themselves as shallow minima. Considering these results it is obvious that the reflectance mode is to be preferred for strongly absorbing materials, whereas the transmittance mode is a possible alternative for weak absorption, especially if distortions by straylight have to be minimized. It should be noticed in this context that also excellent absorption spectra are obtainable with a fluorimeter in the diffuse transmission mode, provided the transmission is reduced by scattering not below 0.1–1% and $K^a \cdot d < 1$.

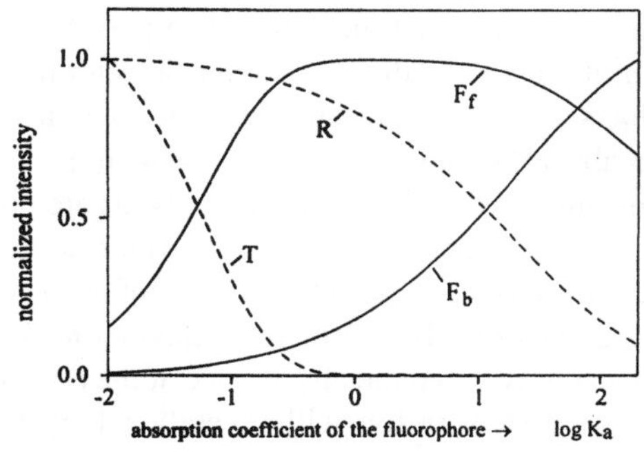

Figure 8.19. Normalized intensities of forward and backward fluorescence, diffuse reflectance and transmittance of a scattering layer ($S = 50$ cm^{-1}, $d = 1$ cm) as a function of the absorption coefficient of the fluorophore.

8.6.2. Inner Filter Effects

In multicomponent systems K^a can be written as a sum of the individual absorption coefficients $K^a_{tot} = \Sigma K^a_i$, where each $K_i(\lambda^a)$ depends in a different way on the wavelength. If one or more of the components are fluorescent, their excitation spectra are mutually attenuated by absorption filters of the other compounds. This effect is included in Eqs. (8.27) and (8.28) so that examples like that of Figure 8.4 can be quantified. The two fluorescent components are monomeric an aggregated pyrene, M_1 and M_n. The fluorescence spectra of these species are clearly different from each other but the absorption spectra overlap strongly. Thus the excitation spectrum of the minority component M_n is totally distorted by the M_1 filter (absorption maxima of M_1 appear as a minima in the excitation spectrum of M_n see Figure 8.4, *top*). In transparent samples this effect can be reduced by dilution. However, this method is not very efficient in scattering media as can be seen by solving Eqs. (8.27 and 8.28) for $bSd \rightarrow 0$. Only the limit $d \rightarrow 0$ will produce the desired relation where fluorescence intensity and absorption coefficient of the fluorophore are linearly proportional to each other in a multicomponent system.

$$\lim_{d \to 0} F_{tot} = q^e \cdot d \tag{8.29}$$

This equation is obtained by inserting first K_{tot} into Eq. (8.23) and then these relations into Eq. (8.15). The practical limit of zero layer thickness depends on the magnitude of the scattering coefficients. In Figure 8.4 (*bottom*), only a few granula of silica were measured in order to obtain the correct spectrum of M_n.

8.6.3. Fluorescence Reabsorption

Reabsorption is a very important error source in thick scattering layers. The fluorescence is suppressed in the region of overlap with the absorption spectrum of the fluorophore and also very often by some background absorption of the substrate. The example of Figure 8.5 shows clearly the possible reabsorption of the NADH-fluorescence by hemoglobin at $\lambda = 580$ nm. The amount of reabsorption is much higher than in transparent samples because fluorescence that is primarily not emitted into the direction of the detector may be scattered backward and finally—perhaps after a long distance of walk—will also find its way to detection. The amount of reabsorption has been calculated within the Kubelka–Munk-approximation for semi-infinite layers.[36–38] The main results are presented in Figure 8.20, where the relation between experimental and true fluorescence intensity is plotted against the diffuse reflectance at λ^e, with the diffuse reflectance at λ^a as parameter. Reabsorption already becomes significant for small deviations from a perfectly "white" reflector ($R^e_\infty = 1$), especially when the absorbance at the excitation wavelength is small, i.e., when R^a_∞ is high. Reabsorption will severely falsify the fluorescence spectra in the region of

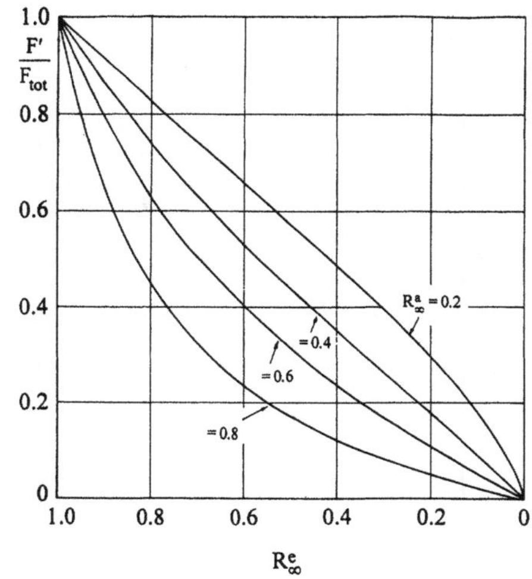

Figure 8.20. Fluorescence reabsorption in a semi-infinite scattering layer. Ordinate: fraction of emitted to true fluorescence intensity. Abscissa: diffuse reflectance at the fluorescence wavelength. Parameter: diffuse reflectance at the absorption wavelength. The scattering coefficient is assumed to be independent of λ ($S^a = S^e$).

overlap with the absorption band, and will give rise to fictitious shifts of the fluorescence maxima of unstructured bands. Figure 8.21 shows the fluorescence of fluorescein on an alumina thin layer chromatographic (TLC) plate. The spectrum shifts with increasing loading significantly to longer wavelengths, but correction for reabsorption according to Figure 8.20 brings back all maxima to the position of the lowest concentration (see bottom of Figure 8.21). A second way of reducing reabsorption is to use thin layers. Calculations start with Eqs. (8.27 and 8.28) and Figure 8.22 presents the relation between experimental and true fluorescence yield in backward and forward direction against the layer thickness. The absorption

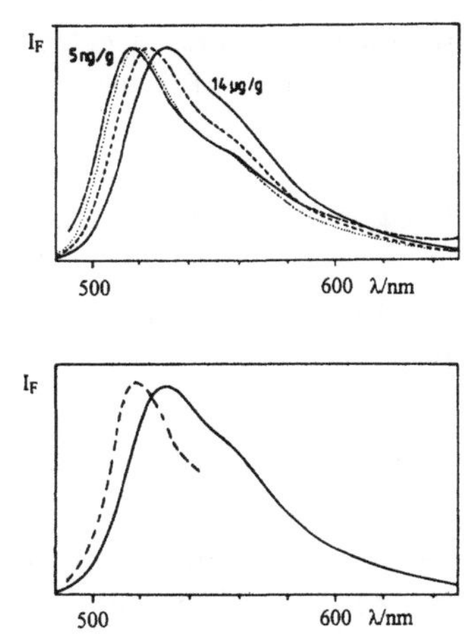

Figure 8.21. Fluorescence spectra of fluorescein on an alumina-TLC plate. *Upper:* redshift of the spectra with concentration according to increasing fluorescence reabsorption. *Lower:* corrected fluorescence spectrum of the most concentrated sample (dashed line) according to Figure 8.20.

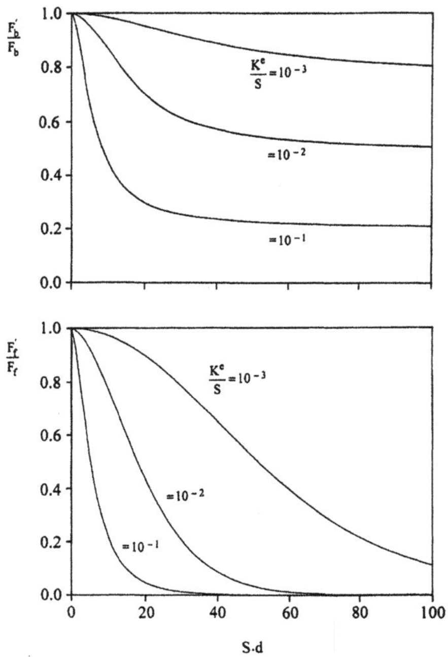

Figure 8.22. Fluorescence reabsorption in a finite scattering layer with $K^a/S = 0.1$ as a function of the scattering thickness Sd. Ordinate: fraction of emitted to true fluorescence intensity in backward (*upper*) and forward direction (*lower*). Parameter: reduced absorption coefficient at the fluorescence wavelength.

coefficient at λ^a is kept constant, and the reabsorption coefficient at λ^e is varied as parameter. One sees immediately that F_f reduces to zero for thick layers, even for low K^e-values, and this explains the total suppression of the 0-0-fluorescence band of pyrene in Figure 8.2 that overlaps strongly with the 0-0-L_b-absorption peak ($K^e/S = 3 \cdot 10^{-3}$), as well as the substantial reabsorption in the region of the much weaker hot sidebands. The calculations show also that very thin layers are necessary in order to reduce reabsorption in F_f or F_b approximately to zero. It is again recommended to test the correctness of the spectra by investigating single granula or very small volumes of tissues.

8.6.4. Fluorescence Quantum Yields

In the semi-infinite layer approximation ($F_b = F_{tot}$) the integral fluorescence quantum yield

$$\phi_f = \int q^e \cdot d\nu^e = \frac{K_f^a + K_0^a}{K_f^a(1 - R^a)} \cdot \int F_b \cdot d\nu^e \tag{8.30}$$

can be determined as easily or difficultly as in nonscattering samples. Experimentally one needs in a first step:

- The fluorescence spectrum, corrected for spectrometer response
- The diffuse reflectance $R = g(K_f, K_0, S)$ in the range of excitation and emission (K_0 is the absorption coefficient of the substrate)
- The diffuse reflectance $R_0 = g(K_0, S)$ of the substrate

Using in detail different, but consistent experimental setups [44, 45] the reflected and emitted intensities have been determined in sum and separately under given I_0. An integrating sphere is not absolutely necessary since the angular distributions of $R(\varepsilon)$ and $F(\varepsilon)$ are both approximately diffuse, but it is necessary to correct the fluorescence intensities for reabsorption (see Section 8.6.3). In a second step one needs the scattering coefficients, S^a and S^e, and the layer thickness. Then also finite layers can be investigated quantitatively. The quantities $F_b(\alpha_0)$, $F_f(\alpha_0)$, $R(\alpha_0)$, and $T(\alpha_0)$ have to be measured under equal geometry and I_0. After correction for reabsorption the quantum yield can be calculated similar to Eq. (8.32), but starting from Eq. (8.15). This procedure is definitely more complex than in nonscattering samples.

The error sources are more numerous than in transparent samples. Especially the absorbed part of light $A = (1 - R - T)$ becomes very uncertain for weakly absorbing samples. Therefore it is recommended to use medium absorbers ($0.3 < A < 0.7$) for quantitative work. It should be also mentioned that the aperture of the spectrometer must high enough to measure reflected, transmitted, and especially emitted light intensities correctly. Usually the spectral intensities of the reflected primary light are much higher than fluorescence. If these differences are balanced with small slit widths that usually create a small area of irradiation and observation, one must be reminded that fluorescence can diffuse very widely, especially when the sample is covered by glass. In Figure 8.23, the sample of Figure 8.2 is irradiated with 0.3 mm monochromator slit width that creates at the sample surface an image of about the same size. Then the fluorescence is scanned laterally from the edge of irradiation. The branches of fluorescence extend to distances of 6–8 mm and it is obvious that the integral intensity will be completely wrong if small slit detection is used.

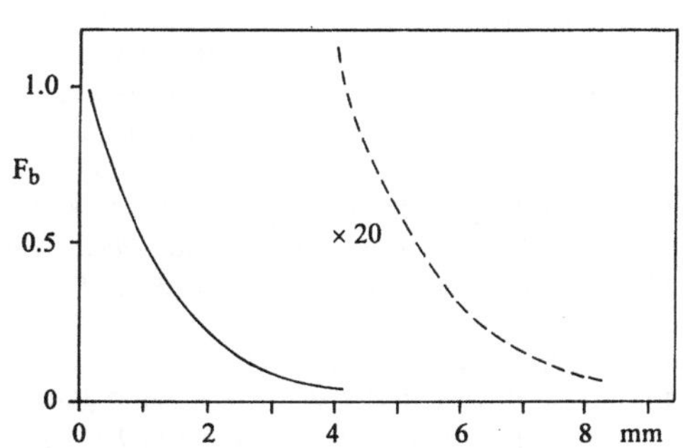

Figure 8.23. Lateral distribution of the fluorescence intensity under line irradiation. Sample: pyrene/silica gel covered with a fused silica plate of 2 mm thickness (the sample is the same as in Figure 8.2).

A few materials are given that can be used as primary fluorescence standards in the adsorbed state, because they have high yields, show little tendency of aggregation provided the surface coverage is low, and are photochemically stable:

- 9,19,-diphenylanthracene on γ-alumina (area = 1 00 m^2g^{-1}) $\phi_F = 0.95$
- perylene on silica gel (area = 500 m^2g^{-1}) $\phi_F = 0.85$
- 2,2′-isopropylene-4,4′*bis*-(dimethylamino)benzophenone on silica gel ϕ_F = 0.95

In addition, the pure solid $(N(C_2H_5)_4)_2MnBr_4$ is recommended as standard. Its phosphorescence yield is high, $\phi_p = 0.8$, the material has only medium absorption despite its high concentration, and there is no long-distance impurity quenching because the luminescence centers are electronically isolated from one another by the large counter ions.

Acknowledgment

This work is supported by projects of the Deutsche Forschungsgemeinschaft "Photochemie an Grenzflächen" and "Chemie in Interphasen."

References

1. G. Kortüm, *Reflectance Spectroscopy*, Springer-Verlag, Berlin (1969).
2. W. Wendtlandt and H. G. Hecht, *Reflectance Spectroscopy*, Wiley & Sons, New York (1966).
3. M. Brun, P. Hubner, and D. Oelkrug, *Fresenius J. Anal. Chem. 344,* 209, (1992).
4. M. Brun, Ph.D. thesis, Tübingen, Germany (1993).
5. D. Oelkrug, M. Brun, and U. Mammel, *J. Luminesc. 60/61*, 422 (1994). Review-articles are, e.g., Refs. 6–11.
6. P. de Mayo, *Pure & Appl. Chem. 54*, 1623 (1982).
7. A. Henglein, *Pure & Appl. Chem. 56*, 1215 (1984).
8. D. Oelkrug, W. Flemming, R. Füllemann, R. Günther, W. Honnen, G. Krabichler, M. Schäfer, and S. Uhl, *Pure & Appl. Chem. 58*, 1207 (1986).
9. J. K. Thomas, *J. Phys. Chem. 91*, 267 (1987).
10. J. K. Thomas, *Chem. Rev. 93*, 301 (1993).
11. M. Anpo and T. Matsuura (eds.), *Photochemistry on Solid Surfaces*, Elsevier, Amsterdam (1989).
12. U. Mammel, Ph.D. thesis, Tübingen, Germany (1992).
13. D. Oelkrug, S. Uhl, F. Wilkinson, and C. J. Willsher, *J. Phys. Chem. 93*, 4551 (1989).
14. D. Oelkrug, S. Uhl, M. Gregor, R. Lege, G. Kelly, and F. Wilkinson, *J. Mol. Struct. 218*, 435 (1990).
15. D. Oelkrug, M. Gregor, and S. Reich, *Photochem. & Photobiol. 54*, 539 (1991) .
16. H. Schneckenburger and W. Schmidt, in: *Fluorescence Spectroscopy* (O. S. Wolfbeis, ed.), p. 94, Springer-Verlag, Berlin, (1992).
17. K. A. Zachariasse, in Ref. (11).
18. R. W. Kessler, S. Uhl, W. Honnen, and D. Oelkrug, *J. Luminesc. 24/25*, 551 (1981).
19. R. W. Kessler, D. Oelkrug, and S. Uhl, *Le Vide, les Couches Minces 209*, 1338 (1981).

20. R. K. Bauer, R. Borenstein, P. de Mayo, K. Okada, M. Rafalska, W. R. Ware, and K. C. Wu, *J. Amer. Chem. Soc. 104*, 4635 (1982).
21. R. K. Bauer, P. de Mayo, W. R. Ware, and K. C. Wu, *J. Phys. Chem. 86*, 3781 (1982).
22. S. Uhl and D. Oelkrug, *J. Mol. Struct. 175*, 117 (1988).
23. A. R. Leheny, N. J. Turro, and J. M. Drake, (a) *J. Chem. Phys. 97*, 3736 (1992); (b) *J. Phys. Chem. 96*, 8498 (1992).
24. H. D. Breuer, in Ref. (11), p. 106.
25. S. H. Kost and H. D. Breuer, *Ber. Bunsenges. Physik. Chem. 95*, 480 (1991).
26. G. Krabichler, Ph.D. thesis, Tübingen, Germany (1986).
27. S. Chandrasekhar, *Radiative Transfer*, Dover Publications, New York (1960).
28. (a) P. Kubelka and F. Munk, *Z. Techn. Pysik 12*, 593 (1931); (b) P. Kubelka, *J. Opt. Soc. Amer. 38*, 448 (1948).
29. L. G. Henyey and J. L. Greenstein, *Astrophys. J. 93*, 76 (1941).
30. A. Schuster, *Astrophys. J. 21*, 1 (1905).
31. B. J. Brinkworth, *J. Phys. D 4*, 1105 (1971).
32. K. Klier, *J. Opt. Soc. Amer. 62*, 882 (1972).
33. L. F. Gate, *Appl. Opt. 13*, 236 (1974).
34. S. Glasstone and M. C. Edlund, *Elements of Nuclear Reactor Theory*, Van Nostrand, New York (1952).
35. M. Brun, Diplomarbeit Chemie thesis, University of Tübingen, Tübingen, Germany (1989).
36. E. Allen, *J. Opt. Soc. Am. 54*, 506 (1964).
37. D. Oelkrug and G. Kortüm, *Z. Phys. Chem. NF 58*, 181 (1968).
38. R. Gade, and U. Kaden, *J. Chem. Soc. Faraday Trans. 86*, 3707 (1990).
39. U. Mammel, M. Brun, and D. Oelkrug, *Fresenius J. Anal. Chem. 344*, 147 (1992).
40. G. Kortüm and D. Oelkrug, *Z. Naturforsch. 19a*, 28 (1964).
41. M. Motamedi, S. Rastegar, G. LeCarpentier, and A. J. Welch, *Appl. Opt. 28*, 2230 (1989).
42. A. J. Welch, J. A. Pearce, K. R. Diller, G. Yoon, and W. F. Cheong, *J. Biomech. Engin. 111*, 62 (1989) and literature cited herein.
43. M. Brun and D. Oelkrug, *J. Opt. Soc. Amer.* (submitted for publication).
44. D. Oelkrug and A. Wölpl, *Ber. Bunsenges. Phys. Chem. 76*, 1088 (1972).
45. J. S. Liu, P. de Mayo, and W. R. Ware, *J. Phys. Chem. 97*, 5995 (1993).

9

Real-Time Chemical Sensing Employing Luminescence Techniques

J. Ricardo Alcala

9.1. Introduction

The behavior of practically all luminescent materials is sensitive to various parameters of physical and chemical origin. The excited state lifetimes and average intensities of the fluorescence and/or phosphorescence of these materials are modulated, for example, by temperature, oxygen, pH, carbon dioxide, voltage, pressure, and ionic strength. Consequently, the luminescence of various materials could be used, in principle, to monitor parameters of interest in medicine, industry, research, and the environment.

Instrumentation and methods currently available provide limited means for real-time measurements of the continuous wave (CW) and transient characteristics of luminescent substances. The measurement, in real time, of the spatial distribution of a parameter of interests, for instance, in cells and in tissue cannot be attained with present technology. Optical fibers are used to monitor the response of a sensor in limited regions in space.

Several books and review articles have been published in the field of sensors, especially in fiberoptic sensors. Some examples are given in Refs. 1–6. However, a chapter that covers the connections between sensor luminescence and parameter sensing is still needed. These connections are of prime importance in the design of probes for chemical sensing applications. The purpose of this chapter is to address this need. It presents a unified and practical approach to sensing employing luminescence. The chapter begins with a brief presentation of the basic concepts that establish the relation between luminescent signals and parameter sensing. The optimum techniques to monitor the response of sensors are subsequently discussed. These techniques are illustrated with the implementation of real-time fiber optic oxygen and temperature

---

J. Ricardo Alcala • Department of Biomedical Engineering, Case Western Reserve University, Cleveland, Ohio 44106.

Topics in Fluorescence Spectroscopy, Volume 4: Probe Design and Chemical Sensing, edited by Joseph R. Lakowicz. Plenum Press, New York, 1994.

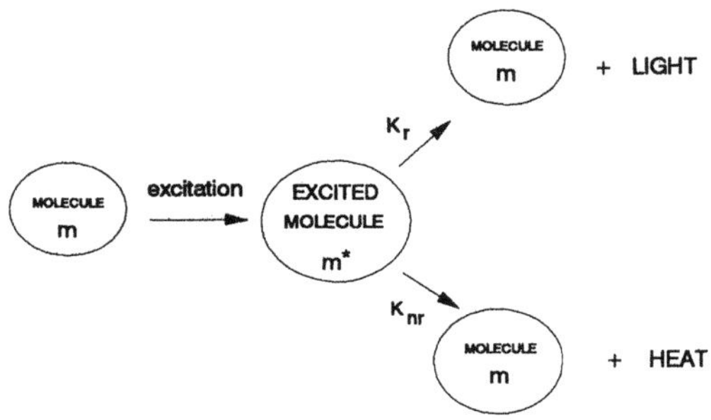

Figure 9.1. Radiative and nonradiative paths to the ground state of excited sensor molecules m^*. All possible mechanisms by which an excited molecule may return to the ground state are sorted in two sets. One set, characterized by the rate constant k_r, is referred as the radiative path. The second set, characterized by the rate function k_{nr}, is referred to in this chapter as the nonradiative path.

sensing systems. Throughout the chapter the words "sensors" or "probes" are used to refer to materials which luminescence is dependent on one or more parameters, either physical or chemical.

9.2. Basic Principles

Consider the process of absorption of excitation by a given sensor molecule m as illustrated in Figure 9.1. Once excited $m*$, the sensor molecule may return to the ground state (m) following several mechanisms. For the purpose of our application, all of these possible mechanisms have been sorted into two groups in Figure 9.1. One group encompasses all processes that emit light. It is referred here as the radiative path. The second group encompasses all mechanisms that do not emit light, referred as the nonradiative path. The combined rate of decay of the radiative mechanisms is denoted by k_r. Similarly, the combined nonradiative decay rate is indicated by k_{nr}.

The sensor molecule probability of remaining in the excited state p_{m^*}, as a function of time, is determined by

$$\frac{dp_{m^*}}{dt} = -(k_r + k_{nr})p_{m^*}, \quad p_{m^*} = e^{-(k_r + k_{nr})t} \tag{9.1}$$

9.2.1. Homogeneous Sensors

A luminescent material in which all excited molecules return to the ground state with the same probability is defined here as a homogeneous sensor. In this case, the number of excited sensor molecules in the excited state, as a function of time $M^*(t)$, is given by a single exponential.

$$M^*(t) = M^*(0)p_{m^*} \tag{9.2}$$

where $M^*(0)$ corresponds to the initial number of excited sensor molecules (at $t = 0$). The sensor transient luminescence $I_L(t)$ is determined by the fraction of excited molecules following the radiative path to the ground state

$$I_L(t) = k_r M^*(t) = k_r M^*(0)e^{-(k_r+k_{nr})t} \tag{9.3}$$

According to Eq. (9.3) the transient luminescence of homogeneous sensors decays exponentially.

The total number of photons P emitted by our sensor is obtained by integrating over all molecules decaying by radiating a photon

$$P = \int_0^\infty I_L(t)dt = \frac{k_r}{k_r + k_{nr}} M^*(0) = QM^*(0) \tag{9.4}$$

Q is commonly referred as the sensor quantum yield. It gives the fraction of excited molecules that return to the ground state by emitting a photon. The steady-state sensor luminescence I_{SS} is given by the number of photons radiated per unit of time by our sensor when excited by steady-state excitation. I_{SS} is proportional to Q:

$$I_{SS} = \frac{dP}{dt} = Q\frac{dM^*(0)}{dt} \tag{9.5}$$

where $dM^*(0)/dt$ yields the rate of steady-state excitation of the sensor molecules. A distinction is made in this chapter between steady-state luminescence I_{SS} and continuous wave luminescence I_{CW}. I_{CW} remains constant with time. I_{SS} is periodic in time. (I_{CW} is a special case of I_{SS}.)

$dM^*(0)/dt$ is determined by the Beer–Lambert equation (when scattering of the excitation is negligible). The rate of sensor excitation is equal to the rate at which photons are absorbed:

$$\frac{dM^*(0)}{dt} = I_O - I_T = I_O(1 - 10^{-\varepsilon CL}) = gI_O \tag{9.6}$$

where I_O and I_T are the incident and transmitted (or not absorbed) excitation light intensities, respectively. I_O and I_T may vary with time as discussed later in this chapter. ε is the sensor extinction coefficient. C is the concentration of sensor molecules per unit volume. L is the average length of the photon path in the sensor. The fraction of absorbed excitation, by $M^*(0)$ molecules, g is given by

$$g = 1 - 10^{-\varepsilon CL} = \sum_{n=1}^{\infty} (-1)^{n+1} \frac{(2.303\varepsilon CL)^n}{n!} \tag{9.7}$$

For the vast majority of sensing applications, the product εCL is less or substantially less than one, for example, in optical fibers L is a few microns, in biomedical applications or research C is usually relatively small. In most practical cases g may be approximated by the first term in the series of Eq. (9.7)

$$g = 2.303 I_O \varepsilon CL, \quad \varepsilon CL << 1 \tag{9.8}$$

Higher terms in the series may be used to increase the accuracy of the approximation. Since g is given by a convergent alternating series (positive and negative terms alternate) in Eq. (9.7), the accuracy attained by including k terms in the estimation of g is higher than the absolute value of the $(k + 1)th$ term. In this chapter Eq. (9.8) is assumed to be a good approximation of g.

When CW I_{CW} or slow varying I_{SS} (see Eqs. 9.10 and 9.11) excitation is used, $dM^*(0)/dt$ is constant (or changes slowly with time) and the CW sensor luminescence is proportional to the quantum yield. For most sensing applications the following equation establishes the relation between the average sensor luminescence and all other relevant parameters:

$$I_{CW} = g I_O Q \simeq 2.303 I_O Q \varepsilon CL \tag{9.9}$$

Continuous wave techniques do not offer the optimum use of luminescence for sensing applications. CW methods, also known as intensity-based techniques, have many inherent limitations. These limitations will be discussed later in the chapter. Many of the limitations of intensity-based methods can be overcome by using steady-state modulated excitation of the form

$$I_O(t) = I_{CW}[1 + M_{I_O} \sin(\omega t)], \quad M_{I_O} \leq 1 \tag{9.10}$$

where M_{I_O} is the modulation of the excitation intensity, ω is the circular modulation frequency. When

$$f_{\text{modulation}} = \frac{\omega}{2\pi} << k_r + k_{nr} \tag{9.11}$$

the sensor luminescence follows the excitation and the cases of I_{SS} and I_{CW} can be treated indistinguishably. At higher frequencies the luminescence, incapable of following the excitation, becomes demodulated and delayed with respect to the excitation.

The use of high-speed modulated excitation ($f > k_r + k_{nr}$) combined with coherent detection methods has resulted in the popular techniques of frequency domain fluorometry, also known as phase-modulation fluorometry. These techniques can be used to determine the temporal characteristics of both fluorescence and phosphorescence and will also be addressed later in this chapter.

9.2.2. Luminescence and Sensing

From the practical point of view, the radiative decay rate k_r may be assumed to be independent of the external parameters surrounding the excited sensor molecule. Its value is determined by the intrinsic inability of the molecule to remain in the excited state. The radiative decay rate k_r is a function of the unperturbed electronic configuration of the molecule. In summary, for a given luminescent molecule, its unperturbed fluorescent or phosphorescent decay rate (or lifetime) may be regarded to be only a function of the nature of the molecule.

In contrast, the nonradiative decay rate k_{nr} may be viewed to be determined by the localized environment of the luminescent molecule. The localized environment perturbs the natural electronic configuration of the sensor molecule increasing the probability of its decay. The functional form of k_{nr} is determined by the nature of the interaction between the excited sensor and its surrounding perturbation. For example, the k_{nr} may be proportional to the concentration, partial pressure, or value of a [*Parameter*] of interest:

$$k_{nr} = \alpha + \beta[Parameter] \quad \alpha, \beta = constant, \tag{9.12}$$

as in the case of oxygen sensing in the absence of diffusion-controlled processes. In such cases the collisional rate, which quenches the luminescence of oxygen sensors, is proportional to the oxygen concentration or partial pressure. The generalized coefficients α and β in Eq. (9.12) determine two independent processes by which the excited molecule is quenched.

The nonradiative decay rate may also show some sort of saturation like in the case of diffusion controlled processes, in which the nonradiative events may be described, for example, by a Langmuir-like function

$$k_{nr} = \alpha + \frac{\beta[Parameter]}{1 + \gamma\,[Parameter]} \quad \alpha, \beta, \gamma = constant \tag{9.13}$$

k_{nr} is bounded between α and $\alpha + \beta/\gamma$. In this case k_{nr} approaches a maximum value (e.g., diffusion determined) regardless of how large the values of [*Parameter*] becomes.

The mechanism of interaction between the excited sensing material and the value of its surrounding [*Parameter*] in equilibrium may be quite complex. The formalisms implemented in the field of catalysis (e.g., see Ref. 7) may find applications in the study of the sensor luminescence response to external agents. From the practical point of view, k_{nr} can be well described by an alternating polynomial function

$$k_{nr} = a_1 + a_2[Parameter] + a_3[Parameter]^2 + a_4[Parameter]^3 + \ldots$$

$$a_i, = \text{constant} \tag{9.14}$$

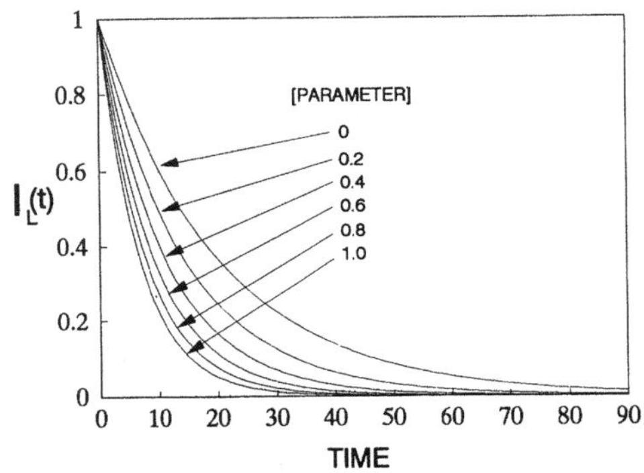

Figure 9.2. Transient luminescence of a sensor as a function of the concentration of a [*Parameter*]. In the figure $k_{nr} = \beta$ (Eq. (9.12) with $\alpha = 0$). Increasing values of [*Parameter*] result in shorter luminescence lifetimes. $k_r = \frac{1}{20}$; $k_{nr} = \frac{1}{10}$ in units of time^{-1}. The curves in the figure are determined by Eq. (9.3).

in which the constants a_i alternate with positive and negative values. The order of the exponent does not imply a multimolecular reaction. The fact that the coefficients of the polynomial alternate between positive and negative values (with the possible exception of the zero and first orders) suggests merely a series expansion of an unknown function, not a kinetic process.

The dependence of k_{nr} on the value or concentration of a [*Parameter*], in the vicinity of the excited sensor, determines both the luminescence intensity and the excited state lifetime of the sensor.

Figure 9.2 illustrates the transient luminescence of a sensor with k_{nr} determined by the ideal case of Eq. (9.12). The figure shows that, as the value of [*Parameter*] increases, the duration (or lifetime) of the luminescence decreases.

9.2.3. Nonhomogeneous Sensors

Sensors are usually attached chemically or physically to other materials here referred as the carrier, like polymers, antibodies, and optical fibers in order to facilitate the sensing process. These carriers generally affect the luminescent characteristics of the sensor molecules. The modification of the luminescent characteristics of the sensor is caused by the creation of more than one microphase or microenvironment for the sensor. Each molecule in its particular microenvironment may return to the ground state following a different set of processes or mechanisms. Alternatively, the nonradiative decay rate of each microphase may be different for each sensor molecule. Depending on the characteristics of the carrier and the sensor, the number of microphases may be one, two, three, or an infinite number.

Consider, for example, a sensor composed of n phases each with a different nonradiative decay rate. The fraction of the excited sensor molecules F_i in phase ith may decay with an overall decay rate $k_r + k_{nri}$. In this case, the average probability of the sensor molecules of remaining in the excited state is given by

$$p_m{}^* = e^{-k_r t} \sum_{i=1}^{n} F_i e^{-k_{nri} t}, \quad \sum_{i=1}^{n} F_i = 1 \tag{9.15}$$

where $k_{nri} \neq k_{nrj}$ for $i \neq j$. The number of molecules in the excited state is given by Eq. (9.2)

$$M^*(t) = M^*(0) p_m{}^* \tag{9.16}$$

and the overall transient intensity of the luminescence is given by Eq. (9.3)

$$I_L(t) = k_r M^*(0) p_m{}^* \tag{9.17}$$

The luminescence contribution of phase ith is also given by F_i when the absorption of the excitation is not phase-dependent. When the absorption of excitation is sensor-phase dependent F_i does not correspond to the actual phase-composition of the sensor–carrier system. In this case, it is useful to refer to the fraction of photons $\mathcal{F}$ emitted by each phase. In our n-component system the total number of photons P is given by integrating Eq. (9.17)

$$P = M^*(0) \sum_{i=1}^{n} F_i Q_i \tag{9.18}$$

where Q_i represents the quantum yield of the ith phase. The steady-state sensor luminescence I_{SS} is given then by

$$I_{SS} = \sum_{i=1}^{n} F_i Q_i \frac{dM^*(0)}{dt} \tag{9.19}$$

where

$$F_i \frac{dM^*(0)}{dt} = g_i I_O, \qquad g_i = 2.303 I_O \varepsilon_i C_i L << 1 \tag{9.20}$$

again, the excitation I_O may vary with time. g_i is the fraction of the excitation absorbed by the ith component. Equation (9.20) assumes negligible attenuation of the excitation. The extinction coefficients can be added linearly only when $g_i << 1$. The steady-state sensor luminescence is then given by

$$I_{SS} = I_O \sum_{i=1}^{n} g_i Q_i \simeq 2.303 I_O L \sum_{i=1}^{n} Q_i \varepsilon_i C_i, \qquad g_i << 1 \tag{9.21}$$

The fraction of photons, or ratio of steady-state intensities, or ratio of CW intensities emitted by phase ith is then

$$\mathcal{F}_i = \frac{F_i Q_i}{\sum_{j=1}^{n} F_j Q_j} = \frac{I_{SS_i}}{\sum_{j=1}^{n} I_{SS_j}} = \frac{I_{CW_i}}{\sum_{j=1}^{n} I_{CW_j}} \tag{9.22}$$

Commonly, the temporal luminescence of sensors in carriers is expressed in terms of preexponential factors and lifetimes

$$I_{\mathcal{L}}(t) = \sum_{i=1}^{n} A_i e^{-\frac{t}{\tau_i}} \tag{9.23}$$

in this case $\mathcal{F}_i$ is given by the well-known equation

$$\mathcal{F}_i = \frac{A_i \tau_i}{\sum_{j=1}^{n} A_j \tau_j} \tag{9.24}$$

Equation (9.24) also holds for the case in which multiple luminescent materials (components), each with a given k_{ri}, are present.

The temporal luminescence of a highly heterogeneous sensor–carrier mixtures cannot be uniquely represented by sums of exponentials (Eq. (9.23)) due to the lack of orthogonality of the exponential function. In this case it becomes appropriate to express equations (9.17) or (9.23) in terms of probability density functions or lifetime distribution functions $\tilde{\mathcal{A}}$ [8–14]

$$I_{\mathcal{L}}(t) = \int_0^{\infty} \tilde{\mathcal{A}}(\tau) e^{-\frac{t}{\tau}} d\tau \tag{9.25}$$

in which the probability of having a preexponential factor of a given value associated with a lifetime τ is given by $\tilde{\mathcal{A}}$. $\tilde{\mathcal{A}}$ is normalized

$$\int_0^{\infty} \tilde{\mathcal{A}}(\tau) d\tau = 1 \tag{9.26}$$

Similarly, the distribution of luminescence intensities (also normalized) is given by

$$\tilde{\mathcal{F}}(\tau) = \frac{\tilde{\mathcal{A}}(\tau)\tau}{\int_0^{\infty} \tilde{\mathcal{A}}(\tau)\tau d\tau} \tag{9.27}$$

Even though the temporal luminescence of a sensor cannot be uniquely represented in terms of lifetime distribution functions, the use of lifetime distributions provides a more convenient way to characterize the transient luminescence of sensors than the use of few discrete exponentials. Lifetime distribution functions require less parameters to describe the sensor luminescence response which is an advantage in the implementation of data analysis for real-time applications.

9.3. Continuous Wave Luminescence Sensing

9.3.1. Homogeneous Sensors

The CW luminescence of ideal sensors is determined by the value or concentration of the external [*Parameter*] of interests. The sensor quantum yield as a function of [*Parameter*] is given by

$$Q = \frac{k_r}{k_r + k_{nr}\,([Parameter])} \tag{9.28}$$

In general, $k_{nr}([Parameter])$ increases monotonically with [*Parameter*] such that Q decreases also monotonically with [*Parameter*]. At a given wavelength, the ratio of quantum yields give, for homogeneous sensors:

$$\frac{I_{\mathrm{CW}}(0)}{I_{\mathrm{CW}}([Parameter])} = \frac{Q(0)}{Q} = 1 + \frac{\beta[([Parameter])}{k_r + \alpha} \tag{9.29}$$

where $I_{\mathrm{CW}}(0)$ refers to the sensor CW luminescence when [*Parameter*] = 0. k_{nr} may be determined by Eqs. (9.12–9.14). In the ideal case when the nonradiative decay rate k_{nr} is linearly related to the parameter concentration according to Eq. (9.12) we obtain the straight Stern–Volmer function

$$\frac{I_{\mathrm{CW}}(0)}{I_{\mathrm{CW}}([Parameter])} = \frac{Q(0)}{Q} = 1 + \frac{\beta}{k_r + \alpha}[Parameter] \tag{9.30}$$

When the relation between then nonradiative decay processes and the concentration or value of the parameter of interest $k_{nr}([Parameter])$ is not linear (e.g., Eqs. (9.13 and 9.14), the intensity ratio of Eq. (9.31) introduces the well-observed problem of curvature in the Stern–Volmer plot (see Fig. 9.3).

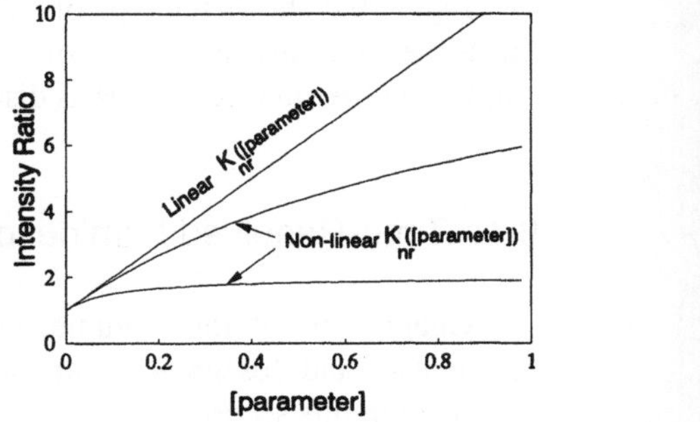

Figure 9.3. Stern–Volmer plot, based on luminescence intensity ratios, of heterogeneous sensor–carrier systems. In general, a straight lines are obtained when the dependence of the nonradiative decay rates for all phases are linear (Eq. (9.12)) regardless of the number of phases in the sensor–carrier preparation. Curvature is found when the value of the nonradiative decay rate k_{nr} does not increase proportionally with [*parameter*] (e.g., Eqs. (9.13, 9.14)).

9.3.2. Nonhomogeneous Sensors

Consider a sensor decay that can be approximated by the superposition of a discrete or continuous set of exponential functions. The CW luminescence intensity is given by Eq. (9.21)

$$I_{CW}[(Parameter]) = 2.303 I_O L \sum_{i=1}^{n} Q_i \varepsilon_i C_i \tag{9.31}$$

from Eq. (9.31) the sensor intensity ratio is determined by

$$\frac{I_{CW}(0)}{I_{CW}([Parameter]} = \frac{\sum_{i=1}^{n} Q_i(0)\varepsilon_i C_i}{\sum_{i=1}^{n} Q_i([Parameter])\varepsilon_i C_i} \tag{9.32}$$

In the case of nonhomogeneous sensors the intensity ratio $I_{CW}(0)/I_{CW}([Parameter])$ introduces significant curvature only when at least one $k_{nri}([Parameter])$ is not linear. When the relation between then nonradiative decay processes and the concentration or value of the parameter of interest is given by Eq. (9.12) the intensity ratio of Eq. (9.32) is practically linear, regardless of the number of components or phases present in our sensor–carrier system. Figure 9.3 illustrates this fact. In the figure the case in which k_{nri} is linearly related to *[Parameter]* (Eq. 9.12) is shown for large number of phases ($n = 50$). The ratio of intensities, in this case, is given by

$$\frac{I_{CW}(0)}{I_{CW}([Parameter])} = \frac{\sum_{i=1}^{n} \frac{k_r}{k_r+\alpha_i}\varepsilon_i C_i}{\sum_{i=1}^{n} \frac{k_r}{k_r+\alpha_i+\beta_i[Parameter]}\varepsilon_i C_i} \tag{9.33}$$

The heterogeneity or homogeneity of a sensor–carrier preparation cannot be established on the basis of the Stern–Volmer plot. Straight or curved behaviors are caused by the relation between the nonradiative processes and the value of *[Parameter]*.

Figure 9.4 shows the type of behavior for an oxygen sensor in which the carrier limits the solubility and diffusivity of oxygen. The data in Figure 9.4 were obtained employing lifetime techniques as discussed in the next section.

9.4. Time-Resolved Luminescence Sensing

Chapter 10 of this volume, entitled "Lifetime-Based Sensing," by Henryk Szmacinski and Joseph R. Lakowicz is an excellent complement to the material presented in this section.

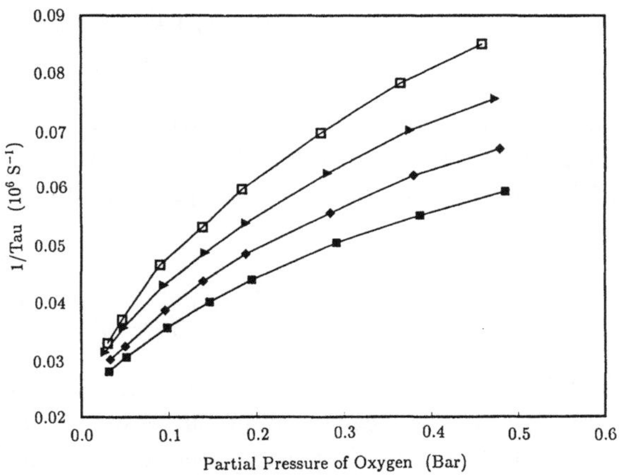

Figure 9.4. Stern–Volmer plot of a fiberoptic oxygen sensor at different temperatures. □, 43°C; ▲, 37°C; ♦, 31°C; ■, 25°C. The sensing capabilities of the fiberoptic sensor are limited by diffusion processes as suggested by the decreasing value of the lifetimes with increasing temperature. (From Ref. 21 with permission.)

9.4.1. Homogeneous Sensors

The time constant or lifetime τ of the sensor luminescence is determined by the value of *[Parameter]*. For sensor–carrier preparations with a uniform composition in which all sensor molecules return to the ground state with the same probability we have:

$$\tau = \frac{1}{k_r + k_{nr}\,([Parameter])} \tag{9.34}$$

Let τ_0 be the lifetime of the sensor when *[Parameter]* = 0, then

$$\frac{\tau_0}{\tau} = \frac{k_r + k_{nr}([Parameter])}{k_r + k_{nr}(0)} \tag{9.35}$$

Again, k_{nr} may be determined by Eqs. (9.12–9.14). In the case when k_{nr} is linearly related to *[Parameter]* Eq. (9.35) reduces to the Stern–Volmer function

$$\frac{\tau_0}{\tau} = 1 + \frac{\beta}{k_r + \alpha}[Parameter] \tag{9.36}$$

9.4.2. Nonhomogeneous Sensors

9.4.2.1. Static Sensors

In the case in which the overall sensor luminescence is the result of isolated phases each with a different nonradiative decay rate, the lifetimes of each phase provides an independent measurement of *[Parameter]*

$$\frac{\tau_{i0}}{\tau_i} = \frac{k_r + k_{nri}([Parameter])}{k_r + k_{nri}(0)} \tag{9.37}$$

where again k_{nri} may be determined by Eqs. (9.12–9.14). The functional dependence of k_{nri} may be different for each lifetime value. In the linear case of Eq. (9.12) the lifetime ratio for each phase of the sensor–carrier preparation yields

$$\frac{\tau_{i0}}{\tau_i} = 1 + \frac{\beta_i}{k_r + \alpha_i}[Parameter] \tag{9.38}$$

In this case, the lifetime ratio is practically linear, regardless of the number of components or phases present in our sensor–carrier system. A similar behavior was obtained in the case of intensity-based sensors as illustrated in Figure 9.3.

Significant curvature may be observed in the case of lifetime- (and intensity-) based sensors, mainly when the relation $k_{nri}([Parameter])$ is not linear. Figure 9.4 shows this type of nonlinear behavior for a fiberoptic oxygen sensor. The figure shows Stern–Volmer-type plots (τ^{-1} versus $[O_2]$) at four different temperatures. The curvature is caused by the inability of the carrier to transport oxygen proportionally to the equilibrium partial pressure of oxygen.

The temperature dependence of the inverse of the lifetime in Figure 9.4 is caused by the effects that temperature has in the transport processes of oxygen in polymers. Typically, as the temperature increases the oxygen solubility in polymers decreases reducing k_{nri}. However, at the same time, oxygen diffuses faster with increasing temperature (k_{nri} increases). The effects of solubility and diffusivity produce opposite effects on k_{nri} with temperature but do not cancel. The increase of k_{nri} (~ $1/\tau$) with temperature suggests that the lifetime of the sensor is controlled by diffusion processes. The data presented in this figure were obtained with the frequency domain instrumentation and techniques presented later in this chapter.

9.4.2.2. Dynamic Sensors

The cases in which the sensor is in dynamic equilibrium between two or more phases or states is not straight forward. In this case the observable lifetime values are dependent upon the dynamics of the sensor–carrier system.

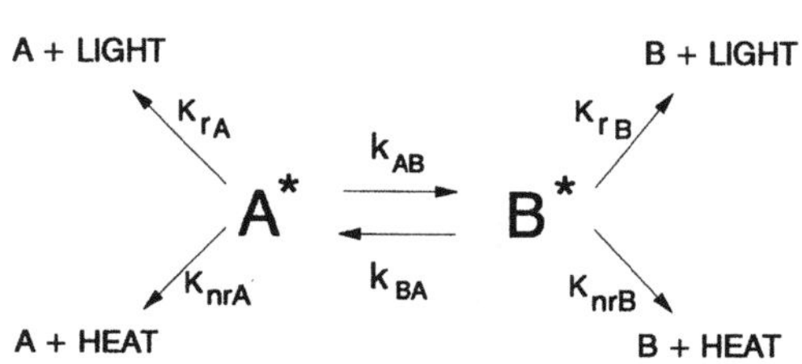

Figure 9.5. Sensor in dynamic equilibrium between two phases or states $A^* \Leftrightarrow B^*$. The measurable lifetime values do not correspond to the lifetimes values of the independent phases or states.

Consider a sensor in dynamic equilibrium between phases or states A^* and B^* as illustrated in Figure 9.5. This situation may occur in pH sensors in which the lifetime of the ionized molecule is different from the lifetime of the neutral molecule. In this case a distinction needs to be made between the lifetimes of the individual states and the lifetimes measured. Let τ_A and τ_B be the lifetimes of states A^* and B^*, respectively. The measured lifetime values, τ_1 and τ_2, are then given by the following equation[12]

$$\tau_{1,2} = \left[\frac{\tau_A^{-1} + \tau_B^{-1} + k_{AB} + k_{BA}}{2} \pm \frac{1}{2} [(\tau_A^{-1} - \tau_B^{-1})^2 + (k_{AB} + k_{BA})^2 + 2(\tau_A^{-1} - \tau_B^{-1})(k_{AB} - k_{BA})^2]^{\frac{1}{2}} \right]^{-1} \quad (9.39)$$

where k_{AB} and k_{BA} are the interconversion rate constants between states A^* and B^* with an equilibrium constant κ

$$\kappa = \frac{k_{AB}}{k_{BA}} \quad (9.40)$$

In the limit when the rates of interconversion k_{AB} and k_{BA} are much slower than the individual state decay rates K_A and K_B, Eq. (9.39) reduces to the static case discussed above and the lifetime ratio of Eq. (9.37) provides useful information on the value of [*Parameter*]. However, in the most general case such lifetime ratio is a complex function of [*Parameter*]as given by Eq. (9.39).

In the limit when the rates of interconversion k_{AB} and k_{BA} are much faster than the individual states decay rates K_A and K_B then a single lifetime is measured:

$$\tau = \frac{\tau_A \tau_B(\kappa + 1)}{\kappa \tau_A + \tau_B} \quad (9.41)$$

and, in this case, the ratio of the measured lifetime values provides a simple relation

$$\frac{\tau_0}{\tau} = 1 + \frac{k_{nrA} + \kappa k_{nrB}}{(\kappa + 1)k_r} \quad (9.42)$$

where k_{nrA} and k_{nrB} may be independently determined by Eqs. (9.12–9.14). When the dependence of the decay rate of both A^* and B^* is linear, the lifetime ratio of Eq. (9.42) also yields also a straight Stern–Volmer plot.

9.4.2.3. Complex Sensors

Nonexponential luminescence decays are not well understood. However, regardless of the lack of understanding, it is a tradition to fit complex decays to sums of exponential functions either discrete or continuous (lifetime distributions). An important limitation of this approach is introduced by the nonorthogonal nature of the exponential function. The practice of fitting nonexponential luminescence decays to sums of exponential functions yields results that are not unique and that cannot be correlated to the [*Parameter*] of interest unless empirical relations are employed. A double exponential fit to the luminescence decay of a sensor may assume, for example, the existence of a two-phase sensor–carrier system, however, the obtained lifetimes may not follow the form of Eqs. (9.37), (9.39), or (9.41). Much work has been done to understand the kinetics of phosphorescent and long lived fluorescent molecules in polymer matrixes. References 15–17 are examples of this research.

A typical example of a more complex system is given by sensors in high concentrations in the carrier polymer. In this case a mixture of first- and second-order decays may be obtained. This situation occurs in fiberoptic sensors when the sensor concentration on the tip of a small fiber is made high with the purpose of obtaining a strong luminescent signal. In this case the sensor molecule probability of remaining in the excited state p_{m^*}, as a function of time, is determined by the following equation

$$\frac{dp_{m^*}}{dt} = -(k_r + k_{nr})p_{m^*} - k_{ex}p_{m^*}^{2} \tag{9.43}$$

which has the following solution

$$p_{m^*} = he^{-(k_r+k_{nr})t} + (1-h)\frac{k_r + k_{nr}}{(k_r + k_{nr} + k_{ex})e^{(k_r+k_{nr})t} - k_{ex}} \tag{9.44}$$

where k_{ex} is the decay rate associated with interactions of two excited sensor molecules. This nonlinear process is considered to be nonradiative in this chapter. h yields the fraction of p_{m^*} that follows the path of an homogeneous decay (first-order decay) while $1-h$ is the fraction of the decay caused by reactions in the excited state due to the high concentration of excited molecules. The sensor transient luminescence is given by Eq. (9.17).

The expected lifetime of sensors that exhibit excited state reactions drifts with time. As the sensor is exposed to excitation the sensor molecules are photodestructed or photobleached. Photobleaching reduces the concentrations of molecules in the carrier and at the same time, the fraction of sensor molecules $(1-h)$ undergoing excited state reactions. A test that a sensor–carrier preparation is not too highly concentrated is the stability of its luminescence lifetime response over long periods of time. Figure 9.6 illustrates this type of test. The figure shows the decay of an oxygen sensor in the sensor luminescence intensity over a period of 100 h (6000 min) of continuous operation. However, the sensor excited state lifetime remains stable over this period, regardless of the amount of photobleaching. This indicates that the sensor preparation was not undergoing excited state reactions. In the figure, the temporal sensor luminescence was determined by the first order (exponential) law. The results presented in Figure 9.6 were obtained from an oxygen sensor bounded to the distal end of a 100-μm fiberoptic probe. The oxygen sensor response was monitored with the frequency domain instrumentation and techniques presented later in this chapter.

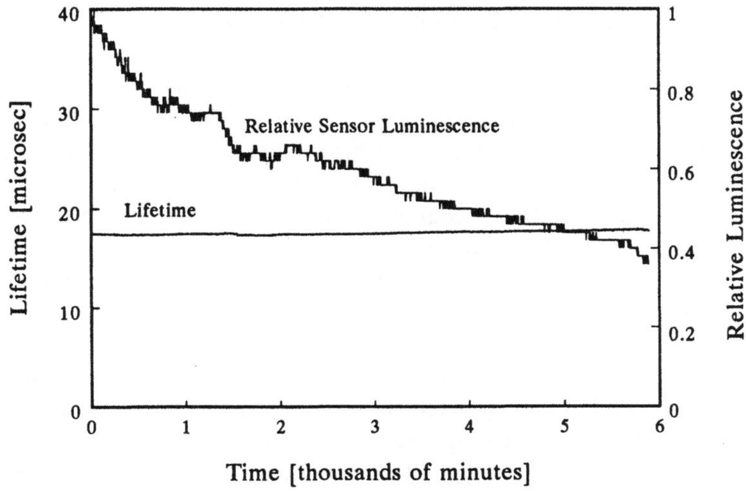

Figure 9.6. Stability of a lifetime-based fiberoptic oxygen sensor over a period of 100 h of continuous operation. Lifetime techniques are insensitive to the process of photobleaching only in the absence of excited state reactions. Excited state reactions of the sensor–carrier system cause drifts in the observed lifetime with photobleaching. They are avoided by limiting the concentration of the sensor in the carrier. (From Ref. 21 with permission.)

9.5. Real-Time Techniques

The goal of luminescent sensing is perhaps to implement methods and instrumentation capable of performing simple, accurate, and precise measurements in real time. These qualities of sensing systems are especially desirable in the biomedical field in which relevant physiological parameters may change constantly during time expands from a fraction of a second to several hours.

The principles discussed in the previous sections of this chapter may be used to monitor the spatial distribution of a given parameter of importance. However, imaging oxygen concentrations, pH values, action potentials, temperature values, and other parameters cannot be attained in real time using the current technology. The real-time applications that are now starting to appear are limited to the sampling of parameters of interest, in limited or localized regions in space, employing optical fibers.

The instrumentation and methods required to monitor the luminescent signal in an optical fiber require the implementation of an spectrophotometer capable of autocalibrating itself, collecting, fitting, and analyzing the data, and displaying the results in a few seconds. These "intelligent" systems are equivalent to a research scientist and a high-quality spectrophotometer capable of performing all tasks in few seconds. The best methods need to be selected to implement these sensing devices. An analysis of the optimum methods currently available is presented in this section.

9.5.1. The Nature of Intensity and Lifetime-Based Sensors

The continuous measurement of weak luminescent signals immersed in orders of magnitude stronger backgrounds is of particular interest. This type of situation occurs in the case of single fiberoptic sensors in which excitation and luminescent response are generally, transported colinearly through one fiber and not at the usual 90° angle commonly found in laboratory spectrophotometers (a less simple alternative, which substantially reduces backscattered excitation, may use two optical fibers, one to conduct the excitation and the other to collect the sensor luminescence). In addition, room background light cannot always be eliminated in the case of real biomedical applications.

Intensity-based sensors are affected by factors that determine the amount of light in optical fibers. Light intensity fluctuations due to fiber microbending, fiberoptic coupling, excitation backscattering, back reflections, sensor photobleaching, and background light are several of the problems commonly found in implementations that require the measurement of absolute luminescence intensities in optical fibers. Continuous wave systems that monitor absolute light intensities do not provide reliable systems for the layperson's use. Dual-wavelength CW methods (intensity ratio approaches) may overcome some of the difficulties.

Lifetime-based systems provide a better solution to the problem of monitoring weak luminescent signals immersed in orders-of-magnitude stronger backgrounds. Lifetime techniques do not monitor absolute light intensities. They measure the rate of change of the sensor luminescence with time. This rate of change is independent of the limitations found in intensity based systems. Consequently, lifetime-based systems are inherently insensitive to light intensity fluctuations due to fiber microbending, fiberoptic coupling, scattering, back reflections, sensor photobleaching, and background light.[18–21]

9.5.2. Time Domain and Frequency Domain Measurements

The transient response of luminescent substances to pulsed excitation can be captured in the time domain by sampling enough data points within the time span of the decay. For example, fast digitizers are commonly employed to store phosphorescence decays. If fast digitizers are unavailable, time-correlated single-photon counting[22] can be used to monitor fluorescence decays.

The transient response of luminescent substances to modulated excitation can be determined in the frequency domain by measuring the phase delay and the demodulation of the luminescence with respect to the excitation.[14, 23–28]

Frequency domain techniques offer advantages over time domain techniques for real-time applications. In the frequency domain the measurements are performed within limited frequency bandwidths. The noise in limited bandwidths is reduced, in most cases substantially. Figures 9.7 illustrates this concept. In Figure 9.7a the

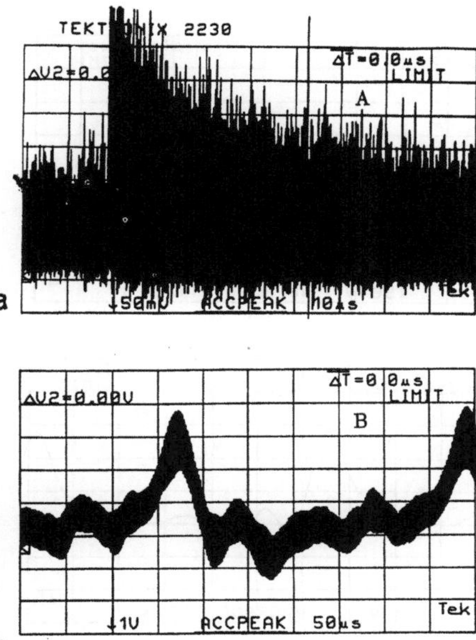

a

b

Figure 9.7. Noise content of a fiberoptic oxygen sensor signal (a) in the time and (b) in the frequency domains. Time domain signals require broad frequency bandwidths. Frequency domain signals require very limited-frequency bandwidths. Noise is reduced by band limiting the signal, an advantage of frequency domain methods.

electrical signal of an oxygen sensor monitored employing a 100-μm core silica fiber is shown. The same electrical signal, frequency band limited to include only the first 16 harmonics, is shown in Figure 9.7b. These signals are averaged one thousand times in Figure 9.8. Let us refer here to the broadband (unlimited bandwidth) signals as the time domain signals (Figures 9.7a and 9.8a). Similarly, the bandlimited signals are referred as the frequency domain signals (Figures 9.7b and 9.8b). The reason for this distinction is due to the fact that the temporal characteristics of the broadband signals can be determined using time domain techniques, while the temporal properties of the bandlimited signals can only be determined using frequency domain methods. Notice the scale differences of the averaged time domain and frequency domain plots. The frequency domain signal remained well defined (highly reproducible) after amplification by a factor of 50. The time base of the signals also shows that the frequency domain signals last, in this case, about seven to eight times longer than their time domain counterparts. The advantages of the band limited or frequency domain approach becomes obvious from these figures when implementing an instrument:

1. The averaged frequency domain waveforms have less noise when compared to the time domain waveforms.
2. The frequency domain signal is easier to capture since it expands over longer time periods.
3. Signals with reduced harmonic content can be determined with fewer samples, in turn, fewer data points translate to faster data processing speeds, an advantage for real-time implementations.

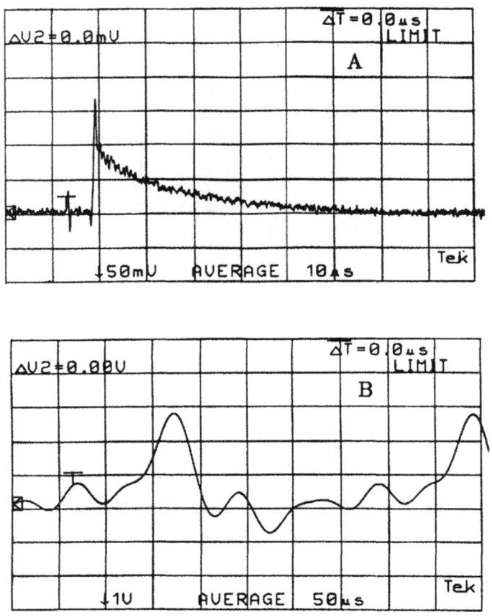

Figure 9.8. Noise content of a fiberoptic oxygen sensor signal (a) in the time and (b) in the frequency domains after 1000 averages. The frequency domain signals from a 100 μm fiberoptic oxygen sensor become practically noiseless while in the time domain noise is still present. The frequency domain signal has been amplified by a factor of 50 to demonstrate the absence of relevant noise.

This comparison between time and frequency domain measurements is performed at submegahertz frequencies in order to avoid the issue of deconvolution of time domain signals. At megahertz frequencies time domain measurements encounter an additional limitation, these signals must be deconvoluted to isolate the sensor response from the instrument response. The need for deconvolutions adds extra software and computation time, which limits the versatility of time domain techniques for real-time applications. No deconvolutions are necessary in the frequency domain as shown below.

9.5.3. The Principle of Frequency Domain Sensing

9.5.3.1. The Homogeneous Case

Going back to Figure 9.1, the probability of finding a sensor molecule p_{m^*} in the excited state is modified in the presence of steady-state excitation ε(t):

$$\frac{dp_{m^*}}{dt} = -\,(k_r + k_{nr})p_{m^*} + \frac{g}{M^*(0)}\varepsilon(t) \qquad (9.45)$$

where $g/M^*(0)$ corresponds to the probability of absorbing an excitation photon by our molecule. Consider $\varepsilon(t)$ to be sinusoidally modulated excitation of the form of Eq. (9.10)

$$\varepsilon(t) = \varepsilon_{\mathrm{CW}}[1 + M_\varepsilon(\omega)\sin(\omega t)], \quad M_\varepsilon(\omega) \leq 1 \qquad (9.46)$$

The solution to the first-order differential equation (9.45) yields

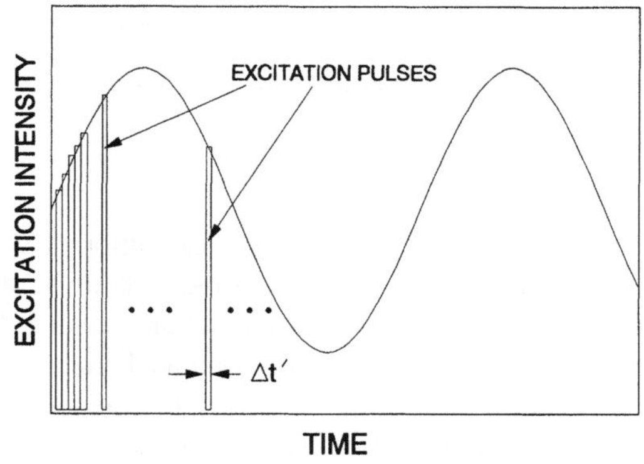

Figure 9.9. Sinusoidal excitation can be regarded as a succession of sufficiently narrow rectangular pulses in time t'. Then, at any particular time t, the fluorescence response to the train of pulses sinusoidally modulated is given by the superposition of all single-pulse responses initiated at times $t' < t$.

$$p_m^* = e^{-(k_r + k_{nr})t} + \frac{g}{M^*(0)} \frac{E_{CW}}{k_r + k_{nr}} \left[1 + \frac{M_\varepsilon(\omega)}{\sqrt{1 + \frac{\omega^2}{(k_r + k_{nr})^2}}} \sin(\omega t - \phi)\right] \tag{9.47}$$

where

$$\phi(\omega) = \arctan\frac{\omega}{k_r + k_{nr}} \tag{9.48}$$

The sensor luminescence response as a function of time $\mathcal{L}(t)$ is then

$$\mathcal{L}(t) = k_r M^*(0) p_m^* \tag{9.49}$$

The steady-state sensor luminescent response ($t \to \infty$) as a time function is written in the following form

$$\mathcal{L}(t) = \mathcal{L}_{CW}\,[1 + M_{\mathcal{L}}(\omega)\sin(\omega t - \phi)], \quad M_{\mathcal{L}}(\omega) \le 1 \tag{9.50}$$

where

$$M_{\mathcal{L}}(\omega) = \frac{M_\varepsilon(\omega)}{\sqrt{1 + \frac{\omega^2}{(k_r + k_{nr})^2}}} \tag{9.51}$$

The $M_{\mathcal{L}}(\omega)$, $M_\varepsilon(\omega)$, and $\phi(\omega)$ are the measurable quantities in the frequency domain. Commonly the modulation ratio $M(\omega)$ is used

$$M(\omega) = \frac{M_{\mathcal{L}}(\omega)}{M_\varepsilon(\omega)} = \frac{1}{\sqrt{1 + \frac{\omega^2}{(k_r + k_{nr})^2}}} \tag{9.52}$$

The CW or intensity-based information is also obtained in the frequency domain and is given by Eq. (9.53)

$$\mathcal{L}_{CW} = gQ\varepsilon_{CW} \tag{9.53}$$

where g and Q are determined by Eqs. (9.6) and (9.4), respectively.

9.5.3.2. *The General Case*

Let us regard the sinusoidal excitation of Eq. (9.46) as a succession of sufficiently narrow rectangular pulses in time t' as illustrated in Figure 9.9. Let us also assume linear regime conditions such that only a small fraction of molecules is excited in the steady state with negligible stimulated emission and excited state reactions.

The luminescence response $\mathcal{L}(t)$ to the continuous-pulse train with amplitude sinusoidally modulated at any particular time t is proportional to the number of photons (or alternatively to the intensity) of the exciting pulse. It is given by the superposition of all single-pulse responses initiated at times $t' < t$.

$$\mathcal{L}(t) = k_r g \sum_i I_{\mathcal{L}}(t - t_i') \, \varepsilon\,(t_i')\, \Delta t_i' \tag{9.54}$$

where $\varepsilon\,(t_i')\,\Delta t_i'$ is the number of photons in the exciting pulse. Equation (9.54) holds only in the linear regime. In the limit when the width of the pulses become infinitely narrow (δ-function excitation) the luminescence is given by

$$\mathcal{L}(t) = k_r g \int_{-\infty}^{t} I_{\mathcal{L}}(t - t')\, \varepsilon\,(t')dt' \tag{9.55}$$

performing the following variable transformation

$$t'' = t - t' \tag{9.56}$$

Eq. (9.55) can be rewritten in the following form

$$\mathcal{L}(t) = k_r g \int_{0}^{\infty} I_{\mathcal{L}}(t'')\, \varepsilon\,(t - t'')\, dt'' \tag{9.57}$$

$\mathcal{L}(t)$ is the convolution of the system impulse response and the excitation waveform. Notice, however, that the lower limit of integration starts at zero not $-\infty$ due to reasons of causality: the sensor luminescence response begins after excitation $t'' = 0$. In our case

$$\varepsilon(t - t'') = \varepsilon_{CW}\,[1 + M_\varepsilon(\omega) \sin \omega(t - t'')] \tag{9.58}$$

substituting Eq. (9.58) into Eq. (9.57) we obtain

$$\mathcal{L}(t) = k_r g \varepsilon_{CW}\,[N + M_\varepsilon(\omega)\,(G(\omega) \sin \omega t - S(\omega) \cos \omega t)] \tag{9.59}$$

where

$$N = \int_0^\infty I_{\mathcal{L}}(t'')\, dt'' \tag{9.60}$$

$$S(\omega) = \int_0^\infty I_{\mathcal{L}}(t'') \sin \omega t''\, dt'' \tag{9.61}$$

$$G(\omega) = \int_0^\infty I_{\mathcal{L}}(t'') \cos \omega t''\, dt'' \tag{9.62}$$

$S(\omega)$ and $G(\omega)$ are the sine and cosine Fourier transforms of the luminescence response to δ-function excitation, respectively. N yields the total number of photons of the response to the δ-function excitation. Equation (9.59) can be rewritten in the following form

$$\mathcal{L}(t) = \mathcal{L}_{\mathrm{CW}}\, [1 + M_{\mathcal{L}}(\omega) \sin(\omega t - \phi(\omega))] \tag{9.63}$$

where $\mathcal{L}_{\mathrm{CW}}$ is determined by Eq. (9.53). $M_{\mathcal{L}}(\omega)$ is the modulation of the luminescence. $\phi(\omega)$ is the phase delay of the luminescence at the particular modulation frequency. The phase delay and modulation ratio between the excitation and the emission constitute the two independent measurable quantities in frequency domain fluorometry.

Comparing Eqs. (9.59) and (9.63) the relation between the measurable quantities and the ideal δ-function pulse response parameters is obtained by

$$\phi(\omega) = \arctan \frac{S(\omega)}{G(\omega)} \tag{9.64}$$

$$M(\omega) = \frac{M_{\mathcal{L}}(\omega)}{M_e(\omega)} = \frac{[S^2(\omega) + G^2(\omega)]^{1/2}}{N} \tag{9.65}$$

For linear systems the emitted fluorescence has the same modulation frequency but is phase-shifted with respect to the exciting light and demodulated such that $M(\omega) < 1$. Knowledge of $\phi(\omega)$ and $M(\omega)$ is equivalent to knowledge of the functions $S(\omega)$ and $G(\omega)$ which correspond to the sine and cosine Fourier transforms of the ideal pulse response $I_{\mathcal{L}}(t)$. Consequently, the measurement of phase and modulation as a function of frequency is equivalent to determining the time evolution of the emitting system to δ-pulse excitation. In frequency domain luminescence no deconvolution for the finite width of the excitation pulse and the time response of the detection system is necessary since the ideal pulse response is obtained. Notice that the derivations performed above hold valid for any functional form of $I_{\mathcal{L}}(t)$. Equations (9.64) and (9.65) reduce to the homogeneous case Eqs. (9.48 and 9.52) in the case of the homogeneous systems. If the luminescence decay is determined by a discrete set of η exponentially decaying components then the sine and cosine Fourier transforms and the normalization factor are given by

$$S(\omega) = \sum_{i=1}^{n} \frac{A_i \tau_i \omega \tau_i}{1 + \omega^2 \tau_i^2} \tag{9.66}$$

$$G(\omega) = \sum_{i=1}^{n} \frac{A_i \tau_i}{1 + \omega^2 \tau_i^2} \tag{9.67}$$

$$N = \sum_{i=1}^{n} A_i \tau_i \tag{9.68}$$

and the observable phase delay and modulation ratio at a given frequency can be obtained from Eqs. (9.64 and 9.65). In the case of continuous distributions of exponentially decaying components the $S(\omega)$, $G(\omega)$, and N functions can be expressed in the following forms:

$$S(\omega) = \int_0^\infty \frac{\tilde{\mathcal{A}}(\tau)\tau\omega\tau}{1 + \omega^2\tau^2} d\tau \tag{9.69}$$

$$G(\omega) = \int_0^\infty \frac{\tilde{\mathcal{A}}(\tau)\tau}{1 + \omega^2\tau^2} d\tau \tag{9.70}$$

$$N = \int_0^\infty \tilde{\mathcal{A}}(\tau)\tau d\tau \tag{9.71}$$

Similarly, the overall observable phase delay and modulation ratio can be obtained from Eqs. (9. 64) and (9.65).

9.5.4. Concurrent Multifrequency Measurements

In order to implement frequency domain based sensing systems capable of monitoring the temporal luminescence of sensors, in few seconds, data must be collected at multiple frequencies simultaneously. Single-frequency techniques have been used to make frequency domain measurements of luminescent decays.[14, 23–28] This approach is unsuitable for real-time applications since data must be acquired at several frequencies in order to precisely and accurately determine the temporal variables of luminescent systems.[11] Each frequency requires a separate measurement, which makes the single frequency approach too slow to monitor the evolution of continuously varying parameters. The need to acquire data at multiple frequencies becomes even more important in fiber optic sensing in which weak and noisy sensor signals, in the subpicowatt range, are common.

9.5.4.1. *Multifrequency Excitation*

Frequency domain measurements require the use of periodic excitation sources. The luminescent molecules respond to the periodic excitation exhibiting the same frequency of modulation. This luminescence exhibits a phase delay and a demodulation with respect to the excitation due to the inability of the sensor molecule to respond to the higher frequencies of the excitation. This inability of the sensor molecules roughly begins at modulation frequencies $f_{\text{modulation}}$ of the same order of magnitude or faster than the decay rate

$$f_{\text{modulation}} = \frac{\omega}{2\pi} \approx k_r + k_{nr}([Parameter]) \tag{9.72}$$

At very high frequencies the luminescence is completely demodulated and only an average intensity is obtained. The maximum frequency of modulation of the response is determined by $k_r + k_{nr}([Parameter])$ of the sensor–carrier preparation.

In the frequency domain, any periodic excitation, $\varepsilon(t)$, can be described by a sum of sinusoidally modulated light waveforms at harmonics of the fundamental frequency of the excitation

$$\varepsilon(t) = \varepsilon_{\text{CW}}\,[1 + \sum_n M_E(n\omega_E)\sin\{n\omega_E t - \phi_E(n\omega_E)\}] \tag{9.73}$$

where ε_{CW} is the average intensity, $\phi_E(n\omega_E)$ and $M_E(n\omega_E)$ are the absolute phase and the absolute modulation of the nth harmonic. $\omega_E = 2\pi f_E$ is the circular frequency. f_E is the fundamental frequency of the modulated excitation.

Sinusoidal excitation provides only one harmonic at the modulation frequency. In contrast, pulsed light provides a large number of harmonics of the excitation repetition frequency. The harmonic content, the number of harmonics and their amplitude, is determined by the pulse width and shape.[25] For example, a train of infinitely short pulses provides an infinite number of harmonics all with equal amplitude. A square wave provides only three modulation frequencies with sufficient amplitude to be usable. Equation (9.74) gives the harmonic content of a train of rectangular pulses $R(t)$ of $\mathcal{D}$ duty cycle (pulse width divided by period) and R_p peak value:

$$R(t) = \mathcal{D}R_p([1 + 2\sum_{n=1}^{\infty} \frac{\sin n\pi\mathcal{D}}{n\pi\mathcal{D}}\cos(n\omega_E t)] \tag{9.74}$$

Notice that the shorter pulse width (or $\mathcal{D}$) the larger the absolute modulation $M_E(n\omega_E) = \sin n\pi D/n\pi D$ (which includes only positive frequencies). The absolute phase delay of symmetric pulses like $R(t)$ may be zero for all harmonics (when the pulse is centered around $t = 0$).

The luminescence $\mathcal{L}(t)$ contains only a subset of modulation frequencies of the excitation light due to the demodulation at higher frequencies caused by the molecule decay rates (see Eq. (9.51)). This subset is given by

$$\mathcal{L}(t) = \mathcal{L}_{CW}\,([1 + \sum_n M_F(n\omega_E)\sin\{n\omega_E t - \phi_F(n\omega_E)\}] \tag{9.75}$$

where $\mathcal{L}_{CW}$ is the average intensity of the luminescent response. $\phi_F(n\omega_E)$ and $M_F(n\omega_E)$ are the absolute phase and the absolute modulation of the nth harmonic of the luminescent response, respectively.

Each sinusoidally modulated light intensity of the response is phase delayed and demodulated with respect to the excitation such that

$$\Delta\phi(n\omega_E) = \phi_F(n\omega_E) - \phi_E(n\omega_E) > 0 \tag{9.76}$$

and

$$M(n\omega_E) = M_F(n\omega_E)/M_E(n\omega_E) < 1 \tag{9.77}$$

The phase delay $\Delta\phi(n\omega_E)$ (also denoted as ϕ as in Eqs. (9.48) and (9.64)), and the modulation ratio (or demodulation), $M(n\omega_E)$, are the direct result of the lifetime of the molecule. Consequently, the measurement of these quantities in the frequency domain provides for the determination of $k_r + k_{nr}$ of the molecule.

The phase delay and the modulation ratio can be determined at each harmonic available for measurement. These harmonics can range from a few hertz to several gigahertz depending on the harmonic content of the excitation and the lifetime of the luminescent molecule.[14, 25, 27, 28] Although all this information is available from the light signals, the instrumentation may limit the number of harmonics that can be measured.

In the case of a single lifetime with no interference from ambient or backscattered light only a single frequency is necessary to determine the lifetime of the luminescence. Determination of the phase and modulation at multiple frequencies is necessary to characterize complex decays in fiberoptic sensors.

Unlike commercial spectrophotometers in which the excitation and the detection systems are perpendicular to one another, in optical fiber sensors the excitation and the response are colinear. Consequently fiber optic sensors are prone to interference from backscattered light that can be orders of magnitude more intense than the dye signal. Although optical filters can be used to minimize the backscattering, it cannot be totally eliminated. In the clinical, industrial, and environmental settings ambient light is an additional source of interference in fiber optics sensors. Furthermore, when immobilized in polymers some dyes exhibit complex decays.[15–17] In highly heterogeneous systems the sensor may decay with a distribution of lifetime values.[9–13] In highly concentrated polymer solutions the sensor may decay by the superposition of first- and second-order decays.

9.5.4.2. *Multifrequency Detection: Analog Parallel Cross-Correlation*

The phase delay $\Delta\phi(n\omega_E)$ and modulation ratio $M(n\omega_E)$ information of the high-frequency signals is transferred to low-frequency signals by amplitude modulation (cross-correlation) of $\varepsilon(t)$ and $\mathcal{L}(t)$ with a periodic train of pulses $C(t)$ given by Ref. 29.

$$C(t) = C_o\,([1 + \sum_m M_C(m\omega_c)\sin(m\omega_c t)] \tag{9.78}$$

of fundamental (circular) frequency ω_c given by

$$\omega_c = \omega_E + \Delta\omega \tag{9.79}$$

where $\Delta\omega$ is a few Hz ($\Delta\omega <<< \omega_o$). The cross-correlation product $\varepsilon(t)\ C(t)$ is given by

$$\varepsilon(t)\ C(t) = \varepsilon_{\mathrm{CW}}\,[1 + \sum_n M_E(n\omega_E)\sin\{n\omega_E t - \phi_E(n\omega_E)\}]$$

$$\times\, C_o[1 + \sum_m M_C(n\omega_c)\sin(m\omega_c t)] \tag{9.80}$$

The lower-frequency terms of the cross-correlation product $\varepsilon(t)\ C(t)$ can be written in the following form

$$\varepsilon(t)\ C(t) = \varepsilon_{\mathrm{CW}}\, C_o[1 + \sum_n \sum_m \frac{M_E(n\omega_E)M_C(m\omega_c)}{2}\cos\{n\omega_E - m\omega_e)t - \phi_E(n\omega_E)\} + \cdots] \tag{9.81}$$

The lowest frequency terms (obtained when $n = m$) are

$$\varepsilon(t)\ C(t) = \varepsilon_{\mathrm{CW}}\, C_o[1 + \sum_n \frac{M_E(n\omega_E)M_C(n\omega_c)}{2}\cos\{n\Delta\omega t - \phi_E(n\omega_E)\} + \cdots] \tag{9.82}$$

Similarly, the lowest frequency terms of the cross-correlation product $\mathcal{L}(t)\ C(t)$ are given by

$$\mathcal{L}(t)\ C(t) = \mathcal{L}_{\mathrm{CW}}\, C_o[1 + \sum_n \frac{M_F(n\omega_E)M_C(n\omega_c)}{2}\cos\{n\Delta\omega t - \phi_F(n\omega_E)\} + \cdots] \tag{9.83}$$

The harmonics of $\Delta\omega$ of the $\varepsilon(t)\ C(t)$ and $\mathcal{L}(t)\ C(t)$ cross-correlation products contain all the phase [$\phi_E(n\omega_E)$ and $\phi_F(n\omega_E)$] and modulation [$M_E(n\omega_E)$ and $M_F(n\omega_E)$] information of the excitation and the system response. This approach, however, is limited

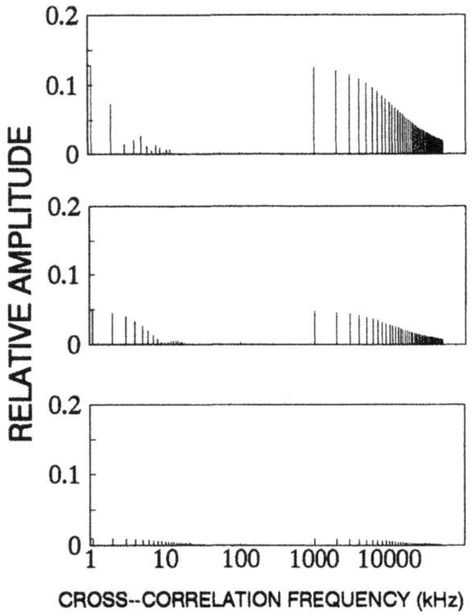

Figure 9.10. Amplitudes of analog cross-correlation frequency components of a 20-nsec lifetime exponential decay and rectangular cross-correlation pulses. From *top* to *bottom* the figure shows the harmonic content of the cross-correlation signal obtained using duty cycles of 30, 10, and 2%, respectively.

to relatively strong luminescence signals not commonly found in the field of fiberoptic sensors.

Analog cross-correlation methods for multiple-frequency acquisition distributes the total power present in the signal among a large set of modulation frequencies as illustrated in Figure 9.10. Hence, the amount of signal available per harmonic is substantially reduced.[30] The use of cross-correlation pulses with high harmonic contents allows for the collection of phase and modulation values at many harmonic frequencies simultaneously with low signal to noise ratios. High signal to noise ratios require the reduction of the harmonic content of the cross-correlation pulse which translates in reduced numbers of harmonic frequencies at which the information can be collected.

9.5.4.3. Multifrequency Detection: Digital Parallel Cross-Correlation

Let us assume two arbitrary periodic waveforms corresponding to an excitation and a response signals as illustrated in Figure 9.11. For the purpose of this discussion the excitation has been illustrated as a train of trapezoidal pulses and the luminescence response as a train of triangular pulses. Using sampling techniques the excitation and response waveforms are shifted to longer time scales as shown in the figure.

Let T_E be the period of the excitation pulses, and T_R be the ideal sampling time required to collect N_p points per period. Let T_S be the sampling time performed by the instrument and N_w the number of waveforms skipped before the next point is sampled. From Figure 9.11 we have

$$T_E = N_p T_R \tag{9.84}$$

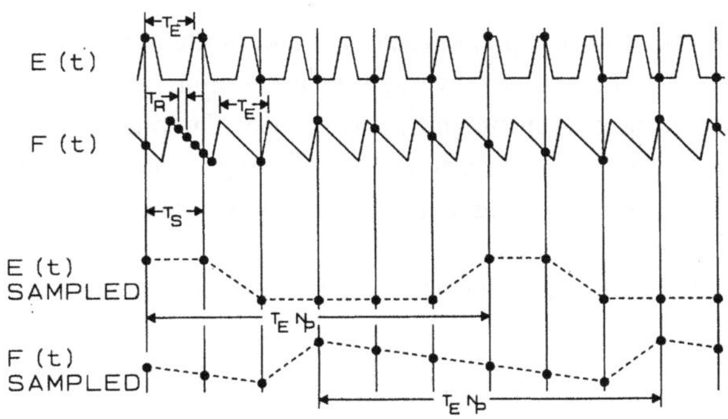

Figure 9.11. Digital sampling: a form of parallel cross-correlation. The phase and modulation information carried by a high frequency signal is shifted to lower frequency signals using synchronously sampling schemes. In this case no external cross-correlation signals need to be introduced. The trapezoidal pulse train illustrates a superposition of harmonics of the excitation and reference signal. The triangular train illustrates a correspondent superimposed set of sinusoidally modulated luminescence responses of the sample. Using sampling techniques, the excitation and response waveforms are shifted to longer time scales as illustrated in the figure (- - - -). T_E illustrates the excitation period. T_R illustrates the ideal sampling time required to collect N_p points per period. T_S illustrates the sampling time required to collect one data point per waveform. $N_w = 1$ in the figure.

and

$$T_S = N_w T_E + T_R \tag{9.85}$$

Equations (9.84) and (9.85) yield the following relation:

$$T_E = \frac{N_p}{N_w N_p + 1} T_S \tag{9.86}$$

Let f_S, f_E be the sampling and the excitation pulse repetition frequencies, respectively. Equation (9.86) can be rewritten in the following form:

$$f_S = \frac{N_p}{N_w N_p + 1} f_E \tag{9.87}$$

Phase and modulation information up to $N_p/2$ harmonics are present when N_p points per waveform are sampled. Let f_M be the maximum frequency sampled, then

$$f_M = \frac{N_p}{2} f_E \tag{9.88}$$

or in terms of the sampling frequency we obtain

$$f_M = \frac{N_w N_p + 1}{2} f_S \tag{9.89}$$

Commercially available sampling and digitizing units provide f_S in the range of fractions of a hertz to tens of megahertz. Values of $N_w N_p$ of the order of few hundreds are easily achievable. In principle, phase delay and modulation ratio information can be captured from a few hertz up to several gigahertz employing synchronous sampling techniques between the excitation and the response. The phase information is lost if the excitation and luminescence response signals are sampled independently without maintaining the phase coherence between the two signals. In this case only the modulation information remains.

In the case of phosphorescence and long-lived fluorescence N_w can be set to zero which yields

$$f_S = N_p f_E \tag{9.90}$$

and

$$f_M = \frac{f_S}{2} \tag{9.91}$$

The value of f_E determines all other variables in the equations above. In turn, f_E is determined by the temporal resolution of interest of the system studied. To resolve an average excited state lifetime τ, the required data sampling rate, in frequency domain techniques is at least an order of magnitude slower than it is in the time domain as stated by the following relation (when $N_p > 32$ and $N_w = 1$):

$$T_E \sim T_S \sim \frac{2\pi}{\tan \Delta\phi_1} \tau \tag{9.92}$$

where $\Delta\phi_1$ is the phase delay of the first harmonic of the response with respect to the first harmonic of the excitation, typically set to 3–5°. Similarly, in the case of phosphorescence and long-lived fluorescence,

$$T_E = N_p T_S \sim \frac{2\pi N_p}{\tan \Delta\phi_1} \tau \tag{9.93}$$

In order to provide the maximum harmonic content for a given number of sampled points, the duty cycle $\mathcal{D}$ of pulsed excitation should be roughly

$$\mathcal{D}_{\text{excitation}} \leq \frac{1}{N_p} \tag{9.94}$$

The smaller the duty cycle the larger the number of harmonics and the less power available per modulation frequency of the excitation.[30] In this case care must be taken to not place to much power in the unused high-frequency region by decreasing the duty cycle of the excitation. A good rule of thumb for externally modulated sources is the following:

$$\mathcal{D}_{\text{excitation}} \sim \frac{1}{N_p} \tag{9.95}$$

Fast Fourier transform algorithms are used to isolate the harmonics sampled. The individual phase delays are obtained from:

$$\Delta\phi_n = \arctan\left(\frac{I_{n,l}}{R_{n,l}}\right) - \arctan\left(\frac{I_{n,e}}{R_{n,e}}\right) \tag{9.96}$$

where $R_{n,l}$ and $I_{n,l}$ are the real and imaginary components of the nth harmonic of the Fourier transform of the luminescence. Similarly, $R_{n,e}$ and $I_{n,e}$ are the real and imaginary components of the nth harmonic of the Fourier transform of the excitation. The modulation ratios are obtained from:

$$M_n = \frac{M_{n,l}}{M_{n,e}} \tag{9.97}$$

where the absolute modulations are given by

$$M_{n,e} = \frac{(R_{n,e}^2 + I_{n,e}^2)^{\frac{1}{2}}}{R_{0,e}} \tag{9.98}$$

$$M_{n,l} = \frac{(R_{n,l}^2 + I_{n,l}^2)^{\frac{1}{2}}}{R_{0,l}} \tag{9.99}$$

9.5.5. The Limit of Fourier Methods in Real-Time Sensing

Fast Fourier transform algorithms are excellent filters for the application of isolating a set of well-defined harmonic frequencies when such frequency values are known. Fourier methods isolate the phase and modulation information at frequencies linearly spaced in frequency space. This is not an ideal situation. To precisely and accurately resolve the temporal characteristics of a luminescent material phase and modulation information need to be acquired at frequency values geometrically spaced over at least three orders of magnitude.[11] Resampling [29] offers a partial solution to this problem. By resampling data are separately collected and isolated using Fourier methods for each order of magnitude. This approach reduces the amount of time required to collect the data but at present is not fast enough to provide real-time results.

9.5.6. Noise in the Time and in the Frequency Domain

The connection between noise in the time and in the frequency domain is given by the linearity property of the Fourier transform (see, e.g., Ref. 31). Let $s(t)$ and $n(t)$ be the signal and noise in the time domain and $S(\omega)$ and $N(\omega)$ their corresponding Fourier transforms, then

$$a_1 s(t) + a_2 n(t) \Leftrightarrow a_1 S(\omega) + a_2 N(\omega) \tag{9.100}$$

According to relation (Eq. (9.100)) the signal to noise ratio is maintained by the Fourier transform in going from the time domain to the frequency domain and vice versa. Consequently, reducing the signal bandwidth in the time domain, by sacrificing the signal temporal resolution, improves the signal to noise ratio (Figures 9.7 and 9.8).

The linearity property of the Fourier transform does not hold for phase delay values. Let σ_t be the intrinsic, single measurement, time domain standard deviation of the filtered signal (in units of amplitude) . Also let $\sigma_\phi(n)$ be the standard deviation of the phase of the nth harmonic averaged j times. The phase noise as a function of harmonic (frequency) is given by

$$\sigma_\phi(n) = \frac{\sigma_t}{\sqrt{j} I_{\mathrm{CW}} M_L(n)} \tag{9.101}$$

which is inversely proportional to the signal modulation. As the frequency increases (the modulation of the luminescent signal decreases) the phase noise increases. The product $I_{\mathrm{CW}} M_L(n)$ is the amplitude of the nth harmonic.

Similarly, the modulation noise is given by the following equation. It is independent of the intensity modulation frequency.

$$\sigma_{M_L}(n) = \frac{\sigma_t}{\sqrt{j}} \tag{9.102}$$

9.5.7. Fiberoptic Sensor Instrumentation

9.5.7.1. Hardware

Figure 9.12 presents the general block diagrams of fiberoptic sensing systems. The instruments require the synthesis of two-phase coherent pulse trains regardless of the cross-correlation mode employed, either analog or digital. In the digital mode, which has the advantages described above, one frequency f_E is used to determine the repetition rate of the excitation source and the other with frequency f_S is used to trigger the data sampling. Both may be generated by coupling two synthesizers to one crystal f_{clock}. The following equations determine the relation among f_{clock}, f_E, and f_S:

$$f_E = \frac{f_{\mathrm{clock}}}{K N_p} \tag{9.103}$$

$$f_S = \frac{f_{\mathrm{clock}}}{K(N_w N_p + 1)} \tag{9.104}$$

where K is a rational number.

The excitation pulse train is used to trigger of the excitation. Mode-locked lasers and cavity-dumped lasers are versatile excitation sources however, they are not ideal to implement portable instruments. Externally modulated solid state microlasers are a

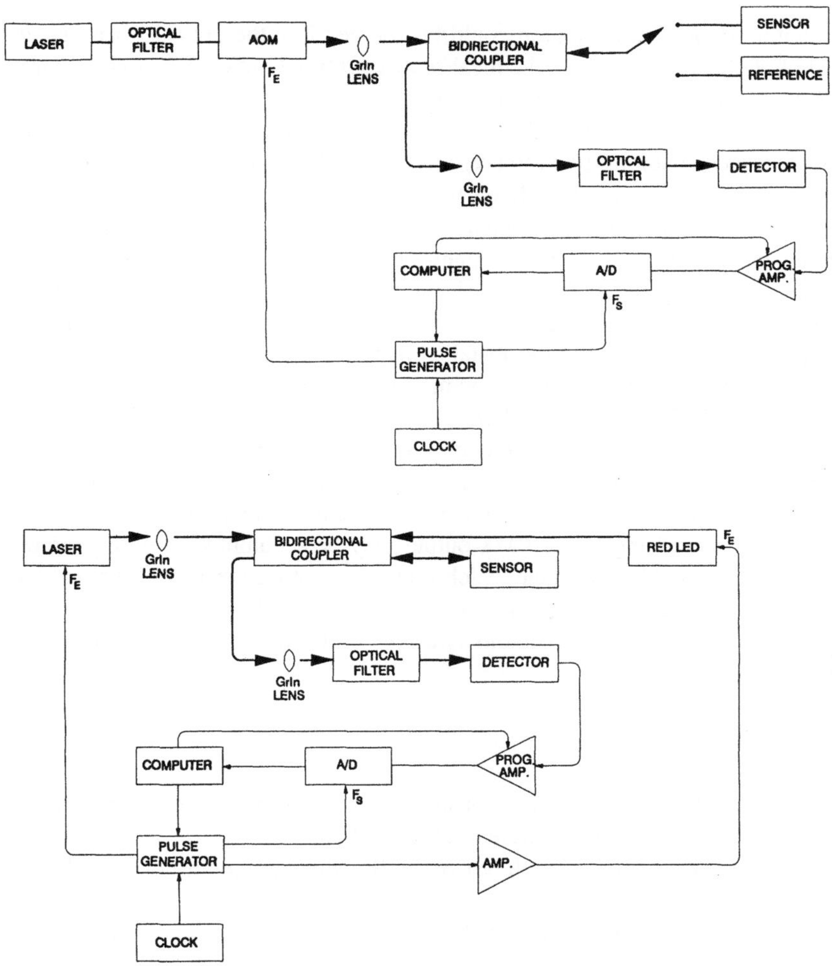

Figure 9.12. Block diagrams of fiberoptic sensing instrumentation. *Top*: externally modulated laser with an acousto-optic modulator. (From Ref. 20 with permission.) *Bottom*: internally modulated laser. (From Ref. 21 with permission.)

better option. Ideal sources would be visible blue or green laser diodes. At present these devices do not exist. Fiberoptic oxygen sensing systems implemented using the principles described in this chapter are externally modulated helium-neon (HeNe) lasers or internally modulated frequency doubled neodymium:yttrium-aluminum-garnet (Nd:YAG) solid-state microlaser.[20, 21] To launch and collect the excitation 2×2 optical fiber bidirectional couplers are ideal. Gradient index lenses are very useful in coupling laser excitation into optical fibers, specially in the case of weak excitation sources. When mainframe lasers are used these lenses should not be employed since substantial amount of excitation could be coupled into the optical fiber causing the rapid photodestruction of the chemical sensor. Optimal optical fibers core diameters, in biomedical chemical sensing, are 100 μ.

The fourth end of the bidirectional coupler may be connected in the case of phosphorescence or long-lived fluorescence to a red light-emitting diode to be used as reference. In the case of short-lived luminescence, a reference needs to be used in connection with a fiberoptic switch. In the case of long-lived luminescence, the spectral mismatch between the red light-emitting diode (LED) and the sensor luminescence does not produce any significant "color effect" errors, which is common in fluorescent lifetime measurements.[32] The red LED pulse readily transmits through the optical long wave pass filter installed just before the photodetector. Because the LED pulses have virtually a zero lifetime for the purpose of phosphorescence measurements, the LED provides a reference for the instrument. The LED and laser may be pulsed alternatively to collect the reference and response signals, respectively. The LED can then track and compensate any drift in the electronics. In this case the phase delay and modulation ratio obtained are true differential measurements.

Analog processing of the PMT output may be accomplished with a circuit that includes an adjustable gain stage which keeps the signal entering the analog to digital (A/D) converter at a constant amplitude. Low-noise, eight-pole, programmable low-pass filters may be used to avoid aliasing, especially in the case of long-lived luminescence.

9.5.7.2. Software

Real-time fiberoptic chemical sensing instrumentation can be automated employing the computer diagram illustrated in Figure 9.13.

The procedures cited in each box of the flux diagram are discussed below.[20, 21]

1. SET AND ENABLE CLOCK FREQUENCY: to be divided by the programmable counters or synthesizers.
2. LOAD COUNTERS: The load and hold registers of the programmable counters 1 and 2 are loaded to produce variable duty cycle pulse trains. The values loaded to the counters are determined by Eqs. (9.103) and (9.104).
3. DISABLE INTERRUPTS: to allow the dedication of the microprocessor to the data acquisition without interruptions in order to maintain synchronization during the data collection.
4. ENABLE COUNTERS: The counters that generate f_E and f_S are armed simultaneously at the beginning of each run to maintain phase coherence between the excitation and sampling signals. The enabling of the counters causes the pulsed excitation to be turned on.
5. DATA COLLECTION: The computer synchronously samples, digitizes, and adds in memory arrays of N_p locations of the reference and of the response signals to complete one scan.

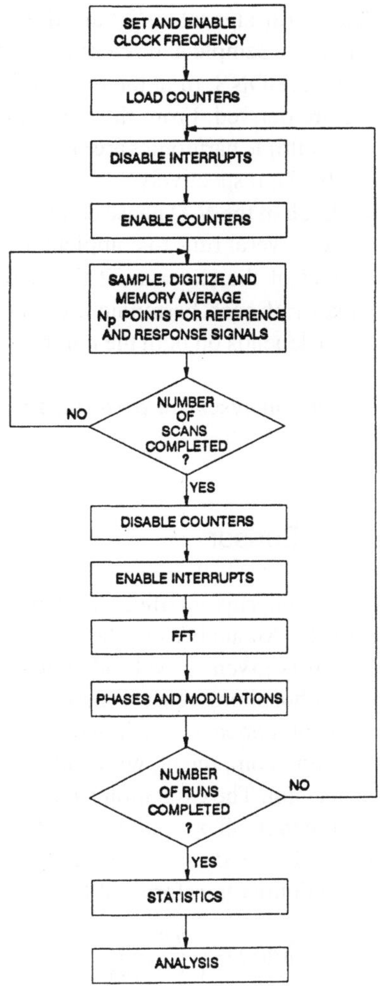

Figure 9.13. Program flow chart. (From Ref. 21 with permission.)

6. NUMBER OF SCANS: Each data point of both the reference and the response signals is memory averaged over several scans to complete a run.
7. DISABLE COUNTERS: The disabling of the counters causes the excitation to be turned off. The sample is not illuminated when data are not collected. In this manner, problems that arise due to long exposures of the sample and reference to excitation (bleaching and heating) are minimized.
8. ENABLE INTERRUPTS: The microprocessor is allowed to direct attention to other tasks after the run is completed.

9. FFT: Fast Fourier Transform algorithms[33] are used to obtain the frequency domain representation of the sample and reference waveforms sampled.
10. PHASES AND MODULATIONS: of at the most $N_p/2$ harmonics of the reference and the response are derived. From this absolute values the phase delay and modulation ratio of multiple harmonics are simultaneously obtained according to Eqs. (9.96) and (9.97), respectively.
11. NUMBER OF RUNS: Each of the $N_p/2$ phase delay and modulation ratio values is memory averaged over several runs to complete the experiment.
12. STATISTICS: At the end of the experiment the statistics (average, standard deviation, and chi-square) of each harmonic are computed. The data quality (precision and accuracy) depends on both the numbers of scans and the number of runs.
13. ANALYSIS: Least-square analyses are performed to monitor changes of the sample in real time.

9.6. Example: an Oxygen Sensor

Platinum and palladium porphyrins in silicon rubber resins are typical oxygen sensors and carriers, respectively. An analysis of the characteristics of these types of polymer films to sense oxygen is given in Ref. 34. For the sake of simplicity the luminescence decay of most phosphorescence sensors may be fitted to a double exponential function. The first component gives the excited state lifetime of the sensor phosphorescence while the second component, with a zero lifetime, yields the excitation backscatter seen by the detector. The excitation backscatter is usually about three orders of magnitude more intense in small optical fibers (100 μ) than the sensor luminescence. The use of interference filters reduce the excitation substantially but does not eliminate it. The sine and cosine Fourier transforms of $I(t)$ yield the following results:

$$S(nf_E) = \frac{\mathcal{F} n\omega_E\tau}{1 + n^2\omega_E^2\tau^2} \tag{9.105}$$

$$G(nf_E) = \frac{\mathcal{F}}{1 + n^2\omega_E^2\tau^2} + (1 - \mathcal{F}) \tag{9.106}$$

where $\mathcal{F}$ is the luminescence fraction from the sensor, $1 - \mathcal{F}$ is the backscattered excitation of zero lifetime, and τ is the excited state lifetime of the sensor. Equations (9.105) and (9.106) yield the relation between the measured phase delays $\Delta\phi_n$ and modulation ratios M, the lifetime τ and its fraction $\mathcal{F}$. For real time applications the number of data points needs to be reduced. Typical number of harmonics needed is about 16.[11] Larger numbers (32, 64, ...) slow the system response without significant gain in sensitivity. Sixteen harmonics correspond to the minimum number required to recover the lifetime values of the oxygen sensitive films over the physiological range of oxygen concentrations. Low concentrations of oxygen cause the sensor to decay with a longer lifetimes and require lower frequencies. High concentrations of oxygen cause the sensor to decay with a relatively short lifetime which requires higher frequencies. The average luminescence intensities seen by the detector are in the 0.3–2.4 p W range (at wavelengths longer than 650 nm). The excitation light employed had a fundamental frequency f_E of 1.5625 KHz and a duty cycle of 0.78%. The sampling frequency f_S was 50 KHz and hence $N_w = 0$ and $N_p = 32$. The relation between oxygen and lifetime is obtained using the following polynomial:

$$[O_2] = a_1 + a_2\tau^{-1} + a_3\tau^{-2} \tag{9.107}$$

where $[O_2]$ is the concentration of oxygen in arbitrary units in equilibrium with the sensor and the calibration constants a_1, a_2, and a_3 are sensor dependent. Figure 9.4 shows a Stern–Volmer type plot (τ^{-1} versus $[O_2]$) at four different temperatures.

Figure 9.6 shows the stability of the lifetime-based system here described over a period of 100 h of continuous operation. The sensor was immersed in saline solution in equilibrium with air at 760 torr. The intensity of the sensor decays with time due to the photodestruction of the porphyrin molecules (photobleaching). However, the lifetime value measured by the instrument remains relatively invariant over this period with a system drift of less than 1%. The instrument adjusts the gain of the programmable amplifier to compensate for the reduction in signal.

Figure 9.14 illustrates the precision and accuracy of the system at various oxygen concentrations as compared to a standard oxygen analyzer (model 570A/580A from

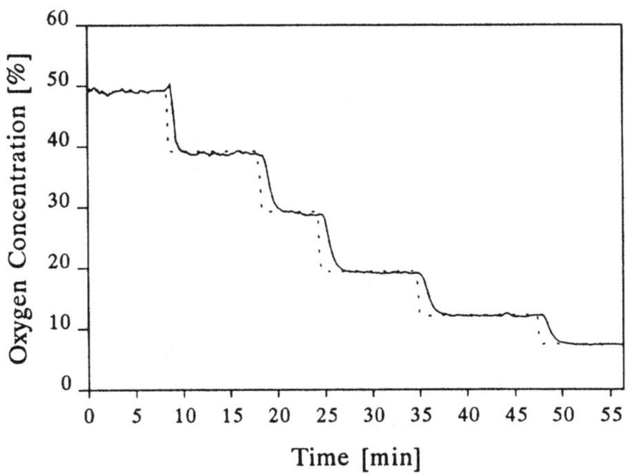

Figure 9.14. Precision and accuracy of the instrument at various concentrations of oxygen as compared to a standard oxygen analyzer (Servomex). The fiber optic sensor monitored the concentration of oxygen in saline solution (continuous stair) in equilibrium with various nitrogen–oxygen gas mixtures monitored by the gas analyzer (dotted staircase). The absolute concentration of oxygen in the gas phase is about 60 times larger than the corresponding equilibrium concentration in the liquid phase. The bath temperature was 37°C. For the purpose of comparison both measurements have been scaled to percent oxygen. (From Ref. 21 with permission.)

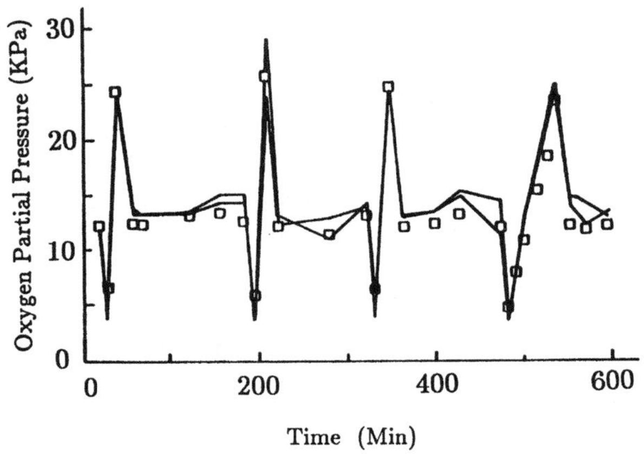

Figure 9.15. *In vivo* partial pressure of oxygen in arterial blood of dogs over the period of 10 h. The solid lines represent the analyses monitor by the instrument. Two different polymer solutions are shown. The dots represent the batch gas analysis measured with a Nova Biomedical blood gas analyzer. (From Ref. 21 with permission.)

Servomex, division of Sybron). Again the fiber optic sensor monitored the concentration of oxygen in saline solution (dashed staircase) in equilibrium with various nitrogen–oxygen gas mixtures. The Servomex measured the amount of oxygen in the gas phase (solid staircase). Roughly, the absolute concentration of oxygen in the gas phase is about 60 times larger than the corresponding equilibrium concentration in the liquid phase. For the purpose of comparing both instruments both measurements have been scaled to percent oxygen. The time delays of the fiber optics sensor includes the time required by the saline solution to reach the new equilibrium after changing the oxygen-nitrogen mixture. Similar performance can be obtained in the gas phase.

Figure 9.15 presents an example of the in vivo measurements of the oxygen content in the arterial blood of dogs over a period of 10 h. The dots represent the batch gas analysis performed with a Nova Biomedical blood gas analyzer. The solid lines represent the analyses monitored by the instrument. Blood oxygen measurements were obtained continuously (not shown in the figure) about every 3 sec. Two different polymer solutions are shown. The measurements performed by the instrument are not affected by the presence of unmetabolized clots and/or anesthetics in the blood stream. Additionally, no deterioration of the signal was found after 10-h periods.

Other applications of the instrumentation and techniques here presented may include (i) temperature sensing employing alexandrite, ruby crystals, and Nd:YAG crystals.[35–41] pH and partial pressure of carbon dioxide (pCO_2) may be determined employing carboxy seminaphtorhodafluor (SNARF)-6.[42]

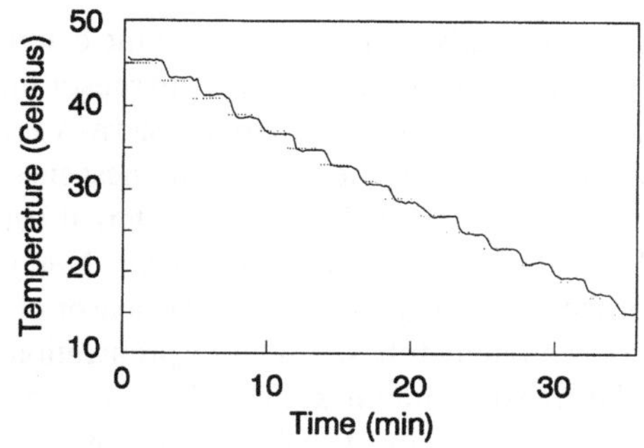

Figure 9.16. Performance of alexandrite based real time optical temperature sensors versus standard (Neslab RTE-111M). Equation 9.107 is used to obtain a working relation between Temperature and τ. The fiber optic sensor monitored the bath temperature (—) in equilibrium with the standard (···).

9.7. Example: a Temperature Sensor

The measurement of temperature is necessary for the calibration of most probes like blood oxygen, pH, ions, voltage, and carbon dioxide sensors. The use of optical methods to invasively measure physiological temperature has the advantage of electrical isolation, when compared to traditional approaches like the use of thermocouplers.

The excited state phosphorescence lifetime of alexandrite crystals may be used to monitor temperature in real time using the principles presented in this chapter. Figure 9.16 illustrates the performance of alexandrite based fiberoptic temperature sensors in the physiological range from 15 to 45 degrees Celsius. Typical precision and accuracy obtained is 0.2 degrees Celsius with about 3 sec. of signal averaging.

9.8. Summary

The fluorescence and phosphorescence of luminescent materials are modulated by the characteristics of the environment to which these materials are exposed. Consequently, luminescent materials can be used as sensors (referred also as transducers or probes) to measure and monitor parameters of importance in medicine, industry and the environment. Temperature, oxygen, carbon dioxide, pH, voltage, and ions are examples of parameters that affect the luminescence of many materials. These transducers need to be excited by light. The manner in which the excited sensor returns to the ground state establishes the transducing characteristics of the luminescent material. It is determined by the concentration or value of the external parameter. A practical and unified approach to characterize the luminescence of all sensors is presented in this chapter. This approach introduces two general mechanisms referred as the radiative and the nonradiative paths. The radiative path, in the general approach, is determined by the molecular nature of the sensor. The nonradiative path is determined by the sensor environment, e.g., value or concentration of the external parameter. The nonradiative decay rate, associated with the nonradiative path, increases monotonically with the values of the external parameter. The radiative and nonradiative mechanisms characterize the sensor luminescence intensity, and sensor luminescence decay rate or lifetime. Stern–Volmer-like functions establish the working relation between the sensor luminescence intensity or lifetime as functions of the concentration or values of the external parameter. Under ideal (linear) relations the Stern–Volmer functions are linear. Nonlinear relations are obtained for complex probes. Complex probes include sensor preparations in dynamic equilibrium, highly concentrated probes, and probe preparations subject to diffusion controlled processes. The transient luminescence of sensors, monitored in the frequency domain, is best suited to implement realtime sensing systems. Frequency domain systems offer the advantaged of enhanced signal to noise ratios even at submegahertz frequencies. The implementation of real-time systems requires the collection of frequency domain data at multiple frequencies concurrently. Analog cross-correlation techniques do not provide the optimum alternative to isolate phase and modulation values at multiple frequencies simultaneously. Sampling techniques are a better solution. Oxygen and temperature sensors are presented as examples of the principles, instrumentation, and methods presented in this chapter.

References

1. O. S. Wolfbeis, *Fiber Optic Chemical Sensors and Biosensors*, Vols. 1 and 2, CRC Press, Boca Raton (1991).
2. D. L. Wise and L. B. Wingard, *Biosensors with Fiberoptics*, Humana Press, Clifton, NJ (1991).
3. D. R. Walt, *Chemical Sensors and Microinstrumentation* (R. W. Murray, ed.), ACS Symposium Ser. 403, American Chemical Society, Washington, D. C. (1989).
4. J. Dakin and B. Culshaw, eds. *Optical Fibre Sensors*, Vols. 1 and 2, Artech House, London (1988).
5. R. Narayanaswamy, Fibre optics for chemical sensing, *Anal. Proc. 22*, 204–206 (1985).
6. J. I. Peteron and G. G. Vurek, Fiberoptic sensors for biomedical applications, *Science 224*, 123–127 (1984).
7. C. G. Hill, Jr. *Chemical Engineering Kinetics & Reactor Design*, pp. 173–192, John Wiley & Sons, New York (1977).
8. E. Gratton, J. R. Alcala, G. Marriott, and F. G. Prendergast, Fluorescence studies of protein dynamics, *Proceedings of the International Symposium on Computer Analysis for Life Science*. Hayashibara Forum'85 (C. Kawabata and A. R. Bishop, eds.), 1-11, Okayama, Japan (1986).
9. D. R. James and W. R. Ware, A fallacy in the interpretation of fluorescence decay parameters, *Chem. Phys. Lett. 120*, 455–459 (1985).
10. D. R. James, Y. S. Liu, P. De Mayo, and W. R. Ware, Distributions of fluorescence lifetimes: Consequences for the photophysics of molecules adsorbed on surfaces, *Chem. Phys. Lett. 120*, 460–465 (1985).
11. J. R. Alcala, E. Gratton, and F. Prendergast, Resolvability of fluorescence lifetime distributions using phase fluorometry, *Biophys. J. 51*, 587–596 (1987).
12. J. R. Alcala, E. Gratton, and F. Prendergast, Fluorescence lifetime distributions in proteins, *Biophys. J. 51*, 597–604 (1987).
13. J. R. Alcala, E. Gratton, and F. Prendergast, Interpretation of fluorescence decays in proteins using continuous lifetime distributions, *Biophys. J. 51*, 925–936 (1987).

14. J. R. Alcala, Ph.D. Thesis, University of Illinois at Urbana–Champaign, Department of Physics (1987).
15. S. A. Bagnich and A. V. Dorokhin, Fractal character of phosphorescence quenching of diacetyl impurity in polymethyl methacrylate, *Opt. Spectrosc.* (*USSR*) *69*, 833834 (1990).
16. E. I. Newhouse and R. Kopelman, Fractal-like triplet–triplet annihilation kinetics in naphthalene-doped poly(methylmethacrylate), *Chem. Phys. Lett. 143*, 106–110 (1988).
17. D. J. Morantz and J. W. Wigley, The contrasting phosphorescence decay kinetics of diacetyl and aromatic ketone phosphors in polymeric matrices, *Polymer Communication 26*, 170–171 (1985).
18. J. L. Gehrich, D. W. Lubbers, N. Opitz, D. R. Hansmann, W. W. Miller, J. K. Tusa, and M. Yafuso, Optical fluorescence and its application to an intravascular blood gas monitoring system, *IEEE Trans. Biomed. Eng. BME-33*(2), 117–132 (1986).
19. B. A. Shapiro, R. D. Cane, C. M. Chomka, L. E. Bandala, and W. T. Peruzzi, Preliminary evaluation of an intra-arterial blood gas system in dogs and humans, *Crit. Care Med. 17,* 455–460 (1989).
20. J. R. Alcala, I. Scott, J. Parker, B. Atwater, C. Yu, R. Fischer, and K. Bellingrath, Real time frequency-domain fiber optic sensor for intra-arterial blood oxygen measurements, in: *SPIE 1885, Proceedings of SPIE Biomedical Optics* (J. R. Lakowicz, ed.), Los Angeles, pp. 306–316 (1993).
21. J. R. Alcala, C.Yu, and G. J. Yeh Digital phosphorimeter with frequency domain signal processing: application to real time fiber optic oxygen sensing, *Rev. Sci. Instrum. 64*, 1554–1560 (1993).
22. D. V. O'Connor and D. Phillips, *Time-Correlated Single Photon Counting*, Academic Press, London (1984) .
23. R. D. Spencer and G. Weber, Measurements of subnanosecond fluorescence lifetimes with a cross-correlation fluorometer, *Ann. N. Y. Acad. Sci. 158*, 361–376 (1969).
24. E. Gratton and M. Limkeman, A continuously variable frequency cross-correlation phase fluorometer with picosecond resolution, *Biophys. J. 44*, 315–324 (1984).
25. J. R. Alcala, E. Gratton, and D. M. Jameson, A multifrequency phase fluorometer using the harmonic content of a mode-locked laser, *Analytical Instrumentation 14,* 225–250 (1985).
26. F. V. Bright and G. M. Hieftje, Rapid-scanning frequency domain fluorometer with picosecond time resolution, *Applied Optics 26*, 3526–3529 (1987).
27. E. Gratton, D. M. Jameson, N. Rosato, and G. Weber, A multi-frequency cross-correlation phase fluorometer using synchrotron radiation, *Rev. Sci. Instrum. 55*, 486–494 (1984).
28. J. R. Lakowicz, G. Laczko, and I. Gryczynski, 2-GHz frequency-domain fluorometer, *Rev. Sci. Instrum. 57*, 2499–2506 (1986).
29. B. A. Feddersen, D. W. Piston, and E. Gratton, Digital parallel acquisition in frequency domain fluorometry, *Rev. Sci. Instrum. 60*, 2929–2936 (1989).
30. J. R. Alcala, Comment on "Digital parallel acquisition in frequency domain fluorometry" [Rev. Sci. Instrum. 60, 2929 (1989)] *Rev. Sci. Instrum. 62*, 1672–1673 (1991).
31. B. P. Lathi, *Linear Systems and Signals*, pp. 474–531, Berkeley-Cambridge Press, Carmichael, California (1992).
32. J. R. Lakowicz, H. Cherek, and A. Balter, Correction of timing errors in photomultiplier tubes used in phase and modulation fluorometry, *J. Biochem. Biophys. Methods 15*, 131–146 (1981).
33. W. H. Press, B. P. Flannery, S. A. Teukolsky, and W. T. Vetterling, *Numerical Recipes: The Art of Scientific Computing*, pp. 381–430, Cambridge University Press, Cambridge (1988).
34. D. B. Papkovsky, J. Olah, I. V. Troyanovsky, N. A. Sadovsky, V. D. Rumyantseva, A. F. Mirinov, A. I. Yaropolov, and A. P. Savitsky, Phosphorescent polymer films for optical oxygen sensors, *Biosensors & Bioelectronics 7*, 199–206 (1991).
35. Z. Zhang, K. T. V. Grattan, and A.W. Palmer, Fiber-optic high-temperature sensor based on the fluorescence lifetime of alexandrite, *Rev. Sci. Instrum. 63*, 3869–3873 (1992).
36. K. T. V. Grattan, R. K. Selli, and A.W. Palmer, Ruby decay-time fluorescence thermometer in a fiberoptic configuration, *Rev. Sci. Instrum. 59*, 1328–1335 (1988).
37. K. T. V. Grattan, R. K. Selli, and A. W. Palmer, Ruby fluorescence wavelength division fiber-optic temperature sensor, *Rev. Sci. Instrum. 57*, 1231–1234 (1987).

38. K. T. V. Grattan and A. W. Palmer, Infra-red fluorescence 'decay time' temperature sensor, *Rev. Sci. Instrum. 56*, 1784–1787 (1985).
39. K. T. V. Grattan, R. K. Selli, and A. W. Palmer, Fiber-optic absorption temperature sensor using fluorescence reference channel, *Rev. Sci. Instrum. 57*, 1175–1178 (1986).
40. K. T. V. Grattan, A.W. Palmer, and C. A. Wilson. A miniaturized microcomputer-based neodymium 'decay-time' temperature sensor. *J. Phys. E. 20*, 1201–1205 (1987).
41. A. T. Augousti, K. T. V. Grattan, and A. W. Palmer, A laser-pumped temperature sensor using the fluorescence decay time of alexandrite, *J. Lightwave Technol. LT-5*, 759–762 (1987).
42. J. R. Lakowicz, Fluorescence lifetime sensing, *Laser Focus World* May, 60–80 (1992).
43. J. R. Lakowicz, H. Szmacinski, and K. W. Berndt, Fluorescence lifetime-based sensing of blood gases and cations, *Proc. SPIE 1648*, 150–163, (1992).
44. S. Motellier, M. H. Michels, B. Direault, and P. Touhoat, Fiber optic pH sensor for *in situ* applications, *Sensors and Actuators B 11*, 467–473 (1993).

10

Lifetime-Based Sensing

Henryk Szmacinski and Joseph R. Lakowicz

10.1. Introduction

At present, there is considerable interest and research activity in the field of chemical sensing, resulting in over 400 papers per year in the scientific literature.[1] Rapid and continuous monitoring of many analytes [pH, partial pressure of carbon dioxide (pCO_2), O_2, metal ions, and other species] is required in many areas of science, including chemistry, biochemistry, environmental sensing, clinical chemistry, and industrial applications. There is increasing interest in optical sensors, particularly fluorescence sensors, because of their advantages over electrochemical sensors.[2–5] These include the absence of electrical interferences (magnetic or electrical fields), no analyte consumption, the possibility of multiple measurements using optical fibers, and the possibility of miniaturization. Fiberoptics can be readily adapted for chemical sensing and be fabricated as miniature single-ended probes suitable for remote sensing and safe operation in chemical environments.[2] Perhaps more importantly, optical detection of analytes can be accomplished without the use of radioactive tracers, and can minimize the need for sample handling and manipulation. Fluorescence sensing provides the additional advantage of good sensitivity, with the specificity being obtained from either the properties of the fluorophore or its fabrication within the sensing element.

At present, most fluorescence sensors or assays are based on intensity measurements, i.e., intensity-based sensing, in which the intensity of the probe molecules change in response to the analyte of interest. Intensity-based methods are initially the easiest to implement because many fluorescent probes change intensity in response to analytes. These intensity changes can be due to changes in extinction coefficient due to probe ionization, changes in quantum yield of the probe on analyte binding, or due

Henryk Szmacinski • Center for Fluorescence Spectroscopy and Department of Biological Chemistry, University of Maryland School of Medicine, and Medical Biotechnology Center of the Maryland Biotechnology Institute, University of Maryland at Baltimore, Baltimore, Maryland 21201. *Joseph R. Lakowicz* • Center for Fluorescence Spectroscopy and Department of Biological Chemistry, University of Maryland School of Medicine, Baltimore, Maryland 21201.

Topics in Fluorescence Spectroscopy, Volume 4: Probe Design and Chemical Sensing, edited by Joseph R. Lakowicz. Plenum Press, New York, 1994.

to inner filtering resulting from the optical density changes of indicators. While intensity measurements are simple and accurate in the laboratory, these are often inadequate in real-world situations. This is because the sample may be turbid, the optical surfaces may be imprecise or dirty, and the optical alignment may vary from sample to sample. In the case of fluorescence microscopy, it is often impossible to know the probe concentration at each point in the image, and in any event, the effective probe concentration changes continually due to photobleaching, phototransformation, and/or diffusive processes. These difficulties with intensity-based sensing appear to be limiting the more widespread use of fluorescence for quantitative clinical chemistry and fluorescence microscopy.

A significant disadvantage of intensity-based sensing is the problem of referencing the intensity measurements. The fluorescence intensity measurement depends on the intensity of exciting light, the extinction coefficient and concentration of the probe, the optical density at the excitation and the emission wavelengths, the optical path length, the fluorescence quantum yield of the probe, and the detector sensitivity. The efficiency of light transmission and collection can vary from one instrument to another, and a constant intensity reference is frequently unavailable. The fluorescence intensity can also vary due to light scattering and/or absorption characteristics of the sample. Moreover, most fluorophores photobleach rapidly, which further complicates the ability to quantitatively use the intensity measurements. These effects result in the need for frequent recalibration and other corrections.

As a result of these difficulties with intensity-based sensing, there have been extensive efforts to develop fluorescent probes and sensing methods which are independent of intensity, in which the instrumental and geometrical parameters are compensated. One approach has been to develop wavelength-ratiometric probes, where ratio of signals at two excitation or emission wavelengths is used for quantitative measurements. In this case the ratio is independent of the probe concentration. This technique is promising, but it has not found widespread use, primarily due to the lack suitable probes. It is difficult to design and synthesize a wavelength-ratiometric fluorophore. Such a probe must display different absorption and/or emission spectra when free in solution and when bound to the analyte. Most of the currently available wavelength-ratiometric probes require ultraviolet (UV) excitation with the associated problems of complex UV laser sources. Moreover, high amounts of autofluorescence are excited at these wavelengths due to NADH, flavin, and other natural fluorophores. The limited availability of wavelength-ratiometric probes can be seen by the scientific literature on Ca^{2+} sensing and imaging, which is dominated by reports of Ca^{2+} measurements using the excitation wavelength-ratiometric probe Fura-2[6] (Figure 10.1). This UV absorbing Ca^{2+} probe is the only practical excitation wavelength-ratiometric probe for Ca^{2+}, and there is only one emission wavelength-ratiometric probe (Indo-1) in current use. Emission ratiometric probes are desirable for use with laser sources, where it is not convenient to change the excitation wavelengths. Attempts to make long-wavelength Ca^{2+} probes have resulted in probes which change

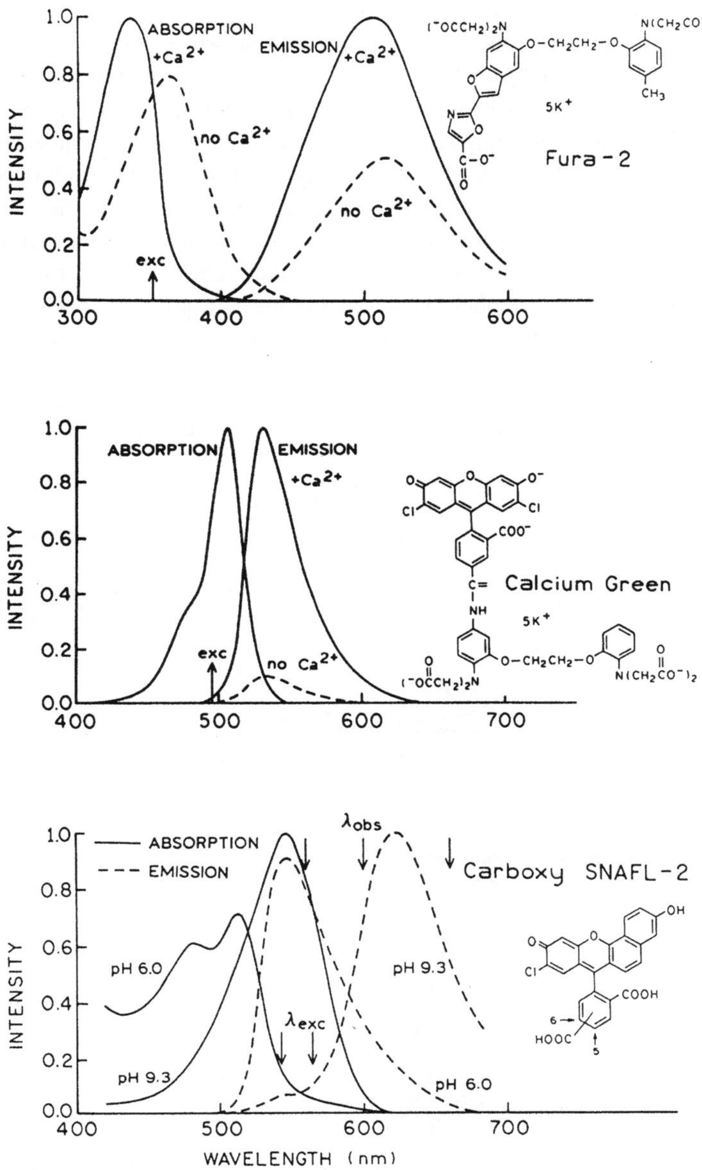

Figure 10.1. Absorption, emission spectra, and chemical structures of representative fluorescence sensors for calcium and pH. The excitation wavelength was 345 nm for Fura-2, 488 nm for Calcium Green, and 543 nm for Carboxy SNAFL-2, $T = 20°C$.

intensity, but do not display spectral shifts in either the excitation or emission spectra, such as Fluo-3, Rhod-2[7,8] and Calcium Color series.[7,9]

While there are relatively few wavelength-ratiometric probes for Ca^{2+}, the choices are still more limited for Na^+ and K^+ which appears to have smaller effects on the spectral properties of probes such as sodium-binding benzofuran isophthalate (SBFI) and potassium-binding benzofuran isophthalate (PBFI).[10] In this case the excitation and emission spectra of SBFI and PBFI show only modest shifts on binding Na^+ or K^+, respectively. In other cases, such as Cl^-, there are no available specific chelating groups, so a wavelength-ratiometric probe for Cl^- does not presently seem possible. Based on these considerations, we decided that it would probably be easier to identify and/or synthesize probes that display changes in lifetime in response to analytes, rather then to design and synthesize probes that display spectral shifts. Our opinion was based

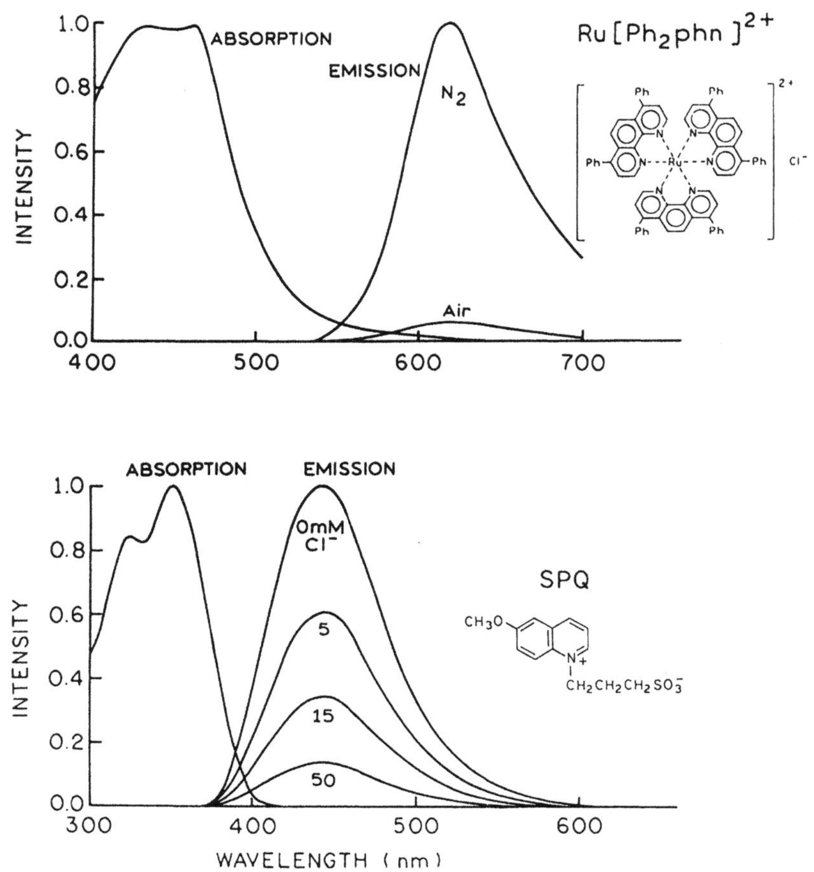

Figure 10.2. Absorption, emission spectra, and chemical structures of fluorescent sensors based on collisional quenching by oxygen (*top*) and chloride (*bottom*).

on the knowledge that a wide variety of quenchers and/or molecular interactions result in changes in the lifetime of fluorophores, while changes in spectral shape were the exception rather than the rule. This prediction proved to be correct, as we now know that long-wavelength excitation probes such as the Calcium Color series,[9] and the analogous Mg^{2+} probes,[11] all display changes in lifetime in response to binding their specific cations even though the absorption and emission spectra are unchanged in the presence of Ca^{2+} or Mg^{2+}. For instance, the Cl^- probes SPQ and SPA display changes in lifetime due to collisional quenching by Cl^-, but there are no shifts in the absorption and emission spectra (Figure 10.2). Additionally, the pH probes of the seminaphtofluorescein (SNAFL) (Figure 10.1) and seminaphtorhodafluors (SNARF) series display changes in their absorption and emission spectra, and also display changes in lifetime on pH-induced ionization.[12, 13] In this case, selection of the excitation or observation wavelength allows the apparent pK_a to be varied by 4 pH units,[13] which increases the dynamic range of the measurements over intensity-based sensing. For these reasons, we are optimistic that lifetime-based sensing can expand the number of measurable analytes as well as the concentration range of the measurements.

10.2. Requirements of a Fluorescent Indicator

There are a number of important criteria in the design of a fluorescent indicator, some of which depend on the particular application.

1. *Sensitivity.* The measured fluorescence parameter of an indicator should be sensitive to changes of analyte in the desired concentration range, as summarized in Table 10.1 for a number of analytes. The indicator should have high extinction coefficient for efficient excitation and high quantum yield for a good signal-to-noise ratio.
2. *Selectivity.* The measured fluorescence parameter should respond selectively to the analyte of interest.
3. *Stability.* The indicator should be chemically stable and insensitive to photobleaching.
4. *Calibration method.* The measured fluorescence parameter should be independent of indicator concentration, geometry of sample, and sensitivity of detection system. Thus, an intensity-based method requires wavelength-ratiometric probes. Lifetime and anisotropy methods do not require wavelength-ratiometric probes, but the lifetime or anisotropy must be sensitive to analyte.
5. *Immobilization or labeling technique.* The polymeric supports should not significantly affect indicator performance, such as response time or selectivity, and if affected the changes should be reproducible. For cellular applications, indicator should be membrane permeable and/or retained by the cells.
6. *Fluorescence spectral properties.* The absorption and emission spectra should be in the visible-wavelength range to avoid autofluorescence from biological systems

Table 10.1. Typical Analyte Concentrations in Blood Serum[14] and in Resting Cells[15]

Analyte	In blood serum (mmol/l)	In resting cells (m*M*)
H^+	35–45 nM	10–1000 nM
(pH)	(7.35–7.46)	(6–9)
Na^+	135–148	4–10
K^+	3.5–5.3	100–140
Li^+	0–2	—
Mg^{2+}	—	0.5–2
Ca^{2+}	4.5–5.5	50–200 nM
Cl^-	103	5–100
HCO_3^-	23–30	—
CO_2	0–15 (% Atm)[16]	—
O_2	5–20 (% Atm)[16]	—

which occurs with UV excitation. The Stokes' shift between spectra should be large for efficient removal of the scatter light and detection of emission.

In our opinion, long-wavelength excitation and emission should be a goal of any attempt at probe design and synthesis. This is because the amount of autofluorescence from biochemical samples and tissues decreases at long wavelengths. Additionally, wavelengths above 650 nm are only weakly absorbed by the skin, opening the possibility of noninvasive transdermal sensing. Additionally, longer absorption and emission wavelength allow the use of inexpensive light sources like LEDs and diode lasers.

Because of recent advances in electronics, electro-optics, acousto-optics, and laser technology, it is now possible to measure the fluorescence lifetime using relatively simple and robust instrumentation. Lifetimes have already been measured using laser diodes,[17, 18] light-emitting diodes (LEDs),[19] and even a simple electroluminescent sheet. [20] In our opinion, an appropriate approach to fluorescence sensing is to reverse the usual procedures of designing the instrument around a known fluorescent sensor. Instead, one should decide on the application, light source, and instrument configuration, and then synthesize the sensor to match the needs of the application. This approach seems practical for lifetime-based sensing because of the comparative ease of obtaining changes in lifetime, as compared to the difficulties of obtained wavelength-ratiometric probes. That is, one can assume with reasonable confidence that a lifetime-based sensor can be designed for use with a desired light source. However, even without the more difficult goal of a wavelength-ratiometric probe, the design and synthesis of a suitable probe will require the cooperative efforts of chemists, biologists, and spectroscopists.

10.3. Molecular Mechanisms for Fluorescence Lifetime-Based Sensing

The *fluorescence lifetime* of a sample is the mean duration of time the fluorophore remains in the excited state. Following pulsed excitation, the intensity decays of many fluorophores are single exponential[21–23]

$$I(t) = I_o \exp(-t/\tau), \tag{10.1}$$

where I_o is the intensity at $t = 0$, and τ is the lifetime. A variety of molecular interactions can influence the decay time, as can be seen from the Jablonski diagram (Figure 10.3). The excited fluorophore can return to the ground state by the radiative (emission) pathway with a rate k_r. The inverse of this rate constant ($\tau_r = k_r^{-1}$) is usually called the intrinsic or radiative lifetime. The radiative decay rate k_r is generally of intramolecular origin, with only a modest dependence on the local environment. There are several formulas to calculate radiative decay rate using the absorption and emission spectra.[24] The simplest of them requires only integration of the first absorption band[25]

$$k_r = 2900 n^2 \bar{\nu}_0^2 \int \varepsilon(\bar{\nu}) d\bar{\nu} \tag{10.2}$$

where n is the refractive index of the solvent, $\bar{\nu}_0$ the wavenumber of the maximum of the absorption band, and $\varepsilon(\bar{\nu})$ in the wavenumber-dependent extinction coefficient. The approximate value of k_r can be estimated from a simplified form of Eq. (10.2), $k_r = 10^4 \varepsilon_{max}$.[24] On binding the analyte the absorption spectrum of the probe in many cases changes (spectral shift and/or change in extinction coefficient), which can result in changes the radiative decay rate and thus affect the fluorescence decay time.

There are many molecular interactions which influence the fluorescence decay times. The measured fluorescence lifetime τ is usually shorter than the radiative lifetime τ_r because of presence of other decay rates which can be dependent on intramolecular processes and intermolecular interactions (Figure 10.3). The measured fluorescence lifetime (τ) is given by the inverse of the total rate of dynamic processes that cause deactivation from the excited (mostly singlet S_1) state

$$\tau = \frac{1}{k_r + k_{nr} + k_Q[Q] + k_T}. \tag{10.3}$$

Nonradiative processes (k_{nr}) can occur with a wide range of rate constants. Molecules with high k_{nr} values display low quantum yields due to rapid depopulation of the excited state by this route. The measured lifetime in the absence of collisional or energy transfer quenching is usually referred to as τ_0, and is given by $\tau_0 = (k_r + k_{nr})^{-1}$.

One easily understood mechanism for changes in lifetime is collisional quenching (Figure 10.3). A variety of substances act as quenchers, including oxygen, nitrous oxide, heavy atoms, Cl^-, and amines, to name a few. By consideration of the lifetime in the absence (τ_0) and presence (τ) of collisional quenchers (no resonance energy

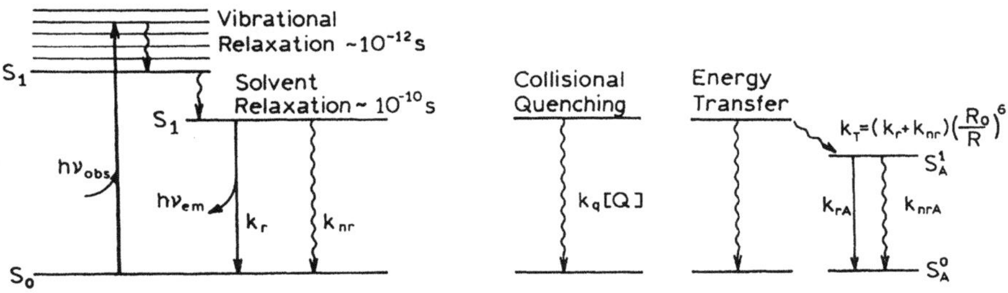

Figure 10.3. Modified Jablonski diagram for the processes of absorption and fluorescence emission (*left*), dynamic quenching (*middle*), and fluorescence resonance energy transfer (FRET) (*right*).

transfer), one can easily see that such quenching decreases the lifetime of the excited state

$$\frac{\tau_0}{\tau} = 1 + \tau_0 k_q[Q] \tag{10.4}$$

where k_q is the biomolecular quenching constant. In fluid solution with efficient quenchers k_q is typically in the range of 10^9–10^{10} M^{-1} sec^{-1}. Collisional quenching by oxygen has been used for intensity-based and lifetime-based sensing of oxygen using the long-lifetime transition metal complexes resulting with large amount of quenching by oxygen.[19, 26–29] Collisional quenching can also be the result of electron transfer processes which are induced by excitation.[30] In some cases, binding of cations can prevent electron transfer from the electron-donating group, preventing intramolecular quenching, which results in an enhancement of quantum yield on binding metal cations without observable spectral shifts.[31]

Another mechanism that decreases the fluorescence intensity and decay time is fluorescence resonance energy transfer (FRET).[32] An energy-transfer-based sensor consists of two kinds molecules, donor (fluorescent) and acceptor (fluorescent or nonfluorescent). For such a sensor the donor need not be sensitive to the analyte, and the acceptor typically displays a change in absorption spectrum due to the analyte. Chemical sensing is achieved by the changes of energy transfer efficiency from donor to acceptor due to the analyte-induced changes in the acceptor absorption. FRET requires partial overlap of the emission spectrum of the donor with the absorption spectrum of the acceptor. There are several approaches for creating energy transfer-based sensors (see Section 10.7).

Lifetime-based sensing can also be based on the existence of two forms of the probe, the fraction of which depends on the analyte concentration. This case is presented in Figure 10.4 where it is assumed that the absorption spectra overlap and both forms can be excited at the same wavelength. The sample displays two lifetimes (τ_F and τ_B) which are characteristic of the analyte-free (F) and analyte-bound (B) forms of the probes. The change in lifetime (from τ_F to τ_B) is due to binding of the analyte

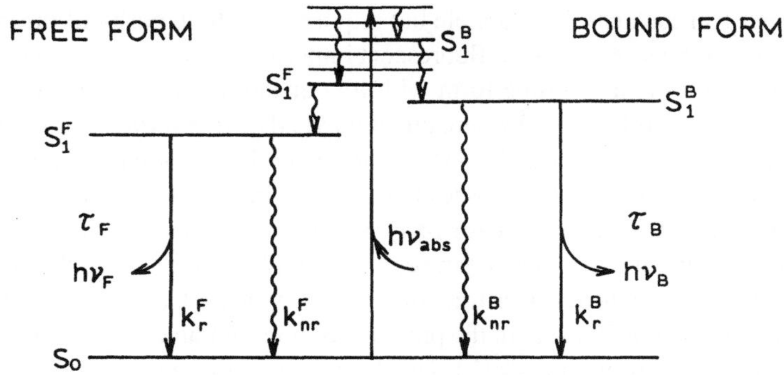

Figure 10.4. Jablonski diagram for two forms of a sensing probe.

may be the result of changes of the radiative and/or nonradiative decay rates. The fluorescence intensity decay of such probe is generally not a single exponential (Eq. (10.1)), but is given as a sum of the exponentials with decay times τ_F and τ_B. Nonetheless, measurement of the mean lifetime can be used to determine the analyte concentration. Complexation may also result in spectral shifts. In Figure 10.4 we have assumed that the complexed form has a blue shifted spectra (higher energy) relative to that of the free form. This type of behavior is known to occur for several Ca^{2+}, Mg^{2+}, K^+, and pH probes, such as Fura-2 (Figure 10.1), Mag-quin-2, PBFI, and the SNAFLs (Figure 10.1) probes, respectively. Some probes display much larger spectral shifts in the absorption spectra than in the emission (Fura-2 (Figure 10.1), Quin-2, Mag-Quin-2). However, there are probes that do not display spectral shifts on complexaton, like Calcium Color Series, but that display significant changes in quantum yield and lifetimes due to different decay rates for free and bound forms (Figure 10.4). For details, see Table 10.2.

It is important to notice that a change in fluorescence quantum yield does not necessarily result in a change in fluorescence lifetime. This is because quantum yield is related to relative decay rates ($\Phi_0 = k_r/(k_r + k_{nr})$) and the lifetime to the absolute decay rates ($\tau_0 = 1/(k_r + k_{nr})$). For instance, the Ca^{2+} probe Rhod-2[8] or Na^+ probe SBFI[10] display significant increase in quantum yield on binding Ca^{2+} (from 0.03 to 0.102) and Na^+ (from 0.045 to 0.083), respectively, but there are essentially no changes in lifetime (2.8–2.9 nsec for Rhod-2 and 0.46–0.48 nsec for SBFI).[33] This effect can be explained that both decay rates (k_r and k_{nr}) change on Ca^{2+} and Na^+ binding for Rhod-2 and SBFI, respectively. A simultaneous decrease of the nonradiative and increase of radiative decay rate will result in an increased quantum yield but may have no effect on the lifetime. The increase of k_r may be compensated by the decrease of k_{nr}. In another case, when quantum yield of one form is much greater than the other it is likely that observed lifetime will be dominated by emission of form with higher quantum yield, which will limit the analyte sensitive concentration range. Lifetime-based sensing in such cases is possible but requires very high purity of the probe and of the sample because emission of fluorescent impurities may be significant compared to very weak emission of probe form with low quantum yield. Such case has been observed for Ca^{2+} probe Fluo-3 where quantum yield of free form is very low (0.005) compared to Ca^{2+}-bound form (0.183).(8) For this probe we observed a longer mean lifetime for Ca^{2+} free form (1.74 nsec) than for bound form (1.01 nsec),(33) which is inconsistent with the large increase of quantum yield. This occurred because the fractional intensity of a long decay time component (2.97 nsec) due to the domination of the fluorescent impurities over fractional intensity of the Ca^{2+}-free form of Fluo-3 with decay time about 30 psec. In the presence of trace amounts of calcium the signal is dominated by emission from Ca^{2+}-bound form with its lifetime of 1.01 nsec. In order to obtain a useful and reliable change in lifetime, the probe must display detectable emission from both the free and bound forms, and the lifetime of each form must be distinct. Then the mean lifetime of the probe reflects the relative fractions of both forms. This reasoning does not apply to collisional quenching, where the intensity decay of the entire ensemble of fluorophores is decreased by diffusive encounters with the quencher.

10.4. Measurement of Fluorescence Lifetimes

There are two widely used methods for measuring fluorescence lifetimes, the time-domain and frequency-domain or phase-modulation methods. The basic principles of time-domain fluorometry are described in Chapter 1, Vol.1 of this series(34) and those of frequency-domain in Chapter 5, Vol. 1 of this series.(35) Good accounts of time-resolved measurements using these methods are also given elsewhere.(36, 37) It is common to represent intensity decays of varying complexity in terms of the multiexponential model

$$I(t) = \sum_i \alpha_i \exp(-t/\tau_i), \tag{10.5}$$

where the α_i are the preexponential factors and the τ_i are the decay times. The fractional intensity of each form to the steady-state intensity is proportional to the preexponential factor and decay time

$$f_i = \frac{\alpha_i \tau_i}{\sum_j \alpha_j \tau_j}. \tag{10.6}$$

This fact can be understood by recognizing that the $\alpha_i \tau_i$ product represents the area under the decay curve which is due to the component with the lifetime τ_i.

We note that a variety of factors can be responsible for a complex multiexponential or nonexponential decays. Depending on the molecular origin of the complex decay kinetics, the values of α_i and τ_i may have direct or indirect molecular significance. For instance, for a mixture of two fluorophores (complexed and free forms or protonated and unprotonated forms) each of which displays a single decay time, the τ_i are the two individual decay times, and f_i are the fractional contributions of each fluorophore to the total emission. If a single fluorophore is in two different environments such as protein-bound and free, then the α_i values often represent the fraction of the molecules in each environment. However, in many circumstances there is no obvious linkage between the α_i and τ_i values and the molecular heterogeneity of the samples. There are a number of cases in which nonexponential decay is encountered such as resonance energy transfer, collisional quenching, and solvent relaxation. While it is of interest to interpret such complex decays in terms of molecular interactions within the sample, it is not necessary to have a complete understanding of the decay processes before a probe can be used for lifetime-based sensing. In lifetime-based sensing, measurement of a mean lifetime is usually adequate to quantify the analyte. The mean lifetime ($\bar{\tau}$) is related to the multiexponential parameters by

$$\bar{\tau} = \sum_i f_i \tau_i. \tag{10.7}$$

The mean lifetime can be used to determine the analyte concentration with a suitable calibration curve.

Instrumentation for Phase-Modulation Lifetime Measurement

The rapid timescale of fluorescence emission results in a significant technological challenge to measure the time-dependent decays. Instrumentation for lifetime measurements, using time-domain or frequency-domain methods, is complex and expensive. Present state-of-the-art instruments use picosecond laser sources and very fast photodetectors to provide resolution of multiexponential and nonexponential intensity decays. Consequently, time-resolved fluorescence spectroscopy has thus been primarily a research tool in biochemistry, biophysics, and chemical physics.[38–41] The instrumentation for chemical sensing may be significantly simplified if the phase-modulation method is used. The phase angle and modulation can be measured with high accuracy, and a single modulation frequency is sufficient for analyte sensing.

In phase-modulation fluorometry, the pulsed light source typical of time-domain measurements is replaced with an intensity-modulated source (Figure 10.5). Because of the time lag between absorption and emission, the emission is delayed in time relative to the modulated excitation. At each modulation frequency ($\omega = 2\pi f$) this delay is described as the phase shift (θ_ω), which increases from 0 to 90° with increasing modulation frequency. The finite time response of the sample also results in demodulation to the emission by a factor m_ω, which decreases from 1.0 to 0.0 with increasing modulation frequency. The phase angle (θ_ω) and the modulation (m_ω) are separate

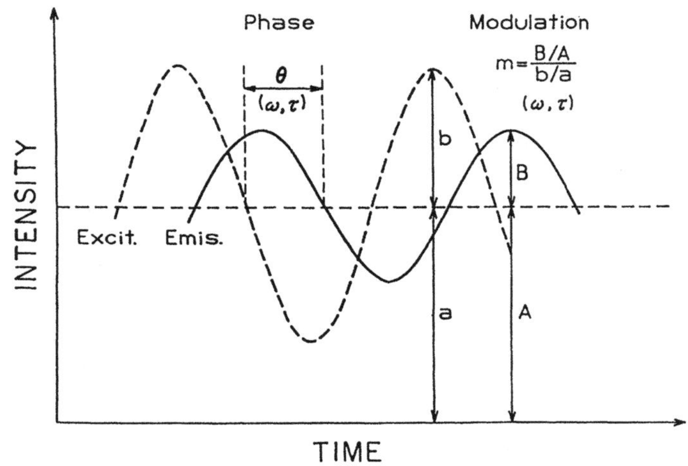

Figure 10.5. Schematic of phase angle and modulation lifetime measurements. ω = Frequency of modulation. τ = lifetime.

measurements, each of which are related to the intensity decay parameters, α_i and τ_i and modulation frequency ω by[42]

$$\theta_\omega = \arctan\,(N_\omega/D_\omega), \qquad m_\omega = (N_\omega^2 + D_\omega^2)^{\frac{1}{2}}, \tag{10.8}$$

where

$$N_\omega = \frac{1}{J}\sum_{i=1}^{n} \frac{\omega\alpha_i\tau_i^2}{1+\omega^2\tau_i^2}, \quad D_\omega = \frac{1}{J}\sum_{i=1}^{n} \frac{\alpha_i\tau_i}{1+\omega^2\tau_i^2}, \quad J = \sum_{i=1}^{n} \alpha_i\tau_i\,. \tag{10.9}$$

Research instrumentation for phase-modulation fluorometry operates over a wide range of light modulation frequencies and excitation wavelengths. Consequently, instruments are somewhat expensive and complex. If the probe characteristics are known, the phase-modulation instrument can be designed with a single wavelength light source and to operate at just one light modulation frequency appropriate for the chosen probe. This specialization can result in decreased complexity and cost since the simple light sources can be used such as HeNe laser and an acousto-optic modulator,[13] or just a laser diode.[17, 18] The acousto-optic modulators are not convenient if a wide range of frequencies are required,[43] but are ideal if only a few frequencies are needed. The electronics for phase-modulation sensing can be based on relatively low-cost components,[20, 44] and phase and modulation at a single frequency are easily measured in a second or less of data acquisition. Also, laser diode sources can allow even simpler instrumentation because they can be modulated by the driving current,[45, 46] and laser diode sources have already been used for frequency-domain measurements to 2 GHz.[17, 18] However, the use of laser diode sources for lifetime-based sensing requires the development of red and/or near-infrared (NIR) lifetime probes.

An advantage of phase-modulation fluorometry for sensing is the use of cross-correlation detection, which results in an increased signal-to-noise ratio and rejection of undesired harmonics. The gain of the photomultiplier tube (PMT) detector is modulated by a small voltage applied to one of its dynodes. The gain modulation frequency is offset from the light modulation frequency by a small amount, typically 25 or 40 Hz. The phase and modulation information are preserved in the low-frequency signal, and θ_ω and m_ω then easily measured with simple electronics. The alternating current (AC) component of the signal is typically passed through a 1-Hz-wide filter, which rejects undesired harmonics. Cross-correlation electronics or numerical filtering results in phase and modulation measurements which are correct irrespective of the form of the light and gain modulation. The measurements do not require sinusoidal modulation, and in fact we frequently make use of pulse-sources for such measurements.[47] The frequency synthesizers for a sensing application need not have a wide range, but the method does require that the two frequencies (f and $f + \Delta f$) be phase-locked. Finally, advances in probe chemistry are likely to result in lifetime-based probes which can be excited with laser-diode sources.[48, 49] This is advantageous because there is less autofluorescence with longer wavelength excitation, and it may become possible to perform noninvasive lifetime-based sensing through the skin. The skin becomes nonabsorbing (scattering, but weakly absorbing) at wavelengths above 600 nm.[50]

10.5. Sensing Based on Probe–Analyte Recognition

The photophysical properties (extinction coefficient, shifts in absorption and emission spectra, quantum yield, and lifetime) of a variety of probes are modified by pH changes, complexation by metal ions, or redox reactions. The resulting changes in photophysical parameters can be used to determine concentration of H^+ and metal cations with suitably designed fluorophores. Most of these resulting sensors involve an equilibrium between the analyte, A, and the free probe (unprotonated or noncomplexed by metal ion), P_F. If the stoichiometry of this reaction is 1:1, the reaction may be represented by

$$AP_B \rightleftharpoons A + P_F. \tag{10.10}$$

The equilibrium is governed by the dissociation constant, K_D, defined as

$$K_D = \frac{[P_F]}{[P_B]}[A]. \tag{10.11}$$

The relative concentration of free, $[P_F]$, and complexed (or bound), $[P_B]$, forms are dependent on analyte concentration, $[A]$, as follows

$$\frac{[P_F]}{[P]} = \frac{K_D}{K_D + [A]}, \qquad (10.12)$$

$$\frac{[P_B]}{[P]} = \frac{[A]}{K_D + [A]}, \qquad (10.13)$$

where $[P]$ is the total concentration of indicator ($[P] = [P_F] + [P_B]$).

The analyte concentration can be determined from knowledge of the dissociation constant K_D, which is a characteristic value for specific probe under specific conditions (temperature, viscosity, ionic strength, pH) and the relative concentrations of the free, $[P_F]/[P]$ (Eq. (10.12)) or complexed forms, $[P_B]/[P]$ (Eq. (10.13)). The analyte concentration range over which probe can be used is determined by the dissociation constant, K_D (Figure 10.6). The useful range of analyte concentrations is typically restricted to $0.1\ K_D < [A] < 10\ K_D$. Concentrations lower than $0.1\ K_D$ and higher than $10\ K_D$ will produce little change in the observed signal. The measured spectra parameters will be dependent on the concentration of complexed form, $[P_B]$ (observed for most metal cation indicators) as well to the concentration of the free form, $[P_F]$ (observed for phenolic pH indicators). The relative concentrations (Eqs. (10.12) and (10.13)) can be determined from steady-state measurements; absorption, emission, and quantum yield, or from time-resolved spectroscopy, i.e., the intensity decay parameters or/and mean fluorescence lifetime.

10.5.1. Intensity-Based Sensing

There are a number indicators (most pH probes) which display changes in the absorption spectra on the complexation, but do not display useful fluorescence.[51] In this case the change in absorbance is due to different extinction coefficients for the free and complexed forms. Because absorbance is proportional to the probe concentration, Eq. (10.11) can be expressed by

$$[A] = K_D \frac{A - A_{\min}}{A_{\max} - A}, \qquad (10.14)$$

where $A_{\min} = \varepsilon_F[P]d$, is absorbance where indicator is only in free form, d is the optical path length, $A_{\max} = \varepsilon_B[P]d$ is the absorbance where indicator is totally complexed, and $A = \varepsilon_F[P_F]d + \varepsilon_B[P_B]d$ is the absorbance where indicator is partially complexed by analyte. The sensitivity of the probe depends on the difference of $A_{\max} - A_{\min}$.

A relation similar to Eq. (10.14) can be obtained for probes which display changes in the emission spectra on complexation. Since the intensity is proportional to the probe concentration, the analyte concentration can be obtained from

$$[A] = K_D \frac{F - F_{\min}}{F_{\max} - F}, \qquad (10.15)$$

where $F_{\min}$ is the fluorescence intensity when indicator is in free form, $F_{\max}$ the intensity when indicator is totally complexed, and F is the intensity when indicator is partially complexed by analyte. Changes in the fluorescence intensity can be only due to different quantum yields of the free and complexed forms without any difference in the absorption spectrum. For some probes, only one form is fluorescent (free or complexed). In this case Eq. (10.13) can be rearranged to give

$$\frac{F_0}{F} = 1 + \frac{[A]}{K_D}, \tag{10.16}$$

where F_0 and F is the intensity of fluorophore in the absence and presence of the analyte, respectively. Eq. (10.16) cannot be regarded as Stern–Volmer relation with $K_{SV} = 1/K_D$ because the concentration of emitting form is not constant (regulated by K_D and analyte concentration).

In order to determine the analyte concentration using Eqs. (10.14)–(10.16), all parameters (absorbancies or intensities) must be determined at the same instrumental configuration, the same optical path length, and the same effective total concentration of the probe. These requirements are often hard to satisfy, especially in microscopy when observing cells. Measurement of $F_{\max}$ and $F_{\min}$ requires lysing the cells and titrating the released indicator, or using ionophores to saturate the indicator. However, such calibration methods do not compensate for dye loss due to photobleaching or leakage during the experiment, and can also alter the spatial distribution of the probe. The use of wavelength-ratiometric probes overcomes these difficulties because the spectral shift, and the intensity ratios, are used to determine the analyte concentration.[52] For excitation-ratiometric probes the analyte concentration can be determined by[52]

$$[A] = K_D \left(\frac{R - R_{\min}}{R_{\max} - R}\right)\left(\frac{S_F(\lambda_2)}{S_B(\lambda_2)}\right), \tag{10.17}$$

where $R = F(\lambda_1)/F(\lambda_2)$ is the measured intensity ratio, $R_{\min}$ and $R_{\max}$ = are ratios for the free and complexed indicator, and $S_F(\lambda_2)$ and $S_B(\lambda_2)$ are factors related to totally free and complexed forms at reference wavelength (λ_2), respectively. Dependent on the kind of probe, the S_F and S_B factors are proportional to the extinction coefficients (absorption-ratiometric), to the extinction coefficients and quantum yields (excitation-ratiometric), and to the fluorescence intensity (emission-ratiometric) of the respective forms of the probe at the reference wavelength λ_2. Then the ratio $S_F(\lambda_2)/S_B(\lambda_2)$ can be rewritten as

$$\frac{S_F(\lambda_2)}{S_B(\lambda_2)} = \frac{\varepsilon_F}{\varepsilon_B}, \quad \frac{S_F(\lambda_2)}{S_B(\lambda_2)} = \frac{\varepsilon_F}{\varepsilon_B}\frac{\Phi_F}{\Phi_B}, \quad \frac{S_F(\lambda_2)}{S_B(\lambda_2)} = \frac{F_F}{F_B}, \tag{10.18}$$

respectively, where ε and F are the extinction coefficients and fluorescence intensities at reference wavelength λ_2 of free (F) and bound (B) forms, and the Φ is the quantum yield of each form. Using Eq. (10.17), the probe concentration and instrumental

Table 10.2. Spectral Properties, Quantum Yields (Φ), and Mean Lifetimes ($\bar{\tau}$) of Free (F) and Bound (B) Forms of Probes for Ion Recognition and Their Dissociation Constants (K_D)

Probe	Absorption $\lambda_F(\lambda_B)^a$ (nm)	Absorption $\varepsilon_F(\varepsilon_B)$ ($\times 10^{-3}$ M^{-1} cm^{-1})	Emission $\lambda_F(\lambda_B)^a$ (nm)	Emission $\Phi_F(\Phi_B)$	$\bar{\tau}_F(\bar{\tau}_B)$ (nsec)	Dissociation constant (K_D)
Ca^{2+} probes						(m*M*)
Quin-2(53, 55–57)	356(336)	5.0(5.0)	500(503)	0.03(0.14)	1.35(11.6)	60,115[b]
Fura-2(33, 52)	362(335)	27(32)	518(510)	0.23(0.49)	1.09(1.68)	135,224[b]
Indo-1(52, 58)	349(331)	33	482(398)	0.38(0.56)	1.40(1.66)	250[b]
CaG(9, 54)	506	74	534	0.06(0.75)	0.92(3.66)	189
CaO(9, 54)	555	80	576	0.11(0.33)	1.20(2.31)	328
CaC(9, 54)	588	108	610(612)	0.18(0.53)	2.55(4.11)	205
BOZ-crown(59, 60)	483(457)	—	642(574)	0.33(0.64)	2.1(3.3)	79
Mg^{2+} probes						(m*M*)
Mag-Quin-2(7, 33)	353(337)	4.2(4.0)	487(493)	0.003(0.07)	0.84(8.16)	0.8
MgG(7, 33)	506	76	532	0.04(0.42)	1.21(3.63)	0.9
BOZ-crown(59, 60)	483(469)	—	642(573)	0.33(0.48)	2.1(3.3)	2.0
K^+ probe						(mM)
PBFI(10, 33)	350(344)	42(47)	546(504)	0.24(0.72)	0.52(0.59)	107
Na^+ probe						(m*M*)
Sodium Green(61)	506	—	536(538)	—[c]	—[c]	6.0
pH probes(12, 13)						pK_a
SNAFL-1	539(510)	51(28)	616(542)	0.093(0.33)	1.19(3.74)	7.7
C. SNAFL-1	540(508)	52(29)	623(543)	0.075(0.32)	1.11(3.67)	7.8
C. SNAFL-2	547(514)	48(25)	623(545)	0.054(0.43)	0.94(4.60)	7.7
C. SNARF-1	576(549)	53(26)	638(585)	0.091(0.047)	1.51(0.52)	7.5
C. SNARF-2	579(552)	60	633(583)	0.110(0.022)	1.55(0.33)	7.7
C. SNARF-6	557(524)	50	635(559)	0.053(0.42)	1.03(4.51)	7.6
C. SNARF-X	575(570)	42(18)	630(600)	0.163(0.069)	2.59(1.79)	7.9
BCECF	503(484)	94(30)	528(514)	—	4.49(3.17)	7.0
Resorufin Acetate(33)	571(484)	67(16)	587(578)	—	2.92(0.45)	~5.7
[Ru(bpy)2-(dcbpy)]$^{2+}$ (62)	458(480)	13(12)	650(675)	—	460(210)	3.9

Abbreviations: BCECF, 2′,7′-*bis*(2-carboxyethyl)-5(and 6)-carboxyfluorescein; [Ru(bpy)2(dcbpy)] $^{2+}$, Bis(2,2′-bipyridine)(2,2′-bipyridine-4,4′-dicarboxylic acid)ruthenium(II); BOZ, Benzoxazine; C., Carboxy; CaC, Calcium Crimson; CaG, Calcium Green; CaO, Calcium Orange; MgG, Magnesium Green; SNAFL, seminaphtofluorescein; SNARF, seminaphtorhodafluor.

[a] λ refers to the wavelength at maximum of respective spectrum.

[b] In buffer with 1 m*M* Mg^{2+}.

[c] New probe, quantum yield, and lifetime changes are expected be similar to Calcium Green and Magnesium Green.

sensitivity do not affect the ratios' values. To avoid any wavelength biases all ratios (R, R_{min}, and R_{max}) should be measured on the same instrumentation.

10.5.2. Lifetime-Based Sensing

Many probes are now known that display changes in fluorescence lifetime on complexation of the analyte, photophysical properties some of them are summarized in Table 10.2. While we have listed the lifetimes of the free and the bound forms of the probes, there is no straightforward equation to calculate the analyte concentration using the mean lifetime as was in the case of the absorbance and intensity (Eqs. (10.14) and (10.15)). The mean lifetime depends not only on relative concentration of the probe species (free and complexed) but also on their decay times, quantum yields, and to some extent on the measurement (method or conditions). While the mean lifetime is independent of total probe concentration, this value generally depends not only on analyte concentration but also on excitation and observation wavelengths.[13]

To illustrate the use of time-resolved data for sensing, assume that each form (free and bound) has a unique decay time, τ_F and τ_B. The intensity decay of a mixture is a double-exponential

$$F(t) = \alpha_F \exp(-t/\tau_F) + \alpha_B \exp(-t/\tau_B). \tag{10.19}$$

The preexponential factors α_F and α_B are related to the concentration of each form $[P_F]$ and $[P_B]$, respectively. The relative intensity observed in the usual steady-state measurement due to each component is given by

$$\frac{f_B}{f_F} = \frac{\alpha_B \tau_B}{\alpha_F \tau_F}, \tag{10.20}$$

Fractional intensities are related to the concentrations of each species by

$$f_i = k I_E \Phi_i \varepsilon_i [P_i] d, \tag{10.21}$$

where Φ is the quantum yield, I_E the excitation intensity, ε the extinction coefficient at excitation wavelength, $[P]$ the concentration, d the optical path length for exciting light, k the geometrical constant, and i refer to each form of the probe. The combination of Eqs. (10.20) and (10.21) yields

$$\frac{\alpha_B}{\alpha_F} = \frac{\varepsilon_B}{\varepsilon_F} \frac{\Phi_B \tau_F}{\Phi_F \tau_B} \frac{[P_B]}{[P_F]}. \tag{10.22}$$

The ratio of preexponential factors (α_B/α_F) represents the ratio of concentrations of the probe in each form ($[P_B]/[P_F]$) only if: (1) there is no change in the absorption spectra or excitation is at the isobestic wavelength ($\varepsilon_F = \varepsilon_B$); (2) only nonradiative processes are affected ($\Phi_F/\Phi_B = \tau_F/\tau_B = (k_r + k_{nr}^B)/(k_r + k_{nr}^F)$; or (3) the value of $\varepsilon_B \Phi_B \tau_F = \varepsilon_F \Phi_F \tau_B$; and (4) there is no shift in the emission spectra or observation is at

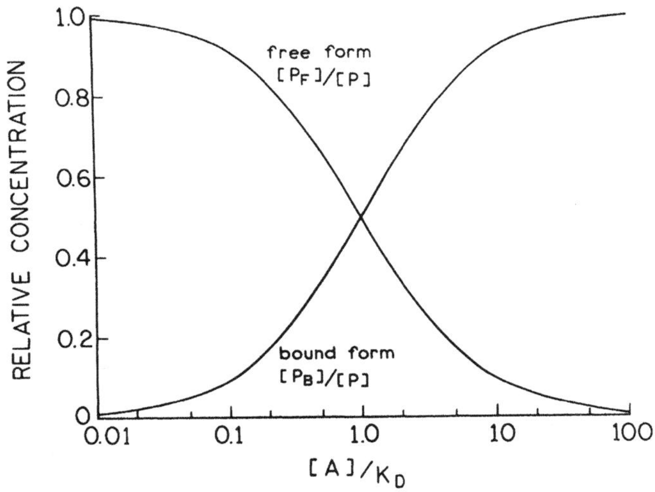

Figure 10.6. Relations between analyte concentration, dissociation constant, and relative concentration of each form of the probe.

the isoemissive wavelength (ratio of Φ_B/Φ_F is independent on observation wavelength). In most experimental cases (Table 10.2), these requirements are not fulfilled. A graphical representation of preexponential factors versus fractional concentration similar to Eq. (10.22) for the calcium probes Quin-2 and Calcium Green has been presented in a recent report.[53] The analytical expression for preexponential factors dependence on analyte concentration can be obtained from Eqs. (10.11) and (10.22) as

$$\alpha_F = \frac{K_D^\alpha}{K_D^\alpha + [A]}, \quad \alpha_B = \frac{[A]}{K_D^\alpha + [A]}, \tag{10.23}$$

where

$$K_D^\alpha = \frac{\varepsilon_F}{\varepsilon_B}\frac{\Phi_F}{\Phi_B}\frac{\tau_B}{\tau_F} K_D. \tag{10.24}$$

Equation (10.23) describes the relations of the preexponential factors to the analyte concentration in the same way as relative concentration of free and bound forms given by Eqs. (10.12) and (10.13). The preexponential factor analyte response function may be shifted toward lower or higher analyte concentrations compared to those obtained from the absorbance or/and intensity measurements (Figure 10.6) because of the apparent dissociation constant (K_D^α) given by Eq. (10.24).

A single modulation frequency is sufficient to measure the fluorescence phase angle and modulation and hence the analyte concentration. These intensity decay parameters (θ and m) are functions of decay times (τ_i), their amplitudes (α_i), and depend on modulation frequency (ω) (Eqs. (10.8) and (10.9)). Simulations for a two-component system (τ_F and τ_B) using Eqs. (10.8), (10.9), and (10.23) resulted in sigmoidal dependences on analyte (similar to Figure 10.6) for the phase angle and modulation. This confirms that intensity decay parameters can be related to the analyte concentration similar to Eqs. (10.14) and (10.15) for the steady-state measurements. In analogy to Eqs. (10.14) and (10.15), one can write

$$[A] = K_D^X \frac{X - X_{\min}}{X_{\max} - X}, \tag{10.25}$$

where X is the analyte-dependent measurable parameter and K_D^X is the apparent dissociation constant associated with the measurable parameter X.

Simulated data show that for $K_D^\alpha = K_D$ (see Eq. (10.24)) the apparent dissociation constants for phase angle and modulation strongly depend on value τ_B/τ_F. If $\tau_B > \tau_F$ (the usual case; see Table 10.2) the phase angle and modulation apparent dissociation constants display significantly lower values than true K_D. For example, if $K_D^\alpha = K_D$, and $\tau_B/\tau_F = 2$, 5, and 10, the apparent dissociation constants are 0.71, 0.44, 0.31K_D for phase angle and 0.33, 0.15, 0.07K_D for modulation, respectively. It should be noted that the apparent dissociation constants from the modulation are significantly lower than from phase angle. The apparent dissociation constants were calculated using a plots of $\log[(X - X_{\min})/(X_{\max} - X)$ versus $\log [A]$ (Hill plot). For sensing, the apparent dissociation constant determines the measurable analyte concentration range that can be determined by the particular measured parameter. The difference in apparent dissociation constants for phase and modulation allows extension of the analyte sensitive concentration range if both parameters are measured. Phase angle and modulation currently seem to be the most convenient parameters in lifetime-based sensing because single frequency is sufficient for direct mapping their values (θ or m) to the analyte concentration. Also, one does not require measurements over wide range of frequencies, which are only needed to recover the preexponential factors and associate decay times.

10.5.2.1. Probes without Spectral Shifts

Figure 10.7 shows dependence of decay time amplitudes (α_i) of Calcium Green on calcium concentration. Calcium Green does not display any spectral shift in absorption and emission spectra. The absorption spectrum of Calcium Green does not change and the emission intensity increases approximately eightfold on binding Ca^{2+} (Figure 10.1).[9] The intensity decay of Calcium Green is heterogeneous and a triple-exponential analysis was required to obtain a reasonable fit for most Ca^{2+} concentrations. The three decay times were essentially constant with $[Ca^{2+}]$, but the α_i values were strongly dependent on $[Ca^{2+}]$.[9] The amplitude of decay time $\tau_1 = 0.05$ nsec decreases and $\tau_3 = 3.66$ nsec increases with increasing calcium concentration

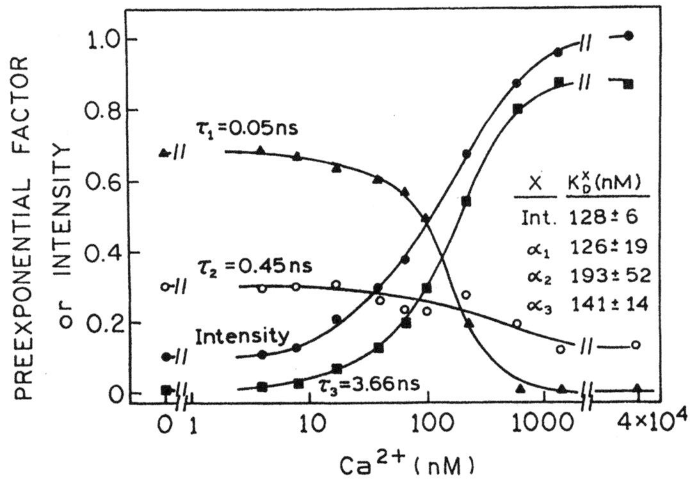

Figure 10.7. Ca^{2+}-dependent steady-state intensity (•) and amplitudes of intensity decays of Calcium Green (in 100 m*M* KCl, 10 m*M* MOPS, 0–10 m*M* Ca-EGTA, pH 7.2, 20°C).

(Figure 10.7). This indicates that the short component (τ_1) is associated with calcium-free and the long (τ_3) with Ca^{2+}-bound form of Calcium Green. The apparent dissociation constants determined using these two components are similar to those of intensity (Figure 10.7). This result is consistent with the absence of the spectral shifts and indicates that mostly the nonradiative rate is affected on Ca^{2+} binding (see also discussion in Section 10.3). For completeness we note there is a third decay time (τ_2 = 0.45 nsec), which is less sensitive to Ca^{2+} (K_d = 193 nM) and is present even at very high Ca^{2+} concentration (40 μM). This component may be associated with impurities or with the less Ca^{2+}-sensitive form of Calcium Green. Molecular Probes have also developed other Calcium Green probes which display weaker binding to Ca^{2+}, Calcium Green-2 (K_d = 574 nM) and Calcium Green-5N (K_d = 3.3 μM).[7, 54]

A complete multiexponential resolution is not needed to use Calcium Green as a lifetime-based probe. This is illustrated in Figure 10.8 where we present the Ca^{2+} response functions of Calcium Green using the phase angle and modulation at a modulation frequency of 151.8 MHz. Plots of $\log[(X - X_{min})/(X_{max} - X)]$ versus $\log[Ca^{2+}]$ resulted in similar slopes for Calcium Green using steady-state intensity and intensity decay parameters (θ and m). The apparent dissociation constants (Figures 10.7 and 10.8) are equal to Ca^{2+} concentration at the zero intercept of the Hill plots using the respective measurable parameter X. However, sigmoidal plots using phase angle and modulation for Calcium Green were slightly distorted at very low and very high calcium concentrations due to a third component (τ_2 = 0.45 nsec) with different apparent dissociation constant (see Figure 10.7). The experimental data for Calcium Green confirm the expected lower values of apparent dissociation constants from phase angle and modulation measurements. For Calcium Green τ_B/τ_F = 3. 98 (Table 10.2),

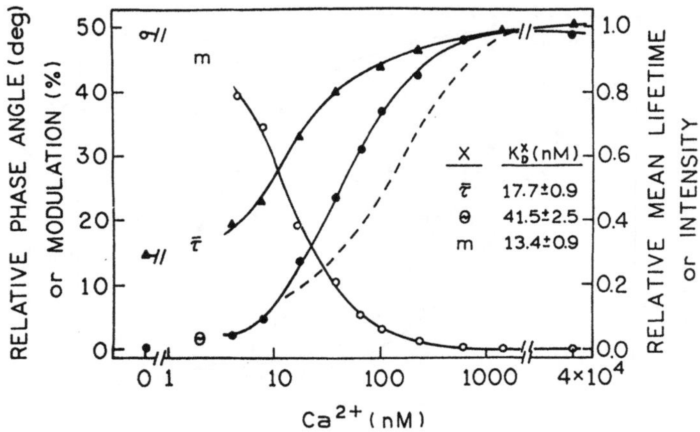

Figure 10.8. Ca^{2+}-dependent phase angle and modulation at frequency of 151.8 MHz and mean lifetime for Calcium Green. Dashed line is a Ca^{2+}-dependent steady-state intensity (from Figure 10.7).

which results in an apparent $K_D^{\theta} = 0.32$ and $K_D^{m} = 0.10$ of that from intensity. Similar data were observed for Quin-2, which also displayed shift in the absorption spectra on Ca^{2+} binding.[55]

10.5.2.2. Probes with Spectral Shifts

If the probe display shifts in absorption and/or emission spectra, the apparent dissociation constant will depend not only on the measured parameter but also on the excitation and emission wavelengths. For probes that display a shift in the absorption spectrum on analyte binding, the value of $\varepsilon_F/\varepsilon_B$ depends on excitation wavelength, which affects the value of K_D^{α} (see Eq. (10.24)). For probes which display shift in emission spectrum the value of Φ_F/Φ_B may depend on observation wavelength. The emission filter (bandpass or long-wave pass) may not transmit the total emission spectra of each form, the sensitivity of detection system may not be the same over the wide range of wavelengths, or observation is through narrow-bandpass filter (e.g., interference filter). In such cases the contribution to the emission signal of each form of the probe will depend on observation method, not only on respective absolute quantum yields (Φ). The free or bound form can be preferentially excited or/and observed, which will result in various values of the apparent dissociation constant K_D^{α} given by Eq. (10.24). The values of K_D^{α} and decay times (τ_F and τ_B) determine the apparent dissociation constants from phase angle and modulation measurements. In Figure 10.9 are presented excitation and observation wavelength-dependent phase angle responses for pH probe Carboxy SNAFL-2. Carboxy SNAFL-2 is a dual-wavelength ratiometric probe (Figure 10.1).[12, 13] If the unprotonated form is preferentially excited or/and observed (longer wavelengths, $\lambda_{exc} = 565$ nm and $\lambda_{obs} = 600$ nm), than

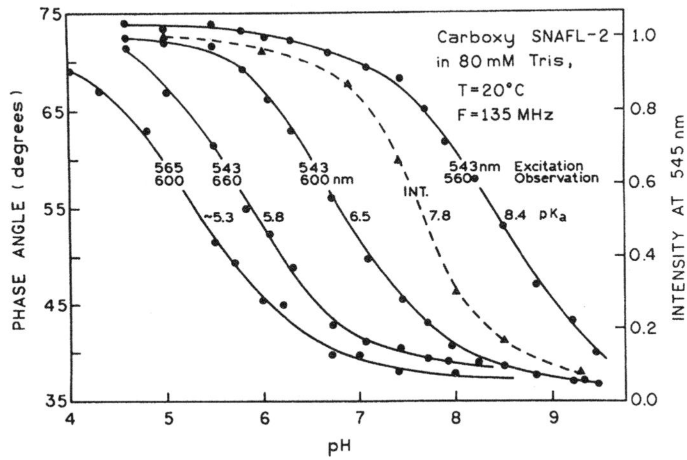

Figure 10.9. Excitation-, emission wavelength-, and pH-dependent phase angle of Carboxy SNAFL-2 (see Figure 10.1 for absorption and emission spectra).

the apparent pK_a's are shifted toward lower values ($pK_a \approx 5.3$), which allows lower pH values to be measured. Using shorter wavelength for excitation ($\lambda_{exc} = 543$ nm) and observation ($\lambda_{obs} = 560$ nm) the pH-dependent phase angle response function is shifted toward higher pH values ($pK_a = 8.4$) because protonated form of Carboxy SNAFL-2 is preferentially excited and observed with its longer lifetime (4.6 nsec, see Table 10.2).

It is important to recognize that the pH range and/or accuracy can be expanded or refined as needed by judicious selection of the excitation and/or emission wavelength. In contrast, intensity-based sensing does not allow expansion of the sensitive analyte concentration range by excitation or/and emission wavelength selection. Using intensity-based sensing, the pK_a for Carboxy SNAFL-2 is 7.65,[12] which typically allows pH determination in the range of $pK_a \pm 1$ (Figure 10.9, dashed line). The pH sensitive range of Carboxy SNAFL-2 for the experimental conditions described in Figure 10.9 is from about 4 to 9.5. This wide pH range is possible, because of the large wavelength shift between the emission (78 nm) and absorption (33 nm) spectra of the acid and base forms.[12, 13]

Table 10.2 summarizes the spectral and lifetime properties of probes for recognition of several cations (Ca^{2+}, Mg^{2+}, Na^+, K^+, and H^+(pH)). The magnitude of the lifetime changes on cation binding determines the sensitivity of the probe, and the phase and modulation accuracy needed for a given accuracy in the analyte concentration (1.5-fold will result in approximately 12° in phase angle). Additionally, probes that display spectral shifts are expected to provide expansion of sensitive analyte concentration ranges, such as for Fura-2 or Mag-Quin-2 using various excitation wavelengths (e.g., 340–380 nm).

10.6. Sensing Based on Collisional Quenching of Fluorescence

An important class of luminescence sensors are those based on the decrease of luminescence intensity and lifetime of the probes as function of analyte concentration. Assume that the probe intensity decays by a single exponential with an unquenched lifetime τ_0. If quenching occurs only by a dynamic (collisional) mechanism, then the ratio τ_0/τ is equal to F_0/F and is described by the classic Stern–Volmer equation

$$\frac{F_0}{F} = \frac{\tau_0}{\tau} = 1 + K_{SV}[Q] \tag{10.26}$$

where $[Q]$ is the quencher concentration, F's are intensities, τ's are lifetimes, and K_{SV} is the dynamic Stern–Volmer constant. The subscript 0 denotes the value in the absence of quencher. The overall dynamic quenching constant K_{SV} can be described by

$$K_{SV} = k_q \tau_0 \tag{10.27}$$

where k_q is the biomolecular quenching rate constant which, for a diffusion-limited reaction can be described by the Smoluchowski equation

$$k_q = 4\pi N' p(D_P + D_Q) \tag{10.28}$$

The D_P and D_Q are the diffusion coefficients of probe and quencher, respectively, N' is the number molecules per millimole, and p is a factor that is related to the probability of each collision causing quenching and to the radius of interaction of probe and quencher. A more detailed treatment of fluorescence quenching including multiexponential intensity decays and static quenching is given elsewhere.(63) There are many known collisional quenchers (analytes) which alter the fluorescence intensity and decay time. These include O_2,(19, 26–29, 64–66) halides,(67–69) chlorinated hydrocarbons,(70) iodide,(71) bromate,(72) xenon,(73) acrylamide,(74) succinimide,(75) sulfur dioxide,(76) and halothane,(77) to name a few.

10.6.1. Oxygen Sensing

The most well-known collisional quencher is molecular oxygen (O_2). Several fluorophores which are potentially useful as O_2 sensors are given in Table 10.3. The frequency responses of several oxygen sensors with a range of lifetimes (τ_0) are shown in Figure 10.10. The arrows indicate the change (decrease in phase and increase in modulation) in the important analytical range of oxygen concentrations from 0% (N_2) to 20% (Air). The magnitude of the phase angle and modulation changes are related with oxygen sensitivity and depend on probe lifetime (τ_0) and modulation frequency (ω). The amount of quenching by oxygen (decrease of the lifetime and quantum yield) increases as the lifetime of the probe increases (see Table 10.3). It should be noted that

Table 10.3. Mean Lifetimes ($\bar{\tau}$) in Solvents Purged by Nitrogen (N_2), Air, and Oxygen (O_2) and Sensing Parameters (Changes in Phase Angle, $\Delta\theta$, and in Modulation, Δm; "Air"–"N_2") of Potential Oxygen Sensors[33]

Probe	$\tau_0(N_2)$ (ns)	$\bar{\tau}$(Air) (ns)	$\bar{\tau}(O_2)$ (ns)	τ_{rel}	$\Delta\theta$ (°)	Δm (%)	F (MHz)
2,5-Dansyl[a]	16.2	11.1	5.7	1.46	–10.5	11.9	10.0
Benzo[g,h,i]perylene[b]	67.8	24.3	6.9	2.79	–27.7	38.0	5.0
Coronene[c]	209	22.8	5.3	9.17	–52.2	67.6	3.0
[Ru(Ph2phn)3]Cl2[a]	5200	282	64.7	18.4	–60.7	78.5	0.2
[Ru(Ph2phn)3]Cl2[d]	5400	625	208	8.6	–53.3	66.0	0.1

[a]In propanol at 20°C,
[b]In dichloromethane at 20°C,
[c]In cyclohexane at 20°C,
[d]In silicone at 20°C.

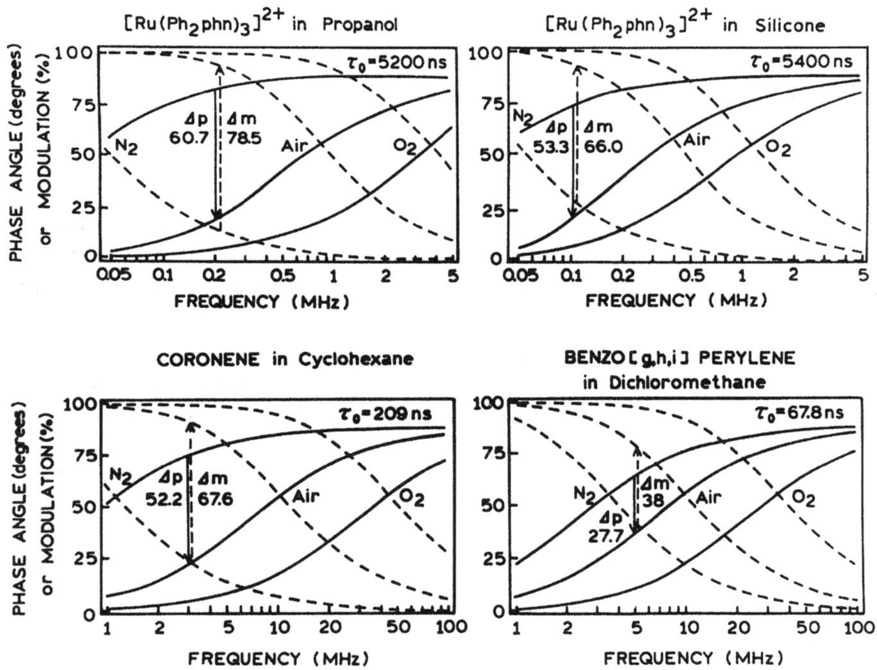

Figure 10.10. Frequency responses of intensity decays of possible oxygen sensors with various lifetimes. The arrows indicate the magnitude of changes in phase and modulation in the range from 0% oxygen (N_2, deaerated solution) to about 20% oxygen (Air, equilibrium with air).

the decrease in lifetime of only 2.79-fold for Benzo[*g*,*h*,*i*]perylene results in quite a sufficient change in phase angle (27.7°) and modulation (38.0%). Such a change should allow accurate oxygen measurements since the phase angle an modulation are usually measured with accuracy of 0.5° and 1%, respectively. These results indicate that phase-modulation method does not require probes with very long lifetimes.

The metal–organic complexes of ruthenium, osmium, iridium, and platinum have been regarded as oxygen probes because of their strong metal-to-ligand energy transitions and long-lived excited states (up to 5 μsec).[26–29, 78] One of the most useful seems is tris(4,7-diphenyl-1,10-phenanthroline)ruthenium(II) ($[Ru(Ph_2phn)_3]^{2+}$), which is highly sensitive to oxygen. In order to obtain an oxygen sensor it is necessarily to immobilize the oxygen-sensitive dye on a rigid support or use a gas-permeable membrane.[79–81] This is because the oxygen-sensitive dyes can be also quenched by a variety of interferents in the solvent. Silicone rubber, in particular, has excellent oxygen permeability, optical properties, can be handled easily and allow the fabrication of very thin films for fast probe response. Additional selectivity to oxygen has been obtained by incorporating of the $[Ru(Ph_2phn)_3]^{2+}$ into silicon, which is highly permeable to oxygen but excludes other chemical species. The possible interferents like CO_2, nitrous oxide, cyclopropane, halothane, Fe^{3+}, and H_2S are all without effect at concentrations well above those normally encountered.[26] Incorporation of oxygen-sensitive dyes into polymeric matrices result in decreased sensitivity compared to probe in solution due to the reduced diffusional coefficient. In case of $[Ru(Ph_2phn)_3]^{2+}$ and silicone rubber there is only modest decrease in sensitivity if phase-modulation method is used (Figure 10.10, *top* and Table 10.3).

10.6.2. Cellular Chloride Sensing

In contrast to metal ions, there are no known specific chelator groups for halides. Fortunately, the halides often act as collisional quenchers. It is interesting to compare the sensitivity and accuracy of a chloride-sensitive probe such as SPQ (6-methoxy-*N*-[3-sulfopropyl]quinolinium). Increasing the sensitivity for chloride is important because, in intracellular applications, a significant decrease of chloride sensitivity was observed (about 10-fold).[68, 82–85] Additionally, the chloride probes typically leak out of the cells. Since all known chloride probes are nonratiometric probes the leakage results in difficulties in calibration. A comparison of Cl^- sensing by intensity and phase angles is shown in Figure 10.11. These plots show the intensity (Eq. (10.26)) and phase angle (Eqs. (10.24), (10.8), and (10.9)) of SPQ decreases due to chloride quenching in aqueous solutions (K_{SV} = 118 M^{-1}) and in the cells (K_{SV} = 13 M^{-1}) for lifetimes of 26.0 and 3.7 nsec, respectively.[85] In aqueous solution the intensity decreases about eightfold and phase angle by 50° (at 20 MHz) in the range of 0–50 m*M* of chloride concentration (Figure 10.11, *bottom*). SPQ in cells display only a 1.7-fold decrease in intensity and 16° in phase angle (at 60 MHz) within the same range of chloride concentrations. This significant decrease of chloride sensitivity of SPQ is due to the

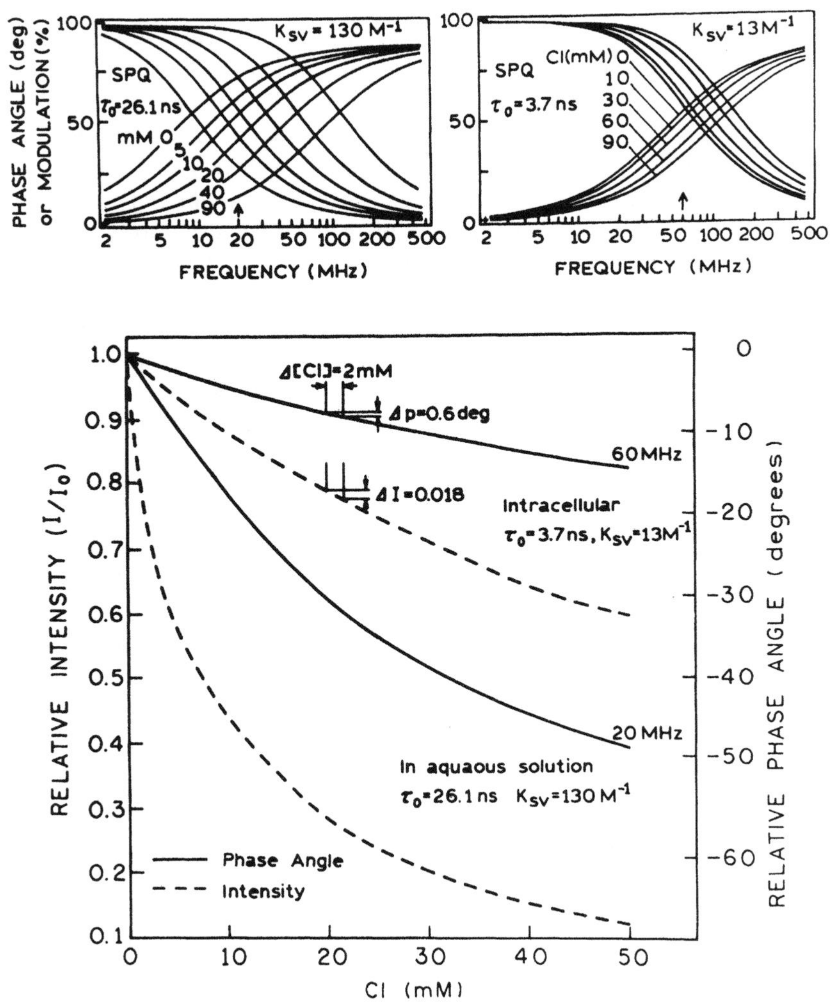

Figure 10.11. Comparison of sensitivity using intensity and phase angle methods for chloride probe SPQ.

non-Cl components like organic anions and proteins.[85] Accuracy of 2 m*M* of chloride requires measurement of intensity with an accuracy of 1.8%, which can be difficult to obtain in microscopic sample. In order to measure chloride to an accuracy of ±2 m*M* one requires measurement of phase angle to an accuracy of 0.6°, which can be easily accomplished even in the presence of probe leakage and photobleaching. In contrast, it is often difficult to measure intensities to an accuracy of 1.8% even in a cuvette, and in a fluorescence microscope using labelled cells such accuracy is not possible. This reasoning supports our claim that lifetime-based sensing is advantageous in cellular imaging and sensing applications. Table 10.4 includes other chloride probes with their

Table 10.4. Photophysical Properties (Quantum Yield (Φ), Lifetime (τ_0), Wavelength at Absorption (λ_{exc}), and Emission Maximum (λ_{em})) of Chloride Probes and Their Quenching Constants (K_{SV})

Probe	Φ	τ_0 (nsec)	λ_{exc} (nm)	λ_{em} (nm)	ε_{exc} ($M^{-1}cm^{-1}$)	K^a_{SV} (M^{-1})
SPQ[68, 69, 85]	0.60	26.0	350	442	3470	118, 13[b]
MQAE[68, 69]	—	21.6	355	440	2800	200
MEQ[83]	0.70	—	344	440	4100	145, 19[b]
TMAPQ[84]	0.73	—	355	450	3700	310, 20[b]
TMAPP[84]	0.50	—	375	440	3600	95
ABQ[84]	0.64	—	355	450	3510	237

Abbreviations: ABQ, 6-methoxy-*N*-(4-aminoalkyl)quinolinium bromide hydrochloride; MEQ, 6-methoxy-*N*-ethylquinolinium chloride; MQAE, *N*-(6-methoxyquinolyl)acetoethyl ester; SPQ, 6-methoxy-*N*-[3-sulfopropyl]quinolinium; TMAPP, 6-methoxy-*N*-(3-trimethylammoniumpropyl)phenanthrindium dibromide; TMAPQ, 6-methoxy-*N*-(3-trimethylammoniumpropyl)quinolinium dibromide.

[a]In aqueous solution;

[b]In cells.

properties. Using probes with higher chloride-sensitivity (larger values of K_{SV}) than SPQ like 6-methoxy-*N*-ethylquinolinium ester (MQAE)[68] or 6-methoxy-*N*-(3-trimethylammoniumpropyl)quinolinium dibromide (TMAPQ)[84] (Table 10.4) one may expect larger change in phase angle and improved accuracy.

10.7. Sensing Based on Fluorescence Resonance Energy Transfer (FRET)

Energy transfer-based probe consists of two parts: donor (D) and acceptor (A). The changes in donor decay time (or intensity) are induced by FRET. The phenomenon of FRET is nonradiative energy transfer from the fluorescence donor to acceptor, without emission and reabsorption of photons. The FRET mechanism is completely predictable based on the spectral properties of the donor and acceptor and operates through space over distances ranging from 10 to 100 Å.[32, 86, 87] The fluorescent donor can be selected for its absorption, emission, quantum yield, and decay time characteristics without concern for its sensitivity to analytes. The acceptor can be fluorescent or nonfluorescent, but must display an absorption in the wavelength range of the donor emission and be photostable. This allows every absorption-based indicator to be made as an fluorescent sensor. It is important to recognize that the spectral properties of the donor and acceptor in an energy transfer-based sensors can be adjusted as desired and the choice of D–A pairs can be modified for use with laser diodes or other desirable light sources. Donor–acceptor probes can provide sensing of analytes for which there are no available direct indicators, e.g., glucose sensing or in immunoassays.

The donor and acceptor molecules can be covalently linked by a flexible or rigid spacer, conjugated to macromolecules, or simply mixed together. To illustrate some features of these approaches, it may be useful to display some simulated results.

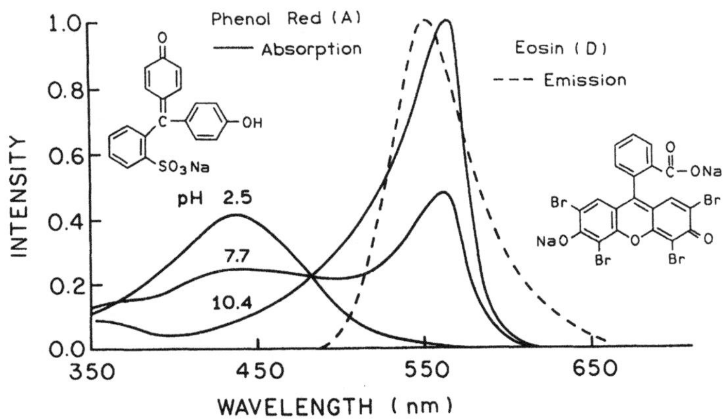

Figure 10.12. Absorption spectra of Phenol Red and emission spectrum of Eosin as potential pH probe based on resonance energy transfer.

10.7.1. Unlinked Donor–Acceptor

Suppose we have a pH indicator like Phenol Red whose absorption spectrum is pH-sensitive with $pK_a = 7.6$ (Figure 10.12). Phenol Red displays two distinct absorption spectra for protonated form (pH 2.5) and for unprotonated form (pH 10.4). One of the possible donors is an Eosin which displays an emission spectrum that overlaps with the absorption spectra of the protonated and unprotonated forms (acceptors) of Phenol Red (Figure 10.12). The critical distances for energy transfer (R_0),[32] calculated from spectral properties of Eosin and Phenol Red, are 28.3 and 52.5 Å for protonated and unprotonated forms of Phenol Red, respectively. For randomly distributed acceptors in three dimensions with no diffusion, the donor decay is

$$I_{DA}(t) = I_0 \exp\left[-\frac{t}{\tau_D} - 2(\gamma_{A1} + \gamma_{A2})\sqrt{\frac{t}{\tau_D}}\right], \tag{10.29}$$

where γ_{Ai} are related to the concentration of acceptors C_{Ai} and the R_{0i}

$$\gamma_{Ai} = \frac{2\pi^{3/2} N R_{0i}^3}{3000} C_{Ai}, \tag{10.30}$$

where N is a number of molecules per mol. The concentration of each form of the acceptor (C_{Ai}) can be calculated from the total acceptor concentration (C_A) and the fraction of each form ($C_A = C_{A1} + C_{A2}$). In case of a pH indicator the fraction of each form can be calculated from the Henderson–Hasselbach equation

$$pH = pK_a + \log \frac{[A^-]}{[HA]}. \tag{10.31}$$

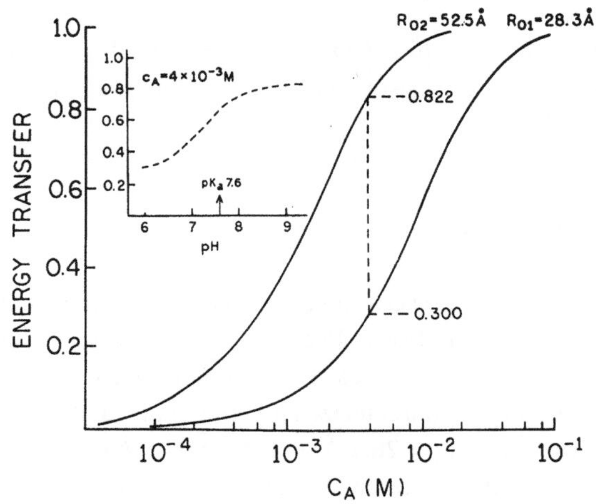

Figure 10.13. Dependence of energy transfer on concentration of Phenol Red, for the protonated (R_{01} = 28.3 Å) and unprotonated form (R_{02} = 52.5 Å). The *inset* shows the pH-dependent energy transfer using an acceptor concentration of 4×10^{-3} *M* (effect of diffusion is omitted).

The energy transfer efficiency for this system can be calculated from relative quantum yield of the donor

$$ET = 1 - \frac{\Phi_{DA}}{\Phi_D} = 1 - \int_0^\infty I_{DA}(t)dt / \int_0^\infty I_D(t)dt, \tag{10.32}$$

where $I_D(t)$ is an intensity decay of the donor with no acceptors. Figure 10.13 (*top*) shows the well-known Förster energy transfer dependence on acceptor concentration and critical distance R_0, in this case for Eosin–Phenol Red (protonated, R_{01} = 28.3 Å) and Eosin–Phenol Red (unprotonated, R_{02} = 52.5 Å). The suitable range of Phenol Red concentrations (significant difference in energy transfer to protonated and unprotonated forms) are from 2×10^{-3} to 8×10^{-3} *M*. pH-dependent energy transfer for this sensor is shown in Figure 10.13 (*inset*) using concentration of Phenol Red 4×10^{-3} *M* and Eqs. (10.29)–(10.32). Such sensors require an acceptor concentration in the range 1–10 m*M*, which results in high optical densities at the excitation and emission wavelengths. At these high acceptor concentrations, there are in addition to FRET, inner-filter effects which depend on the excitation and observation wavelengths, on analyte concentration, and on the detailed macroscopic properties of the sample and detection optics.[88, 89] The phase angle and modulation measurements are not dependent on the total intensity so long as the donor emission is detectable and there is no change in total acceptor concentration. However, it is likely that this kind of sensor will require recalibrations due to washout and leakage of dyes, which affect the acceptor concentration and thus energy transfer efficiency. An advantage is that unlinked donor–acceptor pairs are usually easy to mix together, but can be difficult to chemically couple. Sensing of pH(pCO_2) have been already demonstrated using this type of energy transfer-based sensors with several donor–acceptor pairs and phase-modulation method.[90]

10.7.2. Linked Donor–Acceptor

Donor and acceptors can be covalently linked using a chemical spacer. Assume that we have the same D–A pair Eosin–Phenol Red. In this case we will have a mixture of two linked donor–acceptor species (Eosin–Phenol Red protonated and Eosin–Phenol Red unprotonated) characterized by the same distance distribution and different critical distances (R_{01} = 28.3 Å and R_{02} = 52.5 Å) for FRET. A distribution of D- to -A distances will be present because the linker is typically flexible. The fractional intensity of the first species at time $t = 0$ is g_1 and that of the second species is $(1 - g_1)$. The fractional intensity at time $t = 0$ is equal to fractional concentration of each form, which can be in case of pH indicator (Phenol Red) calculated using Eq. (10.31). The donor fluorescence intensity decay of the mixture is described by the equation

$$I_{DA}(t) = I_D\left[g_1\int_0^\infty P(r)\exp\left(\frac{t}{\tau_{DA1}}\right)dr + (1-g_1)\int_0^\infty P(r)\exp\left(\frac{t}{\tau_{DA2}}\right)dr\right], \tag{10.33}$$

where the decay time of each D–A pair depends on separation r and critical distance R_0 by

$$\frac{1}{\tau_{DAi}} = \frac{1}{\tau_D} + \frac{1}{\tau_D}\left(\frac{R_{0i}}{r}\right)^6. \tag{10.34}$$

The distance-distribution $P(r)$ of the excited molecules of time $t = 0$ is assumed to be Gaussian

$$P(r) = \frac{1}{\sigma\sqrt{\pi}}\exp\left[-\frac{1}{2}\left(\frac{r - R_{AV}}{\sigma}\right)^2\right], \tag{10.35}$$

where σ is the standard deviation of Gaussian (full-width) of half-maximum of the initial distribution (hw = 2.354σ), R_{AV} is the average distance between donor and acceptor. Energy transfer efficiency of the system can be calculated using Eq. (10.32). Assuming that $hw \ll R_{AV}$ one can estimate energy transfer for D–A pair from the equation

$$ET = \frac{R_0^6}{R_0^6 + r^6}. \tag{10.36}$$

Equation (10.36) allows prediction of the optimal length of the spacer based only on values of the critical distances (assuming rigid spacer). Figure 10.14 (*top*) shows the dependence of energy transfer (ET) on the D–A average distance (R_{AV}) (with $hw = 0.1$ R_{AV}) for the critical distances R_{01} = 28.3 Å and R_{02} = 52.5 Å. For a practical sensor, Eosin should be separated from Phenol Red by about 30 to 50 Å. It is difficult to find a rigid chemical spacer longer that 25 Å. Hence, it is important to consider the effect of D–A distance distributions on the photophysical properties of the pH sensor.

Donor and acceptors (Eosin–Phenol Red) separated by an optimal average distance of 40 Å (between R_{01} and R_{02} values) results in very good changes in energy transfer from 0.114 for the protonated to 0.826 for unprotonated form of Phenol Red. It should be noted that this is a significant increase in energy transfer changes compared to unlinked donor–acceptor system (Figure 10.13). It is also useful to discuss the cases where the D–A spacer is shorter (~0.5 × R_{02}, 25 Å) and longer (~2 × R_{01}, 55 Å) than the Förster distance.

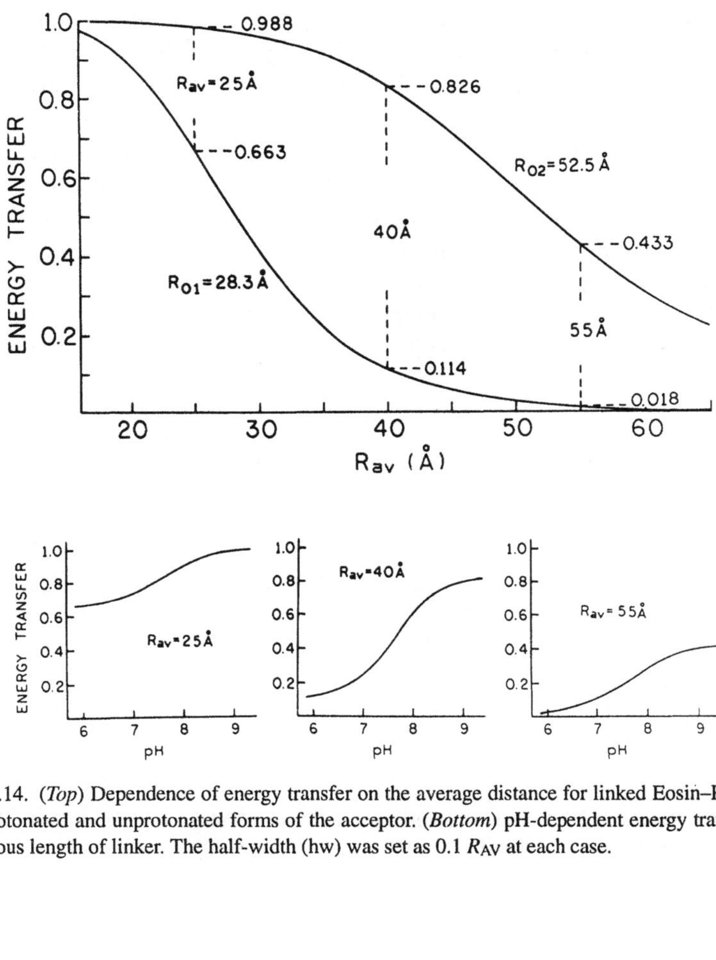

Figure 10.14. (*Top*) Dependence of energy transfer on the average distance for linked Eosin–Phenol Red for the protonated and unprotonated forms of the acceptor. (*Bottom*) pH-dependent energy transfer using three various length of linker. The half-width (hw) was set as 0.1 R_{AV} at each case.

For the D–A pairs with a short spacer, energy transfer to the unprotonated form is very high (0.988) compared to the protonated form (0.663, Figure 10.14). The quantum yield of the protonated form is then about 28-fold greater than unprotonated form. Energy transfer of 0.988 will decrease the lifetime of Eosin from 1.33 nsec (no energy transfer) to about 16 psec. This very weak component with short decay time is likely be dominated at higher values of pH by fluorescent impurities and at lower pH values by fluorescence of D–A pairs with unprotonated form of Phenol Red such as was observed for Fluo-3 (see Section 10.3). Practically, this may result in smaller apparent changes of energy transfer than is presented in Figure 10.14 (*bottom, left*) because of contribution from fluorescence impurities to the emission signal.

If the average distance between donor and acceptors is long (55 Å), then energy transfer to the protonated form (0.018) is negligible compared to that of unprotonated form (Figure 10.14). The pH-induced absolute changes in energy transfer for this case is much lower (0.415) compared to that for R_{AV}= 40 Å (0.718) and comparable with short spacer (0.325) (see Figure 10.14). However, for comparable absolute changes in energy transfer the longer D–A spacer is more promising for lifetime-based sensing than the shorter. This is because for a longer spacer, emission from both forms will have significant contribution over the wide range of analyte concentrations. For Eosin–Phenol Red separated by R_{AV}~ 55 Å the quantum yield of two D–A pairs differs only 1.7-fold.

pH-dependent energy transfer for unlinked and for various average distances between donors and acceptors can be measured as a pH-dependent phase angle and modulation. Eosin-Phenol Red covalently linked at an average distance of 40 Å should display the largest dynamic range in phase angle and modulation (Figure 10.15). The magnitude of phase angle changes is 39.4° and 41.7% in modulation compared with 24.8° and 25.1% for donor and acceptor randomly distributed (unlinked), respectively. For the long donor–acceptor spacer (55 Å) which exhibits rather significant changes

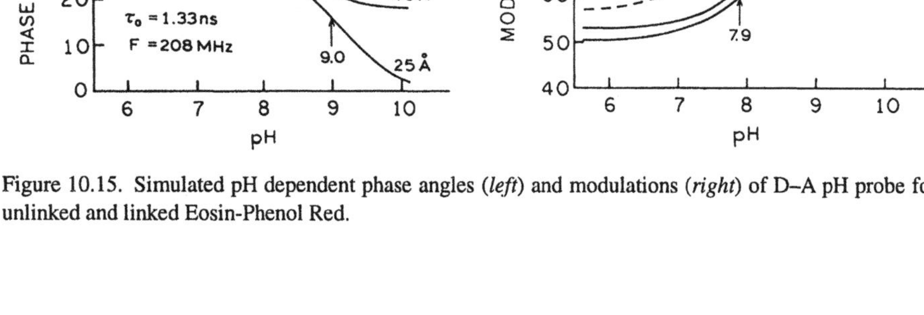

Figure 10.15. Simulated pH dependent phase angles (*left*) and modulations (*right*) of D–A pH probe for unlinked and linked Eosin-Phenol Red.

in energy transfer (from 0.018 to 0.433), we expect only moderate changes in phase angle (14.2°) and modulation (18.9%). The shorter D–A spacer (25 Å) will result in limited range of analyte concentration because of the decreased intensity. In the case of Eosin–Phenol Red at pH 9 the intensity will be 0.06 that at pH 5.5. In the pH range from 5.5 to 9.0, the phase angle will decrease about 15° but with modulation only about 5%.

From the simulations of D–A systems (unlinked and linked) we can conclude that pH-sensitive range can be modulated by using various lengths of spacers. Decreasing the length of the spacer from 55 to 25 Å results in a shift of the apparent pK_a toward higher values using phase angle and modulation (Figure 10.15) as compared with intensity measurements. Generally, if increasing analyte concentration results in increasing the mean lifetime of the probe, then we may expect that phase angle and modulation apparent pK_a will be shifted toward higher values (the apparent K_D will be shifted toward lower values, see Figure 10.8 for Calcium Green). This conclusion does not apply to wavelength ratiometric probes where the apparent value of K_D or pK_a strongly depends on excitation and/or observation wavelengths (Figure 10.9).

It should be mentioned that data for energy-transfer-based probes simulated in this chapter are generic and can apply to any D–A system regardless of the decay time of the donor. The optimal modulation frequency will be determined by the decay time, but the magnitude of changes in phase and modulation will depend only on absolute energy transfer efficiency.

10.7.3. Macromolecules Labeled by Donor and Acceptor

The simulations for the long spacer of R_{AV} = 55 Å is also representative of macromolecules (polymers) labeled by donors and acceptors. Usually, for sensing purpose, one macromolecule is labeled by the donor and the other by acceptor. If the macromolecules are associated, there is some amount of energy transfer. On disassociation, the distance will be too large for significant energy transfer to occur. Such energy transfer-based sensors have been already demonstrated for glucose assays[91, 92] and immunoassays.[93, 94] In glucose assays the mechanism of sensing relies on displacement of acceptor-labeled sugars by the glucose (analyte) from the donor-labeled macromolecule (Concanavalin A (ConA)). Glucose binding to donor labeled Con-A will result donors distant from acceptors with decay time τ_0 (no energy transfer). The dynamic range of such probes depends on the concentration of acceptor-labeled sugar, its affinity of binding to Con A, and D–A distance. It should be noted that in energy transfer-based competitive binding sensors neither the donor nor the acceptor spectral properties need to be dependent on analyte concentration. The important parameter is the relation between value of the critical distance R_0 and the distance between donor and acceptor determined by size of the macromolecules, conjugation sites, and binding sites.

Fluorescence resonance energy transfer has also been used for ionic strength measurements.[95] Fluorescein labeled dextran (donor) and polyethyleneimine-Texas Red (acceptor) were placed behind a dialysis membrane. The polymer association is ionic strength dependent and the ratio of intensities (F_{620}/F_{520}) was used as the measured parameter. Since both the donor and acceptor are fluorescent, this kind of sensor may allow expand the sensitive ionic strength range by shifts in observation wavelength, as was discussed for pH probe Carboxy SNAFL-2 (see Section 10.3).

FRET has also been applied in biological systems using fluorescence microscopy. Donor- and acceptor-labeled lipids can be used for investigation of intracellular structure by examination of its temporal and spatial changes in the distribution of fluorescence molecules in membranes of the living cell.[96] The application of FRET to study binding sites on individual cell surfaces has recently been introduced.[97] Fluorescein covalently conjugated to immunoglobulin E (donor) was bound to Fc-receptors on the cell surface and 2,4-dinitrophenyl (DNP)-haptens were used as an acceptors. The energy transfer between two fluorophores on single cells seen in a fluorescence microscope was determined from the photobleaching kinetics of the donor fluorophore.

Another approach of energy transfer-based probe has been demonstrated for α-Amylase sensing.[98] In this case amylose was doubly labeled, by fluorescein derivative (donor) and Procion Red MX8B (acceptor). As α-Amylase catalyzes the cleavage of the amylose into smaller units, the average distance between fluorescein and Procion Red increases, which reduces the degree of quenching. The rate of increase in fluorescence intensity is proportional to α-amylase activity.

10.8. Summary and Prospects

Lifetime-based sensing is advantageous because the measurements can be independent of fluorescence intensity. This insensitivity offers opportunities in fluorescence microscopy where the local probe concentration generally cannot be controlled, in remote sensing using fiber optics where there are losses of excitation and emission intensity, and in flow cytometry where there is variations in cell volume and extent of probe uptake. Phase angle and/or modulation measurement at a single modulation frequency offers fast temporal resolution and significantly simplified the instrumentation. Amplitude modulation of the excitation can be easily accomplished using a commercially available acousto-optic modulators and continuous-wave (CW) lasers, such as helium–cadmium (HeCd) (354 nm or 442 nm) and helium–neon (HeNe) (543 and 633 nm). Diode lasers can be modulated intrinsically. Lifetime-based sensing (phase-modulation) does not require complete resolution of a multi-exponential decay, and probes with complex intensity decays are suitable for sensing. Phase angle and modulation are very sensitive to intensity decay parameters and a 1.5-fold change in lifetime results in a sufficient change in phase angle (about 12°) and for modulation (about 14%) for accurate analyte measurements. This allows one to use probes that are less sensitive to analyte. Lifetime-based sensing also allows use of the mechanisms of collisional quenching and resonance energy transfer, where there are no spectral shifts for ratiometric sensing. Additionally, syntheses of probes for visible excitation have resulted in probes like Calcium Color series (see Table 10.1), which do not display spectral shifts. Such probes are suitable for lifetime-based sensing.

Many difficulties encountered with intensity measurements can be circumvented by the use of lifetime-based sensing. It is now possible to measure the phase angles on a cell-by-cell basis as they pass through the laser beam using a phase flow cytometer.[99] Fluorescence lifetime imaging, or FLIM, allows image contrast to be created based on the lifetime of the probe at each point in the image. There is a significant interest in creating two-dimensional fluorescence lifetime images, and several techniques have been recently described.[100–105] If the probe lifetime is sensitive to analyte, one now has an ability to create chemical and biochemical images of cells[106] and tissues.

Finally, we note that future instrument for lifetime-based sensing and imaging can be based on laser diode light sources. At present it is desirable to develop specific probes which can be excited from 630 to 780 nm, the usual range of laser diodes. The use of such probes will allow us to avoid the use of complex laser sources, which should result in the expanded use of fluorescence detection in the chemical and biomedical sensors.

Acknowledgments

This work was supported by the National Center for Research Resources (RR-08119), with additional support from National Institute of Health (NIH) grants GM-35154 and RR-07510-01.

References

1. J. Janata and A. Bezegh, Chemical sensors, *Anal. Chem. 60*, 62R–74R (1988).
2. O. S. Wolfbeis (Ed.), *Fiber Optic Chemical Sensor and Biosensors*, Vols. I & II. CRC Press, Boca Raton, Florida (1991).
3. W. R. G. Baeyens, D. De Keukelaire, and K. Korkidis (Eds.), *Luminescence Techniques in Chemical and Biomedical Analysis*, Marcel Dekker, New York (1991).
4. O. S. Wolfbeis (Ed.), *Fluorescence Spectroscopy, New Methods and Applications*, Springer-Verlag, Berlin, (1993).
5. A. W. Czarnik (Ed.), *Fluorescent Chemosensors for Ion and Molecule Recognition*, ACS Symposium Series 538, Washington, D.C. (1993).
6. R. Y. Tsien, Fluorescent and photochemical probes of dynamic biochemical signals inside living cells, in: *Ref. 5*, pp. 130–146 (1993).
7. R. P. Haugland, *Handbook of Fluorescent Probes and Research Chemicals* (K. D. Larison, ed.), Molecular Probes, Inc., Eugene, Oregon (1992–1994).
8. A. Minta, J. P. Y. Kao, and R. Y. Tsien, Fluorescent indicators for cytosolic calcium based on rhodamine and fluorescein chromophores, *J. Biol. Chem. 264*, 8171–8178 (1989).

9. J. R. Lakowicz, H. Szmacinski, and M. L. Johnson, Calcium imaging using fluorescence lifetimes and long-wavelength probes, *J. Fluorescence 2*, 47–62 (1992).
10. A. Minta and R. Y. Tsien, Fluorescent indicators for cytosolic sodium, *J. Biol. Chem. 264*, 19449–19457 (1989).
11. H. Szmacinski and J. R. Lakowicz, Optical measurements of Mg^{2+} Using Fluorescence Lifetimes and Phase-Modulation Fluorometry, *Biophys. J. 64*, A108 (1993).
12. J. E. Whitaker, R. P. Haugland, G. Predengast, Spectral and photophysical studies of benzo[c]xanthene dyes: Dual emission pH sensors, *Anal. Biochem. 194*, 330–344 (1991).
13. H. Szmacinski and J. R. Lakowicz, Optical measurements of pH using fluorescence lifetimes and phase-modulation fluorometry, *Anal. Chem. 65*, 1668–1674 (1993).
14. D. Menisalamani and M. E. Lucas. Fluorescent chemosensors for monitoring potassium in blood and across biological membranes, in: *Ref. 5*, pp. 162–195 (1993).
15. R. Y. Tsien, Fluorescent and photochemical probes of dynamic biochemical signals inside living cells, in: *Ref. 5*, pp. 130–146 (1993).
16. T. Schulte, personal communication, Becton Dickinson Diagnostic Instrument Systems, Towson, Maryland 1989.
17. K. W. Berndt, I. Gryczynski, and J. R. Lakowicz, Phase-modulation fluorometry using a frequency-doubled pulsed laser diode source, *Rev. Sci. Instrum. 61*(7), 1816–1820 (1990).
18. R. B. Thompson, J. K. Frisoli, and J. R. Lakowicz, Phase fluorometry using a continuously modulated laser diode, *Anal. Chem. 64*, 2075–2078 (1992).
19. M. E. Lippitsch, J. Pasterhofer, M. J. P. Leiner, and O. S. Wolfbeis, Fibre-optic oxygen sensor with the fluorescence decay time as the information carrier, *Anal. Chim. Acta 205*, 1–6 (1988).
20. K. W. Berndt and J. R. Lakowicz, Elecroluminescent lamp-based phase fluorometer and oxygen sensor, *Anal. Biochem. 201*, 319–325 (1992).
21. H. E. Zimmerman, J. H. Penn, and C. W. Carpenter, Evaluation of single-photon counting measurements of excited-state lifetimes, *Proc. Natl. Acad. Sci. USA 79*, 2128–2132 (1982).
22. R. A. Lampert, L. A. Chewter, D. Phillips, D. V. O'Connor, A. J. Roberts, and S. R. Meech, Standards for nanoseconds fluorescence decay time measurements, *Anal. Chem. 55*, 68–73 (1983).
23. J. R. Lakowicz, I. Gryczynski, G. Laczko, and D. Gloyna, Picosecond fluorescence lifetime standards for frequency- and time-domain fluorescence, *J. Fluorescence 1*, 87–93 (1991).
24. C. A. Parker, *Photoluminescence of Solutions*, pp. 22–28, Elsevier Publishing Co., New York (1968).
25. E. J. Bowen and F. Wokes, *Fluorescence of Solutions*, Longman Green & Co., London (1953).
26. J. R. Bacon and J. N. Demas, Determination of oxygen by luminescence quenching of a polymer immobilized transition metal complex, *Anal. Chem. 59*, 2780–2785 (1987).
27. E. R. Carraway, J. N. Demas, B. A. DeGraff, and J. R. Bacon, Photophysics and photochemistry of oxygen sensors based on luminescent transition metal complexed to oxygen, *Anal. Chem. 63*, 337–342 (1991).
28. A. Pfeil, Quenching of the phosphorescence of some chromium (III) complexes by molecular oxygen, *J. Am. Chem. Soc. 93*, 5395–5397 (1971).
29. W. W-S. Lee, K-Y. Wong, and X-M. Li, Luminescent dicyanoplatinium (II) complexes as sensors for the optical measurement of oxygen, *Anal. Chem. 65*, 255–258 (1993).
30. R. A. Bissel, A. Prasanna de Silva, H. Q. N. Gunaratne, P. L. M. Lynch, C. P. McCoy, G. E. M. Maguire and K. R. A. S. Sandanayake, Fluorescent photoinduced electron-transfer sensors, in: *Ref. 5*, pp. 45–58 (1993).
31. A. P. de Silva and S. A. de Silva, Fluorescent signalling crown ethers; 'switching on' of fluorescence by alkali metal ion recognition and binding in situ, *J. Chem. Soc., Chem. Commun. 20*, 1709–1710 (1986).
32. Th. Förster, Intermolecular energy migration and fluorescence, *Ann. Phys. (Leipzig) 2*, 55–75 (1948), (Translation 1974 by R. S. Knox).
33. H. Szmacinski and J. R. Lakowicz, unpublished data.

34. D. J. S. Birch and R. E. Imhof, Time-domain fluorescence spectroscopy using time-correlated single-photon counting, in: *Topics in Fluorescence Spectroscopy, Techniques*, Vol. 1 (J. R. Lakowicz, ed.), pp. 1–95, Plenum Press, New York (1991).
35. J. R. Lakowicz and I. Gryczynski, Frequency-domain fluorescence spectroscopy, in: *Topics in Fluorescence Spectroscopy, Techniques*, Vol. 1 (J. R. Lakowicz, ed.), pp. 293–355, Plenum Press, New York (1991).
36. J. N. Demas, *Excited State Lifetime Measurements*, Academic Press, New York (1983).
37. J. R. Lakowicz, *Principles of Fluorescence Spectroscopy*, Plenum Press, New York (1983).
38. D. M. Jameson and G. D. Reinhart (Eds.), *Fluorescent Biomolecules*, Plenum Press, New York (1989).
39. T. G. Dewey (Ed.), *Biophysical and Biochemical Aspects of Fluorescence Spectroscopy*, Plenum Press, New York (1991).
40. J. R. Lakowicz (Ed.), *Time-Resolved Laser Spectroscopy in Biochemistry*, Proc. SPIE, Vol. 1204, (1990) and Vol. 1640, (1992).
41. J. R. Lakowicz (Ed.), *Topics in Fluorescence Spectroscopy*, Plenum Press, New York, Vols. 1 & 2, (1991); Vol. 3, (1992).
42. J. R. Lakowicz, G. Laczko, H. Cherek, E. Gratton, and M. Limkeman, Analysis of fluorescence decay kinetics from variable-frequency phase shift and modulation data, *Biophys. J. 46*, 463–477 (1984).
43. J. R. Alcala, E. Gratton, and D. M. Jameson, A multifrequency phase fluorometer using the harmonic content of a mode-locked laser, *Anal. Instrum. 14*(3&4), 225–250 (1985).
44. B. Barbieri, F. DePiccoli, and E. Gratton, Synthesizers' phase noise in frequency-domain fluorometry, *Rev. Sci. Instrum. 60*(10), 3201–3206 (1989).
45. K. Otsuka, Gigabit optical pulse generation in integrated lasers and application, in: *Picosecond Electronic Devices* (C. H. Lee, ed.), Academic Press, New York (1984).
46. A. Yariv, *Quantum Electronics*, pp. 255–260, John Wiley & Sons (1989).
47. G. Laczko, I. Gryczynski, Z. Gryczynski, W. Wiczk, H. Malak, and J. R. Lakowicz, A 10-GHz frequency-domain fluorometer, *Rev. Sci. Instrum. 61*, 2332–2337 (1990).
48. R. B. Mujumdar, L. A. Ernst, S. R. Mujumdar, and A.S. Wagooner, Cyanine dye labeling reagents containing isothiocyanate groups, *Cytometry 10*, 11–19 (1989).
49. E. Terpetschnig, H. Szmacinski, and J. R. Lakowicz, Synthesis, spectral properties and photostabilities of symmetrical and unsymmetrical squaraines; a new class of fluorophores with long-wavelength excitation and emission, *Anal. Chim. Acta 282*, 633–641 (1993).
50. B. Chance, J. S. Leigh, H. Miyake, D. S. Smith, S. Nioka, R. Greenfield, R. Finander, K. Kaufman, W. Levy, M. Young, P. Cohen, H. Yoshioka, and R. Boretsky, Comparison of time-resolved and -unresolved measurements of deoxyhemoglobin in brain, *Proc. Natl. Acad. Sci. USA 85*, 4971–4975 (1988).
51. E. Koller and O. S. Wolfbeis, Sensor chemistry, in: *Ref. 2*, Vol. I, pp. 303–358 (1991).
52. G. Grynkiewicz, M. Poenie, and R. Y. Tsien, A new generation of Ca^{2+} indicators with greatly improved fluorescence properties, *J. Biol. Chem. 260*, 3440–3450 (1985).
53. K. M. Hirshfeld, D. Toptygin, B. S. Packard, and L. Brand, Dynamic fluorescence measurements of two-state systems: Applications to calcium chelating probes, *Anal. Biochem. 209*, 209–218 (1993).
54. M. A. Kuhn, 1,2-Bis(2-aminophenoxy)ethane-N,N,N′,N′,-tetraacetic acid conjugates used to measure intracellular Ca^{2+} concentration, in: *Ref. 5*, pp. 147–161 (1993).
55. J. R. Lakowicz, H. Szmacinski, K. Nowaczyk, and M. L. Johnson, Fluorescence lifetime imaging of calcium using Quin-2, *Cell Calcium 13*, 131–147 (1992).
56. R. Y. Tsien, New calcium indicators and buffers with high selectivity against magnesium and protons: Design, synthesis, and properties of prototype structures. *Biochemistry 19*, 2396–2404(1980).
57. R. Tsien and T. Pozzan, Measurements of cytosolic free Ca^{2+} with Quin-2, *Methods in Enzymology 172*, 230–261 (1989).

58. H. Szmacinski, I. Gryczynski, and J. R. Lakowicz, Calcium-dependent fluorescence lifetimes of indo-1 for one- and two-photon excitation of fluorescence, *Photochem. Photobiol. 58*, 341–345 (1993).
59. S. Fery-Forgues, M.-T. Le Bris, J.-P. Guette, and J. B. Valeur, Ion-responsive fluorescent compounds. I. Effect of cation binding on photophysical properties of a benzoxazinone derivative linked to monoaza-15-crown-5, *J. Phys. Chem. 92*, 6233–6237 (1988).
60. B. Valeur, J. Bourson, and J. Pouget, Ion recognition detected by changes in photoinduced charge or energy transfer, in: *Ref. 5*, pp. 25–44 (1993).
61. BioProbes 18, Molecular Probes, Inc., Eugene, Oregon, November 1993, p. 20.
62. T. Shimidzu, T. Iyoda, and K. Izaki, Photochemical properties of Bis(2,2′-bipyridine)(4,4″-dicarboxy-2,2′-bipyridine)ruthenium(II) chloride, *J. Phys. Chem. 89*, 642–645 (1985).
63. M. R. Eftink, Fluorescence quenching: Theory and applications, in: *Topics in Fluorescence Spectroscopy, Principles*, Vol. 2 (J. R. Lakowicz, ed.), pp. 53–126, Plenum Press, New York (1991).
64. H. Kautsky, Quenching of luminescence by oxygen, *Trans. Faraday Soc. 35*, 216–219 (1939).
65. W. R. Ware, Oxygen quenching of fluorescence in solution: An experimental study of diffusion process, *J. Phys. Chem. 66*, 455–458 (1962).
66. J. R. Lakowicz and G. Weber, Quenching of fluorescence by oxygen: A probe for structural fluctuation in macromolecules, *Biochemistry 12*, 4161–4170 (1973).
67. E. Urbano, H. Offenbacher, and O. S. Wolfbeis, Optical sensor for continuous determination of halides, *Anal. Chem. 56*, 427–429 (1984).
68. A. S. Verkman, M. C. Sellers, A. C. Chao, T. Leung, and R. Ketcham, Synthesis and characterization of improved chloride-sensitive fluorescent indicators for biological applications, *Analyt. Biochem. 178*, 355–361 (1989).
69. A. S. Verkman, Development and biological applications of chloride-sensitive fluorescent indicators, *Am. J. Physiol. 259*, C375–C388 (1990).
70. J. R. Lakowicz and D. Hogen, Chlorinated hydrocarbon-cell membrane interactions studies by the fluorescence quenching of carbazole-labeled phospholipids: Probe synthesis and characterization of the quenching methodology, *Chem. Phys. Lipids 26*, 1–40 (1980).
71. S. S. Lehrer, Solute perturbation of protein fluorescence. The quenching of other tryptophan fluorescence of model compounds and of lysosome by iodide ion, *Biochemistry 10*, 3254–3263 (1971).
72. M. H. Winkler, A fluorescence quenching technique for the investigation of the configurations of binding sites for small molecules, *Biochemistry 8*, 2586–2590 (1969).
73. A. R. Harrochs, A. Kearvell, K. Tickle, and F. Wilkinson, Mechanism of fluorescence quenching in solution II. Quenching by xenon and intersystem crossing efficiencies, *Trans. Faraday Soc. 62*, 3393–3399 (1966).
74. M. R. Eftink and C. A. Ghiron, Fluorescence quenching of indole and model micelle systems, *J. Phys. Chem. 80*, 486–493 (1976).
75. M. R. Eftink and C. A. Ghiron, Indole fluorescence quenching studies on proteins and model systems: Use of the inefficient quencher succinimide, *Biochemistry 23*, 3891–3899 (1984).
76. O. S. Wolfbeis and A. Sharma, Fiber-optic fluorosensor for sulphur dioxide, *Anal. Chim. Acta 208*, 53–58 (1988).
77. O. S. Wolfbeis, H. E. Posh, and H. K. Kroneis, Fiber optical fluorosensor for determination of halothane and/or oxygen, *Anal. Chem. 57*, 2556–2561 (1985).
78. J. N. Demas, E. W. Harris, and R. P. McBride, Energy transfer from luminescent transition metal complexes to oxygen, *J. Am. Chem. Soc. 99*, 3547–3551 (1977).
79. O. S. Wolfbeis, M. J. Leiner, and H. E. Posh, A new sensing material for optical oxygen measurement, with the indicator embedded in an aqueous phase, *Microchim. Acta*, Part III, 359–366 (1986).
80. J. N. Demas and B. A. deGraff, Interactions of transition-metal-complex photosensitizers with polymers and organized media, *J. Micromol. Sci. Chem. A25*(10&11), 1189–1214 (1988).

81. J. N. Demas and B. A. DeGraff, Applications of highly luminescent transition metal complexes in polymer systems, *Macromol. Chem. Macromol. Symp. 59*, 35–51 (1992).
82. R. Krapf, C. A. Berry, and A. S. Verkman, Estimation of intracellular chloride activity in isolated perfused rabbit proximal convoluted tubules using a fluorescent indicator, *Biophys. J. 53*, 955–962 (1988).
83. J. Biwersi and A. S. Verkman, Cell-permeable fluorescent indicator for cytosolic chloride, *Biochemistry 30*, 7879–7883 (1991).
84. J. Biwersi, N. Farah, Y-X. Wang, R. Ketcham, and A. S. Verkman, Synthesis of cell-impermeable Cl-sensitive fluorescent indicators with improved sensitivity and optical properties, *Am. J. Physiol. 262* (*Cell Physiol. 31*), C243–C250 (1992).
85. A. C. Chao, J. A. Dix, M. C. Sellers, and A. S. Verkman, Fluorescence measurement of chloride transport in monolayer cultured cells. Mechanisms of chloride transport in fibroblasts, *Biophys. J. 56*, 1071–1081 (1989).
86. I. Z. Steinberg, Long-range nonradiative transfer of electronic excitation energy in proteins and polypeptides, *Ann. Rev. Biochem. 40*, 83–114 (1971).
87. L. Stryer, Fluorescence energy transfer as a spectroscopic ruler, *Ann. Rev. Biochem. 47*, 819–846 (1978).
88. D. M. Jordan and D. R. Walt, Physiological pH fiber-optic chemical sensor based on energy transfer, *Anal. Chem. 59*, 437-439 (1987).
89. P. Yuan and D. R. Walt, pH-Dependent fluorescence of merocyanine-eosin-labelled water-soluble polymers, *Macromolecules 23*, 4611–4615 (1990).
90. J. R. Lakowicz, H. Szmacinski, and M. Karakelle, Optical sensing of pH and pCO_2 using phase-modulation fluorimetry and resonance energy transfer. *Anal. Chim. Acta* 272, 179–186 (1993).
91. D. Meadows and J. S. Schultz, Fiber-optic biosensors base on fluorescence energy transfer, *Talanta 35*, 145–150 (1988).
92. J. R. Lakowicz and B. Maliwal, Optical sensing of glucose using phase-modulation fluorimetry, *Anal. Chim. Acta 271*, 155–164 (1993).
93. J. R. Lakowicz, B. Maliwal, A. Ozinskas, and R. B. Thompson, Fluorescence lifetime energy-transfer immunoassay quantified by phase-modulation fluorometry. *Sensors and Actuators B 12*, 65–70 (1993).
94. A. O. Ozinskas, H. Malak, J. Joshi, H. Szmacinski, J. Britz, R. B. Thompson, P. A. Koen, and J. R. Lakowicz, Homogeneous model immunoassay of thyroxine by phase-modulation fluorescence spectroscopy, *Anal. Biochem. 213*, 264–270 (1993).
95. L. M. Christian and W. R. Seitz, An optical ionic-strength sensor based on polyelectrolyte association and fluorescence energy transfer, *Talanta 35*, 119–122 (1988).
96. P. S. Uster and R. E. Pagano, Resonance energy transfer microscopy: Observations of membrane-bound fluorescent probes in model membranes and in living cells, *J. Cell. Biol. 103*, 1221–1234 (1986).
97. U. Kubitscheck, M. Kircheis, R. Schweitzer-Stenner, W. Dreybrodt, T. M. Jovin, and J. Pecht, Fluorescence resonance energy transfer on single living cells. Application to binding of monovalent haptens to cell-bound immunoglobulin E, *Biophys. J. 60*, 307–318 (1991).
98. Z. Zhang, W. R. Seitz, and K. O'Connel, Amylase substrate based on fluorescence energy transfer, *Anal. Chim. Acta 236*, 251–256 (1988).
99. B. G. Pinsky, J. J. Ladaskay, J. R. Lakowicz, K. Berndt, and R. A. Hoffman, Phase-resolved fluorescence lifetime measurements for flow cytometry, *Cytometry 14*, 123–135 (1993).
100. X. F. Wang, T. Uchida, D. M. Coleman, and S. Minami, A two-dimensional fluorescence lifetime imaging system using a gated image intensifier, *Appl. Spectrosc. 45*, 360–366 (1991).
101. G. Marriot, R. M. Clegg, D. J. Arndt-Jovin, and T. M. Jovin, Time resolved imaging microscopy. Phosphorescence and delayed fluorescence imaging, *Biophys. J. 60*, 1374–1387 (1991).

102. R. M. Clegg, B. Feddersen, E. Gratton, and T. M. Jovin, Time-resolved imaging fluorescence microscopy, *SPIE Proc.* (J. R. Lakowicz, ed.), Vol. 1640, pp. 448–460, Billingham, Washington (1992).
103. S. Charbonneau, L. B. Allard, J. F. Young, G. Dyck, and B. J. Kyle, Two-dimensional time resolved imaging with 100-ps resolution using a resistive anode photomultiplier tube, *Rev. Sci. Instrum.* *63*(11), 5315–5319 (1992).
104. J. R. Lakowicz, H. Szmacinski, K. Nowaczyk, K. W. Berndt, and M. L. Johnson, Fluorescence lifetime imaging, *Anal. Biochem.* *202*, 316–330 (1992).
105. C. G. Morgan, A. C. Mitchel, and J. G. Murray, *In situ* fluorescence analysis using nanosecond decay time imaging, *Trends in Anal. Chem.* *11*, 32–41 (1992).
106. J. R. Lakowicz, H. Szmacinski, K. Nowaczyk, W. J. Lederer, M. S. Kirby, and M. L. Johnson, Fluorescence lifetime imaging of intracellular calcium in COS cells, *Cell Calcium* *15*, 7–27 (1994).

11

Fiber Optic Fluorescence Thermometry

K. T. V. Grattan and Z. Y. Zhang

11.1. Introduction

11.1.1. Fiber Optic Temperature Measurement

Temperature is one of the four or five most important parameters in industrial process control and in the chemical industry. Almost all chemical processes and reactions are temperature dependent, and not infrequently in the chemical plant temperature is the only indication of the progress of the process. Where the temperature is critical to the reaction, a considerable loss of product or efficiency may result from operation at incorrect temperatures. In some cases, loss of control of temperature can result in catastrophic plant failure with attendant damage and possible loss of life.

There are many other areas of industry where temperature measurement is essential. Such applications include steam raising and electricity generation, plastics and glass manufacture and moulding, manufacturing processes in metallurgical industries, milk and dairy products, and many other areas of the food industries. Further, in biomedical areas, the taking and the monitoring of patients' temperatures provide the basic diagnostic criterion, and it is essential during some hyperthermia treatments, for the safety of patients.

The importance of temperature measurement can even be seen simplistically by consideration of the financial aspects of the sensors and devices used worldwide. Estimates on the worldwide sales of temperature sensors run to several hundred million dollars per year,[(1)] a figure that could be increased several times when the associated controllers, indicators, and other aspects of the measurement system are added.

A number of environments can present significant difficulties for the determination of temperature. The region to be measured may be moving, extremely hostile, in a position where access is extremely difficult, or where the physical contacting of a

---

K. T. V. Grattan and Z. Y. Zhang • Department of Physics, Measurement and Instrumentation Centre, School of Electrical Engineering and Applied Physics, City University, London EC1 V0HB, United Kingdom.

Topics in Fluorescence Spectroscopy, Volume 4: Probe Design and Chemical Sensing, edited by Joseph R. Lakowicz. Plenum Press, New York, 1994.

sensing probe may even be impossible, where the presence of interference from other electromagnetic noise excludes the use of electronic thermometers. Examples of these situations are the measurement of the temperature in turbine-engines, the monitoring of winding temperatures in electrical transformers, and temperature monitoring during clinical radiofrequency (RF) heat treatment, and so on. To seek alternative means of temperature sensing, one of the most active research and development areas is in thermometry based on the use of fiber optic fluorescent techniques.

It is already over two decades since the first concepts of the use of fiber optic techniques for sensor purposes were discussed. The initial drive for the development of fiber optic sensors came from their potential use in military and aerospace applications where the cost factors of the introduction of new technology were less rigid and the working environment more hostile than is experienced with other areas of application.

The primary reason for interest in fiber optic sensors, in most cases, stems from the fundamental differences between the use of optical fiber and a metal wire for signal transmission.[(2)] These differences give fiber optic sensors the following advantageous characteristics.

1. *Electrical, magnetic and electromagnetic immunity.* This is the dominant attribute of the fiber optic sensor. The materials in the probe are typically good electrical insulators. Since they do not conduct electricity, the probes cannot (in principle) introduce electrical shorting paths or electrical safety problems. Likewise, they do not absorb significant amounts of electromagnetic radiation fields or be heated by such fields with the resultant introduction of thermal errors into the readings.

2. *Small sensor size.* Since the typical sensor does not have to be any larger in diameter than the fiber itself, the sensor can, in principle, be extremely small. This allows its use in such applications as medicine or in microelectronics, where size is critical.

3. *Safety.* The safety issue may be the main reason for the use of fiber optic sensors in some particular areas of application, especially in the chemical industry. Most fiber optic sensors require no electrical power at the sensor end of the system. They generate their own optical signal or they are "powered" remotely by radiation from a light source located within the instrument. Therefore they introduce no danger of electrical sparking in hazardous environments. There is reason to believe that at the normal levels of optical power coupled into fiber optics, i.e., at levels of up to several hundred milliwatts of optical power, there is almost no hazard with any accidental fracture of the cable and possible focusing of the optical radiation by the lensing effects of the broken end.[(2)]

4. *Capability of remote measurements.* The small size of the fiber and its electrical, chemical, and thermal inertness allow long-term location of the sensor deep inside complex equipment and thereby provide access to difficult locations where temperature may be of interest. Beyond this, however, certain of the optical techniques allow noncontact or remote sensing of temperature.

5. *Other advantages.* Beyond these generic advantages, optical techniques exhibit an unusually wide range of operation with precision good enough to meet many requirements. At the same time these techniques provide simplicity of calibration (or, as in the case of the fluorescence lifetime techniques, the absence of the need for calibration for individual probes).

11.1.2. Fiber Optic Sensor Devices for Temperature Measurement

There is perhaps most diversity in the techniques that are used for optically based temperature measurement, as a result of the fact that there are essentially as many ways of making a temperature measurement optically as there are temperature dependent optical properties. As the sensor can either be formed from the fiber itself (termed "intrinsic") or from a material or structure attached to the end of the fiber (termed "extrinsic"), the number of possible fiber optic temperature sensor devices is quite large. Indeed, there has been something of an explosion of device proposals recently seen in the literature. Extensive reviews have been given, e.g., by Grattan[3, 4] and Wickersheim[5] on actual and proposed sensor schemes. Based on the classification methods used by these authors, some typical sensor schemes, among those which have been considered seriously to date, are listed in Table 11.1,[6–23] according to the temperature-dependent properties used. Of these, fluorescence thermometry is among the most versatile and useful.

Table 11.1. Various Fiber Optic Temperature Sensor Schemes

Temperature-dependent property used	Sensor material or structure	Examples of work by
Optical reflection	Liquid crystal, thermochromic material	Rozzell *et al.*,[6] Brenci[7]
Optical scattering	Fiber itself	Dakin *et al.*[8, 9]
Birefringence	Lithium niobate	Knox *et al.*[10]
Fluorescence	Gallium arsenide, various phosphors	Sholes and Small,[11] Wickersheim and Alves,[12] Bosselman *et al.*,[13] Grattan and Palmer[14]
Optical interference	Fabry Perot etalon	Saaski *et al.*,[15, 16] Beheim[17]
Optical absorption	Gallium arsenide crystal, GaAs	Christensen,[18] Kyuma *et al.*[19]
Thermally generated radiation	Blackbody cavity	Dakin and Kahn[20]
Optomechanical schemes	Temperature-induced mechanical displacement	Jones and Spooncer[21]
Interferometric schemes	Temperature-induced interferometric measurement	Bosselmann and Ulrich,[22] Jackson[23]

Abbreviations: GaAs, gallium arsenide

11.2. Fluorescence-Based Fiber Optic Thermometry

11.2.1. Photoluminescence in Fiber Optic Thermometry

Photoluminescence is the phenomenon of light emission, in excess of thermal radiation, from a material which is excited by some form of electromagnetic radiation incident in the ultraviolet, visible, or infrared regions. Such emission is the release of the energy gained from the absorption of the incident photon energy. It may be either fluorescence or phosphorescence, or both, while the distinction between the latter two terms is often somewhat arbitrary.[24] In conformity with current usage in the application field discussed, the term "fluorescence" will be applied generally to all types of photoluminescence, and the corresponding luminescent materials will be termed fluorescent materials or "phosphors" thereafter.

The excitation spectrum of a fluorescent material, i.e., the incident radiation spectrum required for the induction of fluorescence, is determined by the absorption spectrum of the fluorescent material, which it often closely resembles, and by the efficiency with which the absorbed energy is transformed into fluorescence. Normally, the excitation spectrum is of higher photon energy (shorter wavelength) than that of the corresponding fluorescence emission, and in sensor schemes this has an effect in the choice of preferred fluorescent agent, compatible with appropriate optical detection devices.

The initial persistence of fluorescent emission following the removal of excitation depends on the lifetime of the excited state. This emission decays exponentially and the time-constant of such an exponential decay is used as the measure of the lifetime of the excited state, often termed the fluorescence lifetime or fluorescence decay time. In fluorescence thermometry, most materials used have relatively long fluorescence lifetimes ($> 10^{-6}$ sec). It means that fluorescence corresponds to weakly allowed transitions between electronic energy levels of the fluorescent material. The material emits light because that is one of the ways for the electron, once excited by incident radiation, to give up its energy and to return from the excited to the ground state. Any reasonably effective competitive relaxation process can shorten the lifetime of the excited state. A variety of such competitive processes, some radiative and some nonradiative, exist. Hence all luminescent materials can be expected to exhibit a temperature-dependent fluorescence lifetime to a greater or lesser degree, and a temperature-dependent fluorescence intensity over some temperature range. These are the thermal properties used in fluorescence thermometry.

11.2.2. Classes of Fluorescent Materials for Fluorescence Thermometry

Several classes of potentially useful fluorescent materials exist. Most lamp phosphors and many solid state materials are insulating compounds containing ionic "activators." The spectra of the rare-earth activators resemble, to a first approximation,

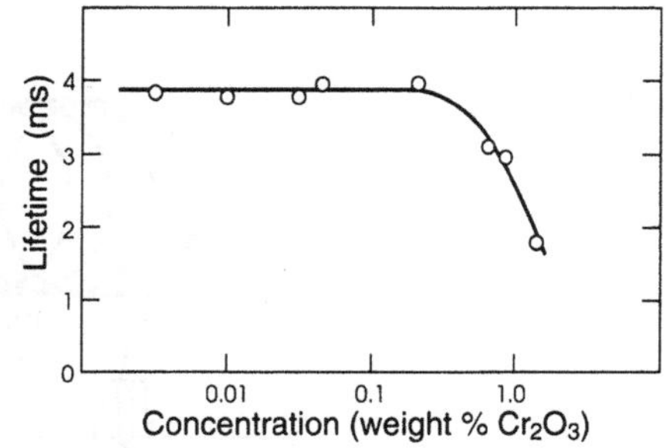

Figure 11.1. An example of concentration quenching. In ruby, the observed decay time of the fluorescence at 77K is independent of chromium concentration up to a concentration of around 0.3 weight% of Cr_2O_3 in the material.[25]

of those of free ions, while the electronic spectra of the transition-metal activators differ from those of free ions significantly due to the strong interaction with the crystal fields of host materials. Many of the early cathode ray tube (CRT) and television phosphors are semiconductors such as ZnS, CdS, or CdZnS, the emission from which corresponds approximately to the semiconductor band edge observed in absorption. There are also organic materials, both liquid and solid, that exhibit "molecular" fluorescence. Finally, there are self-activated materials and materials with charged defects (color centers) that also exhibit fluorescence.

The choice of sensor material determines range, sensitivity, and stability. By considering the latter factors, it is found that inorganic insulating compounds, such as most lamp phosphors and many solid state laser materials, are the most suitable materials for thermometric applications. Indeed, these materials are most commonly used in the existing commercial fluorescence thermometer schemes.

Finally it is worth mentioning the relation between fluorescence properties and the concentration of activator content. The fluorescence intensity increases initially with increases in the concentration of activator content. However, the increase in the concentration above a certain critical value can lead to a reduction in fluorescence intensity. This is called concentration quenching. It can also be observed from the reduction in the lifetime with the increase in the concentration, as shown in Figure 11.1 in the case of ruby, where the activator is the Cr^{3+} ion. Therefore, it is advantageous to select an activator concentration short of the critical value to achieve a good level of fluorescence intensity in a practical sensor system.

11.2.3. Early Fluorescence Thermometer Schemes

A variety of fiber optic thermometry systems using fluorescence sensors have been discussed or become available over the past years. Most of the earliest systems are based on the temperature-dependent fluorescence intensity of appropriate materials. One such example of an early commercial system is the Luxtron model 1000, shown in Figure 11.2, which utilized europium-activated lanthanum and gadolinium

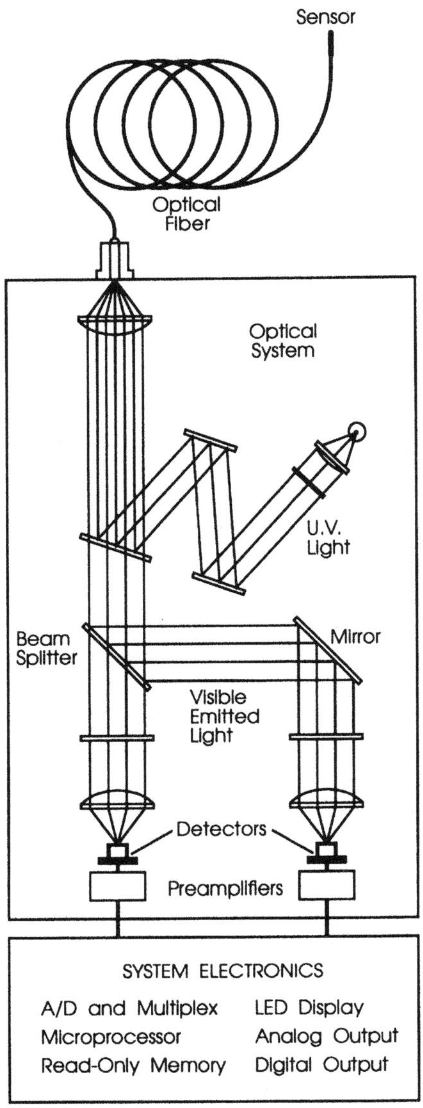

Figure 11.2. Schematic representation of Luxtron model 1000.[12]

oxysulphide as alternate sensor materials.[12] The fluorescence from these rare-earth phosphors consists of sharp lines originating from different excited states of the trivalent europium ion. Since the relative populations of these excited states after excitation are strong functions of temperature, the relative intensities of the emission lines are also quite temperature dependent. By measuring the intensities of two lines originating on different excited states, the system derives the temperature from the ratio of the line intensities.

Another example of an early system is the ASEA model 1010, shown in Figure 11.3, introduced by ASEA AG, a large Swedish automation and energy systems company. In this system, the sensor is a small crystal of gallium arsenide sandwiched between gallium aluminum arsenide layers.[26] The sensor is caused to luminesce in the vicinity of its band absorption edge by radiation from a gallium arsenide light emitting diode (LED). As the temperature of the sensor is raised, its emission broadens and shifts to progressively longer wavelengths. Portions of the luminescence are transmitted through two optical filters with adjacent pass bands, the intensity in each pass band is measured, and a ratio is constructed which can then be correlated with temperature.

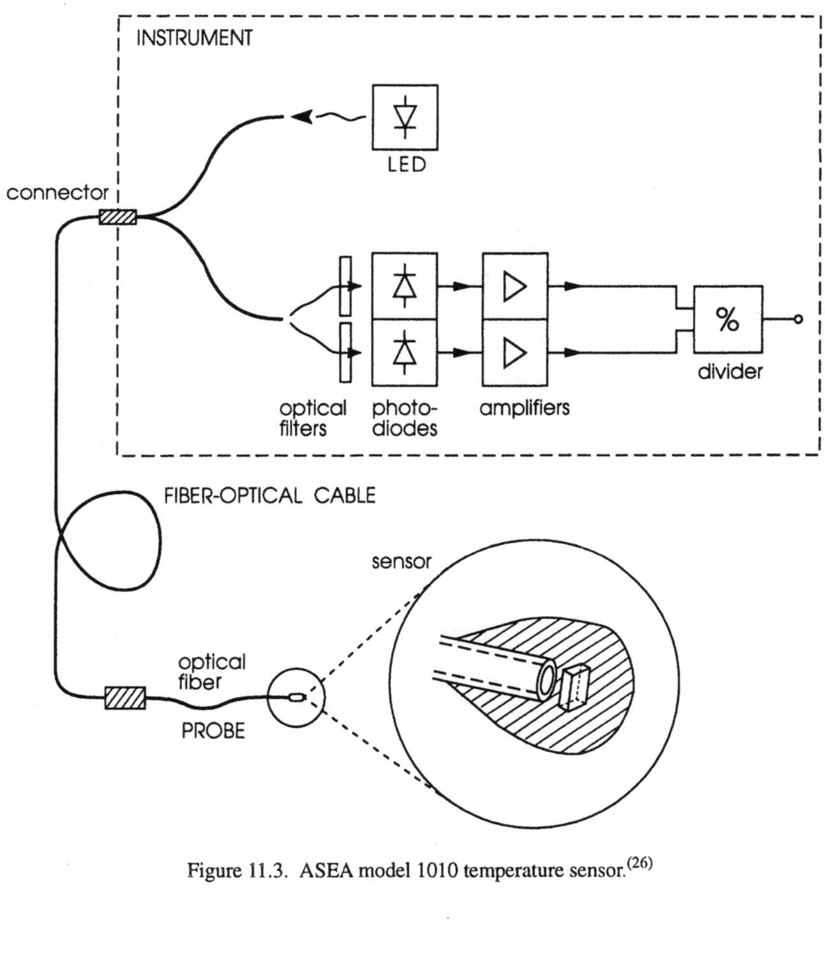

Figure 11.3. ASEA model 1010 temperature sensor.[26]

Although a substantial number of intensity-based systems, such as the Luxtron model 1000, the ASEA model 1010 and the like, were built, it was seen that the technique had limitations in terms of performance and cost.[27] These limitations stem from the need of an additional "referencing channel," i.e., the luminescence intensity at another wavelength, for the separation of the effects of thermally induced changes in intensity from other nonthermal sources of signal change, such as fiber bending, light source, detector degradation, and electronics problems. As a result, the technique based on the measurement of fluorescence lifetimes was developed[11, 13, 14] and has been much preferred in the development of fluorescence-based commercial systems to date.

11.2.4. Fluorescence Lifetime-Based Schemes

In such a category of systems, the temperature-dependent lifetime of fluorescence in an appropriate material is utilized. There are a variety of fluorescent materials potentially useful for temperature sensing, as discussed earlier, and those materials which have been used were often selected with a relatively longer lifetime ($> 10^{-6}$ sec) so that no special high-speed electronic components are needed and the design of lifetime detection electronics could be simple and thus low cost. Since the temperature of an appropriate fluorescent material can be determined from the measurement of a single intrinsic parameter (its fluorescence lifetime), the system in this category is to a first approximation calibration-free. With such system, virtually any probe configuration can be used interchangeably in principle, since the measurement of the fluorescence lifetime does not depend on the exact signal level or on the particular optical configuration.

In addition to choosing the most appropriate fluorescent material to meet the requirements of the desired applications, the achievement of an effective, simple, and economic electronics scheme for the detection of the fluorescence lifetime is also critical in the development of an acceptable sensor system. Quite a few signal processing techniques for the detection of the fluorescence lifetime have been used in fluorescence-based thermometry systems. Indeed, all these systems can be categorized according to the signal processing techniques used, in the following ways.

11.2.5. Pulse Measurement of Fluorescence Lifetime

The common feature of the schemes in this category is that the excitation light applied to the fluorescent material is a high intensity "delta" function pulse (e.g., a laser pulse or that from a flash lamp) or an rectangular pulse, and the measurement is derived from the observation of the fluorescence decay after the removal of the excitation light. The following are the outlines of some typical schemes which are of this type and have been used in thermometry applications.

11.2.5.1. Two-Point Time Constant Measurement

This is a very straightforward method which was used by several groups of workers in the early stage of the development of a fluorescence-based sensor system.[14, 27, 28] The fundamental principle of this technique can be clearly illustrated in Figure 11.4. The principle of this method is to compare the intensity level at two points along the exponential decay curve after the excitation pulse has terminated. The circuitry is designed to measure first the value of decaying signal, I_0, that occurs at a fixed time, t_1, after the termination of the excitation pulse. A second voltage I_0/e is then calculated and established as a reference level. When the decaying signal falls to that level, the time t_2 at which crossover occurs is noted. The interval between t_2 and t_1 is the time constant, τ, of the exponentially decaying signal. In most cases, the fluorescence decay process observes a single or "quasisingle" exponential law, as that shown in Figure 11.4. Thus the time constant, τ, is used as the measure of fluorescence lifetime.

Timing circuits are provided for measuring precisely the time between t_2 and t_1. That time difference can be correlated directly with the temperature of the fluorescent sensor by reference to an empirically determined "look up" table stored digitally within the instrument.

This type of method is simple and inexpensive in relation to the electronic components used. Since the fluorescence signal is measured after the excitation pulse is over, the detector optics do not have to be designed to discriminate strongly against stray signals from the excitation source. However, a significant disadvantage of this type of system is that the signal is only measured at two times and, as a result, precision is greatly limited.

This two-point measurement technique is used in the system reported by Wickersheim and Sun,[27] where a lamp phosphor, tetravalent manganese-activated magnesium fluorogermanate, mentioned above, is used as fluorescent sensor. The excitation

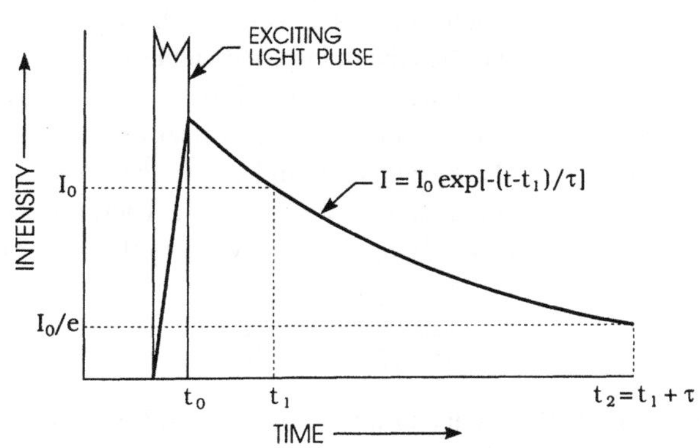

Figure 11.4. Two-point time constant measurement technique.[27]

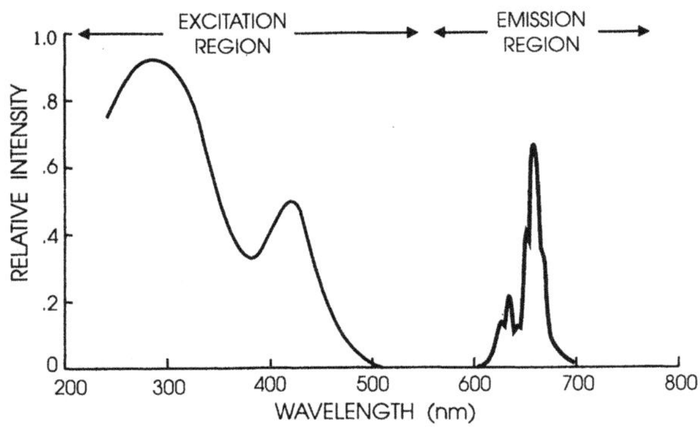

Figure 11.5. Excitation and emission spectra of magnesium fluorogermanate activated with tetravalent manganese.

and emission spectra of this sensor material are shown in Figure 11.5. With a xenon flash lamp used as the excitation source, the observed fluorescence lifetime ranges from approximately 0.5 msec at 450°C to more than 5 msec near liquid nitrogen temperature. It is claimed that an accuracy of ±2°C over the whole range can be achieved without calibration of the instrument. With a single point calibration, ±0.2°C accuracy is reported.

11.2.5.2. Integration Method

To achieve higher precision from the pulse measurement, several techniques have been developed which are based on the integration of the decaying fluorescence signal over different periods of time. One example is the signal processing scheme used by Sholes and Small[11] in their nonfiber study of ruby fluorescent decay. As illustrated schematically in Figure 11.6, when the decaying fluorescence falls below a preset level, the start of the measurement process is trigged. The signal is integrated at two fixed delay times, T_1 and T_2, and then the integration values over these periods of time, A and B are sampled. After the signal has decayed to zero, the integrator is reset and restarted. Integrated noise and direct current (DC) offset levels are then sampled for the same two fixed delays, and are given by C and D, which are equivalent to the noise and DC offset components in A and B, respectively. Therefore, the fluorescence lifetime, τ, could be obtained by solving the following implicit relation:

$$\frac{A-C}{B-D}=\frac{1-e^{-T_1/\tau}}{1-e^{-T_2/\tau}} \tag{11.1}$$

Another example is the balanced integration method described by Sun.[29] This technique is designed to achieve 0.01°C resolution using tetravalent manganese-activated magnesium fluorogermanate, the same sensor material that has been used

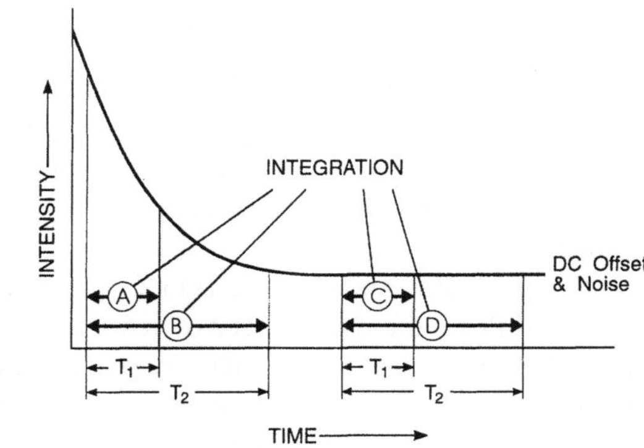

Figure 11.6. Schematic illustration of the integration technique used by Sholes and Small (1980).[11]

with the two-point measurement by Wickersheim and Sun,[27] as mentioned above. The system utilizes a dual-slope integrator to balance the integrals of two sequential areas under the decay curve of the phosphor, as shown in Figure 11.7. The first integration is carried out over a preselected fixed time interval. Integration starts at a predetermined time, t_1, after the excitation pulse has terminated. On completion of this initial integration, i.e., at t_2, polarity is reversed and negative integration (deintegration) begins. This continues until the second integrated area exactly equals the first and the combined integral equals zero. The time of the zero crossing, $t_3 - t_2$, is measured with high resolution (to <20 nsec). The observed value of t_3 can be related to the decay time of the phosphor and hence to its temperature.

The result of this balanced integration system is very sensitive to any background signal, whether optical or electronic. Hence alternating current (AC) noise must be minimized and any DC offsets must be measured and corrected, e.g., in some way similar to that used by Sholes and Small[11] and illustrated in Figure 11.6. It should also be pointed out that the dynamic range of this system is fairly narrow (~100°C for a given choice of the initial integration period) relative to that of the less precise two-point system.[29]

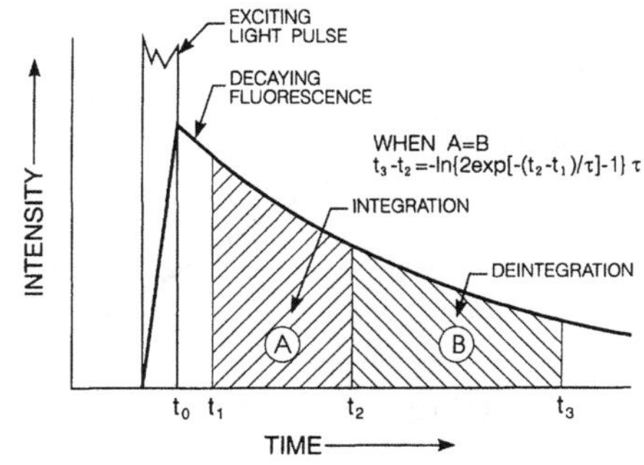

Figure 11.7. Illustration of the balanced integration method.

11.2.5.3. Digital Curve Fit Method

Recently Luxtron has introduced a modular system, WTS-11, designed to monitor the winding temperature in power transformers,[30] utilizing a technique that takes advantage of the newest, commercially available, high-speed digital signal processors (DSPs), where it has been described in detail by Sun[29] and is termed the "digital curve fit method."

In this system, an un-named fluorescent material is used, the excitation spectrum of which allows the use of a convenient excitation source, e.g., a red LED or laser diode. When the sensor is excited by sequential light pulses from an LED or laser diode, a repetitive decaying luminescent signal results. A selected portion of each decay curve is digitized after the detected signal has passed through a low-noise, wide-bandwidth amplifier. The digital samples, after correction for any offset, shown in Figure 11.8, are then processed by the DSP to provide the best exponential by means of a least-squares curve fitting technique. The exponential is first converted to a straight line by taking the natural logarithm of the digitized signal values. The slope of the best-fit straight line is proportional to the decay time of the luminescence.

The result of a number of curve fits are averaged to further reduce the effects of noise. The averaged decay time is then compared with values stored in a digital look-up table to determine the temperature of the sensor.

The use of this technique with a DSP has resulted in a rather compact instrument. It was reported that the development of this technique was aimed to eliminate dependence on many of the drift-prone components and should thereby improve the system stability.[29] However, no assessment of it in respect of these aspects was given in the literature.

However, an investigation into the errors caused by baseline offset and noise in the least-squares estimation of exponential lifetimes has already been carried out by

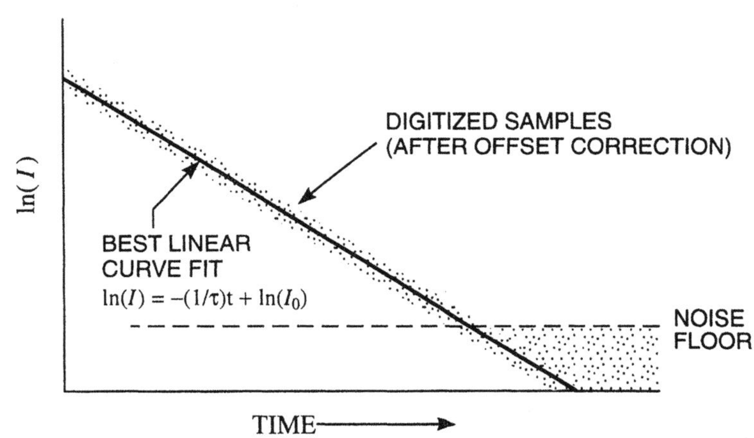

Figure 11.8. Illustration of linearized digital curve fit technique.

Dowell and Gillies.[31] It shows that the two-parameter estimation, which only estimates I_0 and τ and assumes a zero baseline (called DC) offset such as that illustrated in Figure 11.8, is rather sensitive to a small, residual baseline offset. The normalized lifetime-estimation error is proportional to the normalized offsets. The ratio is greater than 3 when the normalized observation time $\beta > 3$, where $\beta = T/\tau$; T is the actual time of observing the signal and τ the lifetime. Therefore, the performance of the digital-curve-fit technique will greatly rely on the effectiveness of the measure utilized to correct the DC offset in the observed signal. The three-parameter least-squares estimation, i.e., one including the estimation of the DC offset, might be a better solution. However, it requires a reiterative numerical algorithm, and this will not only increase the complexity of programming, but also greatly prolong the associated signal processing time.

11.2.6. Phase and Modulation Measurement

A schematic representation of this category of techniques is depicted in Figure 11.9. The intensity of the excitation light is sinusoidally modulated so that the fluorescence response from the sensor material is forced to follow the same sinusoidal law, but lagging behind the excitation light by a phase shift φ, which is expressed as

$$\tan\varphi = 2\pi f\tau \tag{11.2}$$

Thus, the fluorescence lifetime τ can be derived from a measurement of φ.

This technique is of high accuracy and is meant to be used in precision measurement instrumentation, for it is inherently insensitive to the DC-offset and the AC-noise in the sinusoidal signal which can be substantially reduced by a great variety of electronic devices ranging from various electronic analogue filters, and digital filters to the most effective lock-in amplifiers.

At the early stages of development, the lack of a convenient and economic excitation modulation scheme has limited the use of such a phase shift technique in fluorescence thermometry. Now with the wide availability of cheap and easily modu-

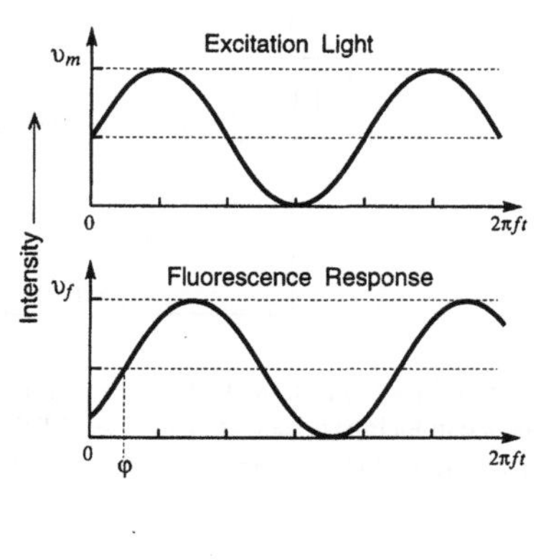

Figure 11.9. Phase and modulation measurement of fluorescence lifetime. $\tan\phi = 2\pi f\tau$; ϕ = phase difference.

lated high-power LEDs or laser diodes, this technique has found its application in the thermometry area. Grattan *et al.*[32] have demonstrated the use of this technique in a ruby fluorescence-decay thermometer, where a high-power green LED was used as the excitation light source. Over the range from room temperature to 170°C, a quite satisfactory measurement deviation was achieved under a poor signal-to-noise ratio condition. A similar system was also reported by Augousti *et al.*,[33] which used alexandrite as the sensor material and was capable of an accuracy of ±1°C over a range 20–150°C, with a response time of 1 sec.

However, since the kind of system requires the measurement of fluorescence during the excitation period, it is highly sensitive to the excitation light leakage to the detector and the optics required to prevent such leakage can make the system very costly.

11.2.7. Phase-Locked Detection of Fluorescence Lifetime

This is a new category of signal processing approach, which has been developed for the achievement of a simple, inexpensive and versatile electronic scheme for the detection of fluorescence lifetime, and has been successfully applied to several fiber-optic thermometer schemes. Though as illustrated schematically in Figure 11.10, these schemes stem from the phase and modulation techniques, but they differ from them in several important aspects. In general, the excitation light source is modulated by a signal from a voltage-controlled oscillator (VCO), either a sine-wave or square wave. The fluorescence response signal from the sensor follows the variation of excitation light in intensity, but lags behind it by a phase shift, φ, which corresponds to the fluorescence decay time as expressed in Eq. (11.2). This signal is sent to a lock-in amplifier to mix with a reference signal derived from the VCO output with a phase lag of a fixed fraction, α, of the period. Then the resulting mixed signal is filtered by a

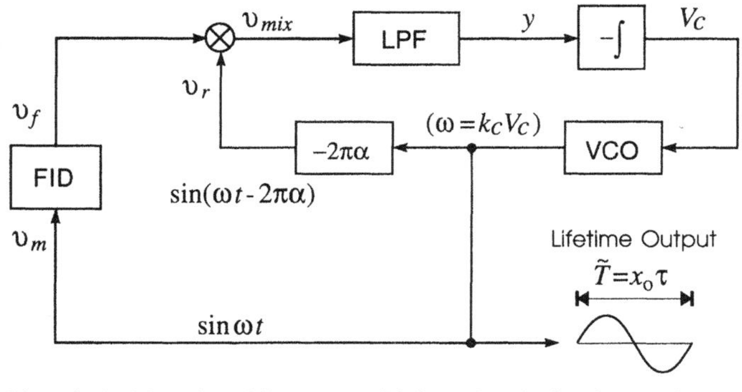

Figure 11.10. Phase-locked detection of fluorescence lifetime using single reference signal. FID = fluorescence inducing and detecting devices; LPF = low-pass electronic filter; VCO = voltage-controlled oscillator; v_m = signal to modulate the output intensity of the excitation light source; v_f = the fluorescence signal.

low-pass filter (LPF) and further integrated. The resultant voltage signal is fed back to control the output frequency of the VCO. The VCO will finally be driven to operate at a frequency at which the integration of the mixed signal is zero. The period of this frequency is directly proportional to the lifetime, and thus the measurement. This allows a high resolution to be achieved over a wide range of lifetimes. It has been found in such systems that while the frequency varies with the fluorescence lifetime, the phase shift between the excitation and fluorescence signals is always kept at a constant value, determined by the reference delay ratio, α. Therefore this category of techniques has been termed the "phase-locked detection" by Zhang *et al.*[34]

11.2.7.1. Simple Oscillator Method

The earliest scheme which is closely related to the *phase-locked detection* technique is that described by Bosselmann.[35] In this scheme, a positive feedback is formed by sending the fluorescence response signal back to control the output intensity of the excitation light source, via a timing circuit which amplifies the signal with an added phase shift, so that the system will oscillate at a frequency determined by the combination of the time-constant of the timing circuit and the fluorescence lifetime. The measurement of the lifetime is derived from the oscillating frequency, which is subject to the parameter drifts of the timing circuit. The sensitivity of the frequency to the lifetime is also limited by the fact that is given by the combination of the lifetime and the time-constant of the timing circuit.

11.2.7.2. Phase-Locked Detection Using a Single Reference Signal

A revision of the simple oscillator method was reported later by Bosselmann *et al.*[13] which gives a primitive form of the phase-locked detection technique, and has been classified as the phase-locked detection using a single reference signal by Zhang *et al.*[34] The operation mechanism of this scheme is similar to that depicted in Figure 11.10 except that a VCO of square-wave type is used, i.e., the excitation light is square-wave modulated. The sensor material used in the thermometer system is a chromium-activated rare earth aluminum borate. In this system, the standard deviation of consecutive measurement is 0.04°C over the temperature region from 0 to 70°C, and no detectable long-term drift is observed.

The conclusion that the measurement sensitivity of using square-wave modulation is greater by a factor of 2 than that of using sinusoidal modulation was made by Bosselmann *et al.*[13] from a comparison between this primitive phase-locked detection scheme and the simple oscillator scheme mentioned before. In fact, the fundamental difference between these two techniques lies in their configurations instead of the modulation modes used. In the simple oscillator system, the fluorescent material is directly used as a timing component in a oscillator by means of optoelectronic coupling, and hence the oscillating frequency is determined by the combination of the fluorescence lifetime and the time-constant of an additional electronic timing circuit, which limits the sensitivity of the oscillating frequency to the lifetime. However, in the phase-locked detection system, the oscillation is generated by a voltage-controlled oscillator with the oscillating frequency only controlled by the fluorescence lifetime, and thereby the sensitivity of the frequency to the lifetime is higher than that in the former case.

Perhaps the first detailed discussion of such a technique in fluorescent thermometry (shown in Figure 11.10) was given by Zhang *et al.* in their work[36] based on both mathematical analysis and experimental simulation. Examples of the electronic design of the corresponding system and the application of the technique in a ruby fluorescence-based fiber-optic sensor system are also listed. This shows that there is no difference in the measurement sensitivity between a system using square-wave modulation and one using sinusoidal modulation. However, the former performs a little better in terms of the measurement resolution.

As demonstrated by the work of several groups,[13, 36, 37] this system is highly suited to high-precision, wide-range applications. However, like the phase and modulation measurement schemes, since the system requires the measurement of fluorescence during the excitation period, even a tiny fraction of excitation light leaking into the photodetector will seriously degrade the performance of the system. An excess of such leakage can even cause the operation of the system to fail.[34] Therefore, high-quality filter optics might be needed to prevent such leakage and this could make the system costly and bulky.

11.2.7.3. Phase-Locked Detection Using Two Reference Signals

The problem of excitation light leakage in a high-temperature sensor scheme based on the fluorescence lifetime of crystalline alexandrite[38] has led to the introduction of a new version of the phase-locked detection technique, termed the phase-locked detection using two reference signals by Zhang *et al.*[34] In that work,[38] two reference signals derived from a VCO output are used to mix successively with the fluorescence signal, so as to delete the excitation leakage component from the final integration of the mixed signal which is used to control the oscillating frequency of the VCO. It has been proved theoretically and experimentally that such a technique is intrinsically immune to the impact of excitation light leakage. Therefore, in the phase-locked detection system, it is no longer necessary to use high cost components in the optical arrangements.

In the system which uses crystalline alexandrite as the sensor material,[38] a measurement reproducibility of ±1 °C is achieved over a wide temperature region from 20 to 700°C. The same technique is applied to another fiber optic thermometer system which is designed for biomedical sensing applications and uses $LiSrAlF_6{:}Cr^{3+}$ as sensor material.[39] The standard deviation of the measurement recorded by this system is better than 0.01°C within the 20°C and 50°C region.

The easing of the requirement for high-quality optics by using this two-reference technique, and employing those characteristics which are essentially inherent to the phase-locked detection techniques, such as high precision, wide measurement range, and essential simplicity, makes this category of technology much to be preferred in the development of cost-effective fluorescence-based fiber-optic thermometers.

11.3. Solid State Materials for Fluorescence Thermometry

11.3.1. Cr^{3+}-Based Material

The transition metal ions, Cr^{3+}, are widely used as active dopants in solid state laser crystals. Unlike rare-earth ions such as Nd^{3+}, Cr^{3+} ions in ionic crystals interact strongly with the crystal-field strength and the lattice vibrations. Thus Cr^{3+}-activated materials are characterized by a wide optical absorption spectrum spanning the ultraviolet (UV) to the red portion of the visible spectrum. This allows the use of cheaper and compact light sources for excitation, such as high-power LEDs and visible laser diodes, highly suitable for practical sensor applications. Again due to the strong crystal-field interaction, the energy gaps of the electronic levels of Cr^{3+} ions vary from one host crystal to another, as do the temperature dependences of the fluorescence lifetimes of Cr^{3+}-doped materials. Therefore, significant variety in such temperature-dependent phenomena is observed and a useful level of control of this temperature dependence is made possible by systematic changes in the crystal field strength, through the variation of the host material or the crystal composition, to cater for differing thermometric needs over various temperature regions.

11.3.1.1. Theoretical Background

The electronic energy levels of the Cr^{3+} ion are well illustrated by the Tanabe–Sugano diagram[40] which plots E/B, the normalized energy of the low-lying excited states as a function of the normalized octahedral crystal-field strength Dq/B. A simplified Tanabe–Sugano diagram is presented in Figure 11.11, for the Cr^{3+} ion in an octahedral crystal field. The ground state is always the orbital singlet, $^4A_2$, irrespective of the strength of the octahedral crystal field. The energy splitting ΔE between the low-lying states, $^4T_2$ and ^2E is defined as:

$$\Delta E = E(^4T_2) - E(^2E) \tag{11.3}$$

It varies strongly with Dq/B and may be positive or negative, as shown in Figure 11.11. The dashed lines indicate the energy level positions of a number of different crystals. In a high-strength crystal field, $Dq/B >> 2.3$ as in ruby (Al_2O_3:Cr^{3+}) and alexandrite[41]; $\Delta E > 0$ (2350 cm^{-1} and ~800 cm^{-1} for ruby[42] and alexandrite,[41] respectively), and the Cr^{3+} emission is dominated by the sharp R lines ($^2E \rightarrow {}^4A_2$ transition). However, by contrast in a low-strength crystal field, $Dq/B << 2.3$, as in LiSAF ($LiSrAlF_6$); $\Delta E < 0$ and the dominant emission is the broad $^4T_2 \rightarrow {}^4A_2$ band transition. The intrinsic

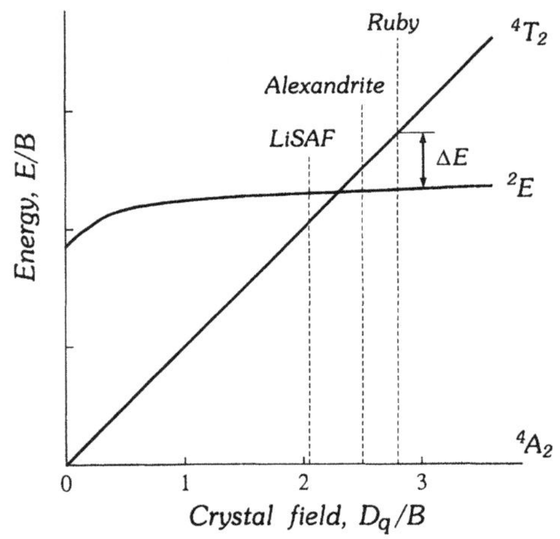

Figure 11.11. Simplified Tanabe–Sugono diagram. The dashed line represents the approximate energy level positions of the Cr^{3+} ion in various host crystals.

lifetime of the Cr^{3+} ions of $^4T_2$ is much shorter than that of the $^4A_2$ state, e.g., 6.6 μsec and 1.54 msec for $^4T_2$ and $^4A_2$, respectively, in alexandrite,(41) and so the magnitude of Dq/B also gives the temperature dependence of the Cr^{3+} fluorescence lifetimes, showing different characteristics. For instance, in the case of $Dq/B >> 2.3$, such a temperature dependence is mainly due to the thermally activated repopulation between the ^2E and $^4T_2$ states, over a wide temperature region (e.g., 300–600K for ruby and 150–700K for alexandrite). In the case of $Dq/B << 2.3$, the ^2E state has little to do with this temperature dependence which is mainly determined by the nonradiative process of the $^4T_2 \rightarrow {}^4A_2$ transition or the so-called thermal quenching of fluorescence.

Therefore, the temperature dependence of the fluorescence lifetimes of Cr^{3+} doped insulating crystals, which have been used as temperature sensing elements or have such potential uses for different temperature regions, are reviewed in terms of the influence of the crystal field strength in light of appropriate choice of sensor element. Modeling work on the temperature dependences of Cr^{3+} fluorescence lifetimes has been carried out by Zhang *et al.*,(43) and the empirical formulas resulting from the above work may thus applied to fit the experimental data of fluorescence lifetimes of Cr^{3+} in different host crystals, showing their promising potential to be used in the assessment, and ultimately for calibration, of thermometers based on such materials.

11.3.1.2. Influence of Crystal Field Strength on Cr^{3+}-Doped Materials

The temperature dependence of the Cr^{3+} fluorescence is strongly characterized by the crystal field strength. Further, the method of the variation of the fluorescence lifetime with temperature can indicate whether the field strength of the host material is high or low. In a host with a low crystal field, the decrease in the Cr^{3+} fluorescence lifetime with increasing temperature changes rapidly only over one continuous temperature region, as is observed in the cases of LiSAF, LiBAF, and LiCAF and shown

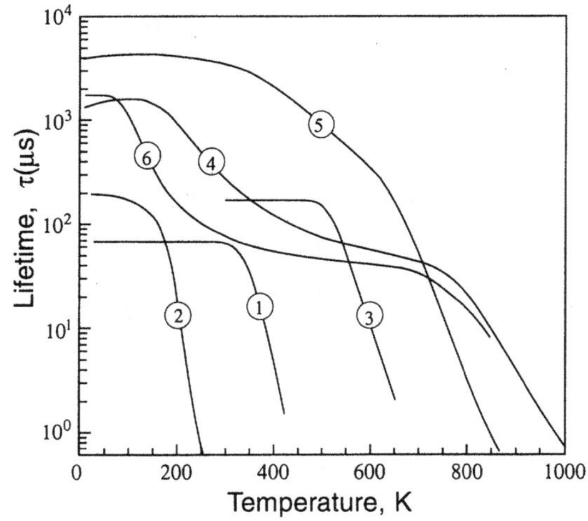

Figure 11.12. Temperature dependences of Cr^{3+} fluorescence lifetime in various host materials. Materials with low strength of crystal field: (1) LiSAF; (2) LiBAF[44]; (3) LiCAF.[44] Materials with high strength of crystal field: (4) alexandrite[41, 45]; (5) ruby (data from ~20 to 300K obtained from Nelson and Sturge,[46] those from 300K and beyond by the author); (6) emerald.[42]

in Figure 11.12. However, in a high-field host, two such regions are found. For instance, in alexandrite, such regions range from ~150 to ~300K and from ~700K and beyond; in ruby, from ~300 to ~500K and from ~600K and beyond; and in emerald, from ~75 to ~600K and from ~650K and beyond as shown in Figure 11.12.

Two configurational coordinate models, presented in Figure 11.13 and 11.14, are sufficient to allow the interpretation of the temperature dependences of the Cr^{3+} fluorescence in crystal materials qualitatively, even quantitatively to an extent of sufficient precision for thermometric applications, as shown in the cases of Cr:LiSAF, alexandrite, and ruby. In high-field-strength host crystals, two mechanisms, the thermal repopulation of the $^4T_2$ and ^2E states and the nonradiative process, dominate the

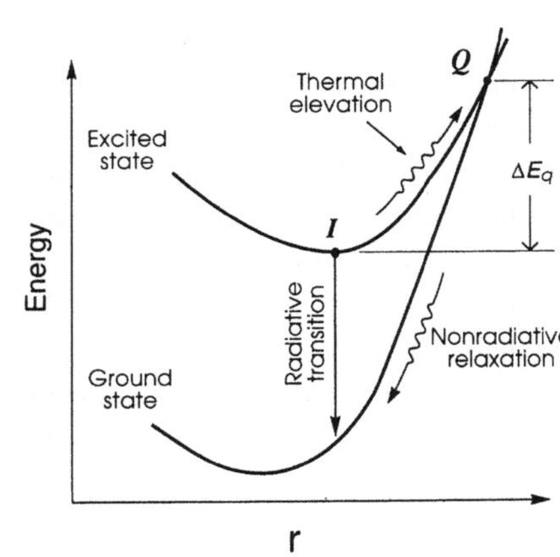

Figure 11.13. The single configurational coordinate model for Cr^{3+} fluorescence in low-field crystals. r = configurational coordinate.

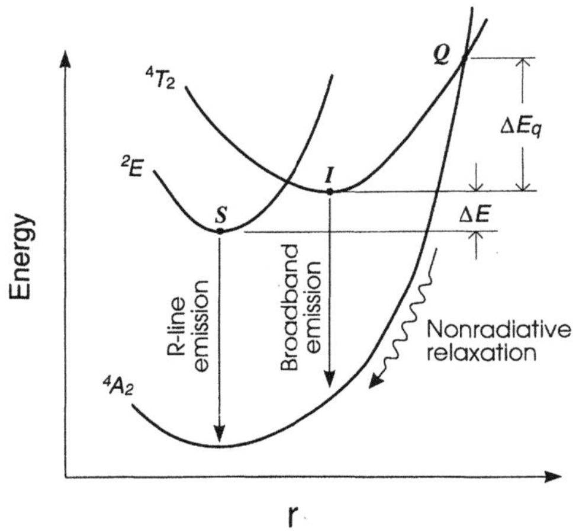

Figure 11.14. The single configurational coordinate model for Cr^{3+} fluorescence in high-field crystals. r = the configurational coordinate.

temperature dependence, alternately over different temperature regions. Thus the fluorescence lifetimes are quite sensitive to temperature variances over a wide temperature region, and are suitable for wide-range thermometric uses. In addition, the total fluorescence efficiency does not decrease with increasing temperature over the region where the thermal repopulation of the $^4T_2$ and ^2E states dominates, and an increase in absorption of the excitation light is observed, as in alexandrite.(45) These features are of significant advantage in achieving an acceptable signal-to-noise ratio in the corresponding optical thermometers. In low-field crystals, only the nonradiative process dominates the temperature dependence, and the fluorescence intensity decreases with temperature increase, as with the lifetime. Therefore the measurement range of the corresponding lifetime-based thermometer is limited, although high measurement resolution could be achieved.

By contrast, in Cr^{3+}-doped low-field crystals, partly due to the lower degree of overlap between the absorption and emission spectra, compared to that in high-field crystals, the self-trapping of fluorescence is rather weak. In some low-field crystals, the dependence of the fluorescence lifetime on the Cr^{3+} concentration is weak even up to very high concentrations, e.g., from 2 to 15% for Cr:LiSAF(47) and from 1 to 10% for Cr:LiBAF.(44) These properties indicate that small changes in the Cr^{3+} concentration of the samples used in the thermometers, the size, and the positioning of the sample materials in the formation of the temperature probes will have no significant effect on the measurements. An important result is that the measurement reproducibility and the exchangeability of the temperature probes made of low-field sensing materials could be better than those in the case of using high-field materials, especially when probes are made by using the sensing materials that are of the same kind but are from different production batches.

11.3.2. Optical Arrangement of Fluorescence Lifetime Thermometers

The optical arrangement of a typical fluorescence lifetime thermometer is one of the simplest, among those of various fiber optic sensor schemes discussed earlier. Examples of typical optical arrangements, used in several Cr^{3+}-based fiber optic thermometer schemes, are illustrated in Figure 11.15. In Figure 11.15a the optical configuration which has been employed by Bosselmann *et al.*[48] in a scheme based on Lu $(Cr_xAl_{1-x})_3BO_3)_4$, a chromium-activated material is shown. The arrangement used by Grattan *et al.*[49] in a ruby fluorescence-based thermometer is depicted in Figure 11.15b, which was aimed to be simple and of low cost, at the expense of a relatively low signal-to-noise ratio. That shown in Figure 11.15c is the apparatus employed by Zhang *et al.* in a Cr:LiSAF-based thermometer for biomedical applications.[50] As seen in these examples, the optical arrangement of a typical Cr^{3+} fluorescence lifetime-based thermometer consists, in general, of the following three parts.

1. An excitation light source with an appropriate emission spectrum to induce a significant fluorescence response from the sensing material, and with some accessories to facilitate the modulation of the output light intensity. Although it is possible to supply high-energy excitation over the absorption band of the material with the use of UV light sources, such as the flash lamp used by Wickersheim and Alves,[12] it is much

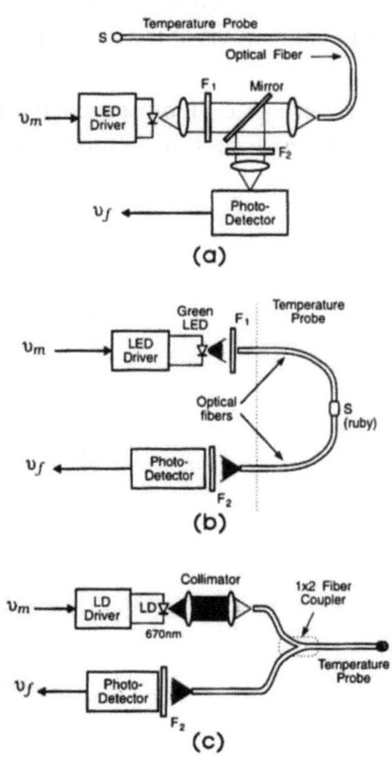

Figure 11.15. Schematics of the optical arrangement and temperature probes for the Cr^{3+} fluorescence lifetime-based fiber optic thermometers. F_1 = short-pass optical filter; F_2 = bandpass or long-pass optical filter; LD = laser diode; LED = light emitting diode; S = the fluorescence material used as sensing element; ν_m = signal to modulate the output intensity of the excitation light source; ν_f = the detected fluorescence response from the sensing element.

simpler to use a visible laser diode or LED, or a helium–neon (HeNe) laser, to pump into the lowest energy absorption band which exists in almost all of Cr^{3+}-activated fluorescent materials, from the green to the red portion of visible spectrum. LED sources can be electronically modulated and now offer nearly as high power output as many HeNe lasers which must, however, be acousto-optically or electro-optically modulated or mechanically "chopped." With both the advantages of LED and HeNe laser sources, laser diodes are the most promising excitation light sources, and they are the ideal light sources for the excitation of Cr^{3+} fluorescence in a sensor system. Rapid advances in laser diode technology and a steady reduction in its price has been seen in recent years, coupled with their wider use in fiber optic sensor systems;

2. A photodetector with a suitable spectral response for the detection of the fluorescence emission. As mentioned above, the fluorescence of most Cr^{3+}-activated materials consists of either the red *R*-lines or a broadband infrared emission with the peak wavelengths ranging from ~750 to 1000 nm, or both emissions. Thus, the widely available, compact and inexpensive silicon PIN photodiodes are the most suitable photodetectors for these applications. In some particular circumstances, where high speed and high sensitivity of detection are required, a silicon-based avalanche diode could be the best alternative to the PIN photodiode, at small price premium;

3. A variety of optical alignment accessories for the launch of the excitation light into the fiber optic temperature probe, the collection of the fluorescence response, and optical filters used to isolate the excitation and fluorescence emission at the detector and in some cases at the excitation source as well.

Fiber optic fluorescence thermometry can provide several quite flexible approaches to access the required measurement regions. The temperature probes can be

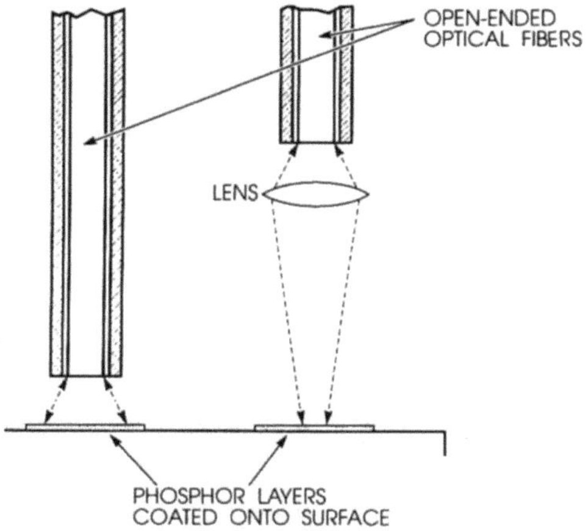

Figure 11.16. Remote measurement probes.[52]

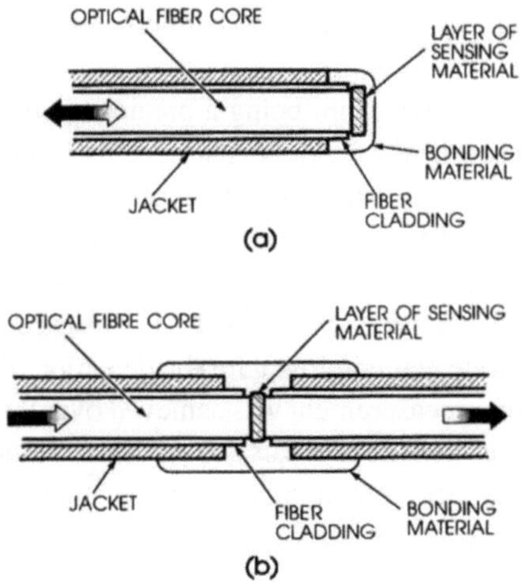

Figure 11.17 Temperature probes for contact measurement. (a) The reflection-type probe, (b) the transmission-type probe.

configured for remote measurement, as shown in Figure 11.16, where the sensing material is detached from the optical fiber and coated onto the surface of the object to be monitored. A unique application of this approach has been demonstrated by Noel *et al.*(51) in the measurement of temperatures on the blades and vanes of an operating turbine-engine, an extremely difficult measurement task for the conventional thermometry technologies, yet highly applicable to fluorescence thermometry.

For more general thermometric applications, the probes are constructed for actual contact measurement, with the sensing materials directly attached to the optical fibers, as shown in Figure 11.17. In the contact measurement configuration, temperature probes may be of either the reflection-type, as that in Figure 11.17a shows, or the transmission-type as in Figure 11.17b. The use of transmission-type probes can avoid the need for optical accessories for the coupling of the excitation light and the fluorescence into the sensor system, as illustrated in Figure 11.15b, while the reflection-type probes are used in the other arrangements shown in the same figure. Thus, the application of transmission-type probes can simplify the design of the optical arrangement and consequently reduce the cost of the device. However, this could cause inconvenience for users in planting the probe in the region where the measurement is to be made. Therefore, the reflection configuration of temperature probes has been more widely used in the development of commercial devices.

Of course, the temperature probes discussed could be further classified according to specific applications, e.g., biomedical temperature measurement, high temperature sensing up to >500°C, and the pyrometry range (≥ 500°C). The development of the temperature probes for such applications cited is discussed where specific applications are concerned.

11.3.3. Ruby-Based Thermometer with Range from 20 to 600°C

Apart from being a precious gemstone, ruby is well-known as the laser crystal used in the world's first successful laser operation. It is also among those earliest materials whose fluorescence lifetime properties were proposed for thermometric applications.(53) The actual use of ruby as the sensor element in a fiber optic fluorescence lifetime thermometer was perhaps first reported by Grattan *et al.*(49) In this thermometer system, an LED was used as the excitation light source and a silicon PIN diode was employed for the detection of the fluorescence signal. As a result, temperature measurement was achieved over the region from room temperature to 170°C. The thermometer system was further developed, as discussed below, in an effort to extend the measurement range of this compact and low-cost device and improve its performance.

11.3.3.1. Experimental Setup

The experimental system setup of the ruby fluorescence thermometer developed by Zhang *et al.*(36) is schematically depicted in Figure 11.18. A green LED, Type TLMP 7513 manufactured by III-V semiconductors with a typical luminosity of 300 mcd, was used as the excitation source which can pump into the strong band centered at ~550 nm, as shown in the absorption spectrum of ruby in Figure 11.19. As the radiation of the green LED also contains a weak emission band in the red portion of visible spectrum, which overlaps part of the fluorescence emission spectrum, a short-pass optical filter, F_1 shown in Figure 11.18, with a cutoff wavelength at 630 nm is used to eliminate this red "tail" of the LED emission. The fluorescence emission spectrum is obvious at longer wavelengths, as shown in Figure 11.20, with the strongest emission on the *R*-lines (around 694 nm).

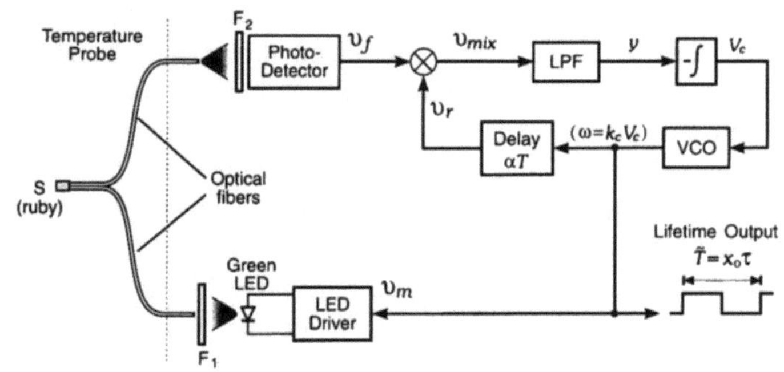

Figure 11.18. The ruby fluorescence lifetime-based fiber optic thermometer system. F_1: short-pass optical filter; F_2: *R*-line band-pass optical filter.

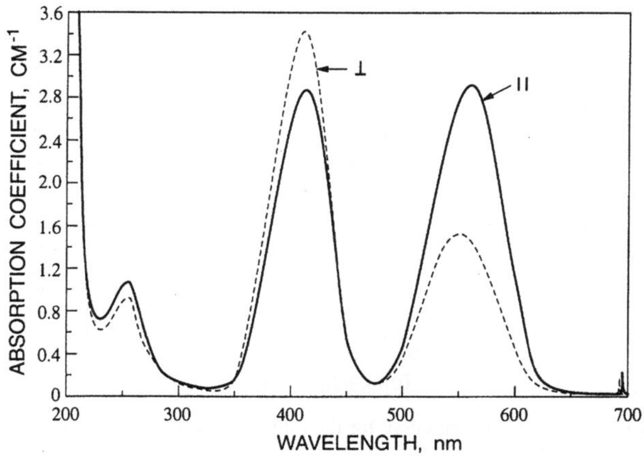

Figure 11.19. The absorption spectrum of ruby.[54] ⊥ = Incident light perpendicular to c-axis; || = incident light parallel to c-axis.

In this system, to achieve high temperature operation, gold-coated silica fibers were first used to fabricate the probe for temperature measurement up to ~600°C. The core diameter of the fiber used was 400 μm. The probe was built in the reflection-type configuration, similar to that shown in Figure 11.17a. Due to the lack of a large diameter fiber coupler, the probe had to be made of a bundle of the gold-coated fibers, so that at the sensor end, the transmission of the excitation light and the collection of the fluorescence response are via separated fibers. The detail of the construction of this special gold-coated fiber probe was discussed by Grattan *et al.*[55]

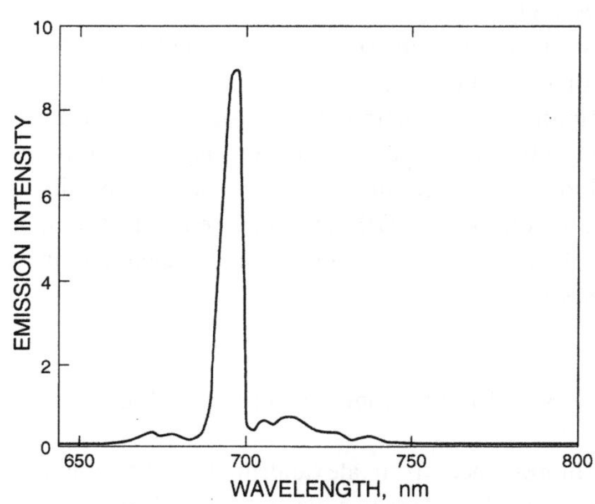

Figure 11.20. The emission spectrum of ruby.[42] Al_2O_3: 0.1% Cr^{3+}.

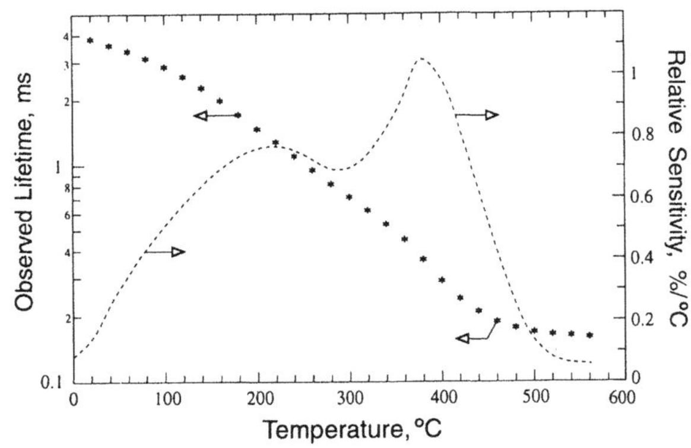

Figure 11.21. Characteristic calibration curve for the ruby fluorescence-based thermometer in the region from room temperature to ~550°C.

11.3.3.2. The Calibration of the Thermometer System

The characteristic calibration curve is shown in Figure 11.21 on a logarithmic scale, over a range from 30 to ~550°C. In the region between 150 and 450°C, the maximum sensitivity is seen. Beyond 500°C, the calibration curve tends to "flatten out" dramatically, and the sensitivity of measurement achievable in this region is limited, as shown by the dashed line in Figure 11.21, which represents the relative temperature sensitivity of the observed fluorescence lifetime, $s_{\Delta\tau/\tau}$ defined as

$$s_{\Delta\tau/\tau} = \left| \frac{(\Delta\tau/\tau)}{\Delta T} \right| \tag{11.4}$$

where τ is the observed lifetime, and $\Delta\tau$ and ΔT are the increments of the lifetime and temperature, respectively.

The intensity of the fluorescence emission detected at the photodetector stage was plotted as a function of temperature over the same range, and this is shown in Figure 11.22. It falls off rapidly with temperature increase over the whole temperature region. This does not contradict the experimental evidence of Burns and Nathan[56] who showed that the fluorescence quantum efficiency of the ruby fluorescence integrated over the entire band from 620 to 770 nm is independent of temperature (to ~5%) in the region from −196 to 240°C, for the emission detected here is only the *R*-line part of the total fluorescence emission.

11.3.4. Alexandrite-Based Thermometer with Range −100–700°C

The use of fluorescence from alexandrite for temperature sensing was first reported by Augousti *et al.*[57, 58] using a low-power LED or a HeNe laser with a rather inefficient modulation accessory made of a bulky, high-voltage controlled Pockels cell,

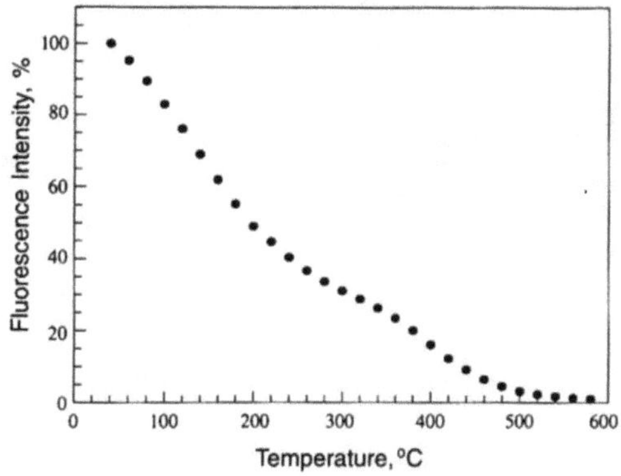

Figure 11.22. The ruby fluorescence intensity recorded in the experiment.

as excitation source. This greatly limited the practical utility of the scheme and a comparatively poor signal-to-noise ratio was obtained. The lifetime measurement technique used also imposed a limit on the consecutive measurement range. Using this limited optical system, an accuracy of ±1°C over a range 20–150°C was achieved.

The subsequent development of laser diode sources at low cost, and improved electronic detection, coupled with new probe fabrication techniques have now opened up this field to higher-temperature measurement. This has resulted in an alexandrite fluorescence lifetime based fiber optic thermometer system,[38] with a visible laser diode as the excitation source which has achieved a measurement repeatability of ±1°C over the region from room temperature to 700°C, using the lifetime measurement technique.

11.3.4.1. Experimental Setup

A schematic representation of the experimental system is shown in Figure 11.23. The modules comprising this system are discussed below.

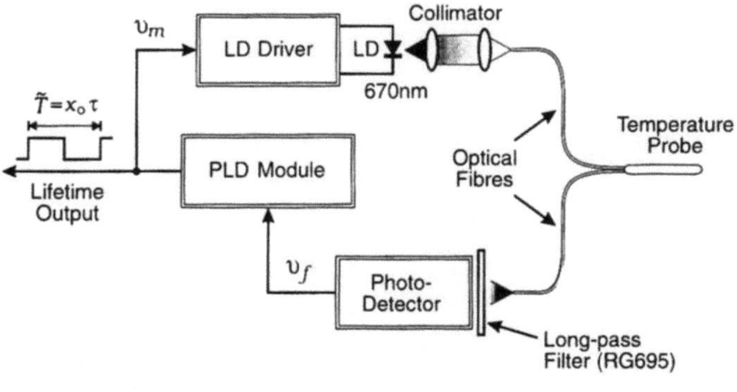

Figure 11.23. Schematic of the optoelectronic setup of the alexandrite fluorescence-based thermometer. LD = laser diode.

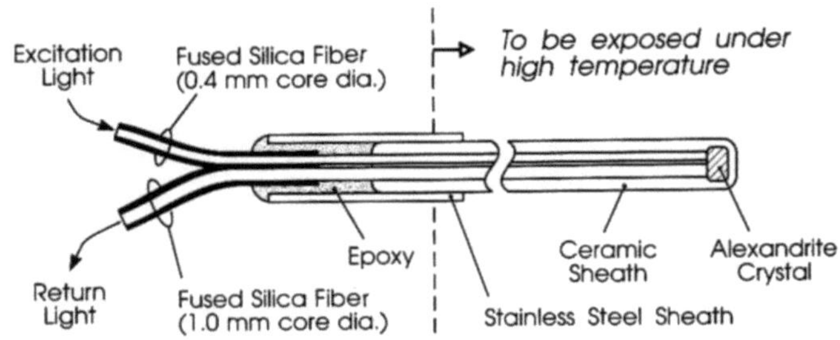

Figure 11.24. Schematic of the optical fiber probe.

The configuration of the probe is as shown in Figure 11.24. The sensor element itself, a small (approximately) rectangular-shaped ($2\times2\times1$ mm) piece of the synthetic crystal alexandrite, is held in a ceramic enclosure at the end of two fused silica fibers. The first fiber, that with a 0.4-mm core diameter, is used to transmit light from the excitation source to the sensor element. The second fiber, with a 1.0-mm core diameter, is used to receive the fluorescence response emitted from the sensor element.

A 3-mW laser diode with peak emission wavelength at 670 nm was used as a pumping source. The output of the laser diode is coupled into the first fiber mentioned above using a collimating and focusing lens, giving a coupling efficiency of more than 50%. Although the alexandrite absorption spectrum spans the wavelength region from about 380 to 680 nm, the absorption coefficient at 670 nm is low compared to that at 633 nm, the HeNe laser wavelength used in the work of Augousti *et al.*[57] However, the convenience of use of such a source, capable of direct modulation, more than compensates.

The output of the system was calibrated against a conventional type K thermocouple and a calibration curve, resulting from using such a laser diode with a 670-nm

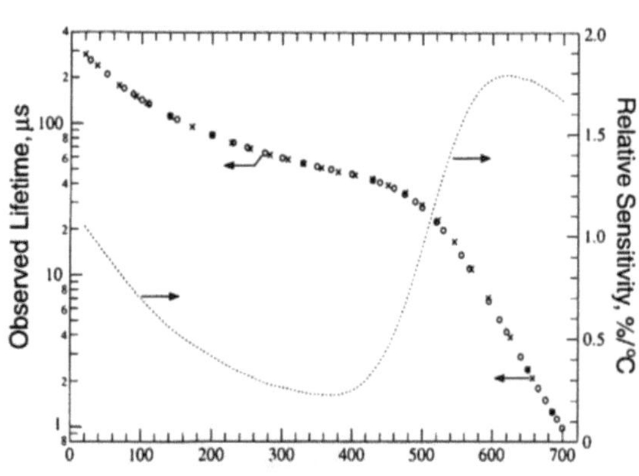

Figure 11.25. Observed alexandrite fluorescence lifetime versus temperature. o o o = data taken in the case of using 670-nm laser diode; x x x = data taken in the case of using 655-nm laser diode; dashed line = relative temperature sensitivity of the observed lifetime.

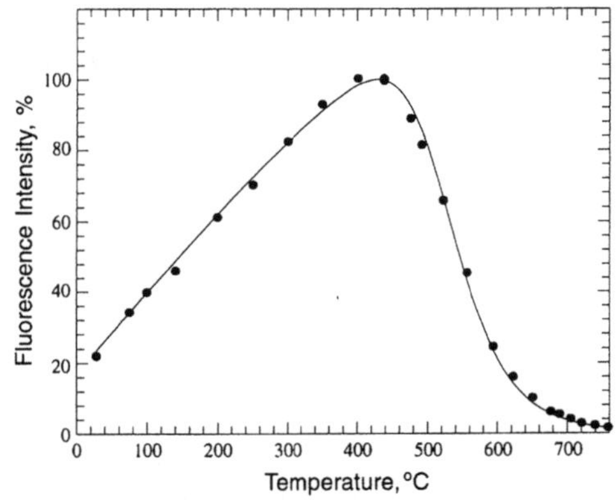

Figure 11.26. Alexandrite fluorescence intensity vs. temperature. • • • = data for fluorescence emission at low-energy broadband (λ > 695 nm) induced by 670-nm laser diode.

lasing wavelength as the excitation source, is depicted in Figure 11.25, drawn with the ordinate having a logarithmic scale. The relative temperature sensitivity of the fluorescence lifetime is calculated[45] based on these observed lifetime data, and is depicted as a function of temperature as the dashed line in Figure 11.25.

The intensities of the fluorescence signal received by the system (using a 670-nm laser diode as excitation source) are also recorded during the calibration process. The data are graphically presented in Figure 11.26 against the temperature. This show that the fluorescence intensity increases dramatically with temperature increase from room temperature or below to 400°C and reaches its maximum at ~430°C before falling off again.

11.3.5. Cr:LiSAF-Based Thermometer for Biomedical Applications

An optical temperature probe, in a fiber optic configuration has been developed[50] for the convenience of planting a temperature-sensing element along the RF applicator for heat treatment for a common condition among the aging male population, benign prostatic hyperplasia (BPH). The trivalent-chromium ion doped material, $LiSrAlF_6$ (Cr:LiSAF) is particularly suitable for this application, in comparison to the others discussed, due to both the temperature sensitivity of its fluorescence lifetime over the biomedical application region (30–50°) and its other optical spectroscopic properties, which give it advantages in the design of the thermometer system itself.

Figure 11.27 is a schematic representation of the Cr:LiSAF fluorescence lifetime-based thermometer. The absorption spectrum of Cr:LiSAF spans the wavelength region from the UV to near 750 nm with a peak falling between 600 and 700 nm, and thus a visible laser diode with a lasing wavelength at 670 nm (and 1 mW of optical power output) can efficiently excite or "pump" the Cr:LiSAF sample used as sensor element, to induce a fluorescence response from it with a sufficiently high signal-to-noise ratio to be detected. As shown in Figure 11.27, the isolation of excitation light

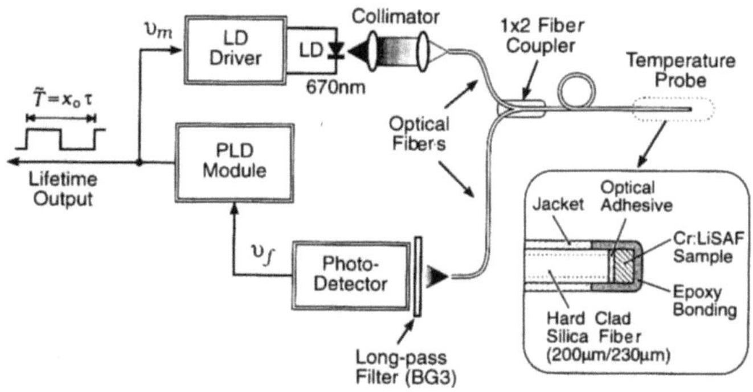

Figure 11.27. Schematic of the Cr:LiSAF fluorescence-based thermometer.

at the fluorescence detection stage is provided by a glass filter. This BG3 glass filter has a long-pass band starting at ~720 nm, and this well matches the Cr:LiSAF emission spectrum. Indeed, due to the clear separation between the excitation light and fluorescence emission, in wavelength terms, little compromise has been made between efficient transmission of fluorescence and good isolation of excitation light at the detector stage. This is a near-perfect optical arrangement and a much higher signal-to-noise ratio is obtained, compared to those in the cases of the ruby- and alexandrite-based systems discussed above.

The data for the fluorescence lifetime for this system are plotted against temperature in Figure 11.28, together with its relative temperature sensitivities calculated using Eq. (11.4). It shows that the Cr:LiSAF fluorescence lifetime decreases monotonically with the temperature increase, although it is rather insensitive to temperature variance around 0°C or below. From about 5°C, the fluorescence lifetime drops more and more sharply with temperature increase. This is seen explicitly from the rapid increase in its

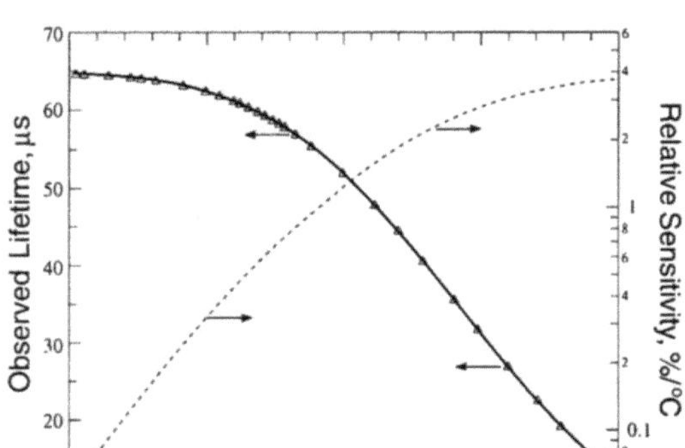

Figure 11.28. The observed lifetime of the Cr:LiSAF fluorescence decay and its temperature sensitivity. △△△ = data for the observed lifetime; the solid line shows curve fitting to the data; the dashed line shows the relative temperature sensitivity of the observed lifetime.

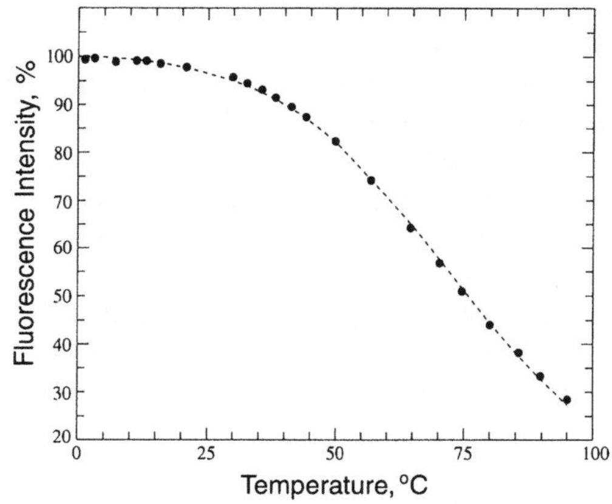

Figure 11.29. The Cr:LiSAF fluorescence intensity recorded in this system. -•-•-•- = data for the fluorescence intensity; dashed line = curve fitting to the data for Cr:LiSAF.

temperature sensitivity with temperature increase. It reaches 0.60%/°C at 35°C, and the significance of such a sensitivity increase is clear from a comparison with the case of ruby, an alternative material for biomedical thermometric uses,(53) whose sensitivity is only ~0.26%/°C at the same temperature. The fluorescence intensity recorded also decreases with temperature increase as shown in Figure 11.29, as would be expected from the configurational coordinate model for the Cr^{3+} fluorescence in low-strength crystal fields.(43)

11.3.6. Discussion of Cr^{3+} Doping Effects in Thermometry

Since the Cr^{3+} impurity is nearly always situated in octahedral sites, the absorption and emission spectra are similar in a large variety of materials, the main distinction being the strength of the octahedral field, rather than its exact nature.(59) This is well demonstrated by the absorption and emission spectra of ruby, alexandrite, and Cr:LiSAF, the materials illustrated in this section. The absorption spectra at lower energies are all characterized by two broad bands corresponding to the ${}^4A_2 \rightarrow {}^4T_1$ and ${}^4A_2 \rightarrow {}^4T_2$ transitions, with the positions of the peaks changing according to the field strength. The emission spectra are formed by broadband emissions in the infrared region mainly due to the ${}^4T_2 \rightarrow {}^4A_2$ transition, with or without the sharp R-line emission related to the ${}^2E \rightarrow {}^4A_2$. The positions of the peak emissions and the presence of the R-lines are again determined by the field strength.

The existence of a strong absorption band, at low energy region that corresponds to the ${}^4A_2 \rightarrow {}^4T_2$ transition, allows the use of a great variety of diode light sources, especially the powerful visible laser diodes recently developed, as discussed. This is one of the most significant advantages of the use of Cr^{3+}-activated materials in the development of fluorescence-based thermometers.

An additional advantage provided by the similarities in the absorption and emission spectra among the Cr^{3+}-doped materials is the potential interchangeability of the optical arrangement of the thermometer system with different sensing materials in the Cr^{3+}-doped family. For instance, an substantial absorption coefficient at around 635 nm is evident in the absorption spectra of ruby, alexandrite, and Cr:LiSAF. Therefore an optical arrangement, which uses a 635-nm visible laser diode as the excitation source and an RG665 long-pass Schott glass filter for the isolation of the excitation light, could be employed effectively for all thermometer systems which use the sensing materials mentioned above.

11.3.7. Cross-Referencing of Fluorescence Thermometry with Blackbody Radiation Pyrometry

So far, in the family of Cr^{3+}-doped materials studied, the highest temperature at which the fluorescence lifetime would be still of thermometric use is around 730°C in the case of alexandrite. The extending of the measurement range toward even higher temperature is severely hampered by difficulties of measuring the ever-shortening lifetimes under a worsening signal-to-noise condition experienced at such high temperatures.

Some rare-earth-activated materials do show strong fluorescence phenomena at temperatures even up to 1000°C with a lifetime long enough to be detected without particular difficulties, as demonstrated in the cases of neodymium:yttrium-aluminum-garnet (Nd:YAG) and $ScPO_4:Eu^{3+}$ by Grattan *et al.*[60] and Bugos *et al.*,[61] respectively.

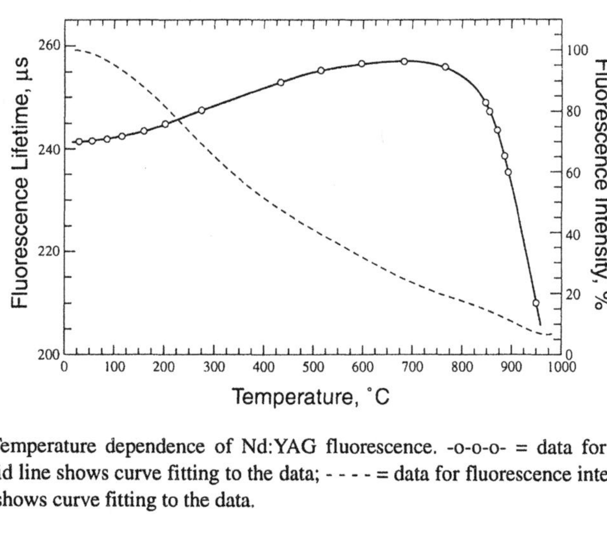

Figure 11.30. Temperature dependence of Nd:YAG fluorescence. -o-o-o- = data for the fluorescence lifetime; the solid line shows curve fitting to the data; - - - - = data for fluorescence intensity at 1064 nm; the dashed line shows curve fitting to the data.

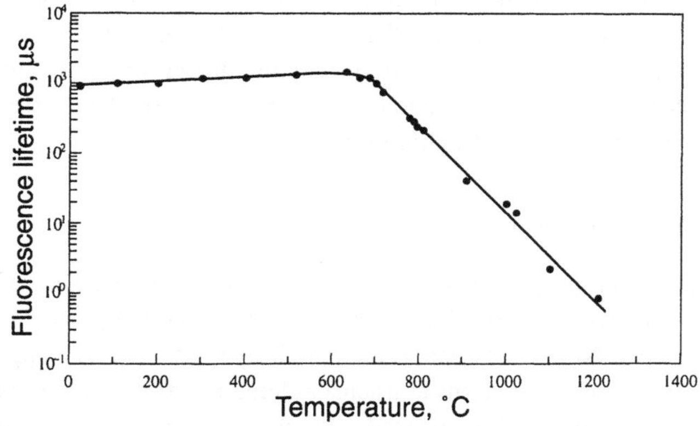

Figure 11.31. Fluorescence lifetime calibration curve for $ScPO_4$:Eu^{3+} from Bugos *et al.*

However, as shown in Figures 11.30 and 11.31, the use of the fluorescence lifetimes of these materials alone can only cover a very limited region of temperature. In both cases, the optimum region for the lifetime information to be used for temperature sensing purposes is from ~700°C and beyond. From room temperature (or lower) to ~600°C, there is some significant increase in the Nd:YAG fluorescence lifetime with temperature increase. This might be of thermometric use, but unless other means can be provided to cover the region from ~600 to ~700°C (and beyond) and to distinguish the lifetime measurements at temperatures higher or lower than this region, a wide measurement region could not be achieved by using this material.

In summary, the use of fluorescence lifetime monitoring for temperature sensing at high temperatures is based on the phenomenon of thermal quenching of fluorescence, while this phenomenon is just the very obstacle that blocks the extending of the measurement further into higher temperatures. Therefore, fluorescence thermometry is intrinsically more effective for measurement within moderate temperature regions, due to this fundamental nature of the fluorescence emission itself.

As a complementary technique to fluorescence thermometry, radiation thermometry is ultimately most effective for high temperature (>500°C) measurement. In fact, the only type of fiber optic system that can cover the range up to 2000°C is based on this technique, such as the Accufiber model 100.[(62)] Though it is claimed that the temperature range can be extended down to 300°C,[(62)] the fundamental nature of the measurement process is such that the system works more effectively in the higher-temperature regions and for many applications 500°C is a realistic threshold. However, a single-fiber optic sensor with extremely wide temperature range might be produced by the combination of above two thermometric techniques, and this is discussed in the work of Zhang *et al.*[(63)] As will be described in the this section, the device reported[(63)] has a relative simplicity of construction, where the fluorescent material used is

Nd:YAG and the lifetime measurement is made by the use of the phase-locked detection of fluorescence lifetime (PLD) scheme. By reference to the use of the fluorescence lifetime measurement, the problems, in pyrometry, of emissivity, and sight path factor in the blackbody radiation measurement could be corrected in such a scheme having an internal "self-calibration."

11.3.7.1. Experimental Setup

The probe used is made of a 1-mm-diameter silica fiber with a piece of neodymium: yttrium-aluminum-garnet (Nd:YAG) crystal cemented on its tip, as shown in Figure 11.32. This piece of Nd:YAG crystal is rectangularly shaped with an approximate size of $1.5 \times 1.5 \times 1.5\ \text{mm}^3$. It functions as a sensor element in the lower temperature range when the fluorescence technique is used. The optical characteristics of Nd:YAG as a fluorescent temperature sensor have been discussed in a previous paper by Grattan *et al.*[60] The blackbody cavity is formed with the cement surrounding the Nd:YAG crystal, as depicted in Figure 11.31, for measurement in the high- temperature region. To strengthen the probe, a quartz tube is used to sheath the silica fiber, as their thermal expansivities match each other very well.

The configuration of the thermometer is presented schematically in Figure 11.33. A Sharp LT010MD0 laser diode, with 5 mW peak emission power and a lasing wavelength at 810 nm, is used to excite the Nd:YAG fluorescence. This is a considerably more powerful source than the LED devices available a few years ago and used in previous work,[64] increasing the optical signals received. A 1×2 bi-directional fiber coupler is used to transmit the input light to the temperature probe, and to collect the resulting fluorescence response, which has a peak emission wavelength around 1064 nm, as well as the blackbody radiation from the probe tip to the detection stage. In addition to the fluorescence response and the blackbody radiation, a fraction of the input light is reflected to the detection stage. Thus, a 1064-nm bandpass interference filter, with a bandwidth of 15.2 nm, is placed before the photodetector, a simple, inexpensive planar 1-mm silicon PIN photodiode (BPX65), to cut off the reflected input light unwanted at the detection stage. The output of the photodetector consists of both the fluorescence response and the thermally generated radiation. The measure-

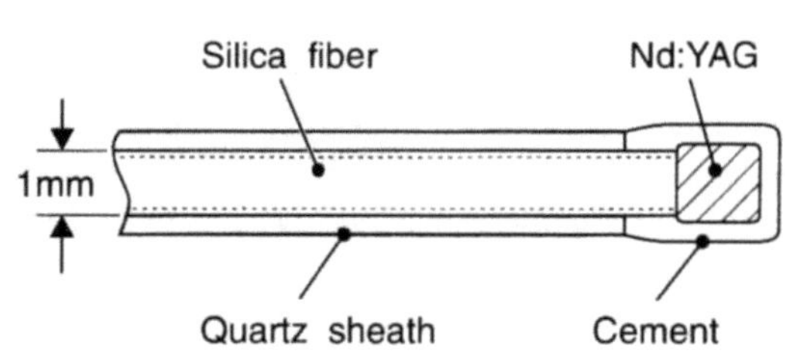

Figure 11.32. Schematic of the optical fiber sensor probe.

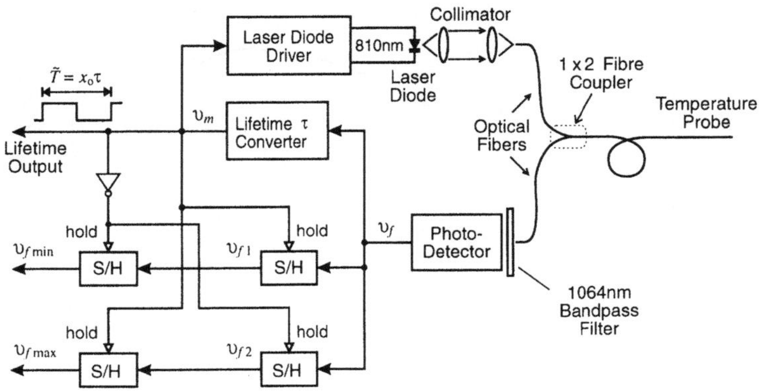

Figure 11.33 Schematic of the optoelectronic sensor scheme. Sample and hold devices.

ment of the fluorescence lifetime was achieved using the PLD scheme discussed above, and blackbody radiation was separated from the fluorescence by the sample and hold (S/H) circuit shown in Figure 11.33.

11.3.7.2. Calibration Results

The temperature probe is calibrated over a temperature range from 50 to 900°C. The calibration curves of the Nd:YAG fluorescence lifetime and the blackbody radiation at 1064 nm against temperature are depicted, as the solid and dashed lines, respectively, in Figure 11.34. As shown in Figure 11.33, the fluorescence lifetime of Nd:YAG increases with the increase of the temperature to ~680°C, and after ~700°C, it decreases more and more sharply with the temperature increase. The temperature resolution that could be achieved from the measurement of the fluorescence lifetime is ±3°C from 100 to 600°C and better than ±2°C, from 750 to 900°C with ±1°C resolution at 850°C. As can be seen, the temperature range from 600 to 750°C is an especially difficult region for the use of lifetime measurement alone. Therefore the radiation technique is best suited to measure the temperature over this region, and upward as well, because it can achieve a higher measurement resolution at temperatures over 600°C. At a temperature lower than 600°C, the lifetime method is generally preferable.

It is recognized that, as with all blackbody radiation, some radiation can be obtained from regions other than from the probe tip. Here, the main concern is the radiation generated along the length of the silica fiber used to transmit the radiation from the probe tip, which may be exposed at high temperature. This unwanted radiation is mostly generated from the fiber surface, and furthermore, only a small amount of such weak radiation can be coupled back into the fiber and transmitted along its transmission direction. Fortunately no significant impact of this radiation on the blackbody radiation measurement was observed.

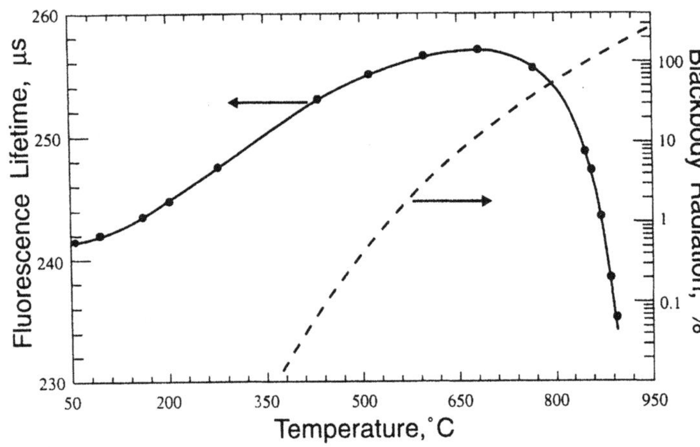

Figure 11.34 Calibration graph of the fluorescence lifetime (solid line) and blackbody radiation (dashed line) versus temperature.

11.4. Discussion and Cross-Comparison of Experimental Devices

11.4.1. Cross-Comparison

A summary of the characteristics of various major experimental devices developed utilizing solid state materials is shown as Table 11.2. The progress made in recent years can be seen. The response times of the devices now are correspondingly shorter than those of previous devices which used the same sensing materials, due to the efficiencies of the new signal processing schemes. The measurement ranges of fluorescence lifetime-based thermometers have been greatly extended into higher temperatures, showing the superiority of using PLD techniques for the detection of fluorescence lifetimes over a wide consecutive range. The improvement (or at worst no deterioration) in the measurement accuracy compared to the corresponding previous devices, shows that the PLD schemes have as high a noise suppression capability as the phase and modulation scheme. For instance, by using the phase and modulation technique, the best measurement deviation achieved was 0.04°C in an LED-excited ruby fluorescence device[67] with a response time of 10 sec; in a similar recent device but using an improved PLD technique,[34] the best deviation achievable was 0.1°C with a response time of only 2 sec. By increasing the response time of the latter device to 10 sec, its resulting measurement deviation could be estimated using the $\sigma_n/\sqrt{m}$ law to be $0.1°\mathrm{C}/\sqrt{5}$ and it is in the same order as that of the former device.

The experimental data on the fluorescence lifetimes of various materials have also been compared to those obtained previously from various processes. Though there are some slight differences, those data obtained from similar materials

Table 11.2. An Assessment of Various Experimental Devices Based on Solid State Fluorescent Materials

Year	Sensing material	Range	σ_n	Response time (sec)	Signal processing technique
1985[65]	Nd:glass (excited by LED)	15–130°C	1°C	6	Pulse measurement
1987[66]	Nd:glass (excited by LED)	–50–200°C	0.7°C	3	Pulse measurement
1987[57]	Alexandrite (excited by He-Ne laser)	20–150°C	0.3°C	3	Phase and modulation
1987[67]	Ruby (excited by LED)	20–160°C	0.3°C	10	Pulse measurement
		20–160°C	0.04°C (at 160°C)	10	Phase and modulation
1988[33]	Alexandrite (excited by LED)	20–150°C	0.3°C	1	Phase and modulation
This work	Ruby (excited by LED)	20–450°C 120–140°C	1°C 0.1°C	2	PLD technique
	Ruby (excited by He-Ne laser)	90–530°C ~370°C	0.3°C 0.03°C	2	PLD technique
	Alexandrite (excited by LD)	20–725°C 20–160°C	1°C < 0.2°C	<1	PLD technique
	Alexandrite (excited by He-Ne laser)	Potentially –100–700°C	1°C	<1	PLD technique
	Cr:LiSAF (excited by LD)	15–150°C 30–100°C	0.06°C 0.02°C	<1	PLD technique
	Nd:YAG (excited by LD)	50–950°C	3°C	<1	Fluorescence lifetime and blackbody radiation

Abbreviations: LD = laser diode; PLD = phase-locked detection of fluorescence lifetime; σ_n = lifetime deviation.

generally share the same temperature profiles. The slight discrepancy among these data originates from differences in concentrations of the impurities in the sensing materials, their geometrical shapes and sizes which would exhibit different levels of fluorescence trapping effects, and other effects due to the physical nature of the material.

11.4.2. Assessment of Fiber Optic Thermometers

Compared to the great diversity in proposed fiber optic sensor ideas, the types of fiber optic thermometers that are commercially available are actually quite few. Though several reports have been given by a major manufacture[5, 68, 69] reviewing various existing commercial systems, cross-comparisons between the performances of these systems are rarely made.

One independent assessment of a few commercial systems that has been published is that carried out by Harmer[70] in 1986. It was undertaken to address the possible use of fiber optic instrumentation for process control in offshore oil rig installations. The report was concerned with investigating whether devices met the manufacturer's design specifications, and tests were performed to consider such characteristics as accuracy, reproducibility, sensitivity, linearity, effects of supply voltage, handling considerations, and aging. In addition to a number of other devices, three prototype commercial temperature sensors were tested. Sensor A was a guide fiber bundle with a bimetallic strip and reflector varying the intensity of the back-reflected light. Sensor B (manufactured by ASEA) used the temperature-dependent fluorescence intensity of a semiconductor crystal, and sensor C (manufactured by Luxtron) depended on the fluorescence lifetime of a phosphor. The comments may be summarized thus: with sensor A, problems associated with intensity modulation were seen, due to connector and cable losses which could be misconstrued to be a temperature error, but sensor B could achieve an accuracy of better than ±0.4°C with individual calibration of the probe. Consistent reliable readings were seen over a period of testing of 2 weeks. For the fluorescence lifetime-type device, errors greater than the specifications were seen. The assessment results for sensors B and C are listed in Table 11.3.

In the test reported, the fluorescence intensity-based sensor did appear to perform well when individually calibrated. On the contrary, the performances of the lifetime-based sensor were extraordinarily poorer than expected in this early (1986) test. The magnitudes of errors induced by the effects of reconnection and microbending in the sensor C, as listed in Table 11.3, were much higher than those in sensor B. It is surprising for the former would be expected to perform intrinsically better than the latter in these aspects. Though sensor C was at an early stage of production at that time, the test did show how differently a device could perform in practice from the specifications given by the manufacturer. Therefore, it is important that commercial devices be subjected to such intensive scrutiny both in the user's interest and for manufacturers to see any device defects that they do not expect or neglect.

Table 11.3. Assessment of Early Commercialized Fiber Optic Thermometer Systems, by Harmer[70]

	Sensor B	Sensor C
Principle	Fluorescence intensity	Fluorescence lifetime
Fiber type	100 μm	305/362 μm
Range	200°C	−35°C to 250°C
Accuracy	±0.4°C	3.5°C (manufacturer's specification: ±0.2°C)
Effect of reconnection	0.7°C (max)	1.5°C
Microbending effect error (5 dB loss)	0.9°C	2.1°C

Of course, if not accompanied by a full understanding of the working principles of these devices, the status of their development, and their potential, the results of this test would be misleading when deciding on the choice of the right type of sensor for further development. It has been shown by later development, as discussed, that the lifetime-based scheme is much to be favored in terms of performance and cost. This was the reason that a major manufacturer, Luxtron, substituted the lifetime-based thermometer for the early, intensity-based one in its commercial production.[5]

The work has shown the diversity of fluorescence-based thermometry schemes, materials, signal processing arrangements, and applications. In addition, a close agreement with theoretical predictions on material performance gives confidence to the assessment of new materials for such uses. Probe sizes are potentially smaller and more convenient due to optical source and detector developments. Thus the future for fluorescence thermometry is bright, with the expectation of a range of new ideas in the field in the future.

References

1. J. S. Johnston, Editorial of special issues on temperature, *Measurement and Control 20*, 5 (1987).
2. K. T. V. Grattan and B. T. Meggitt, *Fiber Optic Sensor Technology*, Chapman and Hall, London, 1994.
3. K. T. V. Grattan, The use of fibre optic techniques for temperature measurement, *Measurement and Control 20*(7), 32–39 (1987).
4. K. T. V. Grattan, Fibre optic techniques for temperature sensing, in: *Fibre Optic Chemical Sensors and Biosensors*, Vol. II (O. S. Wolfbeis, ed.), pp. 151–192, CRC Press, London (1991).
5. K. A. Wickersheim, Fiberoptic thermometry: An overview, in: *Temperature—Its Measurement and Control in Science and Industry*, Vol.6, Part 2, pp. 711–714, American Institute of Physics, New York (1992).
6. T. Rozzel, C. C. Johnson, C. H. Durney, J. L. Lords, and R. G. Olson, A nonperturbing temperature sensor for measurements in electromagnetic fields, *J. Microwave Power 9*(3), 241 (1974).
7. M. Brenci, Thermochromic transducer optical fibre temperature sensor, *SPIE Proc. 515*, 155 (1984).
8. J. P. Dakin, D. J. Pratt, G. W. Bibby, and J. N. Ross, Distributed anti-Stokes ratio thermometry, *3rd Intern. Conf. on Optical Fiber Sensors*, Postdeadline Session, San Diego, February 1985.
9. J. P. Dakin, D. J. Pratt, G. W. Bibby, and J. N. Ross, Distributed optical fibre Raman temperature sensor using a semiconductor light source and detector, *Electron. Lett. 21*, 569 (1985).
10. J. M. Knox, P. M. Marshall, and R. T. Murray, Birefringent filter temperature sensor, *Proc. 1st Conf. Opt. Fibre Sens, (OFS 83)*, p. 1, IEE, London (1983).
11. R. R. Sholes and J. G. Small, Fluorescent decay thermometer with biological applications, *Rev. Sci. Instrum. 51*(7), 882–884 (1980).
12. K. A. Wickersheim and R. V. Alves, Fluoroptic thermometry: a new RF-immune technology, *Biomedical Thermology*, pp. 547–554, Alan Liss, New York (1982).
13. T. Bosselmann, A. Reule, and J. Schröder, Fibre-optic temperature sensor using fluorescence decay time, *Proc. of 2nd Conf. on Optical Fibre Sensors (OFS '84), SPIE Proceeding 514*, 151–154 (1984).
14. K. T. V. Grattan and A. W. Palmer, A fibre optic temperature sensor using fluorescence decay, *SPIE Proc. 492*, 535 (1984).
15. E. Saaski, A family of fibre optic sensors using cavity resonator microshifts, *Proc. of 4th Intern. Conf. on Optical Fibre Sensors*, Tokyo (1986).
16. E. Saaski and J. Hartl, U.S. Patent 4,678,904 (7 July 1987).

17. G. Beheim, Fibre-optic thermometer using semiconductor etalon sensor, *Elec. Lett. 22*(5), 238–239 (1986).
18. D. A. Christensen, A new non-perturbing temperature probe using semiconductor band edge shift, *J. Bioengineering 1*, 541–545 (1977).
19. K. Kyuma, S. Tai, T. Sawada, and M. Nunonshita, *IEEE J. Quan. Electron. QE18*, 676–679 (1982).
20. J. P. Dakin and D. A. Kahn, A novel fibre optic temperature probe, *Opt. Quan. Electron. 9*, 540 (1977).
21. B. E. Jones and R. C. Spooncer, An optical fibre pressure sensor using a holographic shutter modulator with two wavelength intensity referencing, *SPIE Proc. 514*, 223 (1984).
22. T. Bosselmann and T. Ulrich, High accuracy position sensing with fibre coupled white light interferometers, *Proc. of 2nd Conf. on Optical Fibre Sensors (OFS '84), SPIE Proc. 514*, 361 (1984).
23. D. A. Jackson, Monomode optical fiber interferometers, *J. Phys. E 18*, 897 (1985).
24. J. H. Schulman and C. C. Klick, Fluorescence, in: *McGraw-Hill Encyclopedia of Science & Technology*, ed. by S. P. Parker *et al.*, McGraw-Hill, New York (1987).
25. G. F. Imbusch, Inorganic luminescence, in: *Luminescence Spectroscopy* (M. D. Lumb, ed.), Chapter 1, p. 72, Academic Press, London (1978).
26. C. Ovren, M. Adoleson, and B. Hök, Fibre optic systems for temperature and vibration measurements in industrial applications, *Proc. Int. Conf. Optical Techniques in Process Control*, The Hague (1983).
27. K. A. Wickersheim and M. H. Sun, Fiberoptic thermometry and its applications, *J. Microwave Power 22*(2), 85–94 (1987).
28. K. A. James, W. H. Quick, and V. H. Strahan, Fiber optics: The way to true digital sensors?. *Control. Eng. 26*(2), 30–33 (1979).
29. M. Sun, Fiberoptic thermometry based on photoluminescent decay times, *Temperature—Its Measurement and Control in Science and Industry*, Vol. 6, Part 2, pp. 715–719, American Institute of Physics, New York (1992).
30. K. A. Wickersheim, Application of fibre optic thermometry to the monitoring of winding temperatures in medium and large power transformers. *SPIE Proc. 1584*, 3–14 (1991).
31. L. J. Dowell and G. T. Gillies, Errors caused by baseline offset and noise in the estimation of exponential lifetimes, *Rev. Sci. Instrum. 62*(1), 242–243 (1990).
32. K. T. V. Grattan, R. K. Selli, and A. W. Palmer, Ruby decay-time fluorescence thermometer in a fiber-optic configuration, *Rev. Sci. Instrum. 59*(8), 1328–1335 (1988).
33. A. T. Augousti, K. T. V. Grattan, and A. W. Palmer, Visible-LED pumped fiber-optic temperature sensor, *IEEE Trans. Instrum. & Meas. 37*(3), 470–472 (1988).
34. Z. Y. Zhang, K. T. V. Grattan, and A. W. Palmer, Phase-locked detection of fluorescence lifetime and its thermometric applications, in: Int. Conf. Biomedical Optics '93, Los Angeles, January 1993. *SPIE Proc. 1885*, 228–239 (1993).
35. T. Bosselmann, UK Patent Application GB 2,113,837A (1983).
36. Z. Y. Zhang, K. T. V. Grattan, and A. W. Palmer, A novel signal processing scheme for a fluorescence based fibre-optic temperature sensor, *Rev. Sci. Instrum. 62*(7), 1735–1742 (1991).
37. V. Fernicola and L. Crovini, Two fluorescent decay-time thermometers, in: *7th Int. Symp. Temp.*, Toronto, Canada, April 30, 1992, American Institute of Physics, Washington, D.C., 1992.
38. Z. Y. Zhang, K. T. V. Grattan, and A. W. Palmer, Fibre-optic high temperature sensor based on the fluorescence lifetime of alexandrite, *Rev. Sci. Instrum. 63*(5), 3177–3181 (1992).
39. Z. Y. Zhang, K. T. V. Grattan, and A. W. Palmer, Sensitive fibre optic thermometer using Cr:LiSAF fluorescence for bio-medical sensing applications, *Proc. 8th Optical Fiber Sensors Conf.*, Monterey, California, pp. 93–96, IEEE, New York (1992).
40. S. Sugano, Y. Tanabe, and H. Kamimura, *Multiplets of transition-metal ions in crystals*, pp. 107–112, Academic Press, New York (1970).
41. J. C. Walling, O. G. Peterson, H. P. Jenssen, R. C. Morris, and E. W. O'Dell, Tunable alexandrite lasers. *IEEE J. Quan. Electron. QE-16* (12), 1302–1315 (1980).

42. W. H. Fonger and C. W. Struck, Temperature dependences of Cr^{+3} radiative and nonradiative transitions in ruby and emerald. *Phys. Rev. B 11* (9), 3251–3260 (1975).
43. Z. Y. Zhang, K. T. V. Grattan, and A. W. Palmer, Temperature dependences of fluorescence lifetimes in Cr^{3+}-doped insulating crystals. *Phys. Rev. B 48*, 7772–7777 (1993).
44. M. Stalder, B. H. T. Chai, and M. Bass, Crystal growth and spectroscopy of Cr:$LiBaAlF_6$, Paper presented at the Advanced Solid State Lasers Topical Meeting, Hilton Head Island, SC, March 1991.
45. Z. Y. Zhang, K. T. V. Grattan, and A. W. Palmer, Thermal characteristics of alexandrite fluorescence decay at high temperatures, induced by a visible laser diode emission, *J. Appl. Phys. 73*(7), 3493–3498 (1993).
46. D. F. Nelson and M. D. Sturge, Relation between absorption and emission in the region of the R lines of ruby, *Phys. Rev. 137* (4A), A1117–A1130 (1965).
47. M. Stalder, B. H. T. Chai, and M. Bass, Flashlamp pumped Cr:$LiSrAlF_6$ laser, *Appl. Phys. Lett. 58* (3), 216–218 (1991).
48. T. Bosselmann, A. Reule, and J. Schröder, Fibre-optic temperature sensor using fluorescence decay time, *Proc. 2nd Conf. on Optical Fibre Sensors (OFS '84), SPIE Proceedings 514*, 151–154 (1984).
49. K. T. V. Grattan, R. K. Selli, and A. W. Palmer, Ruby decay-time fluorescence thermometer in a fibre-optic configuration, *Rev. Sci. Instrum. 59*(8), 1328–1335 (1988).
50. Z. Y. Zhang, K. T. V. Grattan, and A. W. Palmer, Cr:LiSAF fluorescence lifetime based fibre optic thermometer and its applications in clinical RF heat treatment, Int. Conf. on Biomedical Optics '93, Los Angeles, Jan. 1993, *SPIE Proc. 1885*, 300–305 (1993).
51. B. W. Noel, W. D. Turley, W. Lewis, K. W. Tobin, and D. L. Beshears, Phosphor thermometry on turbine-engine blades and vanes. Proc. 7th Int. Symp. on Temp., Toronto, Canada, 30 April, 1992, American Institute of Physics, Washington, D.C. (1992).
52. K. Wickersheim and M. Sun, Phosphors and fibre optics remove doubt from difficult temperature measurements. *Research & Development* (November), 114–119 (1985).
53. R. R. Sholes and J. G. Small, Fluorescent decay thermometer with biological applications. *Rev. of Sci. Instrum. 51*(7), 882–884 (1980).
54. V. Evtuhov and J. K. Neeland, Properties of ruby as a laser material, in: *Laser, Volume I* (A. K. Levine, ed.), p. 13, Edward Arnold Ltd, London (1966).
55. K. T. V. Grattan, A. W. Palmer, and Z. Zhang, Development of a high-temperature fiber-optic thermometer probe using fluorescent decay, *Rev. Sci. Instrum. 62*(5), 1210–1213.
56. G. Burns and M. Nathan, Quantum efficiency of ruby, *J. Appl. Phys. 34* (3), 703–705 (1963).
57. A. T. Augousti, K. T. V. Grattan, and A. W. Palmer, A laser-pumped temperature sensor using fluorescent decay time of alexandrite, *J. Lightwave Tech. LT-5* (6), 759–762 (1987).
58. A. T. Augousti, K. T. V. Grattan, and A. W. Palmer, Visible-LED pumped fibre-optic temperature sensor, *IEEE Trans. Instrum. and Meas. 37* (3), 470–472 (1988).
59. S. A. Payne, L. L. Chase, and G. D. Wilke, Optical spectroscopy of the new laser materials, $LiSrAlF_6:Cr^{3+}$ and $LiCaAlF_6:Cr^{3+}$, *J. Lumines. 44*, 167–176 (1989).
60. K. T. V. Grattan, J. D. Manwell, S. M. L. Sim, and C. A. Willson, Lifetime investigation of fluorescence from neodymium–yttrium–aluminum garnet at elevated-temperature, *Optics Commun. 62*(2), 104–107 (1987).
61. A. R. Bugos, S. W. Allison, and M. R. Cates, Laser-induced fluorescent properties of europium-doped scandium orthophosphate phosphors for high-temperature sensing applications, *Proc. of IEEE 1991 Southeast Conf.*, 1143–1147 (1991).
62. Accufiber, Inc., Model 100 multi-channel optical fibre thermometer system. Publicity data, Accufiber, Inc., Beaverton, Oregon (1989).
63. Z. Y. Zhang, K. T. V. Grattan, and A. W. Palmer, Fiber optic temperature sensor based on the cross referencing between blackbody radiation and fluorescence lifetime, *Rev. Sci. Instrum. 63*(5), 3177–3181.
64. K. T. V. Grattan, A. W. Palmer, and C. A. Willson, A miniaturized micro-computer-based neodymium 'decay-time' temperature sensor, *J. Phys. E: Sci. Instrum. 20*, 1201–1205.

65. K. T. V. Grattan and A. W. Palmer, Infrared fluorescence "decay-time" temperature sensor. *Rev. Sci. Instrum.* *56*(9), 1784–1787 (1985).
66. K. T. V. Grattan and A. W. Palmer, A miniaturised microcomputer-based neodymium "decay-time" temperature sensor, *J. Phys. E: Sci. Instrum.* *20*(10), 1201–1205 (1987).
67. R. K. Selli, Fibre optic temperature sensors using fluorescent phenomena, Ph.D. thesis, City University, London (1987).
68. K. A. Wickersheim and H. M. Sun, Fiberoptic thermometry and its applications, *J. Microwave Power* *22* (2), 85–94 (1987).
69. K. A. Wickersheim and W. D. Hyatt, Commercial applications of fiberoptic temperature measurement, *SPIE Proc. 1267*, 84–96 (1990).
70. A. L. Harmer, Fibre optic sensors for offshore process control instrumentation, *Proc. Optical Fibre Sensors Conference (OFS '86)*, Informal Workshop at Tusukuba Science City, 1986, VII. Pub: Institute of Electronics and Communication Engineers of Japan, Tokyo, 1986.

12

Instrumentation for Red/Near-Infrared Fluorescence

David J. S. Birch and Graham Hungerford

12.1. Introduction

The phenomenon of fluorescence has been synonymous with ultraviolet (UV) and visible spectroscopy rather than near-infrared (near-IR) spectroscopy from the beginning of the subject. This fact is evidenced in definitive texts which also provide useful background information for this volume (see, e.g., Refs. 1–6). Consequently, our understanding of the many molecular phenomena which can be studied with fluorescence techniques, e.g., excimer formation, energy transfer, diffusion, and rotation, is based on measurements made in the UV/visible. Historically, this emphasis was undoubtedly due to the spectral response of the eye and the availability of suitable sources and detectors for the UV/visible in contrast to the lack of equivalent instrumentation for the IR. Nevertheless, there are a few notable exceptions to the prevalence of UV/visible techniques in fluorescence such as the near-IR study of chlorophyll[7] and singlet oxygen,[8] which have been ongoing for some years.

Only quite recently have the potential advantages of working with near-IR fluorescence probes been realized and this is undoubtedly due to advances in IR instrumentation. A complementary renaissance occurred in Raman spectroscopy with many important materials now studied using IR Raman without interference from background fluorescence.[9] The obviation of autofluorescence in cells, surfactants, and polymers provides a parallel advantage in using IR over UV/visible fluorescence. Other advantages afforded by near-IR over UV/visible fluorescence techniques include reduced Rayleigh scattered excitation (this being proportional to on the inverse fourth power of wavelength), low absorption coefficient in many organic media, e.g., tissue, increased photochemical stability, and compatibility with spinoff technology from recent developments in the communications industry such as optical fibers and

David J. S. Birch and Graham Hungerford • Department of Physics and Applied Physics, University of Strathclyde, Glasgow G4 ONG, Scotland, United Kingdom.

Topics in Fluorescence Spectroscopy, Volume 4: Probe Design and Chemical Sensing, edited by Joseph R. Lakowicz. Plenum Press, New York, 1994.

semiconductor devices, the latter often being inexpensive. At this point in time the primary drawback of fluorescence assay using IR techniques would seem to be the lack of suitable probe molecules for many applications. However, this area is presently receiving considerable attention worldwide and indeed the current state of the art in IR probes is discussed in Chapter 7 (this volume).

In this chapter, we review the instrumentation presently available for studying near-IR fluorescence. This includes modern semiconductor devices such as diode lasers and photodiode detectors and also more conventional devices such as discharge lamps and photomultipliers which are traditionally more usually associated with the study of UV/visible fluorescence. Throughout the chapter emphasis will be placed on the novel red/near-IR aspects of instrumentation and we will assume that the reader has a knowledge of the basics of steady-state and time-resolved techniques to the level consistent with Volume 1 of this series.

Where appropriate we will illustrate the instrumentation with applications demonstrating performance. However, to begin with we will review the red/near-IR implementations of the major system techniques and associated kinetics already in widespread use in fluorescence spectroscopy.

12.2. Techniques

In this section we will review the application of near-IR system instrumentation to the most commonly encountered fluorescence measurements such as steady-state spectra, excited state lifetimes, anisotropy, microscopy, multiplexing, high-performance liquid chromatography (HPLC), and sensors.

12.2.1. Steady-State Spectra

The first measurement we make when starting a fluorescence study is not usually a fluorescence measurement at all but the determination of the sample's absorption spectrum. Dual-beam differential spectrophotometers which can record up to 3 absorbance units with a spectral range of 200–1100 nm are now readily available at low cost in comparison to fluorimeters. The wide spectral response of silicon photodiode detectors has made them preëminent over photomultipliers in this area with scan speeds of a few tens of seconds over the whole spectral range being achieved, even without the use of diode array detection.

Figure 12.1 shows the classic L-format of the most commonly used fluorescence spectrometer configuration which is topologically the same for the measurement of both steady-state spectra and lifetimes. The source and detector options of relevance to IR fluorescence measurements are discussed in Sections 12.3 and 12.4, respectively. The other optical components comprised of the lenses for focusing and collection and monochromators for wavelength selection contain few peculiarities in the near-IR as

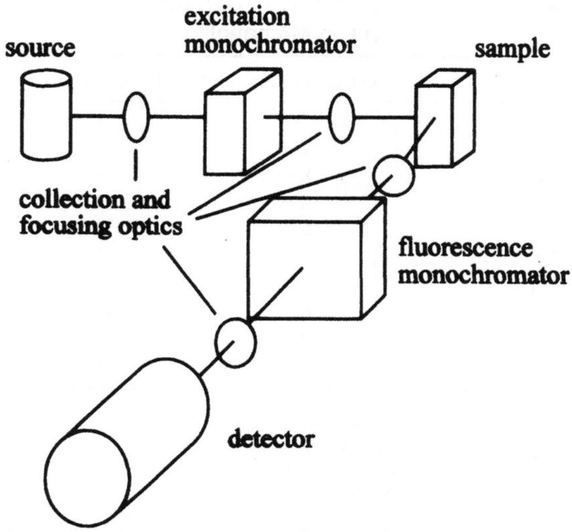

Figure 12.1. L-format fluorescence spectrometer.

compared with the UV/visible. Fused silica and glass lenses transmit light equally well in the near-IR, hence there is no need to incur the significantly higher cost of the former provided the absence from glass of fluorescent ion impurities are assured. The same remarks apply to prism polarizers. Dichroic polymer sheet polarizers are also available for the near-IR. If mirror optics are to be used, coating with gold offers a higher reflectivity in the near-IR than other metal coatings.

Optical monochromators containing ruled diffraction gratings with an appropriate blaze wavelength offer a higher efficiency than holographic gratings in the near-IR. The reason for this is the nature of the photoresist process used in fabricating holographic gratings, and although usually not peaking at the long wavelength end of the near-IR, the response of holographic gratings can be extended with reasonable efficiency (ca. 20%) to beyond 1000 nm. Similar features apply to gratings used in polychromators. However, the multiplexing advantage of higher data accumulation rates afforded by a silicon photodiode array coupled to a polychromator would seem to be outweighed by the greater sensitivity of a photomultiplier and monochromator judging by the predominant use of the latter combination.

Background fluorescence from glass cut-off filters is generally much less when exciting in the near-IR as compared with the UV/visible. Interference filters for the near-IR are also readily available.

Andrews-Wilberforce and Patonay[10] have studied the steady-state metal fluorescence quenching of a series of carbocyanine laser dyes in methanol and listed the peak absorption and emission wavelengths of the dyes. Table 12.1 summarizes the spectral properties of the dyes studied. The small Stokes shift shown in Table 12.1 highlights

Table 12.1. Spectral Peaks of Carbocyanine Laser Dyes in Methanol[a]

Dye[b]	Peak absorption (nm)	Peak fluor. (nm)
IR 125	780	815
IR 132	764	824
IR 140	803	821
IR 144	698	708
DTDCI	647	668
DTTCI	746	777
DODCI	577	605
DOTCI	678	703
HITCP	736	764
HDITCP	771	805

[a]From Ref. 10.
[b]DTDCI 3,3′-diethylthiadicarbocyanine iodide
DTTCI 3,3′-diethylthiatricarbocyanine iodide
DODCI 3,3′-diethyloxadicarbocyanine iodide
DOTCI 3,3′-diethyloxatricarbocyanine iodide
HITCP 1,1′,3,3,3′,3′-hexamethylindotricarbocyanine perchlorate
HDITCP 1,1′,3,3,3′,3′-hexamethyl-4,4′,5,5′-dibenzo-2,2′-indotricarbocyanine perchlorlate

some of the problems of studying such dyes with regard to self-absorption effects and detection of scattered excitation. The narrow spectral bandwidth of laser excitation has obvious advantages with respect to minimizing the detection of scattered excitation.

The advantages of IR over UV/visible fluorescence measurements already mentioned in Section 12.1 are readily realized in the optical configuration of Figure 12.1. Diode lasers emitting in the near-IR offer orders of magnitude enhancement in fluorescence sensitivity as compared with conventional arc lamps and at a cost much less than UV/visible lasers of the same power. For example, by using a 5 mW aluminum-gallium-arsenide (AlGaAs) diode laser at 786 nm and photon counting detection, Imasaka *et al.*[11] have obtained a detection limit of 5 p*M* concentration of the polymethine dye 3,3′-diethyl-2,2′-(4,5.;4′,5′-dibenzo)thiatricarbocyanine iodide (DDTC) in methanol. This is two orders of magnitude lower than they obtained using an Hitachi MPF-4 fluorimeter incorporating a 150 W xenon lamp.

12.2.2. Lifetimes

The most popular technique for measuring fluorescence lifetimes is the pulse method using time-correlated single-photon counting followed by the phase/modulation method. These two approaches, which are to a large extent complementary, are depicted in Figure 12.2. Both usually use the same generic spectroscopic configuration as shown in Figure 12.1, albeit incorporating pulsed or modulated sources, fast

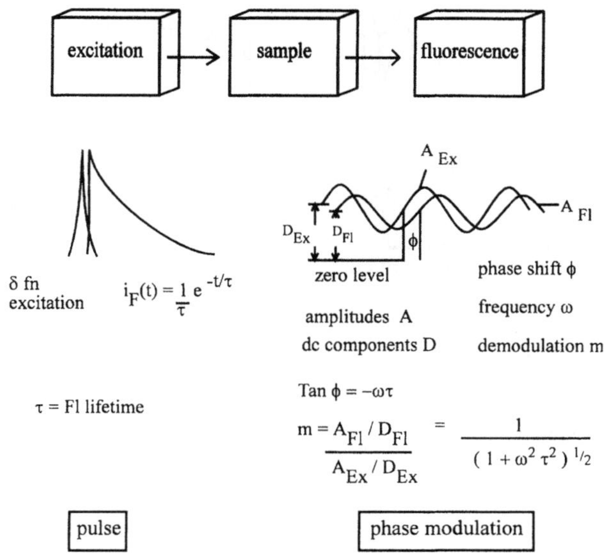

Figure 12.2. Pulse and phase modulation fluorescence lifetime techniques.

response detectors, and interfacing to appropriate data acquisition electronics. Both techniques have been described in detail in an earlier volume in this series.[4]

Until recently there were comparatively few reports of fluorescence lifetime studies of dye molecules in the near-IR, but this situation has changed rapidly. The fluorescence lifetimes of near-IR emitting dyes such as carbocyanines, porphyrins, oxazines, and xanthenes, are usually in the nanosecond region, consistent with the high oscillator strength of the S_1–S_0 transition in such compounds.

Historically the first reported study of the fluorescence lifetimes of near-IR dyes was in 1985.[12] A block diagram of the single-photon timing system used in this work is shown in Figure 12.3. The feature of this instrument which was novel at that time was the use of a diode laser (pulse duration 136 psec, wavelength 823 nm, and power 1.3 W). Imasaka *et al.* used this system to study the fluorescence lifetime of the polymethine dye NK 427 in a range of solvents at 850 nm emission. They observed that the decrease in fluorescence lifetime and quantum yield in a range of solvents was in general in proportion to the decrease in solvent dielectric constant over a range of lifetimes from 1.1 to 0.6 nsec.

Our literature survey indicates that reports of either near-IR excimer emission (as against the ground state aggregation that is well known to occur in many polymethine dyes) or energy transfer from IR fluorophores is nonexistent or at the most sparse. In the authors' laboratory Förster energy transfer from carbocyanine dyes to metal ions in viscous solvents has been studied using single-photon counting with 50 psec diode laser excitation pulses at 670 nm and photomultiplier detection.[13] Table 12.2 illus-

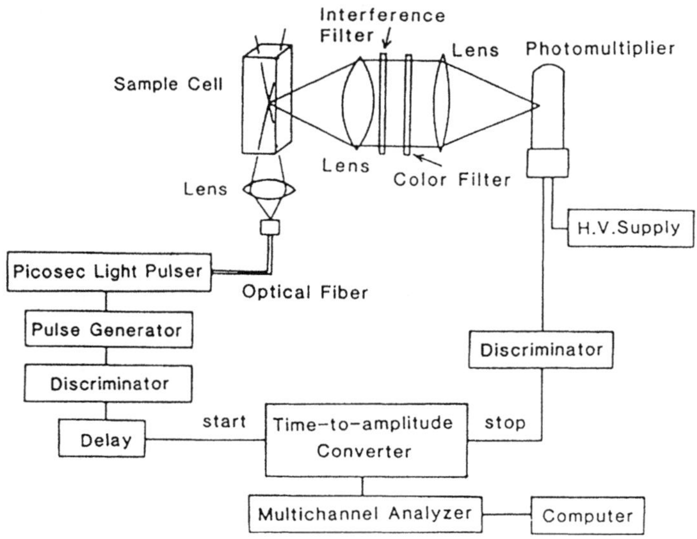

Figure 12.3. Diode laser-based single-photon fluorometer. (From Ref. 12.)

trates this for the quenching of 1,1′,3,3,3′,3′-hexamethylindotricarbocyanine perchlorate (HITCP) by copper ions in propylene glycol at –20 °C. Excellent agreement with Förster's theory is obtained, this data set giving a critical transfer distance of 1.86 nm.

Recently the fluorescence of IR-132 has been determined using single-photon timing at detection levels down to photon bursts from single molecules in 1 picoliter of a 25 f*M* solution of the dye.(14) At these extremely low concentration levels solvent

Table 12.2. Förster Quenching[a] of HITCP by Copper Ions in Propylene Glycol at –20 °C

CuSO4 conc.[4] (× 0.01 *M*)	τ (nsec)	γ	χ^2	χ^2 1 exp
0.0	1.55	—	—	1.12
0.5	1.54	0.069	1.14	1.30
1.0	1.50	0.131	1.22	1.63
1.5	1.53	0.222	1.14	2.47
2.0	1.51	0.287	1.15	2.86
2.5	1.53	0.366	1.14	3.68
3.0	1.56	0.440	1.20	4.18
3.5	1.58	0.519	1.21	5.05
4.0	1.53	0.550	1.18	5.04
4.5	1.55	0.642	1.22	5.91

[a] $i(t) = i(0)\exp(-t/\tau - 2\gamma\sqrt{t/\tau})$.

impurities and Raman scattering become the chief sources of error. The experimental arrangement incorporated an argon ion–pumped self-mode-locked titanium:sapphire (Ti:sapphire) laser giving 120 fsec pulses at 76 MHz with 12 mW at 800 nm. Fluorescence from IR-132 was detected at 840 nm using a single-photon avalanche photodiode (APD). The ability to detect single molecules using low-cost diode laser excitation must surely open up new applications for fluorescence probe methodology, present techniques being limited to studying assemblies of probe molecules, often all in slightly different microenvironments as in the case of a membrane, polymer, or protein.

As well as measuring picosecond to nanosecond lifetimes the single-photon technique can also he used to study much longer lifetimes in the near-IR. Egorov *et al.*[15] have recorded the rise and decay of singlet oxygen emission in water using copper vapor laser excitation of tetra(*p*-sulfophenyl)porphyrin as a sensitizer at 510 nm and single-photon detection at 1270 nm using an S1 photocathode. A singlet oxygen decay time of around 3000 nsec and risetime due to the sensitizer's triplet decay of between 30 and 600 nsec was observed depending on oxygen pressure. Much interest has been devoted to studying singlet oxygen emission in recent years, largely because of its relevance to photodynamic effects. However, more usually singlet oxygen lifetimes have been measured using analog detection with germanium photodiodes[8, 16–18] combined with a sampling oscilloscope or transient digitizing oscilloscope.

Other near-IR techniques that have been used to measure lifetimes, though not to the same extent as the aforementioned methods, include fluorescence up-conversion,[19–21] parametric amplification,[22] streak camera detection,[23] and two-photon excitation.[24] The latter technique is particularly useful as it enables the greater penetration depth of near-IR radiation in organic matter to be used to obtain a well-defined region of excitation, e.g., in single cells or mammalian tissue.

Other near-IR applications which use similar pulse and phase instrumentation as used in lifetime measurements include optical time-domain reflectometry[25] and photon migration in tissue.[26]

12.2.3. Anisotropy

Such is the newness of appreciation of near-IR fluorescence techniques that there is a dearth of examples in the literature of implementations of many of the classic fluorescence methods in the IR. Anisotropy is one striking example of this. However, in a comprehensive study of the anisotropy decay of dyes, including oxazine fluorescence at 720 nm, in mixed isotropic solvents Dutt *et al.*[27, 28] have investigated the effects of viscosity on molecular rotation.

Reported examples of near-IR fluorescence probe studies in anisotropic media are harder to find at present, but no doubt this situation will change as the advantages of excitation in the near-IR, particularly with respect to reducing autofluorescence in biological systems, become more widely appreciated.

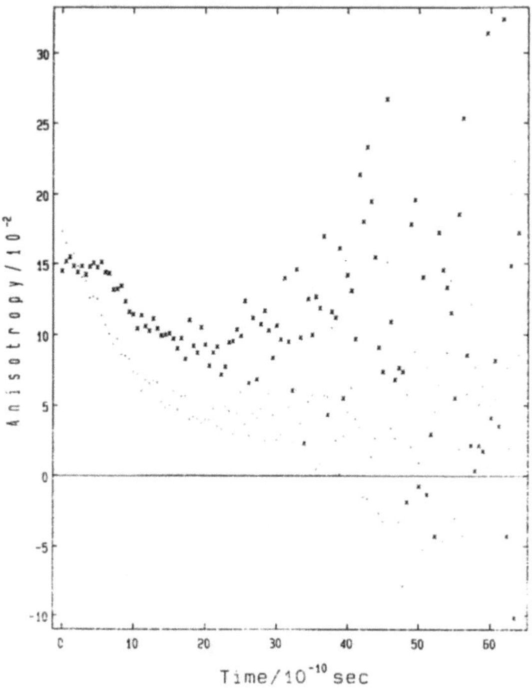

Figure 12.4. Anisotropy decay of IR-140 in liposomes of DPPC above (·····) and below (xxxx) the phase transition.

In the authors' laboratory we have studied the fluorescence depolarization of IR-140 in lipid bilayer membranes of L-α-dipalmitoylphosphatidylcholine (DPPC) and observed similar differences between the gel and liquid crystalline phases as has been widely reported for UV/visible probes such as 1,6-diphenyl-1,3,5-hexatriene (DPH) in this same medium. Figure 12.4 shows some of these results.

12.2.4. Microscopy

The area of fluorescence microscopy is vast and the recent addition of time-resolved techniques has brought a new approach to obtaining contrast other than by using intensity. For a review of fluorescence lifetime microscopy in both the time and frequency domains the reader is referred to the recent articles by Hirayama[29] and Wang *et al.*[30] Again there are few examples in the literature of fluorescence lifetime microscopy studies on near-IR fluorophores. Rodgers and Firey[31] have reported the measurement of the fluorescence lifetime of an individual algal chloroplast of *Mougeotia* using a microscope coupled to single-photon counting at 690 nm emission, obtaining decay constants of 348 and 959 psec.

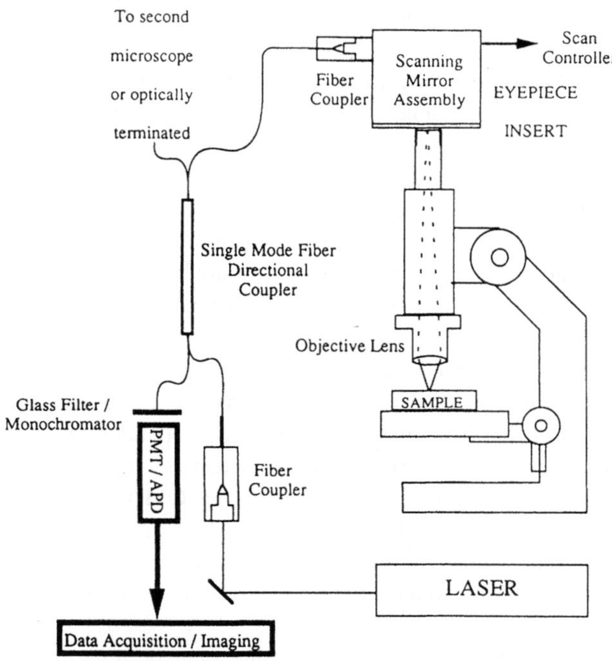

Figure 12.5. Single-photon fluorescence lifetime microscopy. (From Ref. 32.)

More recently Ghiggino and co-workers[(32)] have applied laser scanning confocal fluorescence lifetime microscopy to the study of polyvinyl alcohol films containing rhodamine B (650 nm emission) and cresyl violet (632 nm emission). Synchronously pumped dye laser excitation and APD detection were used with optical fiber coupling. A schematic diagram of their apparatus is shown in Figure 12.5.

Other fluorescence lifetime techniques have also been applied to microscopy including gated image intensifiers and charge-coupled devices (CCDs)[(30)] and a streak-camera based system has been introduced by Hamamatsu Photonics K.K. with a spectral response up to 1200 nm. Phase-modulation methods have also been used to great effect, particularly in displaying two-dimensional (2-D) intracellular fluorescence lifetime images with high time and spatial resolution.[(33)]

Two-photon fluorescence microscopy has also been used with good effect in the near-IR. For example, Ferguson *et al.*[(24)] at the University of Strathclyde have used 270 fsec pulses from a titanium:sapphire (Ti:sapphire) laser at 790 nm to observe visible fluorescence from dyes in zebra fish larvae and erythrocytes. The high depth and lateral definition afforded by the two-photon process and confocal microscopy are useful here. Also, the use of near-IR excitation minimizes photobleaching.

12.2.5. Multiwavelength Array Detection

The use of a linear detector array in the image plane of a polychromator in place of the fluorescence monochromator in Figure 12.1 enables the parallel data accumulation of complete fluorescence spectra. Silicon photodiode arrays, operated in a CCD mode[34] are the most widely used detector elements. The spectral response of the diodes enables fluorescence to be detected from the near-UV up to ca. 1100 nm with a peak response in the near-IR. Up to 8192 elements are now available commercially in a single linear array at low cost. However, the small length of each element (ca. 10 μm) presently limits sensitivity and hence cylindrical lens demagnification is often necessary.

Such diode array arrangements, frequently incorporating internal amplification, have become known as optical multichannel analyzers (OMAs) and incorporate a photocathode and phosphor interspaced with a microchannel plate (MCP) for image intensification. The phosphor emission is then detected using a linear array of silicon photodiodes, sometimes using optical fiber coupling. OMAs have been commercially available for some years[35] and by gating the MCP time-resolved spectra can be determined with nanosecond time resolution. Gratton *et al.*[36] have also applied a cross-correlation method to obtain time-resolved spectra from a commercial OMA system.

Streak cameras and multianode microchannel plate photomultipliers (MCP-PMs) interfaced to a polychromator also permit multiwavelength fluorescence decay measurements, the spectral response of both being determined by the photocathode composition.

In the case of multianode MCP-PMs multiplexed single-photon counting interfaced to a single time-to-amplitude converter is the most efficient and inexpensive way to record multiple fluorescence decay curves simultaneously at different emission wavelengths.[37, 38] Sixteen channels of single-photon timed decay data have recently been accumulated in parallel in the authors' laboratory using a 16-anode MCP-PM and the general multiplexing arrangement shown in Figure 12.6.[39] Such an approach enables global measurements to be combined with global decay analysis.

Farrens and Song[40] have replaced the original spark source with a picosecond diode laser in a multiplexed dual wavelength T-format fluorometer.[41] With an overall instrumental response width of ca. 300 psec full-width half-maximum (FWHM), near-IR fluorescence lifetimes as low as 75 psec in the case of 1,1′-diethyl-4,4′carbocyanine iodide (DCI) (excitation 660 nm) and decay components as low as 48 psec in the case of 124 kDa oat phytochrome (excitation 752 nm) were reported.

12.2.6. Sensors

The near-IR offers many attractions when working with sensors owing to low transmission losses in optical fibers, compatibility with the wealth of optoelectronics

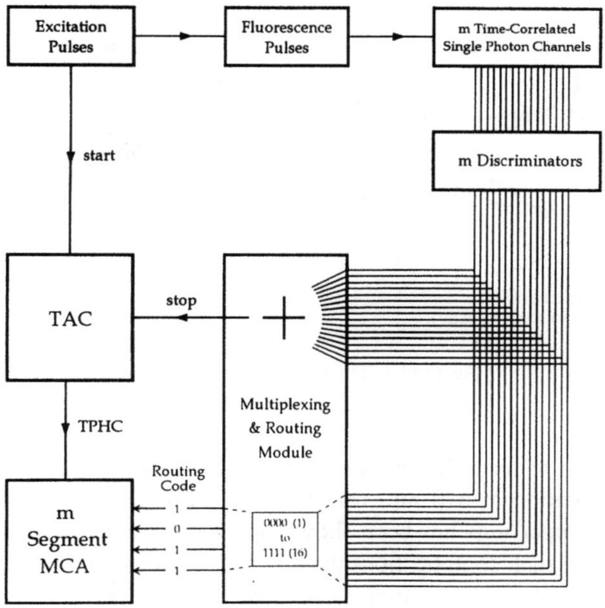

Figure 12.6. Multiplexed single-photon counting arrangement for parallel fluorescence decay measurements (from Ref. 39.)

derived from the communications industry, and the obviation of background fluorescence.

Thompson[42] has recently reviewed some of the techniques and opportunities for fluorescence based fiber optic sensors. Such sensors are based on encapsulating a photophysical phenomenon such as fluorescence quenching in the active element of the sensor tip and relaying the excitation and fluorescence signals using fiber optic coupling. As already mentioned, many of the fundamental photophysical phenomena such as quenching, excimer formation, and energy transfer have been little studied in the near-IR. Hence it is hardly surprising that near-IR fluorescence sensors (which are by definition based on understanding fluorescence phenomena) lag a little way behind the UV/visible equivalents. Sensors operating in the steady-state fluorescence mode are already firmly established, with the potential advantages of the lifetime-resolved mode with respect to simpler calibration and higher specificity becoming more widely appreciated. Figure 12.7 illustrates the general operating principle of a fluorescence lifetime-based sensor using bifurcated optical fiber coupling.

Sensors based on the fluorescence quenching of rhodamine 6G in resins by iodide ions[43] and in Nafion polymer by metal ions in solution[44, 45] have been demonstrated. However, complex fluorescence decay mechanisms often hinder interpretation in lifetime-based sensing and much progress is still to be made in this area before the true potential of lifetime-based sensing becomes a reality. For example, rhodamine 6G in

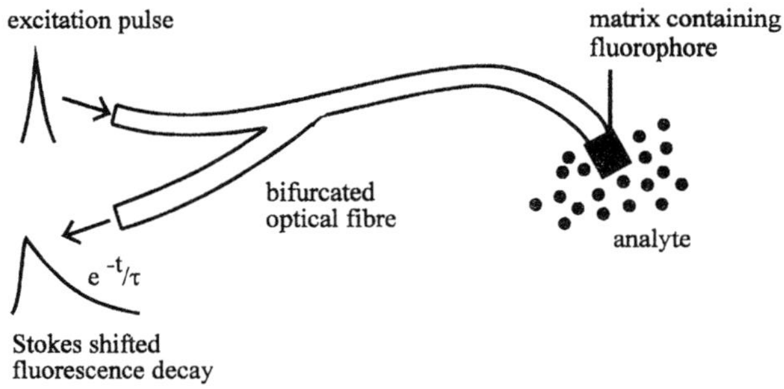

Figure 12.7. Fluorescence lifetime sensor.

Nafion is thought to form dimers giving rise to a biexponential decay.[44, 46] Using diode laser excitation at 670 nm, the fluorescence of oxazine in Nafion and its quenching by copper ions has been shown to give rise to a complex fluorescence decay.[47] Despite such complications there is still room for optimism. For example, Zen and Patonay[48] have demonstrated a pH sensor based on cyanine dye fluorescence intensity in Nafion excited with 30 mW diode laser excitation at 780 nm.

Multilayer Langmuir–Blodgett films doped with a cyanine dye have been deposited on the surface of a quartz multimode optical fiber and the fluorescence properties investigated.[49] The fluorescence intensity of the films was found to be a periodic function of the number of layers due to the waveguide properties of the films.

Nonconventional fluorimeters and sensors that incorporate optical waveguide coupling possess less optical attenuation in the near-IR as compared with the UV/visible because of the reduction in Rayleigh scattering. The temporal dispersion is also reduced in the near-IR for the same reason, e.g., 80 psec/nm/km at 900 nm as compared with 1 nsec/nm/km at 400 nm for a single-mode fiber.

The difference in temporal dispersion in fibers, associated with the Stokes shift and broad spectral content of fluorescence as compared with the excitation, still has to be fully investigated with respect to the effect on convolution analysis.

Care must be taken to avoid the absorption peaks which occur in the near-IR in plastic fibers and plastic-clad silica fibers. Typical transmission characteristics of fibers used in the red and near-IR are shown in Figure 12.8. Single-mode fibers possess the lowest attenuation and minimum temporal-dispersion, but they are inconvenient to work with, having a low acceptance angle and small cross-sectional area (ca. 5 μm diameter) which cause the coupling to be inefficient. Liquid-light guides are now available with transmission up to 900 nm and offer the highest throughput (diameter up to ca. 5 mm with no packing loss), but the poorest multimode temporal dispersion. Gradient index fibers (diameter ca. 50 μm) frequently offer the best all-around performance for fluorescence sensors, because the gradual decrease in refractive index toward the outside of the fiber constrains the light to a more central mode.

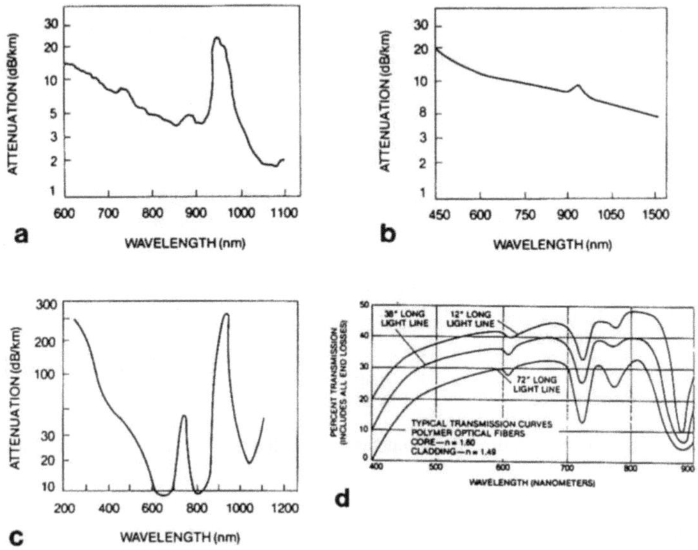

Figure 12.8. Transmission characteristics of some commonly used optical fibers. (a) glass fiber; (b) gradient-index fiber; (c) plastic-clad silica fiber; (d) plastic fiber. (Reproduced with permission from the Ealing Corporation.)

The potential for using multiplexed single-photon counting to acquire fluorescence lifetime data from a distributed array of optical fiber sensors, each sensing a different analyte, has recently been demonstrated by Birch *et al.*[39, 47] Figure 12.9 illustrates such a network to be used in conjunction with the arrangement shown in Figure 12.6.

The opportunities for near-IR fluorescence sensors are of course not only limited to analytical chemistry. Physical parameters such as temperature can also be measured. For example, Grattan and Palmer have used the fluorescence lifetime quenching of neodymium glass fluorescence at 1054 nm, excited at 810 nm with a gallium-alumi-

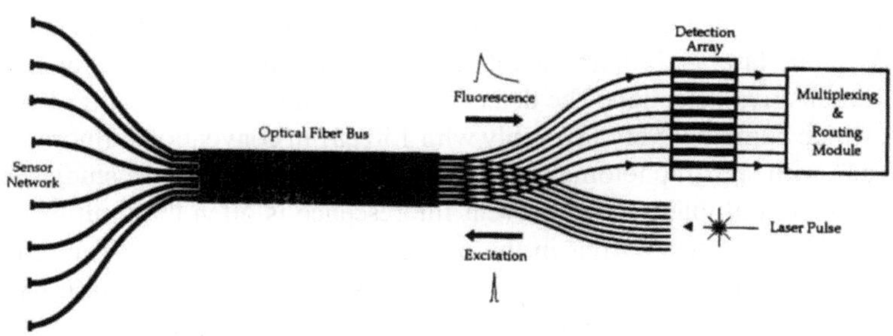

Figure 12.9. Multiplexed single-photon sensor network. (From Ref. 39.)

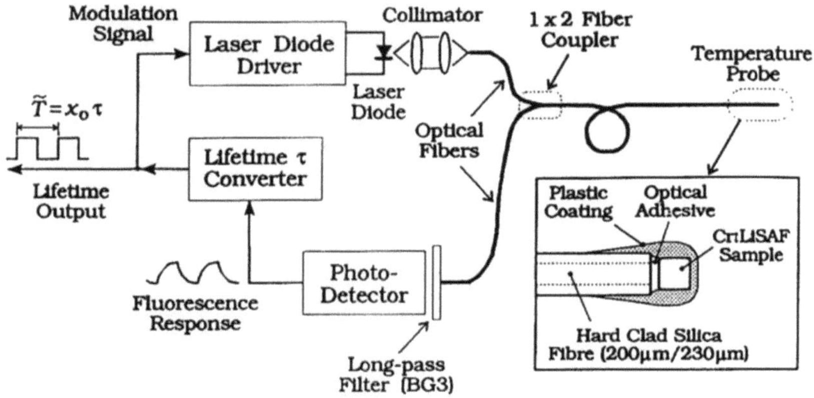

Figure 12.10. Fluorescence lifetime thermometer used in the treatment of benign prostatic hyperplasia. (From Ref. 51.)

num-arsenic (GaAlAs) light emitting diode (LED) and p-i-n (sandwiched intrinsic region between *p*- and *n*-type semiconductor) diode detection, to measure temperature in the range 15–130°C.[50] Subsequent work by the same group has led to fluorescence lifetime thermometry using modulated diode laser excitation and phase-locked detection during radiofrequency heat treatment for benign prostatic hyperplasia.[51] A schematic diagram of the experimental arrangement is shown in Figure 12.10.

In conclusion, it is probably fair to say that the true potential for near-IR fluorescence sensors has yet to be exploited.

12.2.7. High-Performance Liquid Chromatography

The higher sensitivity of fluorescence detection as compared with absorption has been used to good effect in HPLC combined with diode laser excitation in the near-IR. Imasaka *et al.*[52] have labeled serum albumin with indocyanine green and compared the detection limits with near-IR diode laser fluorimetry, conventional arc lamp fluorimetry, and absorption spectrophotometry after separation in a high-performance liquid chromatograph. Figure 12.11 is a block diagram of the experimental layout used. A 15 mW diode laser at 780 nm modulated at 100 Hz was used with photomultiplier and phase sensitive detection. The detection limit with near-IR laser fluorimetry was 1.3 p*M*, which compares very favorably with 150 p*M* in conventional fluorimetry and ca. 250 p*M* with spectrophotometry. The limit on detection in these studies is determined by photomultiplier noise. Solvent fluorescence is often the limiting factor in UV/visible HPLC detection but in the near-IR background solvent fluorescence is negligible. Diode laser excitation at 780 nm and fluorescence detection have been coupled inside a micro-high-performance liquid chromatograph using optical fibers with a view to obtaining better spatial definition and hence higher separation resolution.[53]

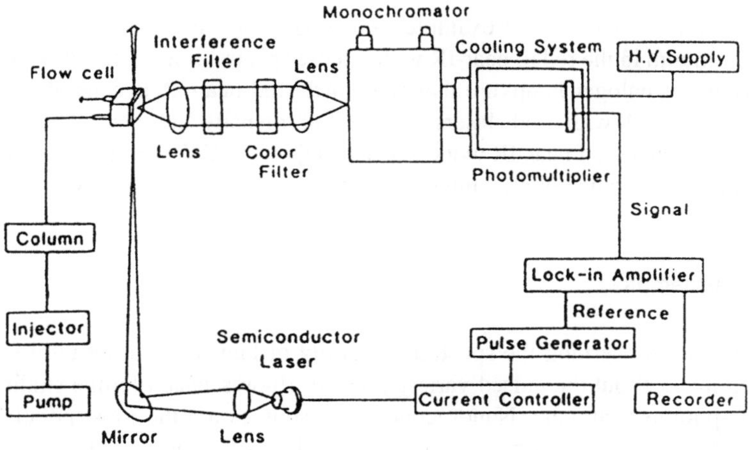

Figure 12.11. HPLC near-IR fluorescence detection. (From Ref. 52.)

The limit of detection in near-IR fluorescence HPLC can undoubtedly be lowered further simply by increasing the laser power, up to 500 mW now being available.

Near-IR fluorescence detection has also been applied to other separation techniques apart from HPLC, e.g., in electrophoresis.[54]

12.3. Sources

We will now discuss the operation and performance of some of the optical sources which are most commonly used for stimulating both steady-state and time-resolved red/near-IR fluorescence. These include lamps, LEDs, and diode lasers as well as a brief discussion of some other types of sources. The fluorescence of IR fluorophores from S_1, the first singlet excited state, can of course still be generated by excitation into S_2, the second excited singlet state, using UV or visible light. The fluorescence from S_1 is then detected following internal conversion from S_2 to S_1. However, this is not good practice given the greater likelihood of self-absorption, distortion of temporal performance due to the effect of a large wavelength difference on the detector, and the propensity of many IR fluorophores to undergo photobleaching. For these reasons we will discuss red/IR sources as a distinct gender.

12.3.1. Lamps

Arc lamps are more usually used in fluorescence studies to provide UV/visible excitation. However, many also possess intense radiation in the red and near-IR, dependent on the filler gas. For example, xenon, krypton, and mercury lamps can all offer a continuum in the red to near-IR with intense spectral lines in the 700–1000 nm region. Tungsten halogen lamps are also useful in the red-near-IR. In general, arc lamps do not compare well with diode lasers in respect of intensity and spectral purity in the near-IR. Nevertheless, if near-IR work is just an adjunct to UV/visible studies then the very broad spectral range of arc lamps makes for much better versatility.

12.3.2. Flashlamps

Comments concerning the spectral versatility of lamps could not until recently have been made about the pulsed variety used for fluorescence lifetime studies with the single-photon technique. Nanosecond spark sources are the most popular light source for fluorescence lifetime measurements. This is because of low cost, ease of use, reliability, and adequate intensity for UV excitation.[4] Hydrogen offers a broad continuum in the UV and nitrogen has intense spectral lines at 316, 337, and 358 nm. The vast majority of spark sources operate by external gating of a hydrogen thyratron which switches a high voltage across a spark gap, the discharge being extinguished by limiting the current flowing by means of a series resistor.[4] Figure 12.12 shows the principal components of the flashlamp used in the authors' laboratory and developed over many years.

The first fluorescence lifetime measurements reported using near-IR spark source excitation where performed in the authors' laboratory using the all-metal coaxial

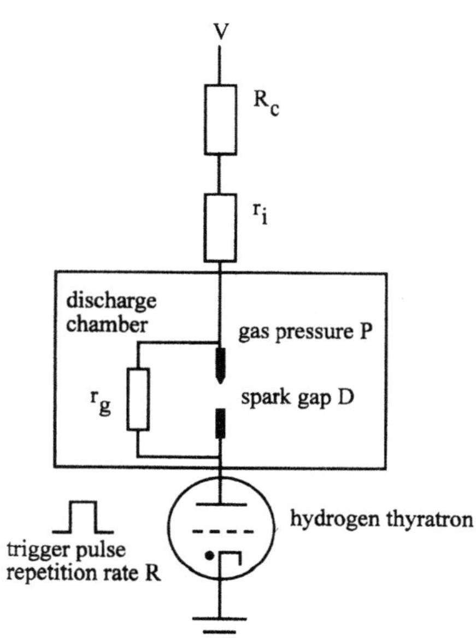

Figure 12.12. Coaxial nanosecond flashlamp components. Typical conditions for red and near-IR operation are: high voltage ($V = 5$ kV), gas pressure $P = 0.8$ B of equal volumes of argon and hydrogen, repetition rate $R = 40$ kHz, and electrode separation $D = 0.6$ mm. The main charging resistor is R_c, the resistor r_i being an isolating resistor for reducing the discharge capacitance. The intergap resistor r_g increases switching stability. Charging resistors are ca. a few megaohms.

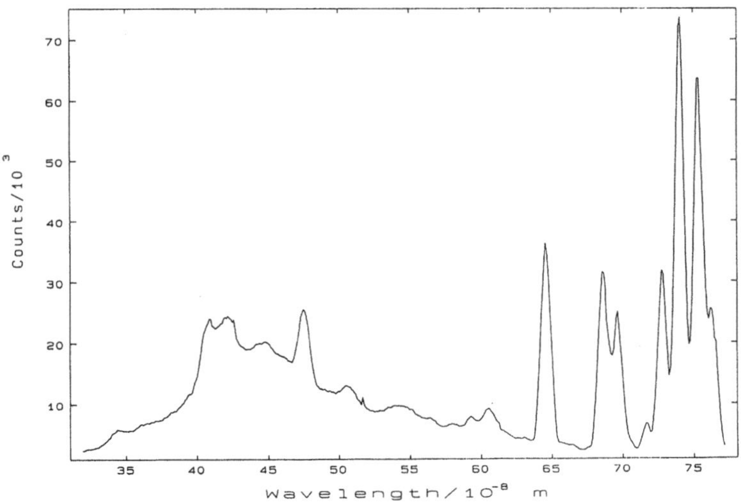

Figure 12.13. Argon nanosecond flashlamp emission spectrum. (From Ref. 55.)

flashlamp filled with a mixture of argon and hydrogen.[55] Figure 12.13 shows the emission spectrum obtained, which includes intense lines at 650 and 750 nm corresponding, respectively, to the $4^3P \rightarrow 3^3P$ and $3^1D \rightarrow 4^3P$ transitions. Many gases have excited states which emit in the red and near-IR, including xenon and neon as well as argon. However, a fact that is not too widely recognized is that changing the gas in a flashlamp has a significant effect on the stable operating conditions. This is because the choice of gas affects the electrical properties of the lamp by virtue of factors such as the dielectric constant, ionization potential, and gaseous diffusion constant. Argon was found to offer the best overall performance in the near-IR in terms of stability and pulse duration. However, it was found necessary to add hydrogen in order to quench the afterglow in argon. Nevertheless, under the conditions shown in Figure 12.12 an overall instrumental pulse full width at half-maximum of 1.0 nsec was obtained at 750 nm, which is comparable to the fastest pulse duration obtained with hydrogen alone. The argon/hydrogen pulse intensity at 750 nm was ca. 15 times more intense than hydrogen at this wavelength. Figure 12.14 shows the fluorescence decay of the dye IR-140 in acetone at wavelengths greater than 850 nm and excitation pulse at 750 nm from the coaxial flashlamp filled with argon/hydrogen.[55] This performance is comparable to that obtained in the UV/visible. The fluorescence lifetime of IR-140 was found to be 1.20 nsec. The argon/hydrogen flashlamp has also been used to study the decay of other polymethine dyes, e.g., DTDCI in acetone excited with the spectral line at 655 nm, obtaining a fluorescence lifetime of 1.46 ± 0.02 nsec.[56] This figure compares very well with a subsequent report of 1.463 ± 0.015 nsec in ethanol recorded using phase fluorometry and diode laser excitation.[57]

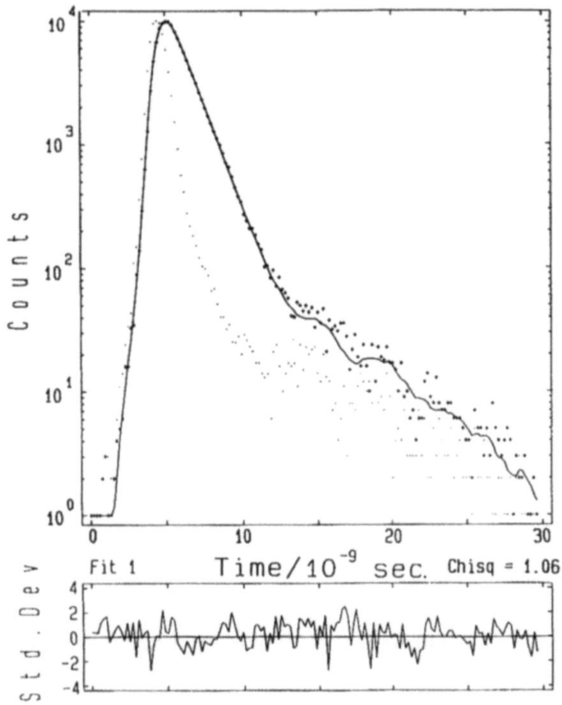

Figure 12.14. Fluorescence decay (+++), excitation pulse (···), fitted function (—) for IR-140 in acetone excited at 750 nm using the argon–hydrogen flashlamp. The decay was measured in 4 min. The fluorescence lifetime is 1.2 nsec. The weighted residuals are shown at the bottom of the figure. (From Ref. 55.)

Two of the main advantages of using near-IR excitation in fluorescence probe studies concern the elimination of background fluorescence and the reduction in scattered excitation. Nowhere are these factors more relevant than in the study of microheterogeneous media such as membranes, micelles, and liquid crystals, all having important applications. In an early study the authors reported the fluorescence lifetime behavior of IR-140 in lipid bilayer membranes of L-α-dipalmitoylphosphatidylcholine (DPPC) and in water in oil microemulsions formed by Aerosol-OT.[58] The latter contains impurities that are extremely difficult to remove and that fluoresce strongly when excited in the UV/visible. Figure 12.15 shows how the fluorescence lifetime of IR-140 in Aerosol-OT inverse micelles dispersed in iso-octane depends on the size of the water pool solubilized inside the micelle. The decrease in fluorescence lifetime with increasing water pool size is interpreted in terms of IR-I40 reporting the reduction in the interfacial viscosity, the insolubility of IR-140 in water and iso-octane causing the dye to lie in the surfactant-water interface. Similar trends in Aerosol-OT microemulsions were previously reported using UV flashlamp excitation to study the anisotropy decay and quenching of more conventional fluorescent probes.[59]

Although nanosecond flashlamps operated in the near-IR possess much less intensity and repetition rate, as well as giving much broader pulses when compared to diode lasers, the recent development of flashlamp operation in the near-IR has

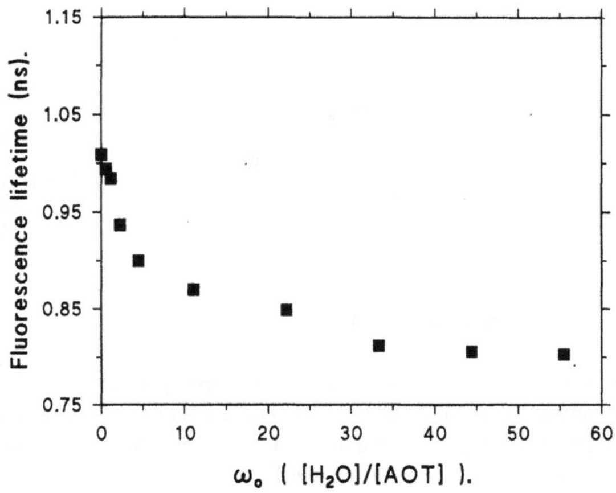

Figure 12.15. Fluorescence lifetime of IR-140 in Aerosol OT (AOT)/iso-octane microemulsions as a function of water pool size ω_0 defined as the molar fraction of water to AOT. (From Ref. 58.)

extended the spectral range of this most popular pulsed source from the vacuum UV to 780 nm.

12.3.3. Light-Emitting Diodes

LEDs are electroluminescent devices fabricated from a semiconductor pn junction and offer inexpensive generation of steady-state or pulsed excitation of low intensity from the near-UV to the near-IR. LEDs epitomize many of the advantages of semiconductor optoelectronics for optical spectroscopy.

The components of an LED are similar to that of the laser diode, the differences between the two devices being largely dependent on the operating conditions and of course the nature of the output.

Figure 12.16a shows the simple band model of a pn junction applied to some of the radiative transitions involved in the operation of different types of LEDs. The most commonly used and simplest radiative mechanism is that of electron-hole recombination directly across the bandgap E_g between the conduction and valence bands. The wavelength of emission then given by hc/E_g. This process can only occur in direct bandgap semiconductors such as gallium-arsenide (GaAs) where the crystal momentum is conserved in the transition, but not in indirect semiconductors such as silicon or gallium–phosphorus (GaP). However, other radiative mechanisms can also be fabricated into pn junctions to vary the emission wavelength such as exciton recombination to an acceptor trap within the bandgap, an energy to which transitions would otherwise be forbidden. Fuller discussions of the physics of LEDs and the theory of pn junctions are available in many texts (e.g., see Ref. 60).

The most common method of generating LED emission is via the injection of minority carriers toward the depletion region at the junction by forward-biasing a pn

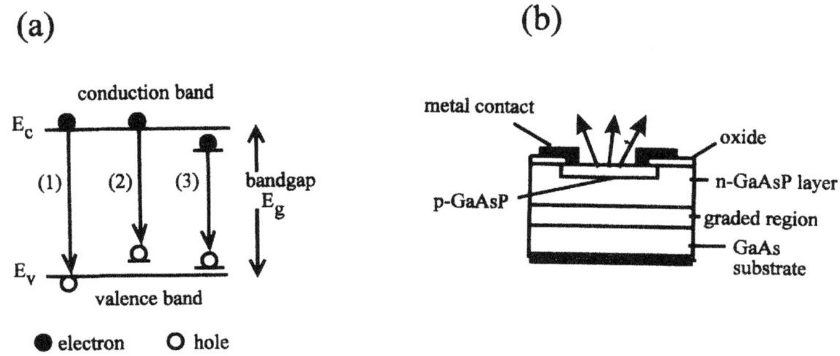

Figure 12.16. (a) Radiative transitions occurring in semiconductors. (1) Intrinsic band to band electron-hole recombination. (2) Extrinsic recombination between electron and neutral acceptor. (3) Exciton recombination between traps in the bandgap. (b) Cross section through a GaAsP LED. The graded region avoids lattice mismatch at the GaAs-GaAsP interface.

junction with as little as ca. 1 V. This is particularly convenient as it involves minimal drive electronics even if modulation and pulsing for time-resolved use is required. Direct current (DC) operation at ca. 100 mW power and modulation frequencies at up to the gigahertz level are readily achievable. It is usual to bias the LED with a series resistor which also limits the current flowing, the power dissipation being low enough for batteries to be used in DC operation.

GaAs LEDs emit in the near-IR, compound alloys of gallium–arsenide–phosphide (GaAsP) in the near-IR to green, silicon–carbide (SiC) in the violet, and gallium–nitride (GaN) and zinc–sulfide (ZnS) in the near-UV. Dopants are required for all these. For example, green light at 565 nm has been obtained in gallium phosphorus (GaP), an indirect bandgap semiconductor, by means of traps introduced by doping with nitrogen in the p and n sides of the junction.

Figure 12.16b shows the a typical cross-section through a GaAsP LED fabricated from layers grown by either liquid phase epitaxy or vapor phase epitaxy.

The overall efficiency of LED emission depends on three factors which vary between the different types of LEDs. These are the efficiency of electron-hole production, the radiative efficiency of recombination, and the efficiency of extraction of the optical signal from the junction.

The emission linewidth of LEDs is typically 10 nm, which is quite broad in comparison to that achievable with laser diodes. Narrower linewidths down to ca. 0.9 nm can be obtained using resonant cavity LEDs.[(61)] Superluminescent LEDs have been produced with low spectral ripple (less than 10% at wavelengths down to 670 nm) by suppressing the optical feedback in laser diode junctions.[(62)]

LEDs have already found a number of applications that demonstrate potential with respect to fluorescence measurements requiring routine operation, low cost, or compact implementations. A 50 mW LED (GaP) with a peak emission at 565 nm has been used in conjunction with HPLC separation and methylene blue fluorescence to detect alcohols down to the microgram level.[63] The high stability of the LED (root mean square (rms) noise ca. 0.0004%) giving a better signal-to-noise ratio than more conventional sources. The same type of LED has also been used to study the fluorescence of Nile Red bound to proteins in a filter fluorometer.[64] The small dimensions of LEDs (a few square millimeters in area) and well-defined spectral bandwidth means they can often be used in close proximity to the sample of interest, without the need for additional focusing or wavelength dispersive optics.

Finally, we should note that LED operation is not limited to inorganic semiconductor materials. Polymer LEDs, which are simpler to fabricate than the inorganic counterparts, have recently been reported.[65, 66]

12.3.4. Diode Lasers

The application of semiconductor lasers to a broad range of areas in spectrometry has recently been reviewed by Imasaka.[67, 68] Topics covered include photoacoustic, absorption, and thermal lens, as well as steady-state and time-resolved fluorescence. Patonay *et al.* have reviewed the application of diode lasers to analytical chemistry.[69] The performance of several commercially available laser diodes for fluorimetry has recently been compared.[70]

In fluorescence studies, diode laser excitation, as compared with that from an LED, offers all the usual advantages associated with lasers, namely, high coherence, narrow spectral linewidth, high intensity, and fast pulse operation. Diode laser output is also highly polarized, facilitating fluorescence anisotropy measurements. Steady-state laser power levels are typically in the 1–100 mW region, higher powers requiring thermoelectric cooling of the diode. High powers are available from ca. 635 nm upward making diode lasers very suitable for studying near-IR fluorescence. Frequency-doubled pulsed diode lasers are commercially available but with < 1 mW power at ca. 400 nm. Optical fiber pigtail coupling is readily available. Pulse durations down to < 50 psec at repetition rates up to 1 MHz are typical.

A diode laser is fabricated from a semiconductor pn junction. The operation of a diode laser is depicted in Figure 12.17. Nearly all diode lasers take advantage of the high efficiency of radiative emission from electron-hole recombination directly across the bandgap (see Figure 12.16a). The lasing requirement for population inversion is obtained by applying a forward bias such that electrons and holes are forced at high concentrations into the depletion layer formed by the space charge at the junction. In the Fabry–Perot construction a resonant laser cavity is formed by optically polishing the opposite faces at the edge of the junction as shown in Figure 12.17b.

Note that spontaneous emission also occurs in diode lasers in the manner of the LED and gives rise to a broadband but weaker optical signal underlying the lasing wavelength. Hence care must be taken in fluorescence work when using cut-off filters to select fluorescence close to the diode laser excitation wavelength. It is this spontaneous emission which occurs first as the bias voltage is increased until, at voltages

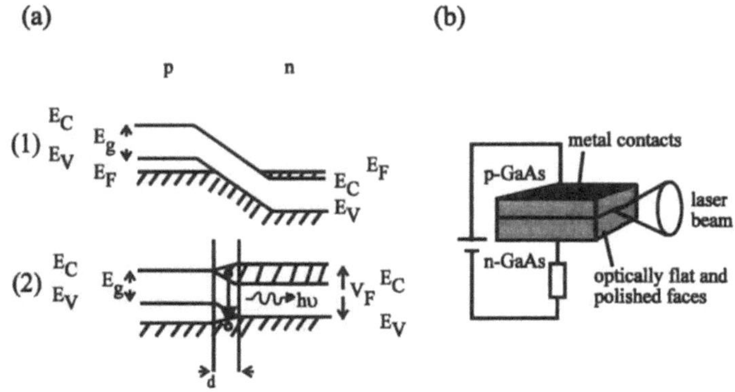

Figure 12.17. (a) Diode laser band structure. (1) In thermal equilibrium. (2) Under forward bias and high carrier injection. E_c, E_v, and E_f are the conduction band, valence band, and Fermi energies respectively. (b) Fabry–Perot cavity configuration for a GaAs diode laser. Typical cavity length is 300 μm and width 10 μm. d is the depletion layer.

greater than that of LED operation, stimulated emission occurs at wavelengths ($\lambda = 2L/n$ where $n = 1, 2, 3 \ldots$ and L is the length of cavity) corresponding to the normal modes of the cavity. Almost instantaneously one mode will dominate and the diode becomes self mode-locking. In GaAs the bandgap is 1.4 eV and this gives rise to the laser wavelength of 880 nm. Linewidths are typically 3 nm, not narrow enough for many types of high-resolution spectroscopy, but acceptable for fluorescence excitation of aromatic fluorophores.

The nature of the optical signal generation and extraction gives a high beam divergence, as high as 30° in many cases. Hence some external focusing is often desirable.

It will be seen that, as in the case of the LED, control of the bias voltage gives simple modulation of the laser output intensity. This is particularly useful in phase-modulation fluorometry. However, a measure of the late awareness of the advantages of IR techniques in fluorescence is that only recently has this approach been applied to the study of aromatic fluorophores. Thompson *et al.*[57] have combined modulated diode laser excitation at 670 and 791 nm with a commercial fluorimeter in order to measure the fluorescence lifetimes of some common carbocyanine dyes. Modulation frequencies up to 300 MHz were used in conjunction with a Hamamatsu R928 photomultipler for detecting the fluorescence. Figure 12.18 shows typical phase-modulation data taken from their work, the form of the frequency response curves is as shown in Figure 12.2 which describes the response to a monoexponential fluorescence decay.

A number of applications of diode lasers to near-IR fluorescence studies have been cited in this chapter.[11–13, 40, 47, 48, 51–54, 57, 67–70] There can be little doubt that the high intensity, repetition rate, and time-resolution afforded by diode lasers combined with

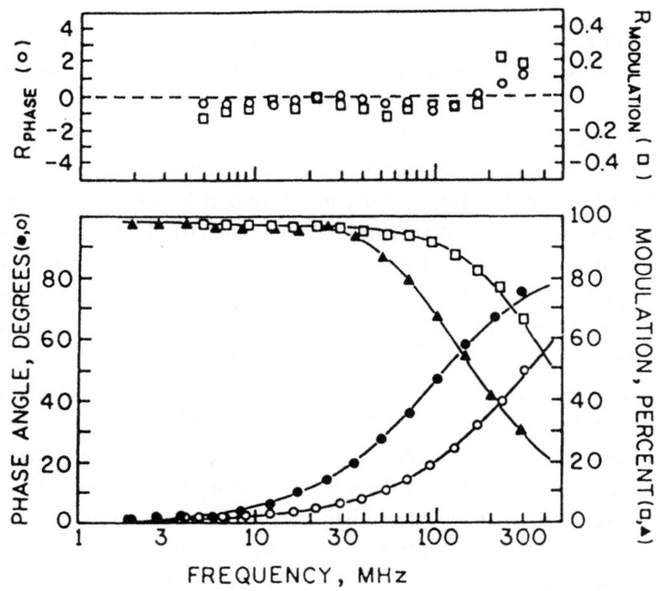

Figure 12.18. The lower portion of the figure shows diode laser phase-modulation fluorescence lifetime measurements of IR-144 (○, □) and DOTCI (●, ▲) in ethanol giving best fit (—) decay times of 0.593 ± 0.007 nsec and 1.71 ± 0.015 nsec, respectively. The residuals between measured data and the theoretical model are shown in the upper portion for IR-144. (From Ref. 57.)

a low cost when compared to mainframe lasers will lead to the increased use of diode lasers in fluorescence spectroscopy in the years ahead. As far as near-IR fluorescence is concerned, diode lasers are at present the nearest there is to an ideal source.

12.3.5. Other Sources

Space does not permit a full discussion of the myriad laser sources available for studying red/near-IR fluorescence. In any case the operation and properties of such sources are well described in many other texts. Nevertheless, for the sake of completeness, we will briefly mention some of the other sources here.

The helium-neon (HeNe) laser immediately comes to mind, having a very useful spectral line at 633 nm for steady-state red/near-IR fluorescence studies. Kessler and Wolfbeis have demonstrated the fluorescence assay of the protein human serum albumin using the probe albumin blue excited with a red HeNe laser.[71] Another useful wavelength available from the green HeNe laser is at 543.5 nm and this has been used with phase-modulation fluorometry by Lakowicz *et al.* to study probes such as carboxy seminaphtorhodafluor-6 (SNARF-6) as a means of measuring pH.[72]

However, unlike diode lasers, an external acousto-optic or electro-optic modulator is required for phase-modulation lifetime fluorometry to be performed with a HeNe laser.

Argon ion lasers, mode-locked to produce pulses in the picosecond domain, are in widespread use, producing power levels generally higher than diode lasers up to ca. 1 W in spectral lines at 488 nm and the green line at 514 nm. Krypton lasers offer comparable performance but into the near-IR with spectral lines up to 799 nm. Argon ion lasers have been widely used in the past to synchronously pump dye lasers, the latter offering picosecond pulses of ca. 76 MHz repetition rate, tunable throughout the visible and near-IR up to ca. 1100 nm using many of the same carbocyanine dyes frequently cited in this chapter. More recently frequency-doubled neodymium: yttrium-aluminum-garnet (Nd:YAG) lasers have replaced argon laser pumping of dye lasers. Some examples of red/near-IR fluorescence lifetime studies using Nd:YAG pumped dye lasers have already been mentioned.[27, 28, 31] The Nd:YAG fundamental at 1064 nm is a little long for most fluorophore absorption, but the second harmonic at 532 nm is useful. A major problem with these alternatives to diode lasers is the higher cost involved, but the underlying trend is undoubtedly toward all solid state tunable laser systems. This has been clearly evidenced in the recent replacement of dye lasers by Ti:sapphire lasers offering a much broader spectral range than any one dye. Some near-IR applications of Ti:sapphire lasers have already been cited.[14, 24]

Very recently, white light continuum pulses of duration ca. 200 fsec, pulse energy ca. 1 μJ, and peak wavelength of ca. 780 nm have been generated at repetition rates up to 250 kHz by commercially available Ti:sapphire regenerative amplified laser systems. Such systems are very expensive, but the expected easier use, as compared with homemade systems, should open up new research applications for time-resolved fluorescence and absorption techniques in the near-IR.

A useful source of continuously tunable radiation from the near UV to the near-IR with unexplored potential in fluorescence studies is the optical parametric oscillator (OPO). These devices have been around since the 1960s[73] and have received a lot of coverage recently in laser and optoelectronic journals.[74] This resurgence of interest in OPOs has been brought about by recent improvements in nonlinear crystals and the development of all-solid-state pump-laser sources with the required levels of coherence and intensity.

The operating principle of an OPO is simply a general parametric effect of which frequency doubling is a special case. OPOs generate the difference frequency (called the idler) between two laser pulses (known as the pump and signal) of different frequencies impinging on a nonlinear crystal. Crystals such as titanyl phosphate, beta-barium borate, and lithium triborate have been used in OPOs. OPOs can be pumped by continuous-wave (CW), Q-switched, mode-locked, or synchronously pumped lasers with concomitant characteristics derived from the pump-laser. Nanosecond, picosecond, and femtosecond pulses have all been generated at mW power levels using OPOs. Frequency-doubled, -tripled, and -quadrupled neodymium: yttrium–aluminum garnet (Nd:YAG) or neodymium:yttrium–lithium fluoride

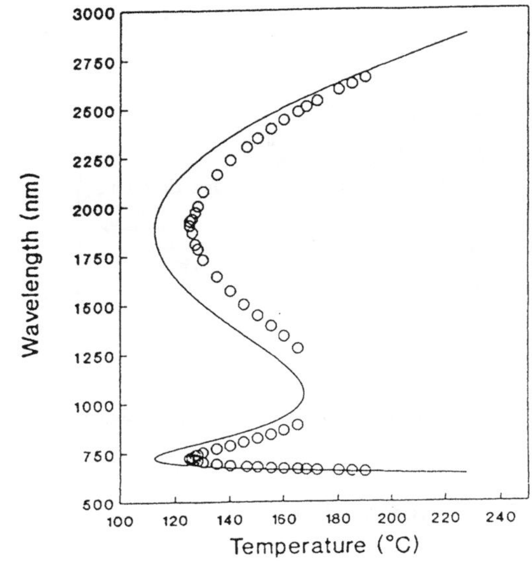

Figure 12.19. Temperature tuning of an optical parametric oscillator. The continuous curve is a theoretical prediction and the circles represent measured data. (From Ref. 75.)

(Nd:YLF) lasers as well as excimer lasers have been used as pump sources, giving OPOs an unparalled tuning range from a single source. Spectral tuning is accomplished by changing the temperature of the nonlinear crystal or by changing the pump laser wavelength. Figure 12.19 shows a typical temperature tuning curve of an OPO and illustrates the very broad tuning range available. A good example of where OPOs might prove useful is in the measurement of time-resolved fluorescence excitation spectra in the red and near-IR for which the narrow spectral bandwidth associated with each diode laser crystal renders these devices unsuitable for excitation studies.

Further important developments in OPO performance undoubtedly lie ahead, in particular with respect to all-solid state systems which are sufficiently easy to operate in analytical instruments.

12.4. Detectors

In this final section, we summarize the operation and characteristics of the principal vacuum tube and solid state detectors that are available for red/near-IR fluorescence studies. These include conventional photomultipliers, microchannel plate versions, streak cameras, and various types of photodiodes. Detector applicability to both steady-state and time-resolved studies will be considered. However, emphasis will be placed on photon counting capabilities as this provides the ultimate sensitivity in steady-state fluorescence measurements as well as permitting lifetime studies.

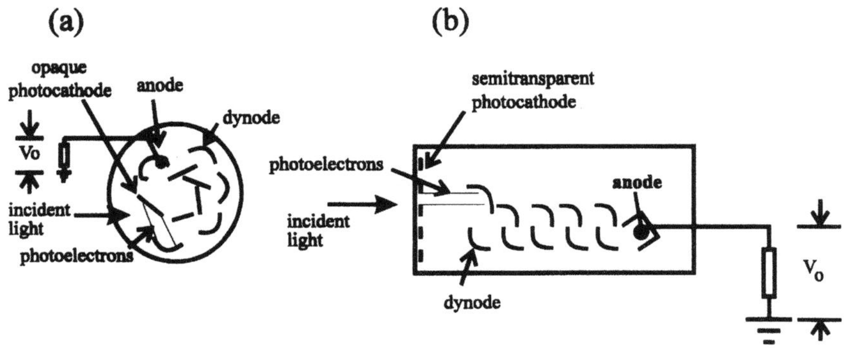

Figure 12.20. Structure of (a) side-window and (b) linear focused photomultipliers.

12.4.1. Photomultipliers

Despite the many advances that have occurred in semiconductor devices, conventional photomultipler vacuum tubes are still by far the most widely used detectors of fluorescence.

Figure 12.20 shows the structure of the side-window circular cage type and linear focused head-on type of photomultiplier which are both preëminent in fluorescence studies. The lower cost of side-window tubes tends to favor their use for steady-state studies, whereas the ultimate performance for lifetime studies is probably at present provided by linear focused devices. In both types internal current amplification is achieved by virtue of secondary electron emission from discrete dynode stages, usually constructed of copper–beryllium (CuBe) alloy, though gallium–phosphide (GaP) first dynodes have been used to obtain higher gains.

The primary concern when choosing a photomultiplier for red/near-IR studies is of course the photocathode spectral response. Figure 12.21 shows the response of some commonly used photocathode compositions.[76] Bialkali compositions (e.g., type D) are used for UV/visible studies up to ca. 650 nm and trialkali compositions (e.g., S20) for near-IR studies up to ca. 850 nm. Silver–oxide–cesium (AgOCs) photocathodes (S1) give a workable response in most fluorescence measurements up to ca. 1000 nm, but are notoriously noisy in DC measurements and at the high gains required for photon counting. Egorov *et al.*[15] have measured the fluorescence lifetime associated with singlet oxygen emission in water at 1270 nm using a single-photon timing photomultiplier with an S1 photocathode. The temporal gate width afforded by time-resolved measurements being useful in reducing the noise recorded. Indium–gallium–arsenide (InGaAs) photocathodes have been used to obtain a spectral response beyond 1000 nm with lower noise than S1 photocathodes. However, in all these characteristics it should be noted there is always a loss in UV/visible response if near-IR performance is to be enhanced. Similarly, consideration should be given to the transmission

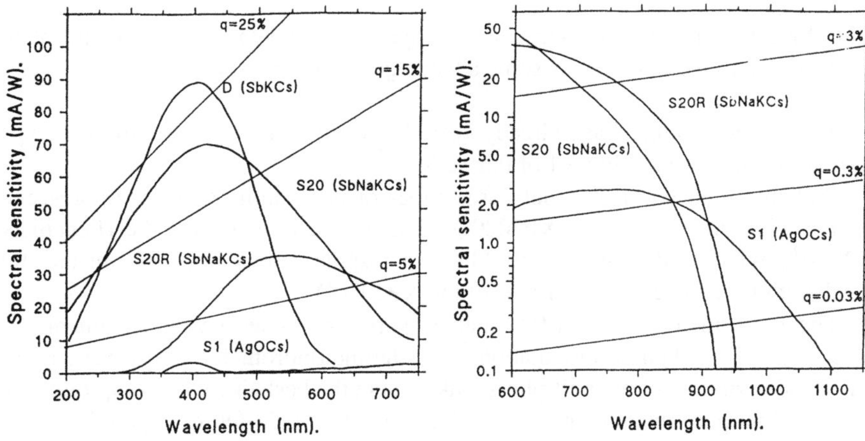

Figure 12.21. Photocathode spectral sensitivity curves in milliamperes of cathode current per watt of radiant power falling on the photocathode. q is the photocathode quantum efficiency, i.e., the probability of photoelectron emission. (Adapted from Ref. 76.)

properties of the window material, fused silica, sapphire, and magnesium fluoride all offering a wide range from the UV to visible, though borosilicate glass is perfectly adequate and less expensive if only visible and near-IR response is required. Cooling of photomultipliers is essential if the spectral response exceeds 850 nm, although small-area photocathode versions have helped to ameliorate the effects of thermionic noise.

The principal requirements for photomultipliers in both pulse and phase-modulation methods of measuring fluorescence lifetimes are as follows.

1. Fast time-response of typically 1.5 nsec risetime and 0.5 nsec FWHM jitter in single-photon transit time.

2. Electron gain > 5 million. Lower gains are workable, but external amplification is then required and the greater is the susceptibility to radiofrequency pickup from the source.

3. Low transit-time dispersion with photon wavelength, i.e., < 0.5 psec/nm. This minimizes the effect on convolution of the difference between the excitation and fluorescence wavelengths. Both side-window and linear focused photomultipliers satisfy this.

4. Low transit-time dispersion with point of illumination on the photocathode, i.e., < 20 psec/mm. Linear focused photomultipliers satisfy this criterion, but side-window devices do not. This again is relevant to successful data analysis.

5. Low noise of ca. < 100 counts/sec.

6. Low afterpulsing due to ion or photon feedback to the photocathode and low prepulsing due to photoelectron generation from dynodes. The lack of a direct line of

sight along the side-window circular cage geometry and the opaque nature of the photocathode in the latter offers some advantages in these respects.

The previous points have already been discussed with respect to different types of photomultipliers in Volume 1 of this series.[77]

Probably the two most widely used types of photomultipliers for fluorescence lifetime work are the Phillips XP2020Q linear focused device and the Hamamatsu R928 side-window device. The R928 has a response up to ca. 930 nm and this wavelength is achieved with variants of the XP2020Q.

The performance of the R928 and its variants has been well documented in pulse[12, 31, 78–81] and phase-modulation[57, 82] lifetime fluorometry. The lower cost and gain of side-window photomultipliers makes them the best all-round compromise for use in dual purpose instruments with both steady-state and lifetime capabilities.

Another side-window photomultiplier often used in red/near-IR fluorescence studies is the Hamamatsu R636 with a gallium–arsenide (cesium) (GaAs(Cs)) photocathode.[11, 53]

Birch *et al.* have used the Philips XP2254B, an S20 version of the XP2020Q, to study the fluorescence lifetimes of a series of aminotetraphenylporphyrins in a multiplexed fluorometer.[83] The extended red response (S20R) version of this device, the XP2257B, has been used with IR spark source excitation to study the fluorescence lifetimes of carbocyanine dyes up to 930 nm emission in isotropic and anisotropic media.[55, 56, 58, 84] An improved voltage divider network has been developed for linear focused photomultipliers which reduces thermionic noise from the photocathode by an order of magnitude by restricting the collection of photoelectrons to the center of the photocathode.[84]

Other linear focused photomultipliers reported in near-IR fluorescence studies include the Hamamatsu R943-02[52] and the RCA C31034,[85, 86] both incorporating gallium-arsenide (GaAs) photocathodes.

12.4.2. Microchannel Plate Photomultipliers

MCP-PMs are widely used with picosecond sources in fluorescence lifetime work in order to make the most of the time-resolution available. However, the greater than an order of magnitude increase in cost of MCP-PMs, as compared with conventional photomultipliers, means that the latter are still more appropriate for use with nanosecond sources such as flashlamps.

The operating principle of an MCP-PM is based on electron multiplication using a continuous dynode structure of ca. 10 μm diameter holes, giving a more compact and hence faster time response when compared with conventional photomultipliers. Rise-times of 150 psec and transit-time jitter (i.e., impulse response) of ca. 25 psec FWHM at 200 counts/sec noise at room temperature have been recorded with the 6 μm channel Hamamatsu R3809 MCP-PM.[87]

The same types of photocathode compositions used in conventional photomultipliers also apply to MCP-PMs and similar considerations apply. Typical performance of MCP-PMs for the UV/visible is demonstrated in several chapters in an earlier volume in this series.[4]

There are at present few reports of the use of MCP-PMs in near-IR fluorescence studies, which reflects more the newness of the application rather than the availability of suitable detectors. Kume *et al.*[88] have used single-photon timing with an MCP-PM to record diode laser pulses of 80 million photons/pulse impinging on an S1 photocathode at wavelengths as long as 1.5 μm. However, the spectral sensitivity of S1 photocathodes is so low at this wavelength that in practice fluorescence signals are often better recorded at such wavelengths using photodiodes (see Sections 12.4.3. and 12.4.4).

Figure 12.22 shows the time-resolved fluorescence spectra of a synthetic bacteria (*R. Sphaeroides*) up to 1000 nm at gate widths of 5 psec, recorded using an MCP-PM with an S1 photocathode.[88]

Multianode versions of MCP-PMs[88] are now available which are particularly suitable for multiplexed single-photon timing measurements (see Sections 12.2.5 and 12.2.6). The potential for MCP-PMs for 2-D imaging has also been demonstrated using

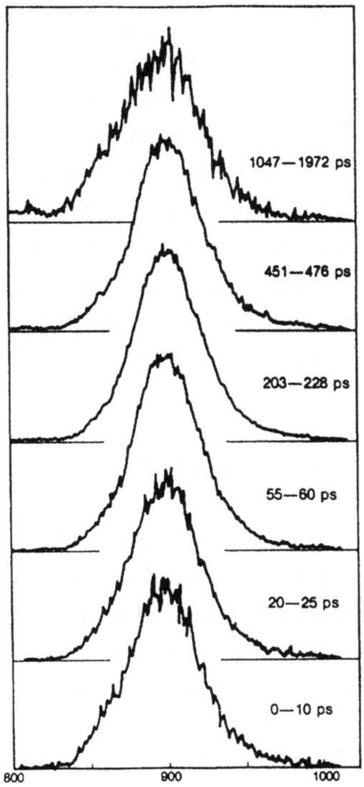

Figure 12.22. Time-resolved fluorescence spectra of synthetic bacteria (*R. Sphaeroides*) recorded using an MCP-PM with an S1 photocathode. (From Ref. 88.)

resistive[89] and delay line[90] anodes. Microchannel plates are widely used for image intensification with CCD cameras[30] and in streak cameras.[23]

12.4.3. Streak Cameras

The operation and application of streak cameras in fluorescence lifetime spectroscopy has been reviewed previously (see, e.g., Refs. 91 and 92). Streak cameras are useful in 2-D time-resolved imaging applications such as microscopy or multiwavelength array fluorometry. The operating principle is based on converting an optical pulse into a photoelectron pulse and spatially dispersing the electron image on a phosphor by means of a synchronized deflection voltage across two plates.

Similar photocathode types such as are used in photomultipliers are available with streak cameras. For example, a streak camera with an S1 photocathode suitable for the detection of near-IR fluorescence has been developed.[23] S20 photocathodes are also readily available.

Major limitations of streak cameras in fluorescence lifetime studies as compared with time-correlated single-photon counting concern the lower dynamic range and higher cost. Nevertheless a 15 psec FWHM response, even faster than the fastest MCP-PM or APD, has been achieved with a streak camera coupled to a CCD camera and analog-to-digital converter (ADC) for digitization of the pulse response.[93] However, the dynamic range in this case was only 105, some four orders of magnitude less than with time-correlated single-photon counting. Streak camera response times even down to the femtosecond region have been demonstrated[94] and might prove useful in fluorescence applications, though the extension to the near-IR places additional limitations on time-resolution.[92] Recently, photon counting with a streak camera,[95] to give a dynamic range approaching that of time-correlated single-photon counting, has been demonstrated at a source repetition rate of only 10 Hz in measurements of the fluorescence decays of rhodamine 6G and 3,3′-diethyloxadicarbocyanine iodide (DODCI) (Figure 12.23).

12.4.4. Photodiodes

Photodiodes occur in many different varieties and are useful in both steady-state and time-resolved fluorescence studies. Photodiodes designed for use in steady-state or on microsecond time-scales are inexpensive and have effective areas up to a few square millimeters, and are capable of efficiently matching to simple focusing optics. However, as the temporal resolution increases so does the cost, and the effective area has to be reduced. For example, APDs with response times in the 50 psec region have effective diameters of ca. 10 μm. Hence, the small active area of high-speed devices is currently the primary drawback in fluorescence studies. Also, photodiodes other

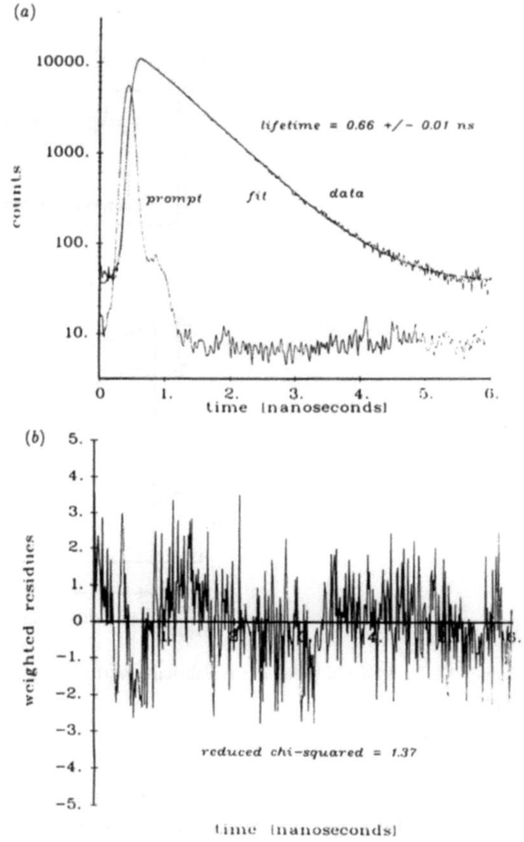

Figure 12.23. Photon counting streak camera measurement of the fluorescence decay of DODCI (10^{-6} *M*) in water. (a) counts; (b) weighted residues. Lifetime = 0.66 ± 0.01 ns. (From Ref. 95.) Note the similarities with single-photon timing data, e.g., see Figure 12.14. Reduced χ^2 = 1.37.

than APDs cannot detect single photons as they lack gain, and they are certainly not as sensitive as photomultiplier tubes.

Photodiodes are based on a pn junction operated at a reverse bias voltage (i.e., the opposite bias to the LED and diode laser). The reverse rather than forward bias voltage has the effect of increasing the voltage across the depletion region such that any photoinduced electron-hole pairs are rapidly swept across the junction, generating a current pulse in the external circuitry (Figure 12.24).

The speed of response of the photodiode depends on the diffusion of carriers, the capacitance of the depletion layer, and the thickness of the depletion layer. The forward bias itself increases the width of the depletion layer thus reducing the capacitance. Nevertheless, some design compromises are always required between quantum efficiency and speed of response. The quantum efficiency of a photodiode is determined largely by the absorption coefficient of the absorbing semiconductor layer. Ideally all absorption should occur in the depletion region. This can be achieved by increasing the thickness of the depletion layer, but then the response time increases accordingly.

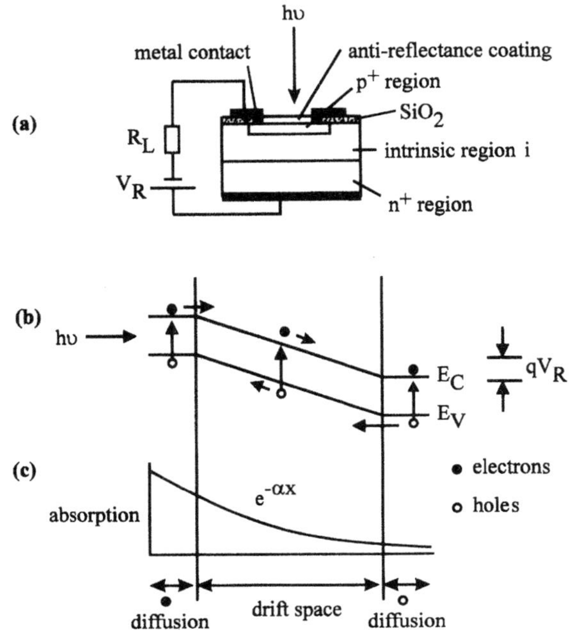

Figure 12.24. PIN photodiode. (a) Fabrication. (b) Energy band diagram. (c) Absorption in the depletion layer.

As a rule of thumb the optimum depletion layer thickness is that defined by half the period of the required modulation response, i.e., at 2 GHz bandwidth and a carrier velocity of 100,000 m/sec the optimum depletion layer thickness is ca. 50 μm.[60]

In one of the most common types of photodiodes used for time-resolved work, the p-i-n photodiode (see Figure 12.24), the depletion layer thickness (i for intrinsic) is fabricated to obtain this optimum performance. Manufacturers usually give full specification sheets detailing, active area, time/frequency response, responsivity amps/watt (A/W) at a given wavelength, dark current, depletion layer capacitance, and bias volts such that with minimal external electronics devices can be made operative.

The upper wavelength response of a photodiode is determined by the bandgap of the semiconductor and the lower wavelength by the absorption of the uppermost semiconductor region (Figure 12.24), wherein recombination occurs rapidly before the carriers reach the depletion region and are dispersed to produce a current. Hence silicon with a bandgap of 1.12 eV produces photodiodes that respond up to the energy-equivalent wavelength of 1100 nm, but germanium with a lower bandgap of 0.66 eV responds up 1800 nm. Silicon has a lower absorption coefficient than germanium in the visible and hence is more frequently used for this spectral region. Silicon photodiodes with a response between 200 and 1100 nm have replaced photomultipliers in many spectrophotometers. GaAs photodiodes are useful up to ca. 870

nm and other compound semiconductors are available for other ranges. Naturally thermionic noise increases as the bandgap is lowered, hence there is no point in working with a photodiode that has a wavelength response significantly longer than is required for the task in hand. Photodiodes often come with Peltier cooling to reduce noise and with the optical viewing surface antireflection coated.

Photodiode detectors have already been cited in this chapter in relation to near-IR fluorescence measurements on singlet oxygen,[8, 16–18] in decay-time temperature sensing,[50] in liquid chromatography,[62] the study of proteins labelled with Nile Red,[64] and diode laser spectrometry.[67] Photodiodes are also conveniently packaged for many applications in an array form enabling rapid data acquisition e.g., in spectrophotometry.[35]

Other types of photodiodes not discussed here include the Schottky barrier, pn homojunction, and heterojunction.

12.4.5. Avalanche Photodiodes

APDs are similar to the photodiodes described in Section 12.4.4, i.e., reverse-biased pn junctions but operate with the reverse bias voltage held slightly above rather than below the breakdown voltage for the junction. In the most commonly used mode the device operation is analogous to a Geiger–Müller counter, the electric field being high enough to sustain an avalanche of carrier multiplication via secondary ionization once a primary electron-hole pair has been photoinduced via absorption in the depletion layer. The diode current can be turned off passively[96] by limiting the current flowing with a suitable resistor or actively[97, 98] by lowering the bias voltage after the onset of the avalanche, the latter leading to reduced dead-time (ca. 50 nsec) and a faster time response. Dark noise in APDs originates in thermal ionization and hence cooling below room temperature is essential in the small bandgap devices used for near-IR studies. Some devices come complete with coupling to an optical fiber pigtail, offering potential applications with fluorescence sensors.

The combination of gain and fast time-response afforded by APDs has found applications in fluorescence lifetime measurements, such devices not being appropriate in steady-state studies. However, these applications have been primarily limited by the extremely small active area of the fastest devices.

The current state of the art in single photon counting APDs has recently been reviewed elsewhere.[99] Single-photon APDs can be plagued by a wavelength-dependent tail in the instrumental response due to the slow diffusion of photoinduced carriers generated outside the depletion layer.[100] This situation is illustrated in Figure 12.25. Such a tail can hinder reconvolution analysis of single-photon timing measurements. This effect has subsequently been reduced using a specially designed single-photon APD and an instrumental response of 20 psec FWHM obtained,[101] comparable to the best so far reported with MCP-PMs.[87] However, in practice such state of the art APD devices only start to find applications in fluorescence when combined with a micro-

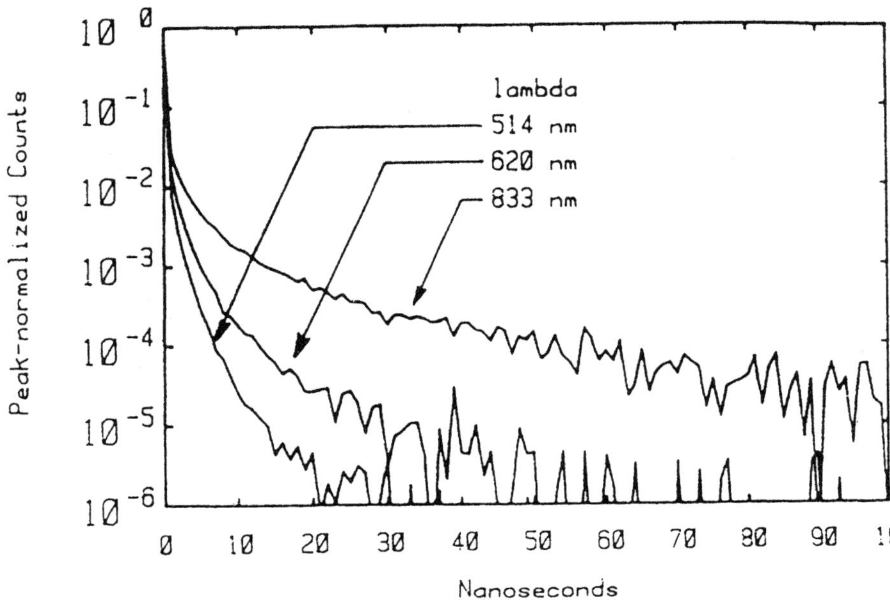

Figure 12.25. Wavelength dependence in the temporal response of a single-photon avalanche photodiode. (From Ref. 100.)

scope[32, 102] or high-powered laser so as to minimize the sensitivity limitations imposed by the ca. 10 μm diameter. The properties of commercially available single-photon APDs with respect to important factors for successful time-resolved fluorescence measurements such as the instrumental response wavelength dependence, and uniformity of response have yet to be fully reported.

Nevertheless, commercially available single photon APDs (RCA C30921S and variants and EG&G SPCM 100/200PQ modules) with a more workable active diameter of 500 μm (0.2 mm square area) have already been proved useful in fluorescence measurements,[14, 32] in spite of having a slower impulse response of ca. 400 psec FWHM.[103]

Single-photon silicon APDs possess a quantum efficiency of ca. 20–40% between 700 and 900 nm which compares very favorably with ca. 3% at best expected from an S20R or S1 photocathode over this range. The lack of late-pulsing in an APD response as compared with a linear focused photomultiplier also has some virtues in the reconvolution analysis of fluorescence decay curves.

A little work has been reported on time-correlated measurements with germanium APDs,[104, 105] showing the potential for extending single-photon APD fluorescence lifetime measurements up to 1.7 μm with picosecond resolution.

Problems still remain in overcoming the intrinsic optical cross-talk in arrays of avalanche photodiodes, which at present precludes equivalent applications to multianode MCP-PMs in such as multiplexed lifetime measurements at different fluorescence wavelengths.

Phase-modulation fluorometry has been performed with APDs to a lesser extent than has single-photon timing, but nevertheless there are some reports of this combination.[106, 107]

There can be little doubt there is still much progress to be made in the application of solid-state detectors to the study of near-IR fluorescence. However, solid-state photomultipliers capable of replacing the conventional photomultiplier tube in mainstream fluorescence applications would still seem to be a long way away.

12.5. Conclusion

As we mentioned at the outset, the study and application of near-IR fluorescence is only at the beginning. The success or failure of this promising approach will depend largely on the synthesis of suitable probe molecules and the performance of the appropriate instrumentation. We can only guess which, if either, will prove the limiting factor, but if the history of UV/visible fluorescence is anything to go by, IR fluorescence will evolve stepwise through the interplay between the different disciplines. The greatest promise in fluorescence instrumentation undoubtedly lies in the development of new semiconductor sources and detectors, combining wide-band spectral properties and fast time-resolution with low cost, reliability, and ease of use. It is very possible that the semiconductor devices discussed in this chapter only give us a snippet of what ultimately might be achieved in terms of device performance and fluorescence applications.

Acknowledgment

The authors would like to thank the Engineering and Physical Sciences Research Council (EPSRC) for research grants relating to this work.

References

1. C. A. Parker, *Photoluminescence of Solutions*, Elsevier Publishing Co., Amsterdam (1968).
2. J. B. Birks, *Photophysics of Aromatic Molecules*, Wiley, London (1970).
3. J. R. Lakowicz, *Principles of Fluorescence Spectroscopy*, Plenum, New York (1983).
4. *Topics in Fluorescence Spectroscopy* (J. R. Lakowicz, ed.), *Vol. 1: Techniques*, Plenum, New York (1991).
5. *Topics in Fluorescence Spectroscopy* (J. R. Lakowicz, ed.), *Vol. 2: Principles*, Plenum, New York (1991).

6. *Topics in Fluorescence Spectroscopy* (J. R. Lakowicz, ed.) *Vol. 3: Biochemical Applications*, Plenum, New York (1992).
7. I. Yamazaki, M. Mimuro, N. Tamai, T. Yamazaki and Y. Fujita, Picosecond time-resolved fluorescence spectra of photosystem I and II in Chlorella pyrenoidosa, *FEBS Lett. 179*, 65–68 (1985) and references therein.
8. See, e.g., M. A. J. Rodgers and P. T. Snowden, Lifetime of singlet oxygen in liquid water as determined by time-resolved infrared luminescence measurements, *J. Am. Chem. Soc. 104*, 5541–5543 (1982).
9. D. B. Chase, Fourier transform Raman spectroscopy, *J. Am. Chem. Soc. 108*, 7485–7488 (1986).
10. D. Andrews-Wilberforce and G. Patonay, Fluorescence quenching studies of near-infrared fluorophores, *Appl. Spec. 43*, 1450–1455 (1989).
11. T. Imasaka, A. Yoshitake and N. Ishibashi, Semiconductor laser fluorimetry in the near-infrared region, *Anal. Chem. 56*, 1077–1079 (1984).
12. T. Imasaka, A. Yoshitake, K. Hirata, Y. Kawabata and N. Ishibashi, Pulsed semiconductor laser fluorometry for lifetime measurements, *Anal. Chem. 57*, 947–949 (1985).
13. I. Downie, An investigation of the suitability of carbocyanine dyes to the time-resolved fluorescence study of microheterogeneous systems, Ph.D. thesis, University of Strathclyde, Glasgow, Scotland (1994).
14. S. A. Soper, Q. L. Mattingly and P. Vegunta, Photon burst detection of single near-infrared fluorescent molecules, *Anal. Chem. 65*, 740–747, (1993).
15. S. Yu. Egorov, V. F. Kamalov, N. I. Koroteev, A. A. Krasnovsky Jr, B. N. Toleutaev, and S. V. Zinukov, Rise and decay kinetics of photosensitized singlet oxygen luminescence in water measurements with nanosecond time-correlated single photon counting technique, *Chem. Phys. Lett.* 421–424 (1989).
16. P. R. Ogilby and C. S. Foote, Chemistry of singlet oxygen. 36. Singlet molecular oxygen luminescence in solution following pulsed laser excitation. Solvent deuterium isotope effects on the lifetime of singlet oxygen, *J. Am. Chem. Soc. 104*, 2069–2070 (1982).
17. J. G. Parker and W. D. Stanbro, Optical determination of the collisional lifetime of singlet molecular oxygen in acetone and deuterated acetone, *J. Am.Chem. Soc. 104*, 2067–2069 (1982).
18. R. W. Redmond, Enhancement of the sensitivity of radiative and nonradiative detection techniques in the study of photosensitization by water soluble sensitizers using a reverse micelle system, *Photochem. Photobiol. 54*, 547–556 (1991).
19. T. A. M. Doust, Fluorescence lifetime and depolarization measurements using frequency up-conversion, in: *Picosecond Chemistry and Biology* (T. A. M. Doust and M. A. West, eds.), pp. 1-34, Science Reviews Ltd, London (1983) and references therein.
20. A. Mokhtari, J. Chesnoy and A. Laubereau, Femtosecond time-and frequency-resolved fluorescence spectroscopy of a dye molecule, *Chem. Phys. Letts, 155*, 593–598 (1989).
21. P. F. Barbara, T. J. Kang, W. Jarzeba and M. Kahlow, Subpicosecond fluorescence measurements—applications in chemistry, in: *Fluorescence Detection II* (E. Roland Menzel, ed.), *Proc. SPIE 910*, 125–130 (1988).
22. D. Hulin, A. Migus, A. Antonetti, I. Ledoux, J. Badan and J. Zyss, Parametric amplification sampling spectroscopy (PASS): A new technique for resolving near-infrared luminescence on a subpicosecond time scale, *Chem. Phys. 46*, 75-77 (1986).
23. Y. Tsuchiya, H. Suzuki, M. Koishi, K. Kinoshita, T. Nakamura, Infrared sensitive universal streak camera for use in 1.0 to 1.6 micron wavelength region, in: *High Speed Photography, Videography, and Photonics III, Proc. SPIE 569*, 181–188 (1985).
24. P. F. Curley, A. I. Ferguson, J. G. White and W. B. Amos, Application of a femtosecond self-sustaining mode-locked Ti:sapphire laser to the field of laser scanning confocal microscopy, *Opt. Quant. Elec. 24*, 851–859 (1992).
25. G. Ripamonti and S. Cova, Optical time-domain reflectometry with centimeter resolution at fW sensitivity, *Elec. Lett. 22*, 818–819 (1986).

26. Time-Resolved Spectroscopy and Imaging of Tissues, (B. Chance, ed.), *Proc. SPIE 1431*, 1–335 (1991).
27. G. B. Dutt, S. Doraiswamy and N. Periasamy, Molecular dynamics of polar dye probes in tertiary-butyl alcohol-water mixtures, *J. Chem. Phys. 94*, 5360–5368 (1991).
28. G. B. Dutt and S. Doraiswamy, Picosecond reorientational dynamics of polar dye probes in binary aqueous mixtures, *J. Chem. Phys. 96*, 2475–2491 (1992).
29. S. Hirayama, Time-resolved fluorescence microscopy, in: *Progress in Photochemistry and Photophysics* (J. F. Rabek, ed.), vol. 6, 1-42, CRC Press, Boca Raton, Florida (1992).
30. Xue Feng Wang, A Periasamy, B Herman and D. M. Coleman, Fluorescence lifetime imaging microscopy (FLIM): Instrumentation and applications, *Crit. Rev. in Anal. Chem. 23*, 369–395 (1992).
31. M. A. J. Rodgers and P. A. Firey, Instrumentation for fluorescence microscopy with picosecond time resolution, *Photochem. Photobiol. 42*, 613–616 (1985).
32. K. P. Ghiggino, M. R. Harris and P. G. Spizzirri, Fluorescence lifetime measurements using a novel fiber-optic laser scanning confocal microscope, *Rev. Sci. Instrum. 63*, 2999–3002 (1992).
33. J. R. Lakowicz, Fluorescence lifetime sensing, *Laser Focus World* 60–80, May (1992).
34. J. V. Sweedler, Charge transfer device detectors and their applications to chemical analysis, *Crit. Rev. Anal. Chem. 24*, 59–98 (1993).
35. D. O. Landon and R. D. Compton, Laser spectroscopy and parallel detection technology, Tracor Northern Technical Report DA-3036-5-4 (1984).
36. E. Gratton, B. Feddersen, M. vandeVen, Parallel acquisition of fluorescence decay using array detectors, in: *Time-Resolved Laser Spectroscopy in Biochemistry II* (J. R. Lakowicz, ed.), Proc. SPIE 1204, 21–25 (1990).
37. D. J. S. Birch, A. S. Holmes, R. E. Imhof, B. Z. Nadolski and K. Suhling, Multiplexed array fluorometry, *J. Phys. E.: Sci. Instrum. 21*, 415–417 (1988).
38. D. J. S. Birch, K. Suhling, A. S. Holmes, A. D. Dutch and R. E. Imhof, Array fluorometry: the theory of the statistical multiplexing of single-photon timing, in: *Time-Resolved Laser Spectroscopy in Biochemistry II*, (J. R. Lakowicz, ed.), Proc. SPIE 1204, 26–34 (1990).
39. D. J. S. Birch, D. McLoskey, A. Sanderson, K. Suhling and A. S. Holmes, Multiplexed time-correlated single-photon counting, *J. Fluorescence 4*, 91–102 (1994).
40. D. L. Farrens and Pill-Soon Song, Subnanosecond single photon timing measurements using a pulsed diode-laser, *Photochem. Photobiol. 54*, 313–317 (1991).
41. D. J. S. Birch, A. S. Holmes, J. R. Gilchrist, R. E. Imhof, S. M. Al Alawi and B. Nadolski, A multiplexed single-photon instrument for routine measurement of time-resolved fluorescence anisotropy, *J. Phys. E: Sci. Instrum. 20*, 471–473 (1987).
42. R. B. Thompson, Fluorescence-based fiber-optic sensors, in: *Topics in Fluorescence Spectroscopy* (J. R. Lakowicz, ed.), Vol. 2, pp. 345–365 , Plenum Press, New York (1991).
43. W. A. Wyatt, F. V. Bright and G. M. Hieftje, Characterization and comparison of three fiber-optic sensors for iodide determination based on dynamic fluorescence quenching of rhodamine 6G, *Anal. Chem. 59*, 2272–2276 (1987).
44. K. S. Litwiler, P. M. Kluczynski and F. V. Bright, Determination of the transduction mechanism for optical sensors based on rhodamine 6G impregnated perfluorosulfonate films using steady-state and frequency-domain fluorescence, *Anal. Chem. 63*, 797–802 (1991).
45. F. V. Bright, G. E. Poirier and G. M. Hieftje, A new ion sensor based on fibre optics, *Talanta 35*, 113–118 (1988).
46. H. Mohan and R. M. Iyer, Photochemical behaviour of rhodamine 6G in Nafion membrane, *J. Chem. Soc. Faraday Trans. 88*, 41–45 (1992).
47. A. Sanderson, A. S. Holmes, D. McLoskey, D. J. S. Birch and R. E. Imhof, A single-photon timing fluorometer with liquid light guide sensing, in: *Advances in Fluorescence Sensing Technology* (J. R. Lakowicz and R. B. Thompson, eds.), Proc. SPIE 1885 466-477 (1993).

48. J. M. Zen and G. Patonay, near-infrared fluorescence probe for pH determination, *Anal. Chem. 63*, 2934–2938 (1991).
49. S. Zhao and W. M. Reichert, Modeling of fluorescence emission from cyanine-dye-impregnated Langmuir-Blodgett films deposited on the surface of an optical fiber, *Thin Solid Films 200*, 363–373 (1991).
50. K. T. V. Grattan and A. W. Palmer, Infrared fluorescence "decay-time" temperature sensor, *Rev. Sci. Instrum. 56*, 1784–1787 (1985).
51. Z. Zhang, K. T. V. Grattan, A. W. Palmer, R. Summers, R. Summan and S. Hughes, Cr:LiSAF fluorescence lifetime based fiber optic thermometer and its application in clinical rf heat treatment, in: *Advances in Fluorescence Sensing Technology* (J. R. Lakowicz and R. B. Thompson, eds.), Proc. SPIE 1885, 300–305 (1993).
52. K. Sauda, T. Imasaka and N. Ishibashi, Determination of protein in human serum by high-performance liquid chromatography with semiconductor laser fluorometric detection, *Anal. Chem. 58*, 2649–2653 (1986).
53. Y. Kawabata, T. Imasaka and N. Ishibashi, Ultramicro flow-cell for semiconductor laser fluorimetry, *Talanta, 33*, 281–283 (1986).
54. L. R. Middendorf, J. C. Bruce, R. C. Bruce, R. D. Eckles, D. L. Grone, S. C. Roemer, G. D. Sloniker, D. L. Steffens, S. L. Sutter, J. A. Brumbaugh and G. Patonay, Continuous, on-line DNA sequencing using a versatile infrared laser scanner/electrophoresis apparatus *Electrophoresis 13*, 487–494 (1992).
55. D. J. S. Birch, G. Hungerford and R. E. Imhof, Near-infrared spark source excitation for fluorescence lifetime measurements, *Rev. Sci. Instrum. 62*, 2405–2408 (1991).
56. G. Hungerford, The application of spark source fluorescence lifetime spectroscopy to the study of infrared fluorescence, transient species and DCM, Ph.D. thesis, University of Strathclyde, Glasgow, Scotland (1991).
57. R. B. Thompson, J. K. Friscoli and J. R. Lakowicz, Phase fluorometry using a continuously modulated laser diode, *Anal. Chem. 64*, 2075–2078 (1992).
58. G. Hungerford, D. J. S. Birch and R. E. Imhof, Spark source infrared fluorometry, in: *Time-Resolved Laser Spectroscopy in Biochemistry III* (J. R. Lakowicz, ed.), Proc. SPIE 1640, 262–270 (1992).
59. P. E. Zinsli, Inhomogeneous interior of aerosol OT microemulsions probed by fluorescence and polarization decay, *J. Phys. Chem. 83*, 3223–3231 (1979).
60. S. M. Sze, *Semiconductor Devices Physics and Technology*, John Wiley & Sons, New York (1985).
61. J. A. Lott, R. P. Schneider Jr, G. A. Vawter, J. C. Zolper and K. J. Malloy, Visible (660nm) resonant cavity light-emitting diodes, *Elec. Lett. 29*, 328–329 (1993).
62. A. T. Semenov, V. R. Shidlovski, S. A. Safin, V. P. Konyaev and M. V. Zverkov, Superluminescent diodes for visible (670nm) spectral range based on AlGaInP/GaInP heterostructures with tapered grounded absorber, *Elec. Lett. 29*, 530–532 (1993).
63. A. Berthod, M. Glick and J. D. Winefordner, Sensitive, indirect photometric detector for high-performance liquid chromatography using a light-emitting diode, *J. Chromatogr. 502*, 305–315 (1990).
64. M. A. Brown, T. E. Edmonds, J. N. Miller, D. P. Riley and N. J. Seare, Novel instrumentation and biomedical applications of very near infrared fluorescence, *Analyst 118*, 407–410 (1993).
65. J. H. Buroughes, D. D. C. Bradley, A. R. Brown, R. N. Marks, K. Mackay, R. H. Friend, P. L. Burns and A. B. Holmes, Light-emitting diodes based on conjugated polymers, *Nature 347*, 539–541 (1990).
66. G. Gustafsson, Y. Cao, G. M. Treacy, F. Klavetter, N. Colaneri and A. J. Heeger, Flexible light-emitting diodes made from soluble conducting polymers, *Nature 357*, 477–479 (1992).
67. T. Imasaka, Semiconductor laser spectrometry, *Anal. Sci. 9*, 329–344 (1993).
68. T. Imasaka, Diode lasers and practical trace analysis, *Anal. Chem. 62*, 363–371 (1990).
69. G. Patonay, M. D. Antoine and A. Boyer, Semiconductor lasers in analytical chemistry, in: *Optical Methods for Ultrasensitive Detection and Analysis*, Proc. SPIE 1435, 52–63 (1991).

70. C. Wersig, W. Finke, E. Handler, H-P Josel and E. Schmidt, Performance of commercial laser diodes in fluorimetric detection, in: *Advances in Fluorescence Sensing Technology* (J. R. Lakowicz and R. B. Thompson, eds.), Proc. SPIE 1885, 389–400 (1993).
71. M. A. Kessler and O. S. Wolfbeis, Laser-induced fluorometric determination of albumin using longwave absorbing molecular probes, *Anal. Biochem. 200*, 254–259 (1992).
72. J. R. Lakowicz, H. Szmacinski and K. W. Berndt, Fluorescence lifetime-based sensing of blood gases and cations, in: *Fiber Optic Medical and Fluorescent Sensors and Applications* (J. R. Lakowicz, ed.), Proc. SPIE. 1648, 150–163 (1992).
73. J. A. Giordmane and R. C. Miller, Tunable Coherent Parametric Oscillation in $LiNbO_3$ at optical frequencies, *Phys. Rev. Lett. 14*, 973–976 (1965).
74. For an up to date coverage, see *J. Opt. Soc. Am. B. 10*, 2147–2243 (1993).
75. M. Ebrahimzadeh, G. J. Hall and A. I. Ferguson, Singly resonant, all-solid-state, mode-locked LiB_3O_5 optical parametric oscillator tunable from 652nm to 2650nm, *Optics Lett. 17*, 652–654 (1992).
76. Philips Data Handbook, Electron Tubes, T9 (1987).
77. D. J. S. Birch and R. E. Imhof, Time-domain fluorescence using time correlated single-photon counting, in: *Topics in Fluorescence Spectroscopy* (J. R. Lakowicz, ed.), Vol 1, pp 1–95, Plenum Press, New York (1991).
78. S. Kinoshita and T. Kushida, High-performance time-correlated single photon counting apparatus using a side-on type photomultiplier, *Rev. Sci. Instrum. 53*, 469–472 (1982).
79. W. R. Ware, M. Pratinidhi and R. K. Bauer, Performance characteristics of a small side-window photomultiplier in laser single-photon fluorescence decay measurements, *Rev. Sci. Instrum. 54*, 1148–1156 (1983).
80. S. Canonica, J. Forrer and Urs. P. Wild, Improved timing resolution using small side-on photomultipliers in single photon counting, *Rev. Sci. Instrum. 56*, 1754–1758 (1985).
81. E. C. Meister, Urs. P. Wild, P. Klein-Bolting and A. R. Holzwarth, Time-response of small side-on photomuliplier tubes in time-correlated single-photon measurements, *Rev. Sci. Instrum. 59*, 499-501 (1988).
82. K. Hadener, S. Bergamasco and G. Calzaferri, Conversion and amplification of photomultiplier anode current in multiple-frequency phase fluorometry, *Rev. Sci. Instrum. 59*, 1924–1927 (1988).
83. D. J. S. Birch, R. E. Imhof and C. Guo, Fluorescence decay studies using multiplexed time-correlated single-photon counting: application to aminotetraphenylporphyrins, *J. Photochem. Photobiol. A:Chem. 42*, 223–231 (1988).
84. D. J. S. Birch, G. Hungerford, B. Nadolski, R. E. Imhof and A. D. Dutch, Time-correlated single-photon counting fluorescence decay studies at 930nm using spark source excitation, *J. Phys. E: Sci. Instrum. 21*, 857–862 (1988).
85. D. C. Stuart and A. D. Kirk, Application of photon counting techniques to measurement of the risetime of chromium (III) phosphorescence in the near infrared, *Rev. Sci. Instrum. 48*, 186–189 (1977).
86. E. Gulari and B. Chu, Photon correlation in the nanosecond range and its applications to the evaluation of RCA C31034 photomultiplier tubes, *Rev. Sci. Instrum. 48*, 1560–1567 (1977).
87. H. Kume, T. Taguchi, K. Nakatsugawa, K. Ozawa, S. Suzuki, R. Samuel, Y. Nishimura and I. Yamazaki, Compact ultrafast microchannel plate photomultiplier tubes, in: *Time-Resolved Laser Spectroscopy in Biochemistry III* (J. R. Lakowicz, ed.), Proc. SPIE 1640 440–447 (1992).
88. H. Kume, K. Koyama, K. Nakatsugawa, S. Suzuki and D. Fatlowitz, Ultrafast microchannel plate photomultipliers, *Appl. Opt. 27*, 1170–1178 (1988).
89. W. G. McMullan, S. Charbonneau and M. L. W. Thewalt, Simultaneous subnanosecond timing information and 2D spatial information from imaging photomultiplier tubes, *Rev. Sci. Instrum. 58*, 1626–1628 (1987).

90. M. R. Ainbund, O. E. Buevich, V. F. Kamalov, G. A. Men'shikov and B. N. Toleutaev, Simultaneous spectral and temporal resolution in a single-photon counting technique, *Rev. Sci. Instrum. 63*, 3274–3279 (1992).
91. Y. Tsuchiya, Advances in streak camera instrumentation for the study of biological and physical processes, *IEEE J. Quan. Elec. QE-20*, 1516–1528 (1984).
92. T. M. Nordland, Streak cameras for time-domain fluorescence, in: *Topics in Fluorescence Spectroscopy* (J.R. Lakowicz, ed.), Vol.1, pp. 183–260, Plenum Press, New York (1991).
93. M. Watanabe, M. Koishi and P. W. Roehrenbeck, Development and characteristics of a new picosecond fluorescence lifetime system, in: *Advances in Fluorescence Sensing Technology* (J. R. Lakowicz and R. B. Thompson, eds.), Proc. SPIE, 1885 155–164 (1993).
94. K. Kinoshita, M. Ito and Y. Suzuki, Femtosecond streak tube, *Rev. Sci. Instrum. 58*, 932–938 (1987).
95. L. M. Davis and C. Parigger, Use of streak camera for time-resolved photon counting fluorimetry, *Meas. Sci. Tech. 3*, 85–90 (1992).
96. R. G. W. Brown, K. D. Ridley and J. G. Rarity, Characterization of silicon avalanche photodiodes for photon correlation measurements. 1: Passive quenching, *App. Opt. 25*, 4122–4126 (1986).
97. S. Cova, A. Longoni and G. Ripamonti, Active-quenching and gating circuits for single-photon avalanche diodes (SPADS), *IEEE Trans. Nucl. Sci. NS-29*, 599–601 (1982).
98. R. G. W. Brown, R. Jones, J. Rarity and K. D. Ridley, Characterization of silicon avalanche photodiodes for photon correlation measurements. 2: Active quenching, *App. Opt. 26*, 2383–2389 (1987).
99. H. Dautet, P. Deschamps, B. Dion, A. D. MacGregor, D. MacSween, R. J. McIntyre, C. Trottier and P. P. Webb, Photon counting techniques with silicon avalanche photodiodes, *App. Opt. 32*, 3894–3900 (1993).
100. S. Cova, G. Ripamonti and A. Lacaita, Avalanche semiconductor detector for single optical photons with a time-resolution of 60ps, *Nucl. Inst. Meth. A253*, 482–487 (1987).
101. S. Cova, A. Lacaita, M. Ghioni, G. Ripamonti and T. A. Louis, 20ps timing resolution with single-photon avalanche diodes, *Rev. Sci. Instrum. 60*, 1104–1110 (1989).
102. T. A. Louis, G. Ripamonti and A. Lacaita, Photoluminescence lifetime microscope spectrometer based on time-correlated single-photon counting with an avalanche diode detector, *Rev. Sci. Instrum. 61*, 11–22 (1990).
103. A. Lacaita, S. Cova and M. Ghioni, Four-hundred-picosecond single photon timing with commercially available avalanche photodiodes, *Rev. Sci. Instrum. 59*, 1115–1121 (1988).
104. W. Fichtner and W. Hacker, Time-resolution of Ge avalanche photodiodes operating as photon counters in delayed coincidence, *Rev. Sci. Instrum. 47*, 374–377 (1976).
105. A. Lacaita, S. Cova, F. Zappa and P. A. Francese, Subnanosecond single-photon timing with commercially available germanium photodiodes, *Opt. Lett. 18*, 75–77 (1993).
106. K. Berndt, H. Dürr and K.-H. Feller, Time-resolved fluorescence spectroscopy of cyanine dyes III. Structure dependence of fluorescence lifetimes, *Z. Phys. Chemie. 268*, 250–256 (1987).
107. K. Berndt, H. Duerr and D. Palme, Picosecond laser spectroscopy with avalanche photodiodes, in: *Time Resolved Laser Spectroscopy in Biochemistry* (J. R. Lakawicz, ed.), Proc. SPIE 909, 209–215 (1988).

13

Application of Fluorescence Sensing to Bioreactors

Govind Rao, Shabbir B. Bambot, Simon C. W. Kwong, Henryk Szmacinski, Jeffrey Sipior, Raja Holavanahali, and Gary Carter

13.1. Introduction

At the heart of a bioprocess is a bioreactor. The bioreactor is typically a glass or stainless steel tank that can culture cells in a liquid medium under aseptic conditions. Air or oxygen is supplied to the cells by sparging the gas into the vessel. The temperature is controlled and agitation is provided to promote homogeneous mixing. The bioreactor employs the biosynthetic pathways of the living cell for the purpose of manufacturing biological products. The primary objective has been to convert inexpensive carbon (or other substrate) to a high-value product that cannot be readily synthesized *in vitro*. This objective, which forms the basis of biotechnology and biochemical engineering, has, in years past, been accomplished primarily by the techniques of mutant isolation, strain development, and genetic manipulation. These approaches are still largely followed today. However, the sophistication achieved in terms of our ability to affect intracellular machinery far exceeds the best available techniques for monitoring/controlling extracellular parameters. For example, virtually all industrial bioreactor production is done with only pH and partial pressure of oxygen (pO_2) measurement and control. Despite all our advances to date which indicate that other parameters such as glucose, CO_2, nitrogen, phosphate, and miscellaneous cation and anion levels in the medium can profoundly affect cellular (and hence bioreactor)

Govind Rao, Shabbir B. Bambot, and Simon C. W. Kwong • Department of Chemical and Biochemical Engineering, University of Maryland Baltimore County, Baltimore, Maryland 21228; and Medical Biotechnology Center of the Maryland Biotechnology Institute, University of Maryland at Baltimore, Baltimore, Maryland 21201. *Henryk Szmacinski* • Center for Fluorescence Spectroscopy and Department of Biological Chemistry, University of Maryland School of Medicine, and Medical Biotechnology Center of the Maryland Biotechnology Institute, University of Maryland at Baltimore, Baltimore, Maryland 21201. *Jeffrey Sipior* • Medical Biotechnology Center of the Maryland Biotechnology Institute, University of Maryland at Baltimore, Baltimore, Maryland 21201. *Raja Holavanahali and Gary Carter* • Department of Electrical Engineering, University of Maryland Baltimore County, Baltimore, Maryland 21228.

Topics in Fluorescence Spectroscopy, Volume 4: Probe Design and Chemical Sensing, edited by Joseph R. Lakowicz. Plenum Press, New York, 1994.

productivities, our capacity to monitor these parameters is limited. This problem has been exacerbated primarily by the lack of technology available to obtain meaningful information in near real time. Although measurement techniques abound for each of the aforementioned analytes, they are specific for each one, generally cumbersome, and not easy to implement online. Despite hopes for breakthroughs using flow injection analysis (FIA) for better measurement, the lack of fast response times and the need to have a recycle loop or a penetration into the aseptic bioreactor environment has dampened industrial enthusiasm for its use. Consequently, this review will focus on the best available on-line sensors for use in fermentation and on the prospects for the near-term future.

Despite all the breathtaking advances in genetic and metabolic engineering, bioreactor technology has changed little since its inception. For several reasons (mostly the regulatory issues involved) it is generally agreed that the stirred tank reactor is here to stay. Most of the sensors used in bioreactors are also improved versions of their original forms (i.e., the glass/calomel electrode for pH sensing, the Clark electrode for pO_2 measurement). There is no commercially available sensor that can continuously and reliably monitor glucose levels in a bioreactor. Furthermore, the principles and instrumentation required for each of these measurements varies considerably, thus adding to their cost and complexity. Since most of these measurements are potentiometric or amperometric, they are also subject to electrical interference and artifacts. There is thus an enormous need for a generic sensing technology that can alleviate these problems and also offer greater versatility.

Fortunately, a recent breakthrough promises just the kind of core technology that can achieve the above goal. The breakthrough is the innovative melding of novel synthetic fluorophores and the use of phase-shift fluorometry to quantitatively measure analytes. The principle is simple: a fluorophore exhibits a phase shift in its emission in response to the analyte in question when excited by an intensity modulated beam of light. The robustness of the measurements can be improved by the simultaneous measurement of the modulation of the emission, which provides a related but independent measurement of the analyte (described in greater detail below). The phase and modulation of the emission are both measurements of the fluorescence lifetime, which changes in response to the analytes. It is conceivable that in a few years, bioreactor monitoring will be primarily implemented through optical means. In this review, we focus on *in situ* pO_2, pH, partial pressure of carbon dioxide (pCO_2), and glucose measurements in bioreactors. We begin with a brief review of the current state of the art for the sensing of these four analytes in bioreactors.

In addition to the above extracellular parameters, cell concentration and cell activity are two important cell-associated parameters that determine how well a fermentation process is performing. The manufacturing of biological products (antibiotics, amino acids, monoclonal antibodies, and other protein products) at large scales requires that cells be cultured at high cell densities and stay metabolically active. Consequently, much effort has been expended to develop techniques that can allow the estimation of cell concentration and cell activity in real time during a fermentation.

For completeness sake, we briefly review the efforts in the area of cell concentration (also referred to as biomass) estimation. Fluorescence measurements have been made of NAD(P)H with the assumption that under balanced growth (i.e., when there are no environmental changes and the cells grow exponentially) the total NAD(P)H content in the bioreactor is proportional to the cell concentration. Under balanced growth this works quite well, however, the problem is that perturbations in the process cause deviations in NAD(P)H concentration. Therefore, alternative methods such as measuring protein fluorescence have been attempted. In addition, there has been significant interest in monitoring cell metabolism on-line. The premise used for this, as in the case of cell mass, is to measure the concentration of NAD(P)H by following its fluorescence. The reasoning behind this is that the reduced pyridine nucleotides are the primary source of reducing power for the cells to carry out various enzymatic reactions. Although it seems hopelessly naive that such a nonspecific measurement could possibly work, many useful correlations to cellular metabolic activity have been reported using this technique. We therefore examine in some detail on the utility of NAD(P)H fluorescence monitoring in bioreactors to follow both biomass and cell activity. Another widely used technique to follow biomass and cell activity is to measure off-gases and compute respiration rates. Although this technique is not optical, we have included it in this review as there is the potential to make inexpensive gas phase pO_2 and pCO_2 measurements using fluorescence lifetimes. We then conclude with the future prospects for bioreactor sensing with an emphasis on using phase-modulation fluorometry.

13.2. Dissolved Oxygen Sensing

One of the most important process measurements made in bioreactors is that of dissolved oxygen tension. oxygen serves as the terminal electron acceptor in aerobic respiration and it is required for the growth of virtually all cells. However, because of its extremely poor solubility in aqueous media it is frequently in danger of limiting the growth of the culture. Consequently, much effort in bioreactor design and operation is directed toward oxygen supply to the fermentor. In doing so, the measurement of oxygen is critical. The current state of the art in oxygen measurement is an improved version of the widely used Clark electrode.

The Clark or polarographic electrode operates on the principle of oxygen reduction at a negatively biased platinum surface. The electrode is constructed with a silver anode, a platinum cathode, and a KCl or AgCl electrolyte. A teflon or silicone membrane separates the electrode components from the surrounding medium (Figure 13.1). When the cathode is polarized with a negative voltage, oxygen diffusing from the surrounding medium through the membrane is reduced at the surface of the cathode. Since the diffusive flux is a function of the partial pressure of oxygen in the fluid, it is possible to calibrate the electrode current versus oxygen tension. It is important to note that the electrode senses oxygen tension (or activity) rather than

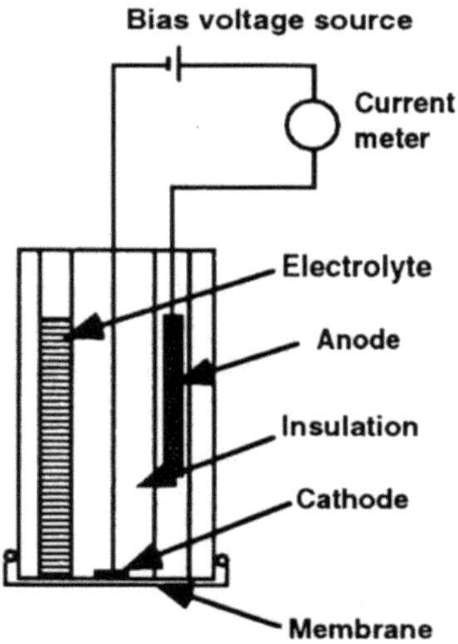

Figure 13.1. A typical Clark electrode for oxygen measurement. (Adapted from Ref. 1.)

concentration. Clark electrodes are calibrated by equilibrating a sample with nitrogen to read zero and followed with air to read 100%. The readings are usually expressed as a percentage of air saturation.

The Clark electrode is widely used because it is the best available technique to date. There are a number of limitations and problems in its use that exist and it is important to be aware of these. We have briefly summarized the most critical ones and for greater details the reader is directed to the excellent review by Lee and Tsao.[1]

1. *Long-term stability.* The polarographic Clark electrode utilizes an electrolyte that is eventually consumed. Consequently, long-term use (several days) results in gradual drift in the electrode signal due tn electrolyte consumption.

2. *Flow dependence.* Since the electrode actually consumes oxygen, measurements in a stagnant sample causes the oxygen tension at the membrane sample interface to deplete resulting in an extension of the diffusion layer into the sample (note that the electrode functions under conditions of membrane controlled diffusion). Thus controlling the sample renewal rate or flow rate at the membrane will affect the accuracy of the reading, the effect decreasing with increasing membrane thickness.

3. *Response time.* In the literature, response time is usually specified as the time taken for the electrode to reach > 90% of the output. Typical response times are around 30 sec. A fast response time is critical when one is measuring transient phenomena such as oxygen respiration rates in tissue or suspended cells and dynamic measurements of the volumetric mass transfer coefficient in bioreactors.

4. *Sensitivity.* Since terminal oxidases have high affinities for oxygen, low-level pO_2 measurements are important. The signal-to-noise ratio may be unacceptable in some situations.

5. *Electrical interference.* When low-level electrical signals are measured, it is very likely that substantial error may be introduced in the measurement by extraneous

interferences such as ground loop currents and interaction with other electrodes (pH and redox). To circumvent this problem one can either "float" the reference ground of the electrode by using an isolation amplifier or a differential amplifier.

The problem with the Clark electrode is that some of these requirements have solutions that are opposing. For instance, flow dependence may be reduced by employing a thicker membrane but this would occur at the cost of increased response time. As a result, most commercially available systems are design compromises that sacrifice a part of some desirable feature. It should he noted that an optical measurement technique where oxygen and/or electrolyte is not consumed will be free of the drawbacks mentioned above.

13.3. pH Sensing

pH sensing is accomplished by means of an electrode with a special glass membrane. The original pH glass was first described in 1930 and is still used today with lithium oxide replacing sodium oxide from the original formulation (Figure 13.2). The principle of operation is based on the presence of a hydrated gel layer at the glass surface that is formed by protonation of the Si-O lattice of the glass. After a day or two, an equilibrium is established and the growth of the gel layer ceases. This layer is only a few nanometers thick, but it determines the performance of the electrode. When the equilibrium at the phase boundary between the solution and the gel layer is disturbed by the migration of protons, a potential is formed. A similar gel layer also exists on the interior of the electrode surface and it has a potential in equilibrium with the reference electrolyte. The difference in potential at the outer and inner gel layers is what is actually measured with respect to a reference electrode and calibrated in pH units. The ohmic connection between the interior and exterior of the electrode is made by means of a slow efflux of electrolyte through a ceramic frit.

The major problems of pH electrodes are the following.

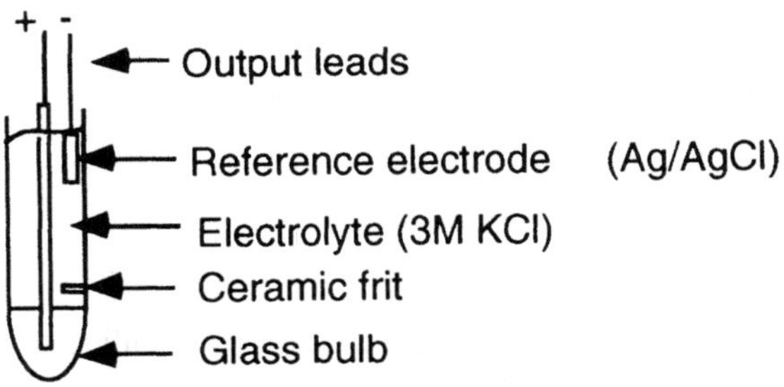

Figure 13.2. A typical pH electrode.

1. The primary source of error is ground loop currents. This is caused by galvanic errors introduced by small potentials resulting from the ionic liquids and dissimilar metals that the electrode is in contact with in a bioreactor. Additional sources of error are interactions with other electrodes. We have frequently found that a pH electrode not connected to an isolation amplifier that floats the reference can show errors of 1–2 pH units. A simple test is to measure the pH of a buffered solution on-line and off-line to check the accuracy of a measurement.

2. Problems with reference electrode/electrolyte. The sample solution can react with the electrolyte. The greatest source of error in biological solutions is through the formation of insoluble silver sulfide, often at the ceramic frit. A blackened spot is usually observed in a pH electrode that has been in service for a few weeks. This precipitate can impede the free flow of electrolyte and cause the probe response to become sluggish and cause large errors in the measured pH.

3. Alkali error occurs when cations (typically Na^+) replace H^+ in the gel layer and thus pH values lower than the actual are measured.

Recent improvements in pH probe design are directed toward alleviating some of the problems mentioned above. For example, the Ingold DPAS prepressurized gel-filled pH electrode performs better than the standard liquid-filled probes, requires minimal servicing since there is no need to refill the electrode with electrolyte, and is relatively simple to install.

Accurate pH measurement is surprisingly complex.[3] A recent review[4] states that if three electrodes measuring pH at one point in an industrial process agree to within 0.1 pH unit, then the electrodes are probably broken, coated, or still covered by their protective caps!

13.4. pCO_2 Sensing

This measurement is actually one of pH. The principle involves the measurement of the pH of a bicarbonate solution separated from the sample by a CO_2-permeable membrane. CO_2 diffuses in (or out) and changes the bicarbonate solution pH in proportion to the pCO_2 in the sample. This method has all the drawbacks of pH measurement with a glass electrode. The problems associated with this method perhaps explain why it is not widely used in biochemical engineering laboratories and process streams despite its considerable metabolic significance.[5]

13.5. Glucose Sensing

Much effort has been expended for some method of online glucose measurement; however, to date there is no real sign of an effective device. Several earlier studies utilized a glucose-permeable membrane with glucose oxidase immobilized and followed the glucose concentration dependent release of hydrogen peroxide by a Clark electrode. This method depends on both the flux of oxygen and glucose as well as the activity of glucose oxidase and this method has not been adopted for online measurements. The elegant work of Schultz and associates based on a displacement technique is the closest demonstration of an on-line measurement technique.[6] The process uses immobilized Concanavalin A (Con A) as the receptor with FITC*-dextran as the bound fluorescent label. Glucose displaces the FITC-dextran whose fluorescence is then proportional to the added glucose.[7] The great advantages of this technique are its reversibility and the lack of consumable reagents. However, the weakness here is the measurement of fluorescence intensity, which becomes a major problem in biological media.[8] Because of these limitations, the measurement of glucose is done offline with an enzymatic analyzer or a spectrophotometric assay.

13.6. Off-Gas Analysis

Off-gas analysis is widely used in many industrial fermentation plants to determine the cellular activity of growing cultures by monitoring respiration. One can measure oxygen uptake and CO_2 production rates and thus measure metabolic activity.[9] In addition, off-gas analysis is also used for monitoring other volatiles, the synthesis of which are strongly dependent on cultivation conditions[10] and product formation.[11] Off-gas estimation and control therefore serves as an indirect method for process analysis and control.

Off-gas analysis is routinely carried out either by mass spectrometry or by gas chromatography (GC), in contrast with traditional infrared and paramagnetic off-gas analyzers which are maintenance intensive and have slow response times and significant calibration drifts.[12]

Mass spectrometry is, in general a fast and robust method for gas analysis that is stable over long periods of time. On line analysis using mass spectrometry is preferably carried out using commercially available capillary inlets. The rapid response times obtained using this inlet allows for sampling at a number of points or a number of fermentors.[13] Alternatively, membrane probes can be used requiring trace amounts of sample gas for analysis and allow for sampling of volatiles from the liquid phase also.[11] However, the diffusion characteristics of membranes are generally nonlinear and have longer response times.

Chromatography is an equally reliable and relatively inexpensive technique for analysis of both volatiles (gas chromatography (GC)) and nonvolatiles (liquid chromatography (LC)). Although somewhat slower (2–3 min analysis time after each injection) the availability of ultrafast capillary GC and ultrafast HPLC (high-performance liquid chromatography) renders this technique equally competitive.[14] With

*Fluorescein isothiocyanate

appropriate molecular probes or immunosensors coupled with lifetime-based sensing, it may be possible in the future to devise optical means for *in situ* analytical monitoring.

The use of off-gas analysis, however, is a relatively expensive proposition as a means of monitoring/controlling extracellular parameters. It is also an indirect method and is generally used to obtain information about the synthesis of products that cannot be analyzed directly. It is possible, however, to use inexpensive phase fluorimetry-based pO_2 and pCO_2 sensors to obtain the same information.

13.7. Biomass Concentration

The end product of a fermentation process is almost always a protein or compound that is produced by the cultivated organism. Maximization of product by monitoring/controlling cell density in a culture is therefore a desirable goal and has traditionally been carried out by taking samples and analyzing them offline. However, these methods are laborious and invasive. Noninvasive indirect online measurements can be carried out by monitoring the NADH fluorescence (see Section 13.8) provided constant intracellular NADH concentration can be assumed.[15] Direct online analysis can be carried out by automatic sampling into a collection chamber and measuring the optical density after appropriate dilutions are carried out. Recently an autoclavable cell mass sensor was developed at Cerex Corporation and The Foxboro Company (U.S. Patent 5,007,740) which can be used online to measure optical density in bubble-free samples.

13.8. Culture Fluorescence

Intracellular reduced pyridine nucleotides NAD(P)H are the primary suppliers of reducing power to anabolic and catabolic pathways. They can be measured because of their fluorescent properties.[16] The fluorescence is caused by the presence of the reduced forms of the pyridine nucleotides NADH and NADPH (jointly referred to as NAD(P)H). These fluorophores absorb light in a wide band around 340 nm, and reemit, or fluoresce, light in a wide band around 460 nm. The phosphorylated and nonphosphorylated nucleotides have essentially equivalent fluorescence properties while the oxidized forms of these nucleotides are nonfluorescent.

It was known that the intracellular concentrations of the reduced and oxidized forms of the pyridine nucleotides vary in different cell types and under different cell culture conditions.[17] Harrison and Chance applied the NAD(P)H fluorescence technique and found that culture fluorescence can be related to the metabolic state of the cells.[18, 19] Since then, more than a hundred papers on NAD(P)H fluorometry have been published. However, they are primarily divided into three major categories:

13.8.1. Biomass Estimation

Most of the studies[15, 20–27] show that a correlation between culture fluorescence and biomass concentration can be obtained mainly in the exponential growth phase. In addition, in order to obtain reproducible correlations, all of the fermentation conditions such as initial substrate concentration, pH, dissolved oxygen level, temperature, and agitation rate have to be the same. However, once the culture is past exponential growth, biomass measurement by following culture fluorescence is no longer accurate.

13.8.2. Substrate Addition/Depletion Responses

Yeast cells have been used as the model system in these studies.[20, 24–29] Typically, a sharp drop in the fluorescence signal was observed during substrate depletion. The fluorescence response caused by substrate addition was found to depend on the metabolic state of the cells as well as on the kind of substrate being used.

13.8.3. Aerobic–Anaerobic Transitions

In these studies, the air supply was shut off during an aerobic fermentation.[19–21, 24, 25, 30, 31] Fluorescence increased abruptly as dissolved oxygen dropped to zero. This effect is due to the increase of the NAD(P)H level which cannot be oxidized through the respiration chain. The goal of these studies was mainly to demonstrate the possibility of using fluorescence to monitor cellular metabolic changes during the fermentation.

In addition to the studies mentioned above, fluorescence measurements have also been applied to many other areas such as mixing time studies,[25, 32] mammalian cell culture,[33] and immobilized cell cultures.[34–36]

One of the main objectives of using culture fluorescence has been to identify metabolic state changes during fermentation and then use this knowledge for process manipulation. Unfortunately, there are very few studies dealing with this aspect. Three research groups,[37–41] however, have worked on an anaerobic bacterium, *Clostridium acetobutylicum*, to study the effects of culture fluorescence. Rao and Mutharasan[37] have demonstrated that culture fluorescence can be used to identify two different production phases: solventogensis and acidogensis. Reardon *et al.*[38] developed a metabolic pathway kinetic analysis method along with the use of fluorescence to determine the pathway rates of *C. acetobutylicum*.

Walker and Dhurjati[42] have used culture fluorescence for on-line discrimination of host and plasmid-carrying strains of *Escherichia coli*. In addition, culture fluorescence has also been used in the control of fed-batch fermentation on yeast cell production.[29, 43]

Although with adequate knowledge of microbial physiology NAD(P)H fluorescence monitoring is quite useful, it has several drawbacks. First, each cell line will have a different amount of average NAD(P)H. Secondly, every species has a different cell wall composition and cellular metabolism and thus exerts a different attenuating effect on the fluorescence. Therefore, the signal must be calibrated separately for each microbial system of interest. This attenuation is known as the "inner filter effect," and it occurs by absorption of both the excited and emitted radiation by the cells and other media components.[44,45] For other factors which can affect culture fluorescence, refer to Li and Humphrey.[46] We note that the inner filter effect can be eliminated by measuring fluorescence lifetimes as described later. Another possible approach that has been described is the use of front-surface detection to observe protein synthesis online,[47] but the front-surface fluorescence intensities are still subject to the effects of absorption and scatter by the sample.[48]

There are many other redox couples within the cell, such as cytochromes or flavoproteins. These other couples usually fluoresce at different wavelengths than those used in NAD(P)H fluorescence. Additionally, they may fluoresce in the oxidized form instead of the reduced form; an example is flavin mononucleotide, which has an excitation range of 436–460 nm and an emission range of 520–570 nm. Therefore, a multiwavelength fluorometer would allow simultaneous monitoring of several intracellular redox couples. An arrangement of this type has been described by Chance *et al.*,[49] and this technique is the topic of current research elsewhere.[46, 50–53] However, as long as fluorescence intensity is used to monitor on-line fermentation, it is impossible to eliminate all background fluorescence, and in any event there is likely to be

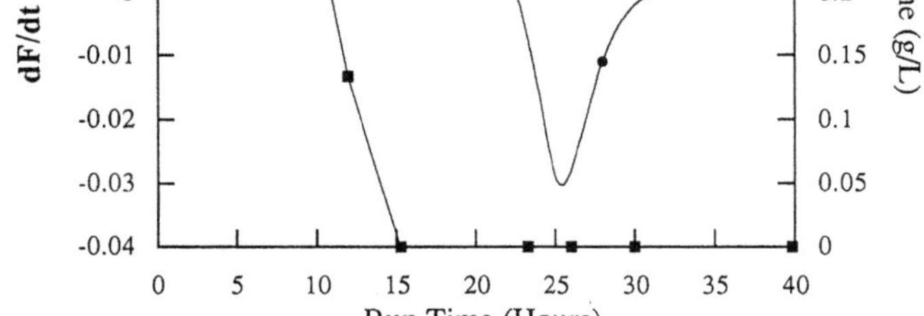

Figure 13.3. Rate of change of NAD(P)H profile (*dF/dt*) in a batch fermentation of *Corynebacterium glutamicum*. First point of inflexion coincides with threonine depletion and the second point marks the end of the fermentation.

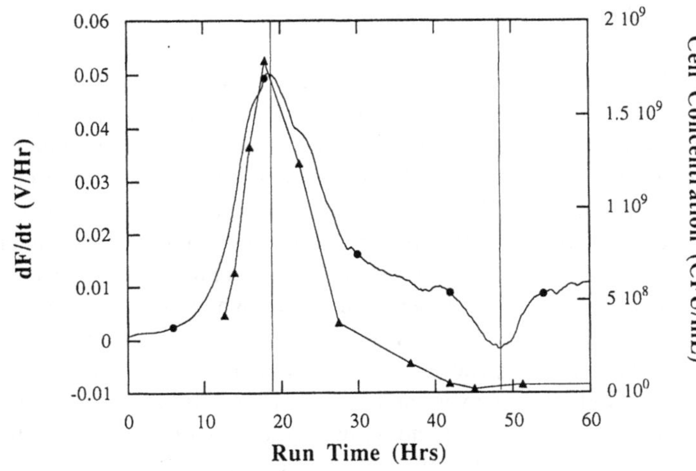

Figure 13.4. *dF/dt* profile correlates with viable cell number of bacteria.

spectral overlap between the emitting species. Therefore, even a multiwavelength fluorometer still has the same drawbacks as a NAD(P)H fluorescence probe. In addition, most of the NAD(P)H fluorescence studies have mainly focused on yeast cells which have very high concentrations of flavoprotein that can cause interference with NAD(P)H measurement. Despite these drawbacks, Kwong *et al.*[(54)] have recently demonstrated that the NAD(P)H fluorescence profile is a useful measurement of the consistency and reproducibility of a batch fermentation. The use of a recycle loop for measuring fluorescence has been found to be useful in eliminating interferences caused by gas hold-up from aeration and agitation.[(55)]

In a recent development, Kwong *et al.*[(56)] describe the utility of following the rate of change of the NAD(P)H fluorescence (*dF/dt*) signal. The advantage that this technique offers is the cancellation of noise and inner filter effects. Furthermore, since this technique measures the rate of change of NAD(P)H fluorescence, it turns out to be very sensitive to metabolic events within the cell as NAD(P)H pools rapidly change depending on the metabolic state of cells. Figure 13.3 shows how the *dF/dt* signal can be used to predict points of threonine depletion on-line in an amino acid fermentation. The first point of inflexion characteristically coincides with threonine depletion, and the second inflexion point with the end of the fermentation (cell death). As shown in Figure 13.4, the overall profile of the *dF/dt* signal strongly correlates with the viable cell count. In the fermentation industry, this is a potentially very useful result as cell viability determination is a cumbersome and laborious off-line technique. Additional results are available in Kwong and Rao.[(57)]

13.9. Other Approaches

We note that this brief review does little justice to monumental work that has been done over the last 40 years to make all the above measurements a reality. It is not possible to present a complete review without writing an extensive treatise on the subject (interested readers are directed to reviews on sensors).[8, 58] In terms of other optical sensing techniques in bioreactors, the sensing of oxygen and pH has been achieved by adding appropriate fluorophores to the fermentation medium and monitoring intensity changes in response to the analyte concentration.[51, 59] The disadvantages of intensity-based sensing were apparent when the fluorescence based measurements agreed only qualitatively with the real values when cells were present. The other disadvantages are the necessity of addition of expensive fluorophores to the broth with no possibility of recovery and the extensive calibration and mathematical corrections necessary for each analyte. However, the loss of expensive fluorophores can be minimized by the use of localized sensing by using fiber optic components.[60, 61]

In terms of enzyme-based sensors for fermentation monitoring, the main problem is the lack of enzyme stability during long-term use. Two recent reviews describe the current state of the art,[62, 63] so we will not dwell on their use.

Additionally, there is a growing use of off-gas analysis for measuring oxygen uptake, CO_2 evolution and respiratory quotient measurement.[9, 64] However, this is not likely to find widespread use despite its usefulness unless the cost involved can be lowered.

13.10. The Future

We believe that these limitations of available conventional as well as fluorescence intensity-based sensors can be circumvented by the use of fluorescence lifetime-based sensing. Phase and/or modulation lifetime measurements (described below) can be accomplished in seconds and are very robust with respect to amplitude changes, electronic noise, and other signal loss problems. The various schemes for fluorescence sensing are illustrated schematically in Figure 13.5. Steady-state intensity measurements (Figure 13.5a) have not found wide acceptance due to lack of stability and the need for frequent recalibration. An additional problem is that intensity measurements work well only in optically clean systems. Wavelength-ratiometric measurements (Figure 13.5b) are desirable for chemical sensing, but there are few available intensity-ratio probes. As a rare example, the Ca^{2+} probe Fura-2 displays a fluorescence spectral shift on binding calcium, which allows wavelength ratiometric measurements independent of probe concentration to be made.[65] We note that probes that display dual emission as the result of an excited state process, such as the light-induced proton loss of naphthols, are not desirable due to the difficulty of controlling the rate processes which depend on buffer concentration as well as the pH.

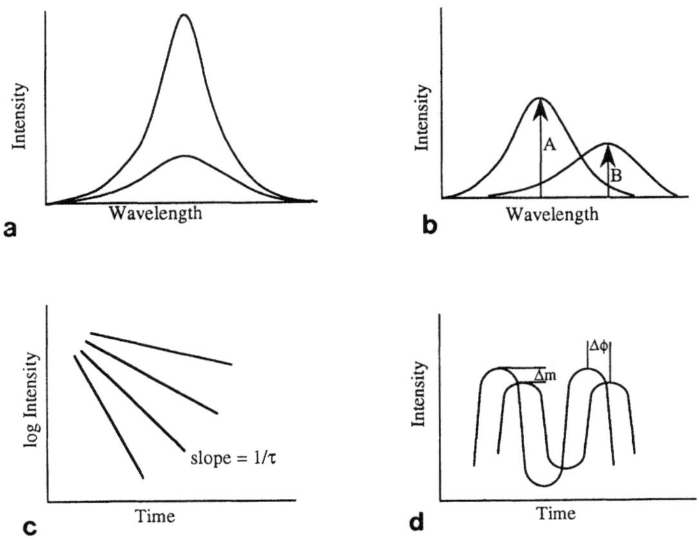

Figure 13.5. Methods of fluorescence sensing. (a) Single excitation/emission wavelength: intensity changes with analyte concentration; (b) wavelength-ratiometric: A/B changes with analyte concentration; (c) liftetime based (time domain): τ changes with analyte concentration; (d) lifetime-based (phase modulation): Δm and $\Delta\phi$ change with analyte concentration.

Decay time sensing (Figure 13.5c and d) is advantageous because such measurements are independent of probe concentration and/or intensity, are mostly independent of absorbance and scattering of the samples, and do not require dual emission from the probe. Among the two possible methods, time-domain or phase-modulation, we prefer the latter because of the possibility of construction of low-cost instruments to measure even subnanosecond lifetimes. In phase fluorometry, sinusoidally modulated excitation light excites a chemical probe. The probe fluoresces in response to the intensity-modulated excitation and is affected by the analyte concentration around it. The emission therefore displays a phase angle shift in relation to the excitation which is dependent on the analyte concentration around the probe. Additionally, the phase and modulation values can provide independent measurements of the decay time, for error checking. This is illustrated in the *upper panel* of Figure 13.6, and the *lower panel* in the figure shows the key elements of a phase fluorometer. The excitation and emission are modulated at the same circular frequency ($\omega = 2\pi f$, f = frequency in Hz). The emission is delayed by the phase angle (Φ). The relative amplitude of the variable portion of the emission (B/A) is smaller than that of the excitation (*b*/*a*). The phase angle (ϕ) and the demodulation factor [m = (*Ba*/*bA*] are both used to *independently* calculate the lifetime (τ) using the equations:

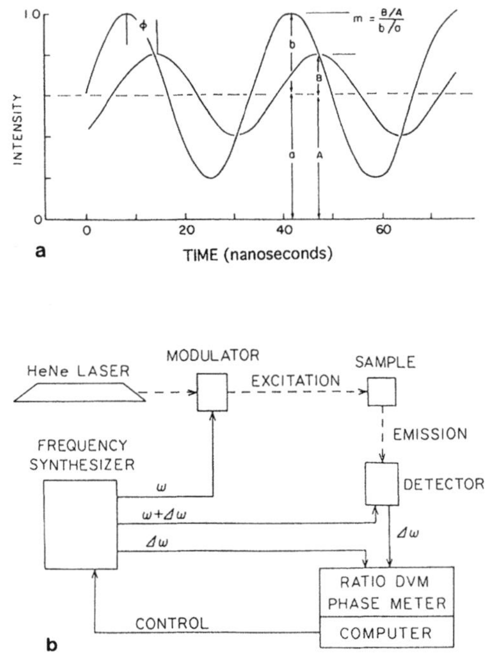

Figure 13.6 Principle of phase-modulation measurement and elements of a phase fluorometer. (a) sample measurement; (b) cross-correlation phase fluorometer.

$\tan\Phi = \omega\tau$ and $m = [1 + \omega^2\tau^2]^{-1/2}$ (*Note*: valid for a single exponential decay)

In the upper panel of Figure 13.6, the emission is drawn assuming a modulation frequency of 30 MHz and a lifetime of 9 nsec. Using the equations above, the phase angle is 59.5° and the demodulation factor is 0.5. (For further details, the reader is referred to Lakowicz[(66)]). Additionally, multifrequency phase and modulation instruments that operate over a range of frequencies have been described[(67, 68)] and simple instruments are possible if only one or several discrete frequencies are required (Figure 13.6, *lower panel*).

The principal advantages of this approach are the following.

1. The base technology is generic and can be applied to many analytes. As just mentioned, the basic idea is to measure the lifetime of the fluorophore. In our case the lifetime is measured using a light source that is modulated at high frequency, and radio-type electronics which measure the phase and modulation of the emission. The most difficult and expensive part of the research-type instruments is obtaining modulated ultraviolet light and a wide range of modulation frequencies. However, these features are not essential for our sensing applications. We have identified chemical probes uniquely sensitive to pH, pCO_2, pO_2, and glucose which can be excited using a 543-nm green helium-neon (HeNe) laser. Also, sensing can be performed using one or a few modulation frequencies, which are easily obtainable with an acoustic-optic (AO) modulator. Furthermore, future sensors that we plan to construct will be based on laser diode excitation, and thus eliminate the AO modulator. Chemical probes are currently being synthesized that use the principle of energy transfer (described below) to shift excitation wavelengths to the red and far-red regions where low-cost laser diodes can be used. Hence, while low-cost instrumentation for bioreactor sensing by phase-modulation methods can be readily constructed with available technology around green HeNe lasers, the prospects for even more inexpensive sensing devices are bright.

2. *The disadvantages of intrinsic (intensity) fluorescence measurements are bypassed.* Intrinsic fluorescence measurements are subject to interference by other medium components (cell mass, complex nutrients, and so on) and this is called the inner-filter effect. Inner filter effects have been successfully modeled and corrected for in cell-free systems.[44, 48] However, correcting for this in a growing culture (or other complex system) is much more complicated. In summary, while other approaches to measure the analytes proposed here have used fluorescence, they have largely failed to work in multicomponent "real" systems because the measurement principle was based on intensity changes. In a real-world system such as a culture broth or blood, intensity measurements have literally faded due to a myriad of interfering components. Our approach offers the potential to avoid such pitfalls by the measurement of phase angles.

13.10.1. Cost Considerations

In terms of cost, in Table 13.1 we compare the costs of the current state of the art in bioprocess measurements.

In contrast, we present in Table 13.2 the potential cost based on phase fluorometry. We have not broken our analysis down by analyte, as one instrument may be sufficient for several analytes. This may be an advantage of our generic core technology whose superior value will become even more compelling as new probes are synthesized for other analytes.

Table 13.1. Costs Associated with Measurements (1992 Base)

Analyte	Hardware	Approximate cost	Vendors
Oxygen	Electrode + meter	$2,000	Ingold, Phoenix
Carbon dioxide	Electrode + meter	2,000	Ingold
pH	Electrode + meter	500	Ingold
Glucose (offline)	Analyzer	6,000	YSI, Beckmann
Off-gases (O_2, CO_2)	Mass spectrometer	30,000–60,000	Dycor, Perkin-Elmer
Total		$40,500–70,500	

Table 13.2. Estimated Cost Using Phase Fluorometry (1993 Base)

Hardware	Cost (estimates)	Footprint size (sq feet)
Phase fluorometer (benchtop)	$100,000	8
HeNe laser-based phase fluorometer	10,000	2–3
Laser diode-based phase fluorometer	2,000	<1

13.10.2. Fluorescence Lifetime-Based Oxygen Sensor

To illustrate the simplicity and stability of phase angle sensing, we have tested an oxygen sensor using the tris [4,7-diphenyl-1,10-phenanthroline] ruthenium (II) complex similar to that described by Demas and co-workers.[69, 70] We tested two configurations, one using a blue light-emitting diode (LED) light source (Figure 13.7), and the other using electroluminescent (ELL) sheet as the modulated light source (Figure 13.8). The ELLs can serve as modulated sources for long decay times of the type displayed by the ruthenium (Ru) complexes.[71] The large area of the ELL allows for substantial total intensity, so that the detector could be a simple PIN photodiode (Figure 13.8) or a large-area photodiode whose size and shape match that of the ELL source.

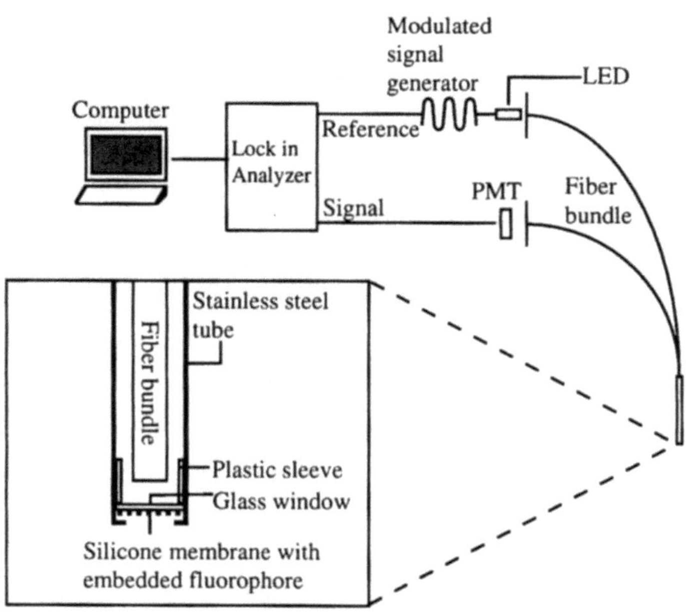

Figure 13.7. LED-based oxygen sensor.

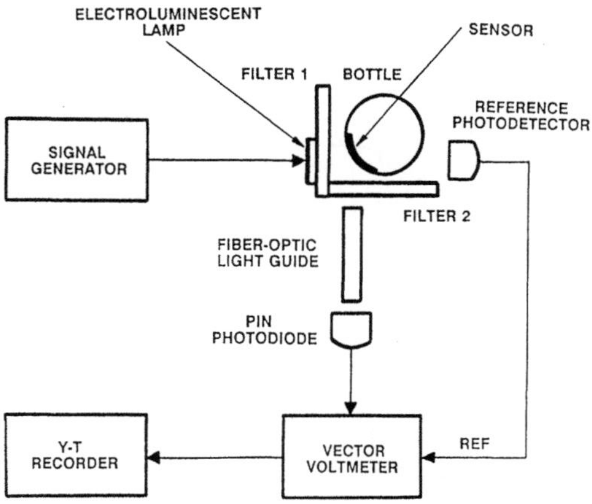

Figure 13.8. Electroluminescent lamp-based oxygen sensor.

In the case of the low-output blue LED, a photomultiplier tube (PMT) detector was used (Figure 13.7).

Remarkably stable phase data were obtained with either configuration. This is illustrated for the LED source in Figure 13.9. Completely comparable results were obtained with an ELL source.[71] The figure shows the phase angles in air and in N_2, where the phase angles are larger when the lifetimes are longer. The sensor was an irregular slab of silicone rubber, containing the Ru complex, inside a scintillator vial. The phase angles were insensitive to removal and replacement of the vial, even though the intensity was highly position-sensitive due to the changes in light collection efficiency and inhomogeneities in the silicone. We also note that the Ru complex is chemically stable and we have observed no deterioration of performance over a few months of use.

Figure 13.10 shows the calibration curve of the LED-based optical oxygen sensor compared with the calibration curve of a commercially available Clark-type sensor (Ingold Electrodes, Wilmington, Massachusetts). While the Clark-type shows a linear calibration, the optical sensor allows a hyperbolic response as predicted by the Stern–Volmer-type equation[72]:

$$\tau_0/\tau = 1 + K\text{pO}_2$$

where τ_0 and τ are the initial and final litetimes, K is the overall quenching constant, and pO_2 is the oxygen tension in the sample. Note that the sensitivity of the optical sensor at low oxygen tensions is significantly higher than that of the Clark-type electrode. This is partly due to the high oxygen permeability of silicone rubber membranes which has the effect of increasing the overall quenching constant, K, in the

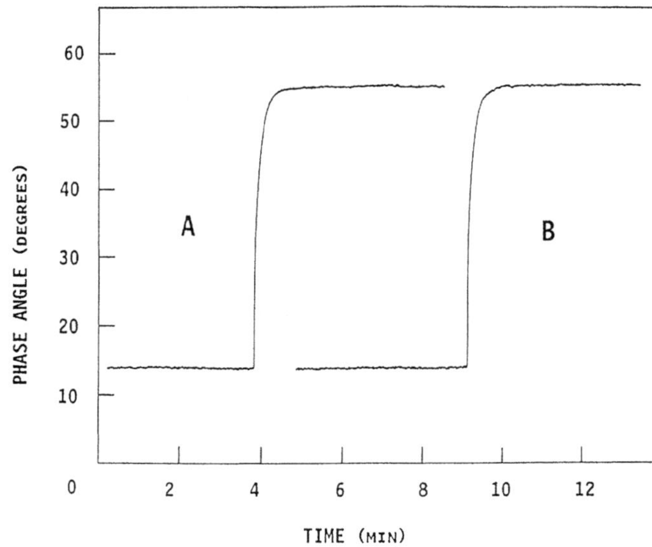

Figure 13.9. Phase angle sensing of oxygen using a LED-based oxygen sensor. The bottle containing the sensor was removed and reinstalled between registrations A and B.

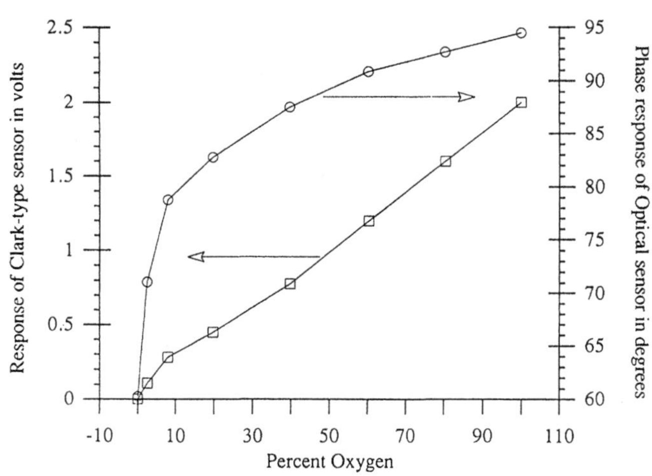

Figure 13.10. Calibration curves for optical sensor and Clark-type oxygen electrode. The phase response of optical sensor and the voltage response of the Clark-type electrode is plotted against percent oxygen in the gas mixture (oxygen and nitrogen) sparged. While the Clark-type electrode shows a linear calibration the optical sensor shows a Stern–Volmer type relationship (see Section 13.10.2).

above equation. As a result higher oxygen sensitivities are obtained at low oxygen tensions which is the operating regime for most bioreactor and waste water treatment facilities.

Pharmaceutical and biochemical applications require sensors that maintain accuracy and reliability through a requisite number of sterilizations (typically 20). To determine the autoclavability of the optical sensor three different sensor membranes with varying probe concentrations were autoclaved at 120°C for 40 min. In each case, an initial "setting" of the silicone rubber membrane occurred after the first autoclaving (as evidenced by a change in texture as well as binding with the glass window on which the membrane is placed) that also caused a small change in dynamic range of the sensor (difference in response between 0% oxygen and 100% oxygen). However, this was a one-time effect and subsequent autoclaving (up to 20 times) did not result in any measurable change in the dynamic range or in the absolute phase angles measured. This, combined with the fact that there is no long-term probe loss from the membrane, allows the sensor to be used in applications which require repeated autoclaving. An additional advantage of the optical sensor is that since it is a nonconsumption-based system the calibration curve is absolute, requiring only a one time calibration after the first autoclaving.

Performance comparisons with a Clark-type sensor demonstrated the applicability of the optical sensor in monitoring dissolved oxygen (DO) levels in a bioreactor.[73] Figure 13.11 shows the response profiles of the optical sensor and Clark-type electrode

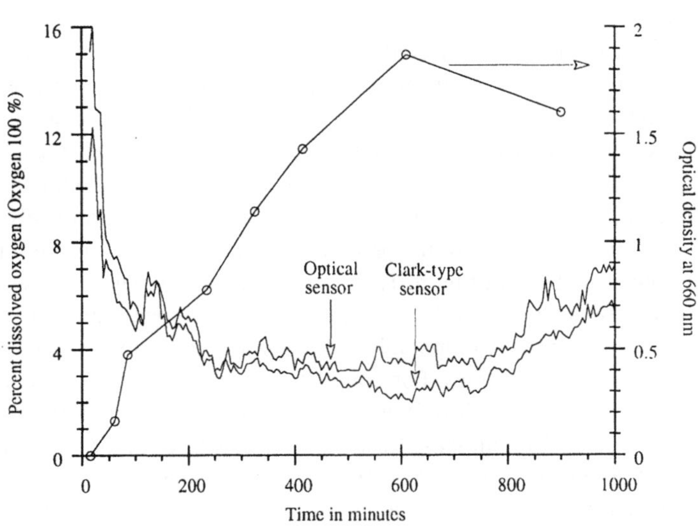

Figure 13.11. Dissolved oxygen profile during a typical *Escherichia coli* fermentation. Using a cubic spline interpolation of the data shown in Figure 13.3, the responses of the two sensors have been converted into percent oxygen and plotted as shown. The optical sensor closely tracks the response of the Clark-type electrode throughout the fermentation.

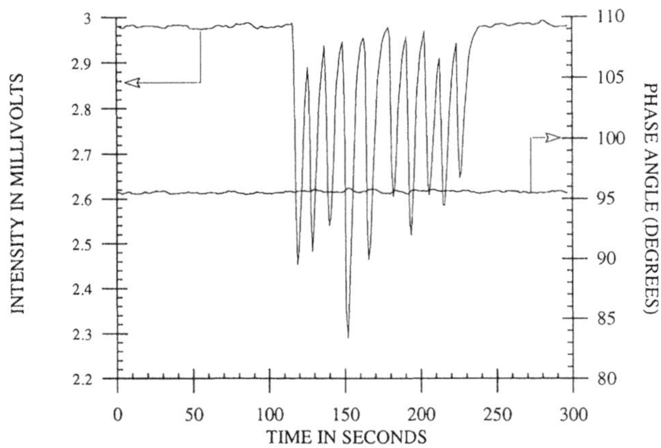

Figure 13.12. Responses of the phase angle and intensity in response to changes in the illumination intensity at fixed oxygen concentration.

to DO in an overnight *E. coli* batch fermentation. Both sensors track the validations in oxygen tension, resulting in a similar DO profile. It must be noted that the local fluctuations in the sensor response is not noise but actual fluctuations in DO. In another experiment, a comparison of the response times of the sensors showed that for low oxygen tensions the optical sensor had a 40% faster response than the Clark-type sensor (data not shown). This is primarily due to the high oxygen permeability of the silicon membrane and well as the non-consumption of oxygen.

Lifetime measurement, as indicated above, is independent of a variety of problems that have hindered the commercialization of sensors measuring fluorescence intensity. In an experiment to compare the phase angle (lifetime) stability with regard to intensity changes, the excitation intensity was varied by waving fingers between the LED source and the detector (Figure 13.12). The intensity trace shows large fluctuations which follow the excitation intensity. In contrast, the phase angles were completely constant. We note that the time-constant for the phase angle measurement was the same as for the intensity measurement. In Figure 13.13, we show the measured phase angles as the source intensity is decreased. No significant change in phase angle is seen for a 10-fold decrease in intensity. Comparably stable phase measurements can be expected even for nanosecond decay times. For completeness, we note that phase sensing of oxygen was first described by Wolfbeis and co-workers,[74] but that this is the only report we know of on lifetime-based sensing. Also, while important, phosphorescence[75] is not likely to provide a general basis for sensing due to the lack of phosphorescence from most chromophores, and the sensitivity of phosphorescence to oxygen quenching which will be present in any sample. Because of the ubiquity of oxygen, and the difficulty of excluding oxygen, we prefer probes with decay times under 10 nsec for which quenching by dissolved oxygen is either negligible or easily accounted for in sensing analytes other than oxygen.

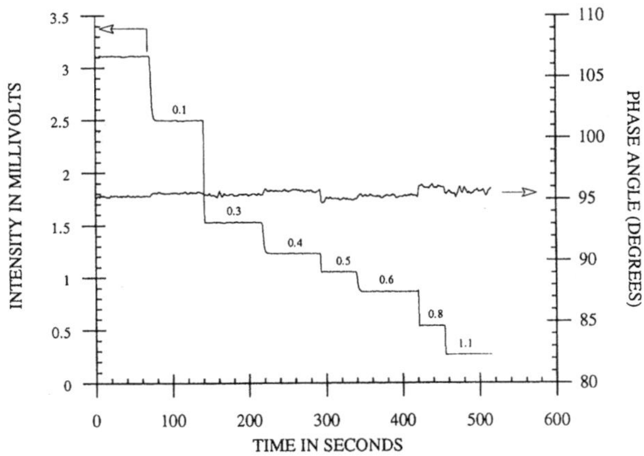

Figure 13.13. Dependence of the phase angle on signal amplitude at fixed oxygen concentration. The optical densities of the filters used to attenuate the excitation are as indicated on the steps. The phase angle readout remained essentially constant except for very low signal amplitudes.

13.10.3. pH Sensors

Lakowicz and Szmacinski[76] have found that the lifetime of the seminaphtorhodafluor/seminaphthtofluorescein (SNARF/SNAFL) series of pH probes are dependent on pH. This is shown for carboxy-SNAFL-2 in Figure 13.14. The changes in phase and modulation are remarkably large and will suffice for good precision measurements of pH. A remarkable feature of lifetime sensing is the ability to change the sensing range (apparent pK_a) by changing the excitation or emission wavelength (data not

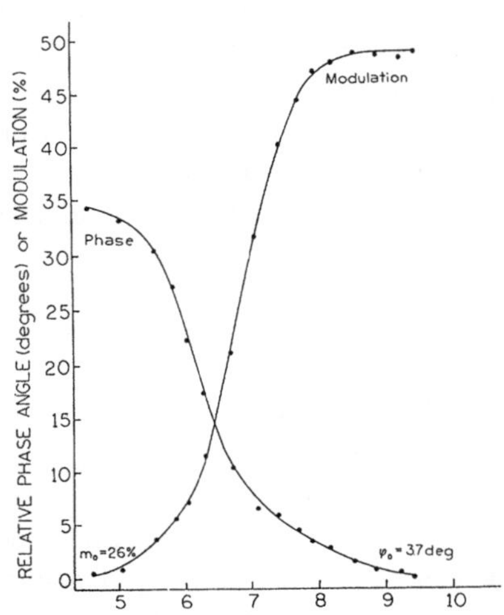

Figure 13.14. pH sensing using the phase and modulation of carboxy SNAFL-2 in Tris (80 m*M*), T = 25°C. λ_{exc} = 543 nm (green He–Ne laser); λ_{obs}: 2–73 Corning filter; f_{mod} = 135 MHz (AO modulator).

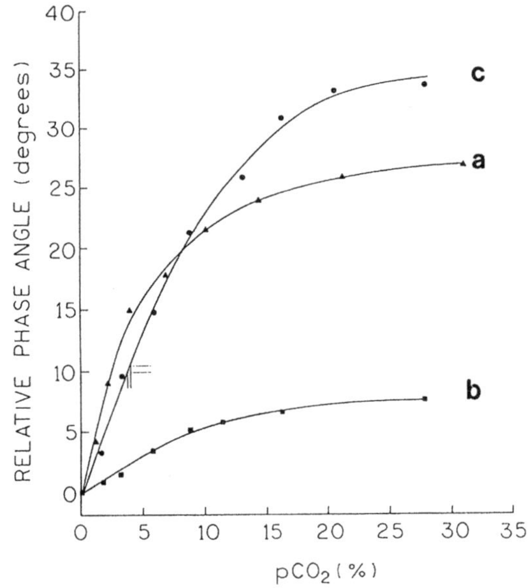

Figure 13.15. CO_2-dependent phase angles. (a) Eosin–Phenol Red (4 m*M*), (b) Rhodamine 6G-Phenol Red (4 m*M*) (c) Texas Red Hydrazide-Bromothymol Blue (2 m*M*). (a) λ_{exc} = 543 nm; λ_{obs} = 580 nm; f = 155 MHz. (b) and (c) λ_{exc} = 543 nm; λ_{obs} = 600 nm, f = 155 MHz.

shown). Alternatively, multiwavelength and/or multifrequency phase and modulation measurements can be used to increase the precision of the measurement, which will be done in testing of these optrodes in bioreactors. In terms of pCO_2 measurements, an approach similar to conventional technology is required. Here, our pH probe is suspended in a bicarbonate solution separated from the bulk sample by a CO_2-permeable membrane. In this case we relied on the principle of energy transfer (below) from pH-insensitive donor to pH-sensitive acceptor.[76] Figure 13.15 is an illustration of pCO_2 sensing with three different probes by this technique. An advantage of this energy-transfer technique is that it can be extended to fluorescent donors at any wavelength by the choice of a suitable acceptor.[76]

13.10.4. Glucose Sensors

Finally, we show results for glucose sensing using lifetimes and energy transfer (energy transfer described later). In this case, we rely on ConA which is labeled with a donor. The acceptor is a dextran-labeled with malachite green (MG). A polymeric acceptor was chosen so that a ConA and competitive binders (dextran) could eventually be fabricated into a sensor with a glucose-permeable membrane.[8] ConA was labeled with three donors (Figure 13.16). In all cases binding of labeled-ConA to the MG-dex-

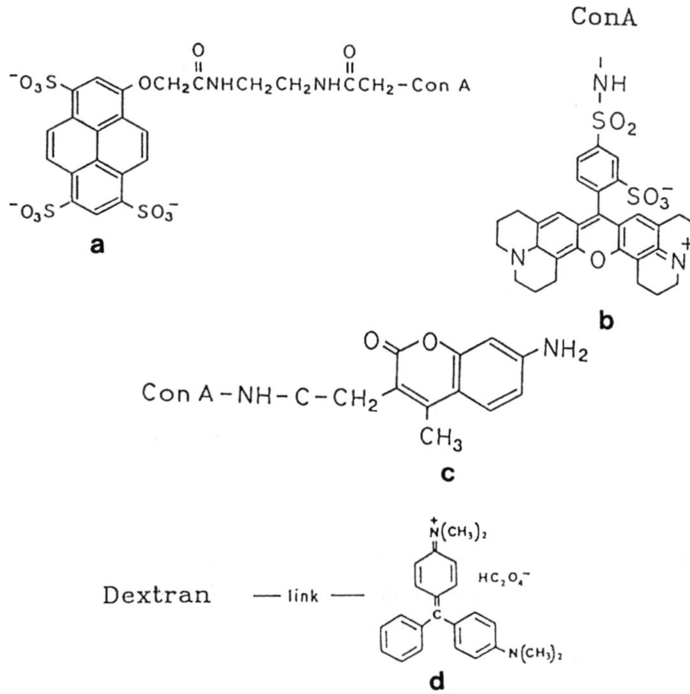

Figure 13.16. Donors and acceptors used for glucose sensing. Donors: (a) Cascade Blue-ConA; (b) Texas Red-Con A; (c) AMCA-ConA. Acceptors: Malachite Green-dextran.

tran resulted in substantial phase angle changes. The use of Texas Red-ConA (Figure 13.17) demonstrates that this donor–acceptor pair could be used with a 543-nm HeNe laser. The data in Figure 13.18 show that the phase angles are sensitive to added glucose, as is required for a competitive glucose sensor. These types of labeled macromolecules can be used to fabricate a glucose sensor for use in bioreactors.

The probes for oxygen, pH, and glucose sensing described here will not show cross-sensitivity to each other. Furthermore, the results shown for O_2 are unique due to the long lifetime of the Ru complex, which allows use of low frequencies. Sensing of pH, pCO_2, and glucose is currently possible using the green (543-nm) HeNe laser and frequencies near 120 MHz. This ensures that the sensing schemes described herein for phase-modulation detection of these analytes in a bioreactor can definitely be carried out using a green HeNe laser as the primary light source. The HeNe laser is simple and widely used as a light source, with a unit cost of around \$2500. The disadvantage is that such a device will still be a tabletop one, and 543-nm HeNe lasers have relatively low optical output. Although the estimated cost of such a unit already is considerably lower than the currently used devices (Table 13.1 and 13.2), we believe

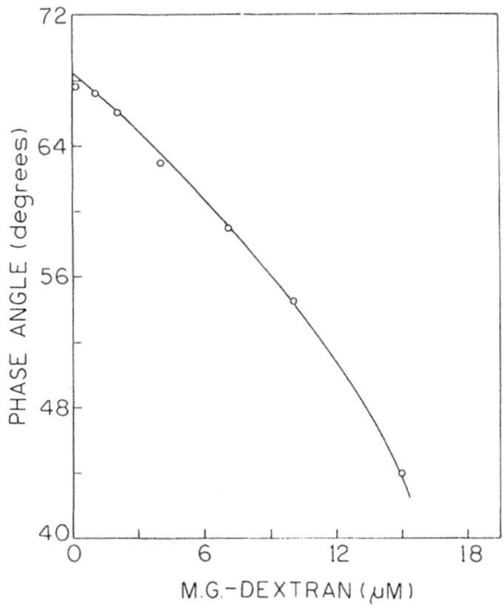

Figure 13.17. Con A binding to dextran measured by energy transfer. Texas Red-ConA; λ_{exc} = 576 nm; f = 315 mHz.

that even smaller and cheaper devices are possible if one can use laser diodes as the light source. Since these devices are currently available commercially only in the far-red range, new probes are required that use energy transfer mechanisms to allow the use of laser diodes.

13.10.5. Utilization of Low-Cost Red LED and Laser Diode Sources

As demonstrated here, one can measure all the four analytes of interest with a green HeNe laser. Importantly, energy transfer concepts allow the use of longer-wave-

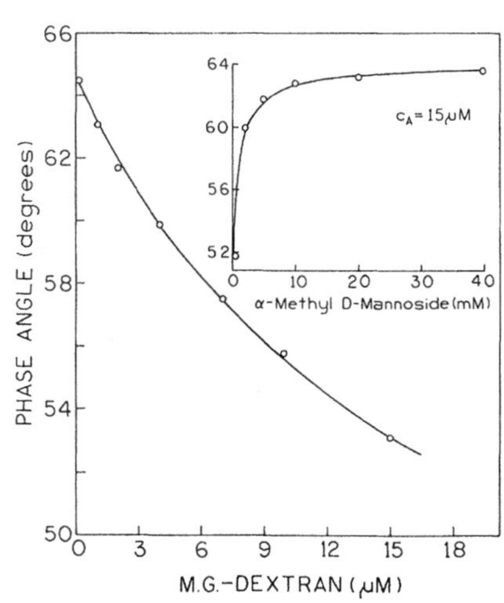

Figure 13.18. Glucose displacement assay using an insoluble polymeric acceptor. AMCA-Con A; λ_{exc} = 356 nm; *f* = 224 mHz. Inset: α-methyl-D-mannoside.

length sources so that low-cost light sources can be used (in the order of a hundred dollars per unit). We also note as an aside that longer wavelengths are relevant to clinical applications, since the human skin is optically nonabsorbing above 620 nm. This allows the possibility of subcutaneous sensors which contain the longer-wavelength probes and the necessary electronics pressed against the skin.

Fluorescence resonance energy transfer (FRET) occurs when the excited state energy is transferred from a donor (D) to an acceptor (A). In general, probes containing two separate molecules can be synthesized, a fluorophore (D) and an analyte-sensitive acceptor (A) as shown below:

D------------------------------A

Covalent link

This energy transfer takes place without the appearance of a photon and the transfer rate depends primarily on the extent of overlap of the emission spectrum of the donor with that of the absorption spectrum of the acceptor, and the distance between the two. The donor can be chosen so that it can be excited by long-wavelength diode lasers. The acceptor should have an absorption spectrum that is sensitive to the analyte (e.g., pH). When the pH changes, the absorption spectrum of the acceptor also changes and should affect the emission of the fluorophore. This in turn should be measurable as a fluorescence lifetime change. Pursuit of energy transfer mechanism sensors have already yielded convincing results in Lakowicz's group. Based on the above considerations, one of our potential pH probes is shown in Figure 13.19.[75]

The pH probe in Figure 13.19 is based on the principle of energy transfer from the pH-insensitive donor (Indocyanine), to a chromophore part which changes absorption in response to pH (Thymol Blue). Such a sensor could be excited with a red HeNe laser or a diode laser. More importantly, the use of a covalently linked energy transfer donor–acceptor pair provides a general approach to making a variety of sensors. While energy transfer pH sensing has been proposed previously,[78–80] the sensor design did

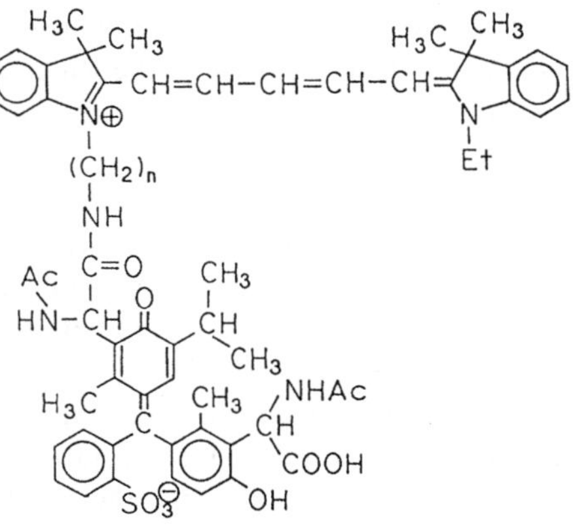

Figure 13.19. Proposed pH probe for use with red laser diode. Expected values: $\lambda_{EX} = 650$ nm; $\lambda_{EM} = 670$ nm.

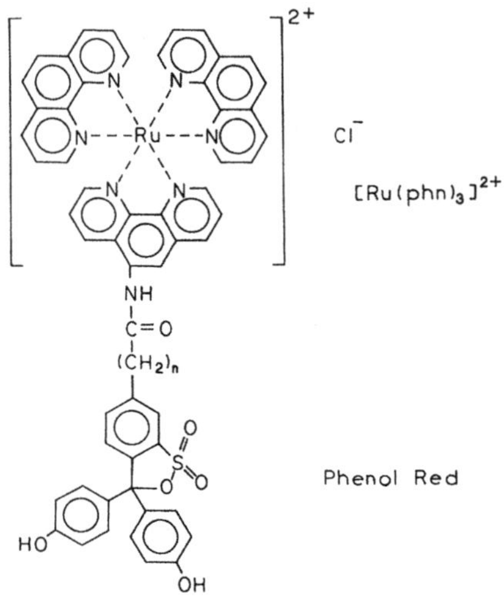

Figure 13.20. Proposed pH(CO_2) probe with long lifetime.

not use covalently linked donors and acceptors, and sensing was based on intensity changes. Consequently, the sensors required frequent calibration to correct for changes in the donor or acceptor concentration, which resulted in varying amounts of energy transfer. Also, lifetime measurements have not been used in any chemical sensing application except for the special case of oxygen sensing using the Ru complexes.[71, 74]

In contrast to earlier sensors, molecules of the type shown in Figure 13.19 are molecular pH sensors that can yield the pH independent of probe concentration. Also, this concept is extendible to other D–A pairs with different excitation wavelengths. One can also conceive of long-lifetime pH sensors, for use with the LED or ELL sources, as shown in Figure 13.20. The precise probe shown in Figure 13.20 would be sensitive to oxygen quenching, which could be avoided by the use of a complexed lanthanide donor in place of the Ru complex. Such complexes are not quenched by oxygen, and are widely used in immunoassay,[81, 82] and have recently been used in fluorescence microscopy.[83]

The results shown in Figure 13.14 represent intrinsic changes in the decay time, which is not easily predictable from the steady state fluorescence nor is it extendible to longer wavelengths. We note that the upper excitation wavelengths for the SNAFL and SNARFs are about 610 and 630 nm, respectively, so laser diode sources cannot be used and use of a red HeNe laser is questionable. However, energy transfer pH sensing is easily extended to excitation with 633- or 670/760-nm sources. To date, a covalently linked pH donor–acceptor pair has not been synthesized. Fortunately, this linkage can be mimicked by using a high concentration of acceptor immobilized with the donor in a suitable matrix. While the O_2 probe was easily entrapped in silicone rubber, the pH probes requires a more hydrophilic matrix. In a recent report we have described an effective immobilization method using Sol-Gel glasses.[84] The Sol-Gel is a unique inorganic polymer similar to glass that is water permeable and has good optical properties. Entrapment of the pH energy transfer probes in the Sol-Gel matrix resulted in a novel, chemically active, optical sensing material. Sol-Gel matrices have previously been used to entrap organic dyes (for review see Avnir[85]) and have been found to provide an environment that is optically and chemically inert. The phenomenon of pH-dependent energy transfer was studied both in solution and after entrapment in a thin (submicron thickness) Sol-Gel film coated on a glass substrate. Figure 13.15 shows the pH-dependent phase angles in three different donor–acceptor pairs in solution. The concentrations of acceptors used are indicated. The pH was varied by using a CO_2-permeable membrane and a bicarbonate buffer. In the case of the Texas Red–Bromothymol Blue pair, the ionization of Bromothymol Blue at higher pH (lower pCO_2) values results in increased spectral overlap and energy transfer from the Texas Red donor, and thus shorter lifetimes and smaller phase angles at lower pCO_2. The high acceptor concentration, however, resulted in an optically dense solution where most of the incident light was absorbed by the acceptor. This problem was eliminated in the Sol-Gel environment because of the small optical path (thin coating). Figure 13.21 shows the intensity response to pH of the Texas Red–Bromothymol Blue energy transfer system in a Sol-Gel matrix. The pH was varied by transferring the Sol-Gel matrix between pH buffers. The change in fluorescence intensity was reproducible and reversible. Figure 13.22 shows the phase response of pH of the same system. Again, reproducible and reversible changes in phase angles were observed. It is important to note that the Sol-Gel sensor was stable and retained its pH sensing capabilities even after repeated autoclaving.

Further miniaturization and lower-cost sensors will put an instrumented bioreactor in every scientist's reach. If this effort succeeds, it should be possible to instrument shake flasks at low cost. Since this is the most widely used "bioreactor" of any kind and the bulk of basic research as well as the annual screening of hundreds of thousands

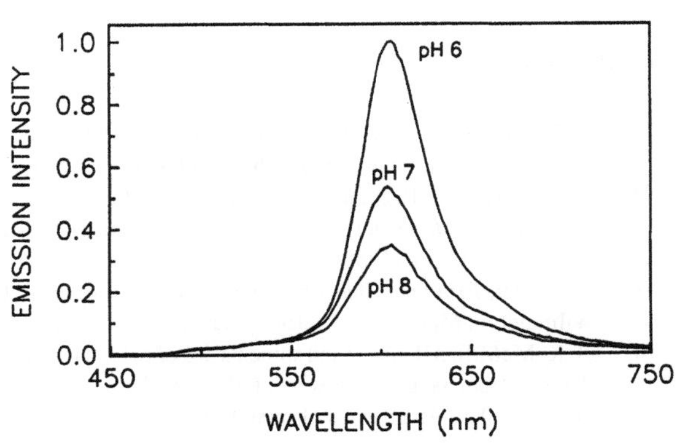

Figure 13.21. The pH-dependent emission spectra of Texas Red Hydrazide and Bromothymol Blue in Sol-Gel (80 mM Tris). Excitation = 442 nm.

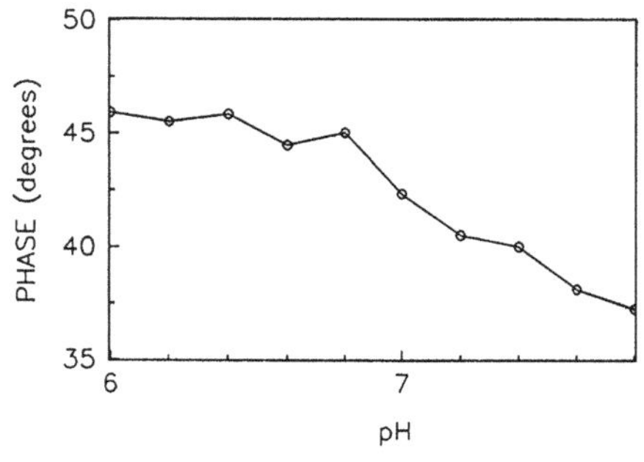

Figure 13.22. Dependence of the Texas Red Hydrazide/Bromothymol Blue phase angle on pH.

of soil samples for new drugs is done by drug companies with this device, it is likely that a lot cost means to monitor the environment within shake flasks will enable greatly improved success rates.

Looking toward the future, we believe that probes for sensing several other analytes by phase-modulation fluorometry will follow rapidly. It is feasible that in a few years, many analytes will be monitored at low cost by phase fluorometry. Together with miniaturization of computers and advances in software, sophisticated computer control algorithms can be readily implemented. It is possible that by making instrumented bioreactors inexpensive through this technology and with further advances in genetic engineering, the ripple effects of new drug discovery, low-cost cell culture, and low-cost sensors for biomedical use will help cut down on the enormous fraction of gross national product (GNP) (14% as of 1991) devoted to health care.

Acknowledgments

GR acknowledges funding from the National Science Foundation (BCS-8911957 and BCS-9157852). HS acknowledges grants DIR-8710401 from the National Science Foundation and RR4800 01 and RR0710 01 from the National Institutes of Health.

References

1. Y-H. Lee and G. Tsao, Dissolved oxygen electrodes, in: *Advances in Biochemical Engineering*, Volume 13 (T. K. Ghose, A. Fiechter, and N. Blakebrough, eds.), Springer-Verlag, Berlin (1979).
2. S. P. Srinivas, G. Rao, and R. Matuarasan, Redox potential in anaerobic and microaerobic fermentation, in: *Handbook on Anaerobic Fermentation* (L. E. Erickson and D. Y. C. Fung, eds.), Marcel Dekker, New York (1985).
3. H. Buhler, *Principles and Problems of pH Measurement*, (available from Ingold Electrodes, Inc., Wilmington, Massachusetts) (now Mettler Toledo, Inc.) (1980).
4. G. K. McMillan, Understand some basics of pH measurement, *Chem. Eng. Prog. 87*, 30–37 (1991).
5. P. J. F. Gommers, B. J. van Schie, J. P. van Dijken, and J. G. Kuenen, Biochemical limits to microbial growth yields: An analysis of mixed substrate utilization, *Biotechnol. Bioeng. 32*, 86–94 (1988).

6. J. S. Schultz, S. Mansouri, and I. Goldstein, Affinity sensor: A new technique for developing implantable sensors for glucose and other metabolites, *Diabetes Care 5*, 245–253 (1982).
7. K. R. Srinivasan, S. Mansouri, and J. S. Schultz, Coupling of concanavalin A to cellulose hollow fibers for use in glucose affinity sensor, *Biotechnol. Bioeng. 23*, 233 (1986).
8. J. S. Schultz, Design of fibre-optic biosensors based on bioreceptors, in: *Biosensors: Fundamentals and Applications* (A. P. F. Turner, I. Karube, and G. S. Wilson, eds.), Oxford University Press, New York (1987).
9. S. J. Coppella and P. Dhuljati, Low cost computer coupled fermentor off-gas analysis via quadrupole mass spectrometer, *Biotechnol. Bioeng. 29*, 679–689 (1987).
10. R. O. Thomson, Bacterial toxins, in: *Secondary Products of Metabolism* (A. H. Rose, ed.), Academic Press, New York (1979).
11. E. Heinzle, H. Kramer, and I. J. Dunn, State analysis of fermentation using a mass spectrometer with membrane probe, *Biotechnol. Bioeng. 27*, 238–246 (1985).
12. B. Buckland, T. Brix, H. Fastert, K. Gbewonyo, G. Hunt, and D. Jain, *Bio/Technol. 3*, 982 (1985).
13. E. Heinzle, K. Furukawa, I. J. Dunn, and J. R. Bourne, Experimental methods for online mass spectrometry in fermentation technology, *Bio/Technol. 1*, 181–188 (1983).
14. J. K. Mclauglllin, C. L. Meyer, and E. T. Papoutsakis, Gas chromatography and gateway sensors for online state estimation of complex fermentations, *Biotechnol. Bioeng. 27*, 1246–1257 (1985).
15. T. Scheper and K. Schugerl, Culture fluorescence studies on aerobic continuous cultures of *Saccharomuces cervevisiae*, *Appl. Microbiol. Biotechnol. 23*, 440–444 (1986).
16. L. N. M. Duysens and J. Amesz, Fluorescence spectrophotometry of reduced phosphopyridine nucleotide in intact cells in the near-ultraviolet and visible region, *Biochem. Biophys. Acta 24*, 19–26 (1957).
17. J. London and M. Knight, Concentrations of nicotinamide nucleotide coenzymes in micro-organisms, *J. Gen. Microbiol. 44*, 241–254 (1966).
18. D. E. F. Harrison, Undamped oscillations of pyridine nucleotide and oxygen tension in chemostat cultures of klebsiella aerogenes, *J. Cell Biol. 15*, 514–521 (1970).
19. D. E. F. Harrison and B. Chance, Fluorimetric technique for monitoring changes in the level of reduced nicotinamide nucleotides in continuous cultures of microorganisms, *Appl. Microbiol. 19*, 446–450 (1970).
20. W. B. Armiger, J. F. Forro, L. M. Montalvo, and J. F. Lee, The interpretation of on-line process measurements of intracellular NADH in fermentation processes, *Chem. Eng. Commun. 45*, 197–206 (1986).
21. W. Beyeler, A. Einsele, and A. Fiechter, On-line measurements of culture fluorescence: Method and application, *Eur. J. Appl. Microbiol. Biotechnol. 13*, 10–14 (1981).
22. J. H. T. Luong and D. J. Carrier, On-line measurement of culture fluorescence during cultivation of *Methylomonas mucosa*, *Appl. Microbiol. Biotechnol. 24*, 65–70 (1986).
23. H.-P. Meyer, W. Beyeler, and A. Fiechter, Experiences with the on-line measurement of culture fluorescence during cultivation of *Bacillus subtilis*, *Escherichia coli* and *Sporotrichum thermophile*. *J. Biotech. 1*, 341–349 (1984).
24. T. Scheper, A. Gebauer, A. Sauerbrel, A. Niehoff, and K. Schugerl, Measurement of biological parameters during fermentation processes, *Anal. Chim. Acta 163*, 111–118 (1984).
25. T. Scheper, T. Lorenz, W. Schmidt, and K. Schugerl, On-line measurement of culture fluorescence for process monitoring and control of biotechnology processes, *Ann. NY Acad. Sci. 506*, 431–445 (1987).
26. D. W. Zabriskie, Use of culture fluorescence for monitoring of fermentation systems, *Biotechnol. Bioeng. Symp. 9*, 117–123 (1979).
27. D. W. Zabriskie and A.E. Humphrey, Estimation of fermentation biomass concentration by measuring culture fluorescence, *Appl. Environ. Microbiol. 35*, 337–343 (1978).
28. A. Einsele, D. L. Ristroph, and A. E. Humphrey, Substrate uptake mechanism for yeast cells. A new approach utilizing a fluorometer, *Eur. J. Appl. Microbiol. Biotechnol. 6*, 335–339 (1979).

29. D. L. Ristroph, C. M. Watteeuw, W. B. Armiger, and A. E. Humphrey, Experience in the use of culture fluorescence for monitoring fermentations, *J. Ferment. Technol. 55*, 599–608 (1977).
30. B. Chance and B. Thorell, Fluorescence measurements of mitochondrial pyridine nucleotide in aerobiosis and anaerobiosis, *Nature 184*, 931–934 (1959).
31. E. L. Winter, G. Rao, and T. W. Cadman, Relationship between culture redox potential and culture fluorescence in *Corynebacterium glutamicum. Biotechnol. Tech. 2*, 233–236 (1988).
32. A. Einsele, D. L. Ristroph, and A. E. Humphrey, Mixing times and glucose uptake measured with a fluorometer, *Biotechnol. Bioeng. 20*, 1487–1492 (1978).
33. S. A. Siano and R. Mutharasan, NADH fluorescence and oxygen uptake responses of hybridoma cultures to substrate pulse and step changes, *Biotechnol. Bioeng. 37*, 141–159 (1990).
34. P. M. Doran and J. E. Bailey, Effects of immobilization on the nature of glycolytic oscillations in yeast, *Biotechnol. Bioeng. 29*, 892–897 (1987).
35. W. Muller, G. Wehnert, and T. Scheper, Fluorescence monitoring of immobilized microorganisms in cultures, *Analytica Chim. Acta 213*, 47–53 (1988).
36. K. F. Reardon, T. Scheper, and J. E. Bailey, *In situ* fluorescence monitoring of immobilized *Clostridium actobutylicum, Biotechnol. Lett. 8*, 817–822 (1986).
37. G. Rao and R. Mutharasan, NADH levels and solventogenesis in *Clostridium acerobutylicum*: new insights through culture fluorescence, *Appl. Microbiol. Biotechnol. 30*, 59–66 (1989).
38. K. F. Reardon, T.-H. Scheper, and J. E. Bailey, Metabolic pathway rates and culture fluorescence in batch fermentations of *Clostridium Acetobutylicum. Biotechnol. Prog. 3*, 153–167 (1987).
39. A. K. Srivastava and B. Volesky, Measurement and regulation of culture reduction state in *Clostridium acetobutylicum, Biotechnol. Bioeng. 38*, 181–190 (1991a).
40. A. K. Srivastava and B. Volesky, NADH fluorescence in a carbon-limited fermentation, *Biotechnol. Bioeng. 38*, 191–195 (1991b).
41. A. K. Srivastava and B. Volesky, On-line fluorescence measurements in assessing culture metabolic activities, *Appl. Microbiol. Biotech. 34*, 450–457 (1991c).
42. C. C. Walker and P. Dhurjati, Use of culture fluorescence as a sensor for on-line discrimination of host and overproducing recombinant *Escherichia coli. Biotechnol. Bioeng. 33*, 500–505 (1989).
43. C. M. Watteeuw, W. B. Armiger, D. L. Ristroph, and A. E. Humphreys, Production of single cell protein from ethanol by fed-batch process. *Biotechnol. Bioeng. 21*, 1221–1237 (1979).
44. S. P. Srinivas and R. Mutharasan, Inner filter effects and their interferences in the interpretation of culture fluorescence, *Biotech. and Bioeng. 30*, 769–774 (1987).
45. N. S. Wang and M. B. Simmons, Effect of background fluorophores on the NADH fluorescence probe signal, *Biotechnol. Tech. 5*, 241–246 (1991).
46. J. K. Li and A. E. Humphrey, Use of fluorometry for monitoring and control of a bioreactor, *Biotechnol. Bioeng. 37*, 1043–1049 (1991).
47. A. J. Lipton and M. M. Domach, Observing protein synthesis, export, and tryptophan incorporation by front-surface fluorometry, *Biotechnol. Bioeng. 39*, 13–19 (1992).
48. J. Einsinger and J. Flores, Front-face fluorometry of liquid samples, *Anal. Biochem. 94*, 15–21 (1979).
49. B. Chance, B. Schoener, R. Oshino, F. Itshak, and Y. Nakase, Oxidation-reduction ratio studies of mitochondria in freeze-trapped samples. *NADH and flavoprotein fluorescence signals, *J. Biol. Chem. 254*, 4764–4771 (1979).
50. W. Halangk and W. S. Kunz, Use of NAD(P)H and flavoprotein fluorescence signals to characterize the redox state of pyridine nucleotides in epididymal bull spermatozoa, *Biochim. Biophys. Acta 1056*, 273–278 (1991).
51. B. H. Junker, D. I.C. Wang, and T. A. Hatton, Fluorescence sensing of fermentation parameters using fiber optics, *Biotechnol. and Bioeng. 32*, 55–63 (1988).
52. J. K. Li, E. C. Asali, A. E. Humphrey, and J. J. Horvath, Monitoring cell concentration and activity by multiple excitation fluorometry, *Biotechnol. Prog. 7*, 21–27 (1991).

53. S. A. Siano and R. Mutharasan, NADH and flavin fluorescence responses of starved yeast cultures to substrate additions, *Biotechnol. Bioeng. 34*, 660–670 (1989).
54. S. C. W. Kwong, L. Randers, and G. Rao, Consistency evaluation of batch fermentations based on online NADH fluorescence, *Biotech. Prog. 8*, 410–412 (1992).
55. S. J. Coppella and G. Rao, Practical considerations in the measurement of culture fluorescence, *Biotechnol. Prog. 6*, 398–401 (1990).
56. S. C. W. Kwong, L. Randers, and G. Rao. On-line detection of substrate exhaustion using NAD(P)H fluorescence, *Appl. Environ. Microbiol. 59*, 604–606 (1993).
57. S. C. W. Kwong and G. Rao, Metabolic monitoring by using the rate of change of NAD(P)H fluorescence, *Biotechnol. Bioeng.*, in press.
58. R. C. Hughes, A. J. Ricco, M. A. Butler, and S. J. Martin, Chemical microsensors, *Science 254*, 74–80 (1988).
59. W. S. Kisaalitas, P. J. Slininger, R. J. Bothast, J. F. McCarthy, and R. L. Magin, Application of fiber-optic fluorescence measurements to on-line pH monitoring of a pseudomonad fermentation process, *Biotech. Prog. 7*, 564–569 (1991).
60. O. S. Wolfbeis, Fiber-optic sensors in biomedical sciences, *Pure Appl. Chem. 59*, 663–672 (1987).
61. O. S. Wolfbeis, in: *Fiber-Optic Chemical Sensors and Biosensors,* Vols. I and II (O. S. Wolfbeis, ed.), CRC Press, Boca Raton, Florida (1991).
62. W. H. Mullen and P. M. Vadgama, Microbial enzymes in biosensors, *J. Appl. Bact. 61*, 181–193 (1986).
63. S. L. Brooks, I. J. Higgins, J. D. Newman, and A. P. F. Turner, Biosensors for process control, *Enzyme Microb. Technol. 13*, 946–955 (1991).
64. R. D. Kiss and G. N. Stephanopolous, Metabolic activity control of the L-lysine fermentation by restrained growth fed-batch strategies, *Biotechnol Prog. 7*, 501–509 (1991).
65. R. Y. Tsien and T. Pozzan, Measurement of cytosolic free Ca^{2+} with Quin-2, *Methods. Enzymol. 172*, 230–262 (1989).
66. J. R. Lakowicz, in: *Principles of Fluorescence Spectroscopy*, Plenum, New York (1983).
67. J. R. Lakowicz and B. P. Maliwal, Construction and performance of a variable-frequency-phase-modulation fluorometer, *Biophys. Chem. 21*, 61–78 (1985).
68. J. R. Lakowicz, G. Laczko, and I. Gryczynski, 2-GHz frequency-domain fluorometer, *Rev. Sci. Instrum. 57*, 2499–2506 (1986).
69. J. R. Bacon and J. N. Demas, Determination of oxygen concentrations by luminescence quenching of a polymer-immobilized transition-metal complex, *Anal. Chem. 59*, 2780–2875 (1987).
70. E. R. Carraway, J. N. Demas, B. A. DeGraff, and J. R. Bacon, Photophysics and photochemistry of oxygen sensors based on luminescent transition-metal complexes, *Anal. Chem. 63*, 337–342 (1991).
71. K. W. Berndt and J. R. Lakowicz, Electroluminescent lamp-based phase fluorometer and oxygen sensor, *Anal. Biochem. 201*, 319–325 (1992).
72. C. A. Parker, *Photoluminescence of Solutions*, Elsevier, Amsterdam (1968).
73. S. B. Bambot, R. Holavanahali, J. R. Lakowicz, G.M. Carter, and G. Rao, Phase fluorometric sterilizable optical oxygen sensor, *Biotechnol. Bioeng. 43*, 1139–1145 (1994).
74. M. E. Lippitsch, J. Pusterhofer, M. J. P. Leiner and O. S. Wolifbeis, Fiber-optic oxygen sensor with the fluorescence decay time as the information carrier, *Anal. Chim. Acta. 205*, 1–6 (1988).
75. J. M. Vanderkooi, G. Maniara, T. J. Green, and D. F. Wilson, An optical method for measurement of dioxygen concentration based upon quenching of phosphorescence, *J. Biol. Chem. 262*, 5476–5482 (1987).
76. J. R. Lakowicz, H. Szmacinski, and M. Karakelle, Optical sensing of pH and pCO_2 using phase-modulation fluorometry and resonance energy transfer, *Analytica Chim. Acta 272*, 179–186 (1993).
77. J. R. Lakowicz, H. Szmacinski, and K. W. Berndt, Fluorescence lifetime-based sensing of blood gases and cations, *SPIE Proc.* 1648-17 (1992b).
78. G. Gabor and D. R. Walt. 1991. Sensitivity enhancement of fluorescent pH indicators by inner filter effects, *Anal. Chem. 63*, 793–796 (1991).

79. P. Yuan, D. M. Jordan, F. P. Milanovich, and D. R. Walt, Fiber-optic chemical sensors based on energy transfer, *SPIE Proc. 906*, 28–29 (1988).
80. P. Yuan and D. R. Walt, pH-Dependent fluorescence of merocyanine-eosin-labeled water-soluble polymers, *Macromolecules 23*, 4611–4615 (1990).
81. E. P. Diamindis, Immunoassays with time-resolved fluorescence spectroscopy: Principles and applications, *Clin. Biochem. 21*, 139–150 (1988).
82. T. Lovgren, I. Heemila, K. Pettersson, and P. Halonen, in: *Time-Resolved Fluorometry in Immunoassay, Alternative Immunoassays*, John Wiley & Sons, New York (1985).
83. E. J. Soini, J. Pelliniemi, A. Hemmila, V-M Mukkaka, J. J. Kankare, and K. Frojdman, Lanthanide chelates as new fluorochrome labels for cytochemistry, *J. Histochem. Cytochem. 36*, 1449–1451 (1988).
84. S. B. Bambot, J. Sipior, J. R. Lakowicz, and G. Rao, Lifetime-based optical sensing of pH using resonance energy transfer in sol-gel sensors, *Sensors & Actuators B*, in press.
84. D. Avnir, S. Braun, and M. Ottolenghi, The Encapsulation of Organic Molecules and Enzymes in Sol-Gel Glasses: Novel Photoactive, Optical, Sensing and Bioactive Materials. A Review. National Technical Information Service (NTIS) Publication No. AD-A224 154, 21 pp. (1991).

14

Principles of Fluorescence Immunoassay

Alvydas J. Ozinskas

14.1. Introduction

Immunoassays are preferred for the quantitation of many clinically relevant analytes. The high binding affinities and the specificity of analyte recognition displayed by antibodies greatly simplify the accurate determination of analytes, despite the presence of many other substances in a sample of interest. Useful immunoassay strategies require appropriate levels of analyte sensitivity. The sensitivity of the earliest immunoassays was dramatically improved by the introduction of radioisotopes as tracers. Radioimmunoassays (RIAs) became a benchmark against which subsequent immunoassay technologies were measured. Moreover, RIAs provided guiding principles that were extended for immunoassay development with tracers other than radioisotopes. The pursuit of alternative tracers gained momentum due to the hazards and precautions required in working with radioactive materials, their relative instability, and due to the mounting problems and expense associated with their manufacture and disposal. Thus, reporters based on direct and enzyme-generated fluorescence and absorbance became more widely used to fulfill the growing need for nonradioisotopic tracers. As a result, RIA technology is declining in commercial use, and although the absorbance-based enzyme immunoassay (EIA) is a rather mature technology, RIAs and EIAs are still widely used.

Fluorescent labels have long been used for immunological staining in fluorescence microscopy, both prior to and since the introduction of the RIA. Fluorescence technology has been extensively applied to instrumented immunoassay development, not only since high detection sensitivities are possible, but also since fluorophores display a variety of measurable properties. The emission intensity, orientation, waveform, lifetime, and the interrelationships between these properties contribute to the continuing evolution of fluorescence immunoassays and to their commercial development. The sophistication of fluorescence technology for conducting clinical immu-

Alvydas J. Ozinskas • Becton Dickinson Diagnostic Instrument Systems, Sparks, Maryland 21152.
Topics in Fluorescence Spectroscopy, Volume 4: Probe Design and Chemical Sensing, edited by Joseph R. Lakowicz. Plenum Press, New York, 1994.

noassays is evident in instrumentation and immunoassay kits using advanced methods, and further advances are indicated by the rapid pace of new technology development.

The success of a particular immunoassay technology is often measured to the extent that it is useful in commercial and clinical settings. Fluorescence polarization has been used for immunoassays for over two decades, and has realized significant clinical applications during the last decade. Time domain methodology has also been successfully used in clinical settings, with commercially available kits for many analytes. These so-called time-resolved methods make somewhat limited use of time domain techniques, but deliver highly sensitive results with virtually all analytes. Several studies that exploit frequency domain methods have emerged, and may have potential for clinical utility. Frequency domain methods can be largely independent of relative fluorescence intensities and associated fluorescence background problems. Methods based on phase-modulation spectroscopy are responsive to the fluorescence lifetime, an absolute property, making them very robust in terms of accuracy and real-time monitoring.

Developments in fluorescence methods are merged with advances in instrumented technologies, as virtually all instrument components become improved, simplified, and more economical. Laser diodes are powerful, compact, and inexpensive sources of monochromatic light. Photomultiplier tubes can be obtained commercially in small, unified packages that contain the amplifier and high-voltage divider. Photodiodes capable of detecting photon bursts from single molecules have been reported. Coupled with these developments, advances in fluorescent probes, especially at near-infrared (near-IR) and IR wavelengths, indicate that fluorescence technology for immunoassay applications has the potential for continued rapid development.

The aim of this chapter is to discuss fluorescence concepts that are used in selected immunoassay applications. The primary focus is on fluorescence topics of recent interest that provide insight into the characteristic properties of antibodies and antigens in immunoassays, or that describe enhancements in immunoassay technologies. The basic reagents and instrumentation required for immunoassay purposes are discussed first, followed by a brief description of immunoassay formats. The principles that are utilized in various fluorescence immunoassay technologies are outlined with specific examples and their significance. Since it is beyond the scope of this chapter to review all of the applications of fluorescence immunoassays, apologies are extended to authors that this chapter fails to cite. A number of comprehensive treatments of fluorescence immunoassay (FIA) applications and related topics are available.[1–8]

14.2. Fluorescence Immunoassay Reagents

Fluorescence immunoassays require labeling of antibodies, antigens, or both. The structure and some of the properties of antibodies that are important to the construction of immunoassays are briefly discussed first in this section, followed by a general discussion of probes and some of their characteristics.

14.2.1. Antibodies

Antibodies of the IgG class are globular glycoproteins with molecular weights of about 150,000 to 160,000 Da, comprising two heavy (H) chains (50,000 Da) and two light (L) chains (25,000 Da), joined to form a Y-shaped molecule that is stabilized by a number of disulfide bonds (Figure 14.1). Three major antibody domains are distinguished, based on enzymatic fragmentation of antibodies. The two identical analyte- or antigen-binding domains are known as Fab regions, named after the fragment with the antigen binding site, and the stem is known as the Fc region, named after the fragment that crystallizes. Antibodies can be cleaved by the enzyme papain just above the disulfide hinge region, where the Fab domains join the Fc region, releasing Fab fragments.[9] Enzymatic cleavage with pepsin also occurs in the hinge region, and yields $F(ab')_2$ fragments, in which the Fab′ fragments are joined by one or more disulfide bonds.[10, 11] The disulfides can be reduced under mild conditions to yield individual Fab′ fragments with thiol functions that are suitable for binding reporter groups. The different structural types of immunoglobulin (Ig) molecules are classified as immunoglobulin A (IgA), immunoglobulin D (IgD), immunoglobulin E (IgE), immunoglobulin G (IgG), and immunoglobulin M (IgM), based on differences in their heavy chains. The IgG class is most often used for immunoassays, and is further categorized into various subtypes. The IgM class, which is occasionally used for immunoassays, comprises five antibody molecules joined into a decavalent structure by a J-chain and disulfide bridges. The J-chain is also found in IgA dimers and trimers.

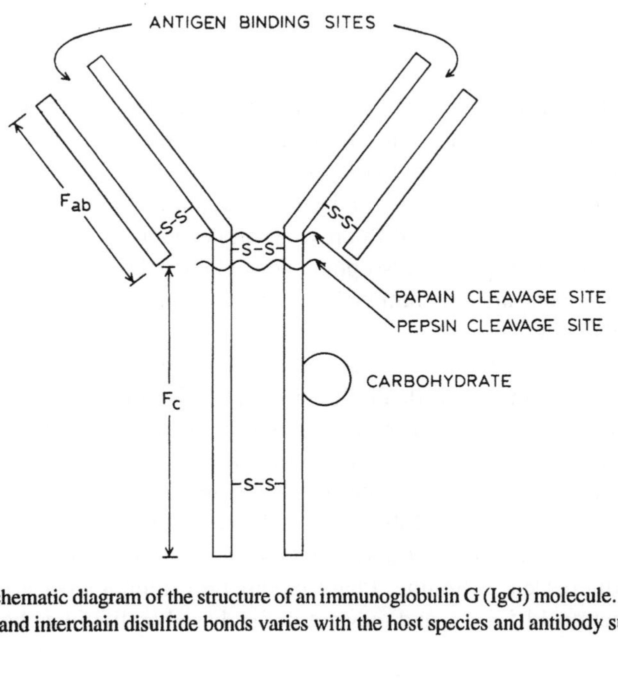

Figure 14.1. Schematic diagram of the structure of an immunoglobulin G (IgG) molecule. The location and extent of intra- and interchain disulfide bonds varies with the host species and antibody subclass.

Antibodies are formed in response to stimulation of a host animal with an immunogen. The immunogen must be relatively large, with a molecular weight in excess of about 10,000 Da. Small-molecular-weight antigens, or haptens, are not large enough to elicit an immune response, and must be bound to a suitable carrier protein such as bovine serum albumin. Polyclonal antibodies are purified from the sera of immunized animals, and have a heterogeneity of structures with varying specificities and affinities. In many cases, polyclonal antibodies can be fractionated into subpopulations with higher binding affinities.

Monoclonal antibodies are produced by hybridoma cell line clones.[12] Individual spleen cells from immunized mice or rats each produce antibodies with a uniform structure. The cells are fused with myeloma cells that are capable of continuous propagation. The hybridoma clones resulting from the fusion are isolated by limiting dilution and selected for those that produce antibody in reasonable yield, with desirable characteristics, such as high affinity and low cross-reactivity. Hybridomas are commonly grown intraperitoneally in suitably prepared mice and rats. The implanted tumor cells secrete the antibody into the ascitic fluid. The clones can also be grown in bioreactors where large scale *in vitro* production is possible. The uniform structures of monoclonal antibodies and the continuous propagation of the clones that produce them make them ideal reagents for immunoassays. Polyclonal and monoclonal antibodies are available from a large number of commercial suppliers with specificities for hormones, therapeutic drugs, drugs of abuse, disease markers, antibodies, and other analytes.[13]

14.2.2. Fluorescent Probes

In FIAs, the antibody, antigen, or both may be labeled with reporter groups to transduce a binding interaction into a quantifiable change. The use of fluorescent probes in immunoassays has previously been reviewed,[1, 14, 15] including a detailed account of conjugation chemistry,[16] and general reviews of protein modification also exist.[17, 18] Desirable properties of fluorescent probes include gentle reactivity, high water solubility, low nonspecific adsorption, good photostability, as well as high molar extinction coefficients and quantum yields. For certain applications, the fluorescence lifetimes are also critical.

The fluorescence lifetime can be measured by time-resolved methods after excitation of the fluorophore with a light pulse of brief duration. The lifetime is then measured as the elapsed time for the fluorescence emission intensity to decay to $1/e$ of the initial intensity. Commonly used fluorophores have lifetimes of a few nanoseconds, whereas the longer-lived chelates of europium(III) and terbium(III) have lifetimes of about 10–1000 μsec (Table 14.1). Chapter 10 (this volume) describes the advantages of phase-modulation fluorometers for sensing applications, as a method to measure the fluorescence lifetime. Phase-modulation immunoassays have been reported (see Section 14.5.4.3.), and they are in fact based on lifetime changes.

Table 14.1. Physical Data of Some Fluorophores Used in Immunoassays

Fluorophore	Σ(molar)	Ex (nm)	Em (nm)	τ (nsec)
Fluorescein isothiocyanate, dichlorotriazinoaminofluorescein	66,000	492	520	4.5
Rhodamine B isothiocyanate	103,000	545–560	585	3
Tetramethylrhodamine isothiocyanate	105,000	540	575	2
Lissamine rhodamine B sulfonyl chloride	80,000	555	580	1
Dansyl chloride	4,300	345	535	14
Anilinonaphthalene sulfonic acid		375	475	16
Pyrenebutyrate	40,000	340	385	100
B-phycoerythrin	2,400,000	545	576	2.5
Allophycocyanin	700,000	634	650	—
Eu^{3+}-β-naphthoyltrifluoroacetone	36,000	340	613	500,000

Relatively few probes have been widely applied to FIAs. Fluorescein and rhodamine, their derivatives, and to some extent phycoerythrin have dominated immunoassays that are based on steady-state measurements. Dansyl, umbelliferone, coumarins, and pyrenes are among others that have been used. Several more commonly used dyes are shown in Figure 14.2. Shorter-wavelength dyes tend to be less useful since they excite at wavelengths where endogenous chromophores, fluorophores, and

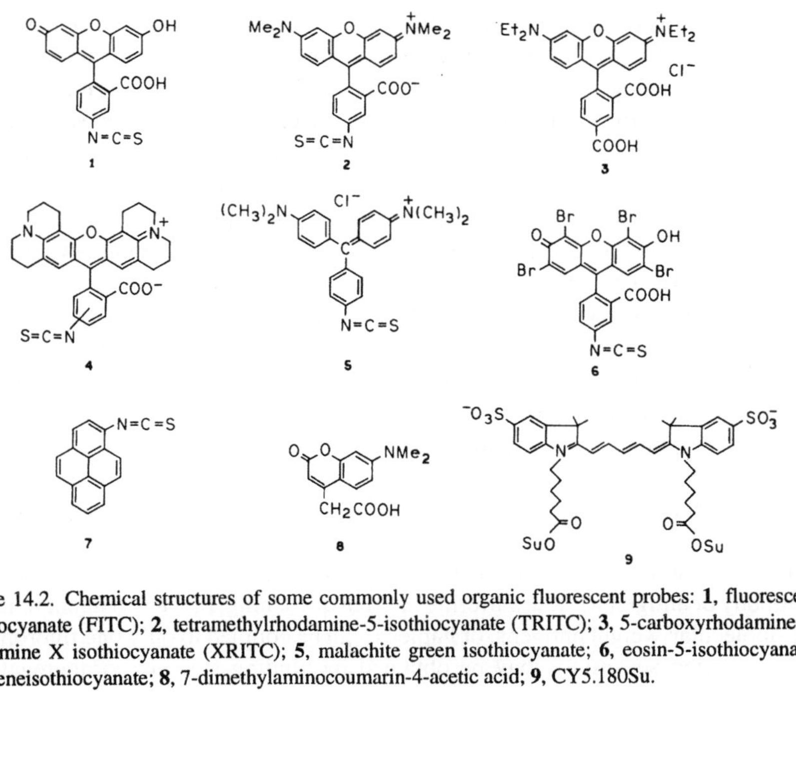

Figure 14.2. Chemical structures of some commonly used organic fluorescent probes: **1**, fluorescein-5-isothiocyanate (FITC); **2**, tetramethylrhodamine-5-isothiocyanate (TRITC); **3**, 5-carboxyrhodamine B; **4**, rhodamine X isothiocyanate (XRITC); **5**, malachite green isothiocyanate; **6**, eosin-5-isothiocyanate; **7**, 1-pyreneisothiocyanate; **8**, 7-dimethylaminocoumarin-4-acetic acid; **9**, CY5.180Su.

some drugs in patient samples can absorb and emit, and also where glass and plastic attenuate the light. Similarly, the intrinsic fluorescence properties of proteins are virtually useless for immunoassay purposes due to the short excitation wavelengths that are required and the lack of sensitivity. Chelates of Eu^{3+} and Tb^{3+} have been extensively used in time-resolved immunoassays .

Virtually all fluorophores that are used in immunoassays have characteristics that, if improved, would lead to more ideal immunoassay performance. Such efforts have been expended notably in the development of improved probes for fluorescence energy transfer immunoassays (Section 14.5.3.) and time-resolved immunoassays (Section 14.5.2.). The total synthesis of new fluorophores for targeted immunoassay purposes is laborious, and there is a risk that new probes may not meet performance specifications. Yet further probe development is essential, particularly in the near-IR (near-IR probe development is the topic of Chapter 7 in this book), with functional groups that make covalent attachment simple enough for general use. Relatively few such probes are available that excite at near-IR and IR wavelengths. A low-molecular-weight near-IR fluorophore that is presently readily available (from Biological Detection Systems, Pittsburgh, Pennsylvania) is CY5 (Fig. 2), which has properties that make it well suited for general immunochemical applications.(19) Allophycocyanin is an example of a proteinaceous near-IR fluorophore that is useful for immunoassays. Reports have appeared that detail the use of a new class of proprietary phthalocyanine fluorophores for immunoassay purposes.(20, 21) The availability of more near-IR fluorophores will likely lead to significant improvements in FIA development. Whole blood FIAs will become increasingly more feasible, since near-IR fluorophore use is not complicated by the fluorescence of endogenous fluorophores nor by inner filter effects due to endogenous chromophores such as hemoglobin. Furthermore, the sensitivity of detection can be increased in comparison with visible probes, since near-IR fluorophores tend to have large molar extinction coefficients (> 200,000), and signal-to-noise ratios can be increased at longer wavelengths. Direct evidence for the photon burst from a single molecule of the near-IR fluorophore IR-132 was acquired with a single photon avalanche photodiode, which has a high quantum efficiency in the IR.(22) Another example of the need for probe development is the general lack of fluorophores that enable fluorescence polarization immunoassays (FPIAs) for large-molecular-weight analytes. As discussed in Section 14.5.1., fluorescence lifetimes of commonly available probes are long enough only for the FPIA determination of relatively small-molecular-weight analytes. The development of longer lifetime probes has the potential to enable simple, homogeneous immunoassays for large-molecular-weight analytes.

Direct attachment of some probes to reagent components can result in limited detection sensitivity, since there is a distinct limitation to the number of small-molecular-weight fluorophores, with the notable exception of lanthanides, that can be attached to an antibody or antigen without suffering internal quenching. In many reagent schemes, small-molecular-weight antigens, or haptens, can be coupled to only one fluorophore. Better detection sensitivity may be obtained by binding antigens to aggregates of fluorophores. The large proteinaceous fluorophores, phycoerythrin and allophycocyanin, actually comprise a number of fluorophore moieties. These proteins have large extinction coefficients and high quantum yields. Probe amplification is possible with various reagents, such as liposome-encapsulated fluorophores, latex beads impregnated with fluorophores, other microparticles, and enzymes. The use of enzymes with fluorogenic substrates, for example, typically results in 5- to 1000-fold greater sensitivity than is achievable with chromogenic substrates.

Various functional groups on the antibody surface are available for coupling to probes, although amino groups are most widely used.[16] Amino group labeling can result in compromised antibody binding activity, since both the heavy and light chains at the antigen combining sites terminate in α-amino groups, and since ε-amino groups of lysine may occur near the binding sites. Thus, if antibody molecules are inactivated during the labeling process, then reporter groups that are attached to inactivated antibodies can falsely indicate larger contributions from the unbound fraction. Steric problems can also occur with large-molecular-weight probes such as phycoerythrin.[23, 24] To avert some of these problems, site-specific labeling at more remote locations is possible. Vicinal diols of Fc region carbohydrates can be oxidized and used for coupling to a label.[25, 26] Labels can be bound to hinge region thiols of intact antibodies after partial reduction.[24, 27] Hinge region disulfides that join $(\mathrm{Fab}')_2$ fragments following pepsin digestion (see Section 14.2), and the thiols that result from reductive cleavage to form Fab′ fragments are suitable for remote, site-specific labeling.[28, 29] The use of Fab′ fragments also avoids potential cross-reactive interactions of the Fc region. Univalent antigen recognition afforded by using Fab and Fab′ fragments prevents formation of precipitable large binding complexes, although this is at the expense of the enhanced binding affinity, or avidity, that is possible with two binding sites on the same antibody molecule.

Immunological interactions can be mediated through the use of a nonimmunological strong binding pair, such as a biotinylated reactant to a binding partner that is labeled with avidin or streptavidin. The association constant for the biotin–avidin interaction is on the order of $10^{15}\ M^{-1}$, and since each avidin molecule has four biotin binding sites, the binding interaction is quite strong and can be amplified.

Quantitation of the ratio of covalently bound fluorophores to antibody molecules is relatively straightforward, requiring the use of dye extinction coefficients and simple algorithms.[17, 19, 30] Caution must be exercised in cases where aggregation of conjugated dye results in additional blue-shifted absorbance maxima in the absorbance spectra of fluorophore-protein conjugates, which can in turn result in erroneous estimation of incorporation ratios and also in reduced fluorescence intensities. In this case, the dye concentration is more accurately determined spectrophotometrically after dispersing the aggregated dye with an organic solvent, such as dimethylformamide. The antibody is more accurately quantitated by a protein assay. In cases where the ratio of label to protein or other bound substance cannot be quantitated, it is customary to report a dye concentration determined by spectrophotometry, together with the molar ratio used for the conjugation reaction.

14.3. Fluorescence Instrumentation

Fluorometers designed for research purposes[31] are typically equipped with a xenon arc lamp, monochromators, one or more photomultiplier tubes, cuvet holders, and a computer interface. Some research level fluorometers, such as the Perkin-Elmer LS50, have optional microtiter plate reading accessories with fiber optic bundles. This is convenient since 96-well microtiter plates are commonly used for immunoassay development, and many commercial immunoassays are based on the use of microtiter plates. Fluorometers designed for commercial immunoassay purposes are generally dedicated instruments with few, if any, data acquisition and reduction parameters that can be manipulated by the user.

In instruments for immunoassays, the light source may be as simple as a common tungsten or halogen lamp with inexpensive filters for wavelength selection. Developments in lasers are making monochromatic light available at relatively low cost, with adequate power for most applications. Infrared diode lasers have been available for some time, and seem to have largely preceded the development of probes that make them useful for FIA. Diode lasers at wavelengths as low as 630 nm are now inexpensive and are produced in quantity. Sample measurement may be done in cuvettes, tubes, microtiter plates, flow cells, and also in specially designed containers, with fiber optic probes and evanescent waveguides. A novel cuvette has been reported to more than double the efficiency of fluorescence measurements, for both steady-state and phase-modulation measurements.[32] The excitation is reflected back through the sample, which excites more fluorophore molecules, thereby improving the signal-to-noise ratio. The most common photodetectors for practical use have been photomultiplier tubes (PMTs). However, developments in charge-coupled device (CCD) and photodiode detectors have resulted in significant improvements in detection sensitivity. An important characteristic of these detectors is their red to IR wavelength sensitivity. These factors may make CCDs and photodiodes more attractive for increased immunoassay use in the future.

Fluorometers can be distinguished according to whether they are capable of making steady-state intensity measurements or lifetime measurements. Steady-state fluorometers can be operated in the analog mode, where the output current of the PMT has a linear response to the number of incident photons over a wide range. They can also be operated in the photon counting mode, in which the PMT produces pulses in response to the incidence of single photons. This mode is useful for low concentrations of fluorophores where high sensitivity is required, but is limited by a short dynamic intensity range. Fluorescence lifetimes can be measured by time-resolved, or pulse fluorometry; and by phase-modulation, or harmonic fluorometry. Time-resolved fluorometry involves intensity measurements where the probe lifetime does not change appreciably. For immunoassay measurements, the fluorescence intensity is integrated over a specific time interval (Section 14.5.2.), such that these are in fact gated intensity measurements, whereas the fluorescence lifetime is measured as the time required for the fluorescence intensity to decay to $1/e$ of the original intensity. With time-resolved instrumentation and long lifetime probes, the sensitivity of gated intensity detection is about 10^{-16} *M*, as compared with the 10^{-11} to 10^{-12} *M* limit of steady-state analog instrumentation. In the phase-modulation method, incident light is intensity modulated in a sinusoidal manner, and provision is required for measuring a reference sample. The detector quantitates the phase angle and the demodulation factor, which change in response to fluorescence lifetime changes. This method can distinguish very small lifetime differences in fluorophore mixtures. These concepts are considered in more detail in Section 14.5.4, and have been comprehensively described.[31]

14.4. Immunoassays

Immunoassays are classified in ways that distinguish whether antibody or antigen is labeled, that depend on the number of different antibody reagents, and that indicate whether or not separation and wash steps are required. Heterogeneous immunoassays generally utilize a solid phase binder to facilitate separation and washing steps to remove unbound sample and reagents (Figure 14.3). A solid support can be coated with one or more antibodies, or with the analyte. Analyte binding or its absence is deduced from measurement of a labeled antibody or a labeled analyte reagent that is supplied

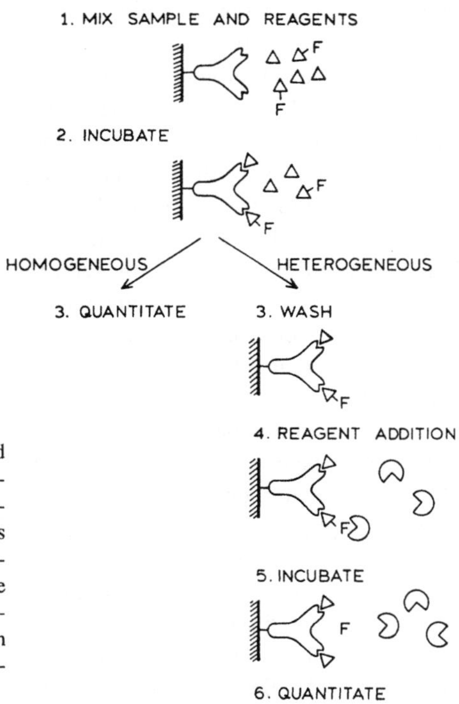

Figure 14.3. Distinction between homogeneous and heterogeneous immunoassay formats. Haptenic analytes are indicated as triangles, and conjugated fluorescent probes are indicated by the letter "F." In this hypothetical depiction, the homogeneous immunoassay is quantitated in the original reaction mixture. The heterogeneous immunoassay requires removal of unreacted tracer, further addition of reagents such as an enzyme to release a fluorescent molecule "F," followed by quantitation.

in the liquid phase. Once the analyte in the sample is bound to the solid phase during a suitable incubation period, the solid phase is subjected to washing steps. The bound solid phase can then be exposed to other reagents that enhance or amplify detection, followed by signal measurement. In homogeneous immunoassays, the sample to be analyzed is mixed with the reagents, and then following an appropriate incubation time, the signal can be measured directly within the original reaction mixture without a requirement for separation and wash steps. In so-called true homogeneous assays, the antibody–antigen binding interaction occurs and is measured within the liquid reaction phase. Distinguished from true homogeneous assays are apparent homogeneous assays, which may actually be effected by separations that are induced within the reaction vessel, although no reagent addition or removal is required once the sample is added. Homogeneous immunoassays are especially desirable for commercial applications since separation and wash steps are not required, which simplifies instrumentation, reduces sources of error, shortens time to results, and decreases biohazards.

The dynamic range and sensitivity of homogeneous assays tend to be more limited in comparison with heterogeneous FIAs. The presence of the sample within the reaction mixture to be measured results in vulnerability of analyte quantitation to matrix interference that is exaggerated over the matrix effects commonly found with heterogeneous assays. For this and other reasons, the amount of sample that can be used is usually limited. Interference with the fluorescence signal from a serum sample can be caused by endogenous fluorophores such as bilirubin, and by light-scattering particles like proteins and lipids. Reagents may also be a source of interference. For example, anilinonaphthalene sulfonic acid (ANS) and furosemide, which may be used to displace sites in proteins that bind analytes such as the thyroid hormone thyroxine and steroids,[33, 34] can fluoresce in the visible spectrum. Nevertheless, as will be seen in subsequent sections, a number of opportunities exist for configuring homogeneous immunoassays using fluorescence. Homogeneous immunoassays and the dedicated analyzers for their measurement should be easier and more economical to manufacture. It seems likely, however, that most homogeneous strategies known to date will not easily exceed or even compare with the sensitivities of heterogeneous immunoassays.[35] Many of these problems will be eased with development of near-IR probes, sources, and detectors. Long wavelengths will avoid fluorescence from endogenous sources and reagents. Sensitivities and signal-to-noise ratios also tend to improve favorably at long wavelengths.

Various immunoassay formats have been devised within the framework of homogeneous or heterogeneous types. Several basic formats are discussed here (Figure 14.4), of which many variations are possible. Note that Figure 14.4 depicts formats with solid phase binders, which facilitate the separation and washing steps utilized in heterogeneous assays. Similar principles are applied in homogeneous assays. A fundamental distinction is between competitive and noncompetitive immunoassays. In a common competitive format, the antibody is immobilized on a solid phase such as a plastic tube, bead, or microtiter plate (Figure 14.4a). In this example, the tracer is a fluorophore-labeled analyte, which competes with analyte added in the sample for a

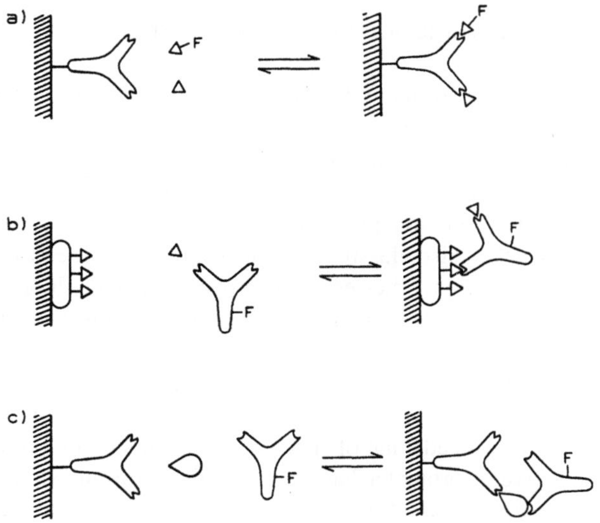

Figure 14.4. Diagram of three basic immunoassay formats. (a) Common competitive format; (b) competitive assay with immobilized antigens bound to a carrier protein (immunometric assay); (c) two-site immunometric assays. Haptenic analytes are indicated as triangles, whereas larger-molecular-weight analytes are shown as teardrop shapes. Conjugated fluorescent probes are denoted by the letter "F."

limited number of antibody binding sites. The resulting signal response is inversely proportional to the analyte concentration, or dose. Another type of competitive assay employs immobilized antigens which, for small-molecular-weight antigens, are often covalently bound to a carrier protein coated on a solid phase (Figure 14.4b). The immobilized antigen then competes with antigen added in the sample for binding sites on the labeled antibody, also resulting in a signal response that is inversely proportional to the analyte concentration. Assays in which the antibody bears the label are known as immunometric assays. Two-site immunometric assays, also referred to as double antibody or "sandwich" assays (Figure 14.4c), are used for antigens with large molecular weights that are on the order of 20,000 Da or greater. Sandwich formats are possible with antigens that have two or more distinct antibody recognition sites, or epitopes. Two antibodies are used. Generally one of the antibodies is labeled, and for some assays both antibodies might be labeled, but each recognizes a different epitope on the antigen. Since the presence of increasing quantities of analyte produces greater numbers of binding complexes, the signal response is directly proportional to analyte concentration. Sandwich assays tend to be more sensitive than competitive immunoassays. Quantitative immunoassays require the use of calibrators that contain varying amounts of analyte to generate a dose-response curve (concentration curve) for analyte quantitation. Qualitative immunoassays require establishment of a threshold value to indicate whether an analyte is present in a sample of interest.

14.5. Fluorescence Immunoassay Applications

FIAs are classified according to the identities and performance characteristics of the probes that are used. Immunoassays in which fluorophore-labeled reagents are used for steady-state intensity measurements may simply use the generic term, FIAs. If a fluorophore property is altered as a result of the immunoreaction, then the assay may be named after the transduction principle, such as quenching or energy transfer immunoassays. Assays are also classified by the technologies used for measurement, as exemplified by polarization, time-resolved, or phase-resolved fluoroimmunoassays (PRFIAs). The distinctions may be blurred when more than one principle or technology is used. In these cases, classification may be rather arbitrary.

FIAs can be based on steady-state intensity measurements without probe amplification, owing to the sensitivity of detection that is possible with fluorescence instrumentation, which exceeds that of spectrophotometers by two or three orders of magnitude. A sensitive fluorometer has been described for an estradiol assay[36] in which the limit of estradiol detection is 3×10^{-11} *M*. Estradiol antibody labeled with rhodamine B is reacted with estradiol samples. Unreacted labeled antibody is removed with Sepharose-estradiol-casein beads, and the remaining fluorescence is directly proportional to the analyte concentration. The detection limit of rhodamine B on the same fluorometer is 5×10^{-12} *M*. This instrument uses a 0.75 mW green helium-neon (HeNe) laser to irradiate the sample from above, at the air-liquid interface, to increase the light path and to decrease surface reflections. The sample compartment has a top-mounted photon trap, and a mirror mounted on the side of the sample compartment opposite the PMT to enhance detection.

In space-resolved immunoassays, a smooth metal slide is coated with an antibody monolayer, and a parallel laser beam is used to quantitate surface bound fluorophore.[37,38] Scattered light is low since the excitation is reflected into a different space, although scatter still remains the principal source of background.

In an assay to detect antibodies against HIV, a common problem in direct multiple labeling of proteins is circumvented by using reversible coupling chemistry.[39] Self-quenching of fluorophore labels above a specific incorporation ratio is a problem that must be avoided to maximize sensitivity. Multiple fluorescein moieties are incorporated into antibody against human immunodeficiency virus (HIV) using disulfide bridges, such that the fluorescein is highly quenched. Following the immunoreaction, the disulfides are disrupted by dithiothreitol. Fluorescein is released into solution and is dequenched, where it can be quantitated with a 20-fold increase in signal over that achievable with conventional fluorescein labeling. Fluorescein fluorescence can be further enhanced with micelles and cyclodextrins.[40]

Enzymes can be linked to immunoassay reagents to amplify detection by the use of fluorogenic substrates. Enzyme-linked fluoroimmunoassays (ELFIAs) are very similar to photometric EIAs in format and workflow. EIAs are widely used, and many commercial ELFIA assays and systems are available.[1] The most commonly used enzymes in ELFIAs are horseradish peroxidase, alkaline phosphatase, and β-Dgalactosidase, with various available substrates. Using the same enzyme, a FIA can be as much as 1000-fold more sensitive than a corresponding EIA.[1]

Fluoroimmunoassays based on microparticles generally use spherical latex particles with mean diameters of 100 nm–50 μm. These are coated with antibody or antigen, and a fluorescent label. In some instances, viable cells have been used as both solid phase and antigen source, particularly in flow cytometry. Following the immunoreaction, microparticles used in liquid phases can be separated by centrifugation, filtration, or magnetic manipulation. Flow cytometry is capable of counting particles and distinguishing them by size and by emission wavelength, and can be used with four-color detection.[1]

In a particle concentration fluoroimmunoassay system that is based on microtiter plates with 0.22-μ filters on the bottoms, microparticles are used with a flowthrough wash system. Either fluorescein-labeled antibodies or cells serve as labels.[41] Carboxyfluorescein diacetate, which crosses the cell membrane, is hydrolyzed by cytoplasmic esterases, thereby trapping carboxyfluorescein within the cell. The method is also useful for screening and isotyping antibodies.

Phase-separation immunoassays have been reported, in which the solid phase particles are formed after the immunoreaction is completed.[42] Phase-separation immunoassays are advantageous since the unstirred layer of solution near a solid surface alters diffusion and binding kinetics at the surface in comparison with the properties of the bulk solution. In phase-separation assays for IgG and IgM, capture antibodies are bound with monomers suitable for styrene or acrylamide polymerization.[42] Monomer-labeled capture antibodies are reacted with analyte and with fluorescein- and/or phycoerythrin-labeled antibodies in a sandwich assay, followed by polymerization of the monomers. Fluorescence of the resulting particles is quantitated in a FACS IV flow microfluorometer, and is directly proportional to analyte concentration.

14.5.1. Fluorescence Polarization Immunoassays

FPIAs were among the earliest homogeneous assays. Since establishment of the theory and method by Dandliker,[43] a large number of FPIA reports has been generated for therapeutic drug monitoring, determination of hormones, drugs of abuse, and some proteins and peptides.[1] Much of the recent activity is due to the commercial availability of complete assay kits and dedicated instruments from various manufacturers, some of which are fully automated,[1] such as the Abbott IMx™.[44] A significant component of the commercial interest is the relative ease of manufacturing homogeneous immunoassay systems and the potentially short time to results. Assay times can be as short as 2 min.[45] For the most part, FPIAs are based on a competitive format, and are conducted in free solution. The fluorophore-labeled antigen is directly measured. Despite some limitations, most notably the practical restriction to analytes with low molecular weights, considerable current interest is indicated by the continuing appearance of many research reports.

According to fluorescence selection rules, only molecules that have their absorption dipoles properly oriented with the electronic dipole of polarized light can be excited.[31] Consequently, the emission is also initially polarized. The polarization, P, is defined by the expression

$$P = \frac{I_{\parallel} - I_{\perp}}{I_{\parallel} + I_{\perp}} \tag{14.1}$$

where $I_{\parallel}$ is the parallel intensity, and $I_{\perp}$ is the perpendicular intensity. The initial orientation of the emission dipoles is such that the fluorescence intensity is maximal if the emission is monitored through a polarizer that is parallel to the source, whereas through a perpendicular polarizer, the intensity is minimal (Figure 14.5). For completely polarized light, the polarization is 1, since $I_{\perp} = 0$. Scattered light is completely polarized, and can cause large inaccuracies in polarization measurements. In random solution, the emission dipoles become depolarized due to rotational motion. Since in this case the intensities observed through parallel and perpendicular orientations of the emission polarizer are theoretically equal, the polarization is zero.

It must be noted that for a randomly oriented tracer in solution, the polarization measurement may not provide a zero value as one would expect, due to potential bias

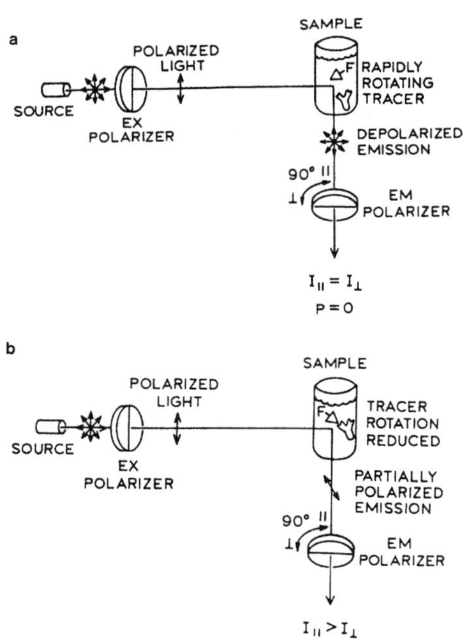

Figure 14.5. Representation of fluorescence polarization immunoassay measurements. (a) Unbound tracer; (b) bound tracer.

in the optical paths of the parallel and perpendicular polarizer orientations.[31] Consequently, in many FPIAs, a control polarization value is determined to correct for optical path bias and for background.[46] Not all FPIA methods require a control measurement,[45] simplifying the workflow.

Polarization equations are based on a planar description of light, which is reasonable when considering the excitation source. Polarization, however, fails to account for the symmetrical distribution of fluorophore emission.[31] The anisotropy, r, more accurately describes the emission by

$$r = \frac{I_{\parallel} - I_{\perp}}{I_{\parallel} + 2I_{\perp}} \tag{14.2}$$

in which the factor $2I_{\perp}$ reflects the symmetrical emission. Polarization is usually used for immunoassays, although at least one immunoassay report based on anisotropy measurements has appeared.[47]

The Perrin equation[48] describes the relationship between the rotational correlational time, ϕ, and the fluorescence lifetime, τ, for steady-state measurements:

$$(1/P - 1/3) = (1/P_0 - 1/3)(1 + 3\tau/\phi) \tag{14.3}$$

where P_0 is the polarization value in the absence of rotational diffusion, and P is the polarization after complex formation. The relationship between the volume, V, of the tracer and the rotational correlation time is

$$\phi = 3\,\eta\;V/kT \tag{14.4}$$

where the rotational correlation time, ϕ, is defined as the time required for rotation through 68.5° and η is the viscosity of the medium. The volume term, V, is dependent on the size and shape of the molecule, k is the Boltzmann constant, and T is the absolute temperature.

Antibody molecules and antibody binding complexes have rotational correlation times that usually exceed 100 nsec.[47] For small-molecular-weight analytes (≤10,000 Da), fluorophores used for FPIA tracers should have lifetimes that are considerably shorter than the tumbling times of antibodies so that the antibody-bound tracer can emit prior to randomization of the emission dipoles. The tracer is usually comprised of a fluorophore-hapten conjugate. The fluorophore lifetime must be greater than the rotational correlation time of the hapten-tracer conjugate due to the need for rapid depolarization of the unbound tracer emission. Fluorescein is the most widely used probe, with a lifetime of about 4 nsec.[47, 49] The rotational correlation time of free fluorescein, and of fluorescein bound to a small-molecular-weight analyte such as phenytoin, is about 1 nsec.[47] The dynamic range of an FPIA tends to be rather narrow. Most FPIAs span fewer than 0.3 polarization units. The units are expressed in mP, so that at a typical resolution of 1 mP, 300 points of resolution can be expected over a dynamic range of 0.3 P. Note that the total possible range is from 0 to 0.5 P, if the probe has $P_0 = 0.5$.

Since FPIAs are conducted as homogeneous immunoassays, they are susceptible to effects from endogenous fluorophores and from intersample variations. Such problems and others due to the sample matrix are largely avoided by sample dilutions of several hundredfold. Low-affinity, nonspecific binding of tracers to sample proteins, when present in sufficiently high concentrations, can result in a falsely elevated polarization signal. Interference from sample proteins can be eliminated when warranted, by proteolytic hydrolysis with pepsin.[46]

Free rotation of the fluorophore in the antibody-bound tracer can also limit FPIA sensitivity. Fluorophore rotation in bound tracer can be affected by the length of the linker, or bridge, between the antigen and the fluorophore of the tracer conjugate, as found in a single reagent polarization assay of methamphetamine.[50] Varying bridge lengths between fluorescein and covalently bound methamphetamine were examined in this study. In the case of a tracer with a seven-carbon bridge, the antiserum titer (or dilution of antiserum required to achieve a specific value or result) was equivalent to that of tracers with four or five carbon bridges, but the maximum polarization value was lower. This result suggests that the longer bridge allows more free rotation, which results in greater fluorescence depolarization of the bound tracer. The single-reagent assay of methamphetamine is possible since the selected antibody has an appreciable dissociation constant, so that methamphetamine added in the sample is capable of displacing and equilibrating with the antibody-bound tracer reagent. The tracer and antibody are commonly supplied as separate reagents, as in a FPIA of astromycin.[51]

Limitations in assay sensitivity can result not only from free rotation of the bound probe, but also from local rotation of the Fab regions of the antibody. Evidence for this has been shown by a comparison of theoretical and experimental anisotropy measurements and by phase-modulation measurements (see Section 14.5.4.) of antibody binding to fluorescein-labeled phenytoin.[47] In this way, the fluorescence lifetimes and anisotropies of local subpopulations of the antibody-bound tracer are resolved within a single binding mixture to extract rotational information. This study shows that 47% of bound tracer has a rotational correlation time of 1.1 nsec. This indicates rapid rotation of the bound tracer at a rate equivalent to that of free tracer. The lower polarization limit is thereby increased, resulting in less sensitivity. Also shown is that about 53% of total antibody-bound probe has a rotational correlation time of 28 nsec, a significant increase over free probe, but not as high as the expected 100-nsec range for the overall tumbling time of an antibody. The inferred reason is that local motion of the Fab domain of the antibody is faster than the tumbling motion of the entire complex. Thus the upper polarization limit is reduced, further limiting the dynamic range.

Since local motions of flexible protein domains can randomize the emission dipoles in a manner that does not reflect the overall tumbling time of the entire protein, and since there is an apparent lack of fluorophores with the appropriate properties for larger analytes, most FPIAs are useful only for determination of relatively low-molecular-weight analytes. Nevertheless, a few studies of larger-molecular-weight analytes have been reported. A competitive assay for human chorionic gonadotropin (hCG) has been reported with FITC-labeled hCG.[52] Human and bovine insulin, labeled with fluorescein, were used in competitive FPIAs with sensitivities approaching 10^{-10} *M*, and with dynamic ranges of more than two logs of insulin concentration.[53] The latter study also reported that antibody binding to the tracers induces negligible intensity and lifetime changes. Smaller proteins, protein fragments such as Fab fragments of antibodies,[54] and peptides that bear representative epitopes are likely to be useful for FPIAs of larger analytes. Longer-lifetime probes may be more useful for larger molecules. Derivatives of dansyl and umbelliferone, which have longer lifetimes than fluorescein, have been investigated in FPIAs of creatine kinase–MB (CK–MB).[55]

The Abbott IMx™, a dedicated commercial immunoassay analyzer that employs FPIAs for small molecules, can also determine larger analytes by a fluorescence-based microparticle capture enzyme immunoassay (MEIA).[44] In this system, antibody-coated 0.47-μm latex particles are used for both sandwich and competitive assays, and alkaline phosphatase conjugates that bind to the particles cleave 4-methylumbelliferyl phosphate to generate the fluorophore.

A report of a particle-based FPIA[56] details improvement in the limit of detection by two orders of magnitude over a conventional FPIA format for rabbit IgG, to 10^{-10} *M*. Goat anti-rabbit capture antibody (binder) is immobilized on polystyrene microspheres or colloidal gold particles, and the sample competes with a fluorescein-labeled rabbit IgG probe. The observed increase in sensitivity is apparently due to the large increase in effective mass of the binding complex. The polarization range is rather limited, though, reportedly due to low concentrations of immobilized antibodies.

14.5.2. Time-Resolved Fluorescence Immunoassays

Time-resolved fluorescence immunoassays (TRFIAs) are among the most sensitive of the FIAs. In time-resolved measurements, the source is briefly pulsed before the long-lived lanthanide chelate tracer emission intensity is measured. Scattered light and background fluorescence from sources such as endogenous fluorophores in reagents and samples, hardware components such as glass or plastic cuvettes, and fluorometer optics are allowed to decay over time before a selected part of the exponential intensity decay curve of the lanthanide tracer is integrated (Figure 14.6). This is achieved by gating the phototube, either by turning on the power after decay of the prompt fluorescence background, or by electronic gating, where a portion of the decay curve is selected for integration.[6, 57–59] Gating is possible because of the long decay times of lanthanides and the short decay times of sample autofluorescence. Thus, the combination of background rejection and the long time interval for quantitation of the emission intensity accounts for the high sensitivities of TRFIAs.

Since lanthanide ions themselves are very weak absorbers, they must be chelated with ligands that capture light and transfer the light energy to the chelated ions (Figure

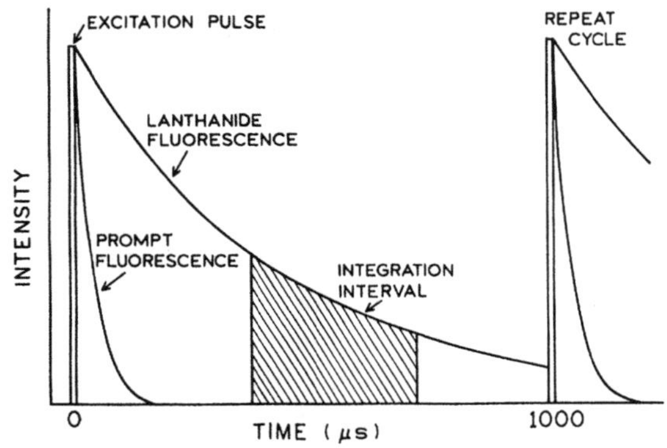

Figure 14.6. Time-resolved fluorescence immunoassay measurement cycle.

14.7). Lanthanide chelates display long lifetimes, large Stokes shifts, narrow emission maxima, and can have relatively high quantum yields. Chelates of europium and terbium are among the most luminescent, contributing to the interest in developing new complexes of these ions.[60–65] Chelates of europium(III) and terbium(III) also are the most commonly used lanthanide probes for immunoassays, with lifetimes that are typically 10 to 1000 μsec.[6, 65–67] The fluorescence of europium chelates tends to be quenched by water, which can be excluded from contacting the chelated ions by using enhancers such as β-diketonate ligands (Figure 14.7), tri-*N*-octylphosphine oxide, and micellar reagents such as Triton-X 100.[65, 67, 68] Terbium chelates are more stable in water, and usually do not require protection from water to fluoresce.[69] However, terbium chelates excite maximally below 300 nm. This can be problematic, since glass and plastic attenuate light below 320 nm.[69]

The chelates that are required for optimally strong binding of europium in immunoassays are often only weakly fluorescent themselves, whereas optimally fluorescent europium conjugates tend to dissociate more readily during immunocomplex formation. Consequently, a dissociative enhancement approach is used in numerous assays manufactured by Wallac Oy under the trade name DELFIA®, monitored using the 1230-Arcus™ or 1232-DELFIA®.[1] The assays are conducted in strips of microtiter plate wells coated with antibodies or antigens. Immunoreagents labeled with the weakly fluorescent, tight binding chelates of ethylenediamine tetraacetate (EDTA), diethylenetriaminepentaacetic anhydride (DTPA), or comparable derivatives are immunologically bound to the microtiter well. The lanthanide ions are then acid-dissociated to form more brightly fluorescent chelates for signal measurement. An early example of the dissociative enhancement method is a TRFIA for rabbit antibody, which uses polystyrene beads.[66]

A number of dissociative enhancement immunoassays have been developed using Eu^{3+} or Sm^{3+} complexed to a proprietary chelator, N^1-[*p*-isothiocyanatobenzyl]-diethylenetriamine-N^1,N^2,N^3,N^3-tetraacetate (Figure 14.7).[1, 70] This chelator is cou-

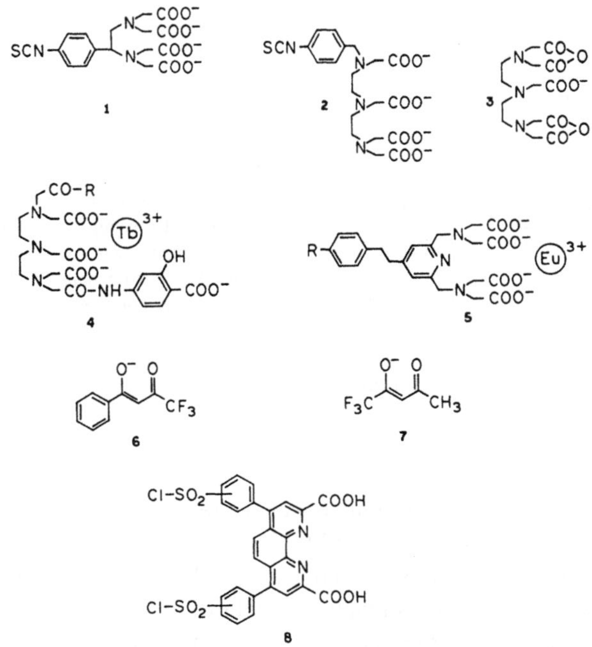

Figure 14.7. Chemical structures of some chelators and enhancers used in time-resolved fluoroimmunoassays: **1**, isothiocyanatophenyl-EDTA; **2**, N^1-*p*-(isothiocyanatobenzyl)-diethylenetriaminetetraacetic acid; **3**, diethylenetriaminepentaacetic anhydride (DTPA); **4**, Tb^{3+}-DTPA-*p*-aminosalicylic acid; **5**, Eu^{3+}-pyridine-bis(methylaminodiacetic acid); **6**, benzoyltrifluoroacetone (BTA, a β-diketone); **7**, trifluoroacetylacetone (a β-diketone); **8**, 4,7-bis(chlorosulfonylphenyl)-1,10-phenanthroline-2,9-dicarboxylic acid (BCPDA).

pled to anti-digoxin antibody in a competitive immunoassay to measure serum digoxin, and a digoxin-albumin conjugate is adsorbed to microtiter strip wells.(71) Another application of this chelator is in a TRFIA of vascular permeability factor.(72) Some additional examples of the dissociative enhancement approach are a sensitive time-resolved immunoassay for free thyroxine (T_4) evaluated against a RIA,(73) and assays for thyrotropin (TSH), luteotropin (LH), testosterone, and cortisol.(74)

A different commercial methodology uses europium–BCPDA (BCPDA is 4,7-bis(chlorosulfophenyl)-1,10-phenanthroline-2,9-dicarboxylic acid, Figure 14.7) chelates, which are of medium stability, covalently conjugated to avidin or streptavidin, and saturated with excess Eu^{3+}.(75) The chelator conjugates bind to biotinylated antibody, which has been immunologically bound to microtiter plates. These chelate complexes fluoresce brightly on a dry solid phase or in solution.(6, 75, 76) This approach has been developed by CyberFluor, Inc., using the CyberFluor 615™ time-resolved fluorometer for numerous analytes, examples of which are immunoassays of carcinoembryonic antigen (CEA),(77) cortisol,(78) human chorionic gonadotropin (hCG),(79) and alpha-fetoprotein (AFP).(80) The europium–BCPDA-labeled streptavidin conju-

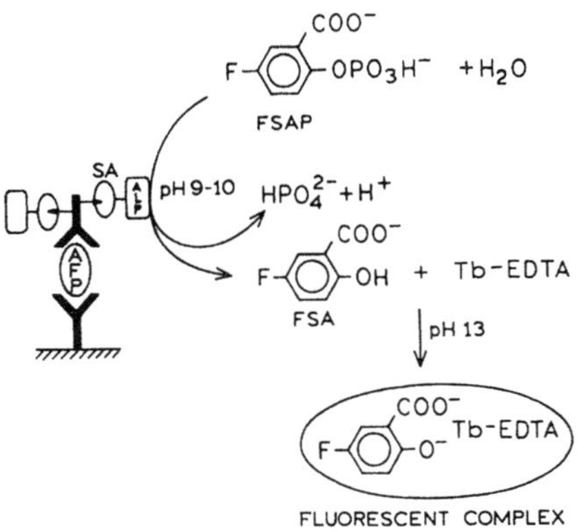

Figure 14.8. Enzymatically-amplified TR-FIA of α-fetoprotein (AFP), using streptavidin-alkaline phosphatase conjugate (SA-ALP) and 5-fluorosalicyl phosphate (FSAP) to generate 5-fluorosalicylic acid (FSA). The FSA combines with terbium-EDTA to form a bright, fluorescent complex. (Adapted from Ref. 83 with permission.) (©1992, American Chemical Society).

gate also has been applied in a similar manner to the detection of IgG and IgM antibodies against rubella virus.(81) The BCPDA approach has been proposed as part of an alternative to Western Blots.(82) In this approach, antigens are first separated by high-performance liquid chromatography, immobilized on microtiter plates, and then detected by time-resolved fluorescence of a europium chelate tracer antibody, with sensitivities in the 10^{-17} *M* range.

A more recent methodology reported by the CyberFluor group uses enzymatic amplification to generate chelators for terbium. In a time-resolved model immunoassay for AFP,(83) the biotinylated second antibody of a sandwich assay is exposed to a streptavidin-alkaline phosphatase conjugate (Figure 14.8). The enzyme cleaves 5-fluorosalicyl phosphate (FSAP) to form 5-fluorosalicylic acid (FSA), which together with EDTA results in a long-lived, brightly fluorescent Tb^{3+} chelate. The substrate, FSAP, is incapable of forming such a chelate. This amplification method has also been applied in a highly sensitive assay for TSH.(84) The method also has been demonstrated with other enzymes.(85)

The large Stokes shift of lanthanide chelates likely accounts for the high labeling efficiency that is possible with lanthanides, as opposed to organic fluorophores. Organic fluorophores are characterized with relatively small Stokes shifts, resulting in self-quenching when the number of fluorophores incorporated in each molecule is great enough. Self-quenching becomes noticeable when more than about eight fluorescein moieties are incorporated into IgG.(86) This does not seem to be a problem with lanthanide chelates. In the case of a TRFIA for human growth hormone using a dissociative enhancement method, 11 molecules of europium chelate per antibody molecule were incorporated.(87) Detection of rubella antibodies by TRFIA was done using a tracer with 14 BCPDA molecules bound per streptavidin molecule.(81) As many as 50 europium ions can be incorporated into a single protein molecule while maintaining immunoreactivity and stability.(66)

Time-resolved approaches for multi-analyte immunoassays have been described recently. Simultaneous determination of LH, follicle stimulating hormone (FSH), hCG, and prolactin (PRL) in a multisite manual strip format has been reported.[88] Four microtiter wells are attached to a plastic strip, two-by-two and back-to-back, such that the wells can be read on a microtiter plate reader. In a quadruple-label format, the simultaneous quantitative determination of four analytes in dried blood spots can be done using europium, samarium, dysprosium, and terbium.[89] In this approach, thyroid-stimulating hormone, 17-α-hydroxyprogesterone, immunoreactive trypsin, and creatine kinase MM (CK-MM) isoenzyme are determined from dried blood samples spotted on filter paper in a microtiter well coated with a mixture of antibodies. Dissociative fluorescence enhancement of the four ions using cofluorescence-based enhancement solutions enables the time-resolved fluorescence of each ion to be measured through four narrow-band interference filters.

Homogeneous TR-FIAs have been reported in which proprietary lanthanide chelates are used. In a homogeneous immunoassay for T_4, a fluorescent europium chelate coupled to thyroxine is quenched by antibody binding.[90] A similar approach is used for estrone-3-glucuronide.[91] TRFIAs based on homogeneous methods have not yet become widely used.

14.5.3. Fluorescence Energy Transfer Immunoassays

Nonradiative fluorescence resonance energy transfer (FRET) is a mechanism that can transduce antibody–antigen binding interactions into measurable parameters. This mechanism is utilized in fluorescence energy transfer immunoassays (FETI). The early development of FETI was done by Ullman *et al.*,[92] and was commercialized by Syva Co. for use with the Syva Advantage™ instrument. One of the immunoreagents in FETI is labeled with a donor fluorophore. Another immunoreagent is labeled with an acceptor chromophore, which may or may not be fluorescent, to quench the donor. FETI measurements are based on fluorescence intensity changes of one of the bound species. Because the fluorescence properties of the unbound species essentially remain unchanged, homogeneous immunoassay formats are possible. Besides intensity changes, energy transfer also induces fluorescence lifetime changes. Thus the energy transfer mechanism is suited for technologies that reflect lifetime changes, such as phase-resolved immunoassays and phase-modulation fluorescence immunoassays, as discussed in Section 14.5.4.

Energy transfer occurs through space as described by the rate expression

$$k_T = (r^{-6} J \kappa^2 n^{-4} \phi_d/\tau_d) \times 8.71 \times 10^{23} \text{ sec}^{-1} \tag{14.5}$$

in which k_T is the rate of energy transfer, r is the distance between donor and acceptor, J is the overlap integral between the donor emission and the acceptor absorbance spectra, κ^2 is a term that relates the relative orientation of the donor and acceptor transition dipoles, n is the refractive index of the medium, ϕ_d is the quantum yield of the donor, and τ_d is the lifetime of the donor. Energy transfer expressions are highly sensitive to the distance between donor and acceptor, varying as the sixth power of the distance. The efficiency of energy transfer is given by

$$E = \frac{\phi_0 - \phi_q}{\phi_0} = \frac{r^{-6}}{r^{-6} + R_0^{-6}} \tag{14.6}$$

where ϕ_0 and ϕ_q are the quantum yields of the donor in the presence and absence of acceptor, respectively. The Förster distance, R_0, is the distance between the donor and acceptor at which energy transfer is 50% efficient, and is expressed in Å as

$$R_0 = 9.79 \times 10^3 (\kappa^2 n^{-4} \phi_d J)^{-6}. \tag{14.7}$$

The energy transfer efficiency is therefore increased with a larger acceptor extinction coefficient, better spectral overlap between the donor emission and the acceptor absorbance, and higher quantum efficiency of the donor. The orientation term κ^2 can vary from 0 to 4, and for randomly oriented molecules is 2/3. Random orientation is, in fact, generally assumed when calculating the R_0.

The predominant donor–acceptor pair for FETI measurements is fluorescein-rhodamine, for which R_0 is 58 Å.(86) A highly efficient donor–acceptor pair that was used in a phase fluorescence immunoassay (see Section 14.5.4.2) is B-phycoerythrin (BPE) and CY5, with an R_0 of 72 Å,(34) which is a long distance in comparison with commonly used donor acceptor pairs.(93) This is due to the good spectral overlap (Figure 14.9), the excellent quantum yield of BPE (0.98), and the high extinction coefficient of CY5 (200,000). The theoretical maximum value of R_0 for a hypothetically perfect donor–acceptor pair is 84 Å.(86) A detailed analysis of energy transfer distance and other parameters that affect energy transfer immunoassays has been described in a time-resolved energy transfer approach by Morrison.(23)

The high molar absorptivities and quantum yields of the large protein fluorophore phycoerythrin (240,000 Da) have been exploited in energy transfer assays. Phycoerythrin has been used as both donor and acceptor, with several bound antigen molecules per phycoerythrin molecule.(86, 94) The usefulness of BPE is indicated in competitive assays for human IgG that use fluorescein-labeled antibody as donor to

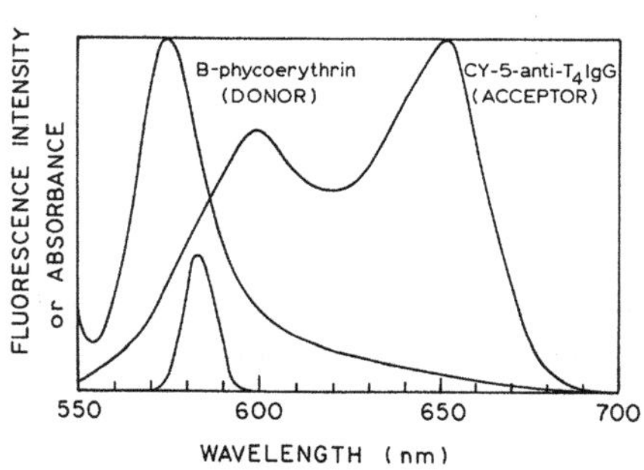

Figure 14.9. Overlap between the emission spectrum of a BPE–thyroxine donor conjugate and the absorption spectrum of the acceptor conjugate CY5-anti-thyroxine IgG. Donor emission through a 580-nm interference filter is shown below the full emission spectrum. (From Ref. 34 with permission.)

B-phycoerythrin-labeled antibody (BPE–IgG), and that use Texas Red-labeled antibody as an acceptor for BPE–IgG.[24]

Most FETI measurements are of changes in the donor emission, so that any chromophore with the proper absorption characteristics is suitable as an acceptor. A nonfluorescent acceptor may be preferred in cases where the donor and acceptor emission spectra show overlap in the region of interest. Thus unwanted contributions and complications from the fluorescence signal of the acceptor can be prevented.[86] Khanna and Ullman[86, 95, 96] have reported the development of nonfluorescent acceptors, one of which is an almost nonfluorescent 4′,5′-dimethoxyfluorescein derivative that has been used successfully in FETI.

Changes in the emission of a fluorescent acceptor also can be used to monitor FETI, and may be preferred in some instances, particularly when there is concern that donor quenching may occur by mechanisms other than nonradiative energy transfer. A detailed study describes monitoring of acceptor fluorescence using time-resolved energy transfer.[23] Pyrenebutyrate, a relatively long-lifetime fluorophore ($\tau = 100$ nsec), is conjugated to $F(ab')_2$ fragments of anti-human IgG, and serves as the donor. The acceptor, BPE, has a relatively short lifetime ($\tau = 2.5$ nsec), and is covalently bound to Fab′ fragments of human IgG. In this system, time-delayed (time-resolved) acceptor measurement ensures that only emission from the long-lived donor stimulates acceptor fluorescence. Fluorescence arising from direct excitation of the acceptor by the source and from the prompt background are thereby eliminated.

The concentrations of both the donor and acceptor must be maintained at sufficiently low levels to prevent inaccuracies due to inner filter effects, in which the excitation is excessively attenuated by absorption, and emitted light from the donor is trivially absorbed by the acceptor. In concentrated solutions, these undesirable effects may be overcome to some extent by making front face measurements.[31] Thus, when the excitation and emission are at the same surface, the light is not required to traverse the full length of the absorbing medium. An indication of whether inner filter effects are significant can be obtained by monitoring the steady-state emission spectra of a donor in the presence of increasing amounts of acceptor. Observation of an essentially unchanging emission spectrum when the acceptor concentration is increased beyond saturation suggests that the decrease in fluorescence intensity with subsaturating amounts of acceptor is not due to trivial absorption (Figure 14.10). Besides concentration effects, immune complex formation may cause signal changes due to effects other than nonradiative energy transfer, such as relief of internal quenching on binding fluorophore–hapten conjugates. The fluorescence of a conjugate can be partially quenched by electron transfer, as in a fluorescein-gentamicin conjugate.[86] In the case of a fluorescein–thyroxine conjugate, heavy atom-induced quenching is caused by the iodine atoms of thyroxine, which are in close proximity to the dye.[86] The use of a rigid piperazine linker, and substitution of two iodine atoms by bromines, largely alleviates this effect.

The donor and acceptor should be brought as close together as possible to maximize energy transfer. In the case of donor or acceptor labeling of small-molecular-

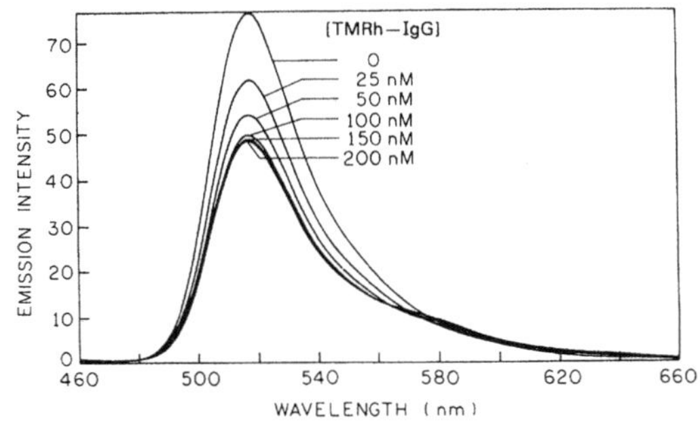

Figure 14.10. Emission spectra of fluorescein-goat anti-mouse IgG (DTAF-IgG) donor in the presence of increasing amounts of tetramethylrhodamine-mouse IgG acceptor (TMRh-IgG), with excitation at 400 nm. (Reproduced from Ref. 107.)

weight analytes, there are relatively few degrees of freedom for label attachment. For large-molecular-weight analytes and antibodies, random labeling with donor or acceptor molecules results in a heterogeneity of energy transfer distances. For example, if antibody is randomly labeled with several fluorescein moieties for each antibody molecule, one or more of the fluorescein labels may be too far away to result in significant energy transfer to an immunologically bound rhodamine-labeled antigen. This limits the maximum possible amount of energy transfer, and also limits the dynamic range of an immunoassay. Measurement of donor fluorescence that is not quenched in a donor–acceptor complex falsely indicates a smaller bound to unbound donor ratio. When labeling antibodies with BPE, the efficiency of energy transfer is improved by using antibody Fab fragments, thereby shortening the maximum possible distance. The effect of distance has been considered in a competitive assay of human IgG, using BPE-labeled Fab′ fragments of human IgG as acceptor, and pyrenebutyrate-labeled $F(ab')_2$ fragments of anti-human IgG as donor.[23] When several Fab′ molecules are covalently bound to BPE, energy transfer is less efficient than when only one Fab′ is covalently bound. Decreased energy transfer efficiency presumably is due to an increase in the effective radius of the multiple Fab′ fragments incorporated into the BPE conjugate.

Since best results are obtained when all donor moieties are capable of efficient energy transfer, donor labeling of the analyte may be more effective than donor labeling of antibody, when measuring small molecular weight analytes. In a double antibody sandwich format, where it is necessary to use a donor-labeled antibody, it is important to recognize that antibody binding sites may be inactivated by the labeling reaction. Inactive donor antibody would elevate the donor background, thereby reducing the assay dynamic range. Similarly, the dynamic range may also be reduced when monitoring changes in the fluorescence intensity of the acceptor, since less acceptor would be bound to donor. Another disadvantage of energy transfer sandwich assays, and of other formats where there are multiple binding sites, is the biphasic nature of calibration curves for these assays.[92] One possibility in this case is to analyze the sample at two different dilutions, to determine at which point on the biphasic curve the sample occurs. An alternative is to analyze the sample at two different reagent conditions, one condition where the reagent concentrations emphasize sensitivity, and the other where dynamic range is emphasized at the expense of sensitivity.

An energy transfer time-resolved immunoassay approach in which a lanthanide is used as the donor has been patented.[97] The potential usefulness of lanthanide chelates as donors for FETI has been questioned in earlier studies.[92] More recently, a terbium chelate could not be quenched in attempts to develop a FETI.[65] It is suggested that the chelated emitting center is too well protected from its environment to undergo nonradiative energy transfer, which does not seem consistent with the Förster mechanism. Conversely, enhancement of terbium chelate fluorescence by salicylates has been observed,[65] and a coumarin derivative has been found to transfer energy to europium in a thorium phosphate xerogel.[98]

14.5.4. Phase-Modulation Fluoroimmunoassays

Immunoassays based on phase-modulation spectroscopy have been implemented by two distinctly different approaches. Phase-resolved immunoassays rely on fluorescence intensity measurements, in which the emission of one fluorescent species in a mixture is suppressed, and the remainder is quantitated. Phase fluorescence immunoassays utilize measurements of the phase angle and modulation, which change in response to fluorescence lifetime changes. Common aspects of the theory and instrumentation are discussed in this section, followed by individual discussions of the different approaches.

14.5.4.1. Phase-Modulation Fluorescence Spectroscopy

Phase-modulation immunoassay measurements are made with sinusoidally modulated light. Since the emission is a forced response to the excitation, the emitted light has the same periodicity as the excitation. Due to the time lag between absorption and emission, the emission is delayed in comparison with the excitation. The time delay between the zero crossing of one period of the excitation and of the emission is measured as the phase angle (Figure 14.11). The emission is also demodulated, due to a decrease in the alternating current (AC) component of the AC to direct current (DC) ratio.

Instrumentation for phase-modulation spectroscopy has been described.[31, 99–101] Immunoassay measurements have been made using modulation frequencies of about 10 to 250 MHz, generated by frequency synthesizers. Commercial instruments (ISS, Inc. and SLM, Inc., both in Champagne-Urbana, Illinois) are typically equipped with a Pockels cell or a Debye–Sears modulator. The Pockels cell is an electro-optical device made from a crystal of potassium deuterium phosphate. Relatively high voltages and highly collimated light are required to use this modulator. The Pockels cell can be driven at selectable frequencies up to about 400 MHz, with practical modulation efficiencies approaching 70%. The Debye–Sears modulator is an acousto-optical

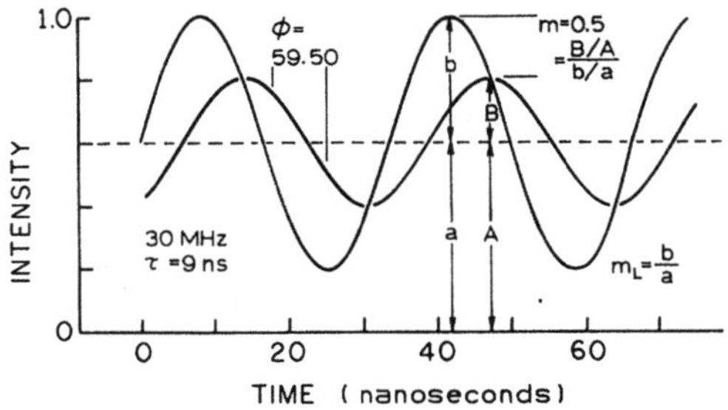

Figure 14.11. Diagram of phase-modulation fluorometry with sinusoidally modulated excitation, with demodulated and delayed, or phase-shifted emission. (From Ref. 31 with permission.)

device, comprised of a quartz crystal immersed in a tank of either ethanol-water or methanol-water. A standing wave is generated in the tank between the vibrating crystal and a reflector plate, forming a grating consisting of sinusoidal oscillations in the refractive index of the medium. This modulator is limited to essentially three frequencies, the fundamental and two harmonics, and has a modest modulation efficiency of about 50%. Diode lasers can be directly modulated for use in phase-modulation fluorometry,(102) and are readily available commercially at wavelengths as low as 630 nm. Pulsed lasers can be used at the fundamental frequency and its harmonics. In phase modulation spectroscopy, signal detection is most commonly achieved using PMTs. The Hamamatsu R928 PMT can be used at frequencies up to about 250 MHz. Microchannel plate detectors have a much faster response, due to their fast rise times.

In phase fluorescence immunoassays and fluorescence lifetime immunoassays (see Section 14.5.4.3), the phase angle and modulation need not be measured at high frequencies to obtain meaningful information. For an immunoassay it is not necessary to measure the entire frequency response, but only at one or several light modulation frequencies. Technically, the actual measurements can be simplified by converting the high-frequency signals to much lower frequencies by cross-correlation.(103) Both the sample PMT and the reference PMT are gain-modulated at the incident light frequency, w, plus a very low offset frequency, $\delta\omega$, typically 25 Hz. The high-frequency components can be removed by low-pass electronic filters. The remaining 25-Hz signal carries the information on the phase angle and modulation, which are readily measured with simple equipment consisting of a time interval counter and a digital voltmeter. Phase-resolved measurements require selection of the detector phase angle to measure the intensity of the desired component(s), and to suppress the emission of the undesirable components (discussed in Section 14.5.4.2).

14.5.4.2. Phase-Resolved Fluoroimmunoassays

Phase-resolved, or phase-suppression technology was investigated for FIA development by McGown and co-workers(49, 53, 105, 106) to test the prediction that phase fluorometry would be useful for monitoring two-state equilibria in the analysis of macromolecular binding interactions.(31, 104) Fluorophores with different fluorescence lifetimes can be quantified by phase-resolved fluorescence intensity measurements within a single mixture with this methodology. A single fluorophore tracer in two environments, such as bound and free, will often display different lifetimes in the different environments. These can be analyzed for the fractional contribution that each makes to the total intensity. This approach was applied by McGown in PRFIA. Thermodynamic binding parameters of immunoassay components have also been studied by phase-resolved fluorometry.[105]

For a single fluorophore with an exponential decay, the phase and modulation are related to the fluorescence lifetime by the expressions

$$\tan \phi = \omega\tau \tag{14.8}$$

$$m = (1 + \omega^2\tau^2)^{-1/2} \tag{14.9}$$

where ϕ is the phase angle, ω is the sinusoidal modulation frequency, m is the modulation intensity ratio, and τ is the fluorescence lifetime. If the sample being measured has two or more exponential decays, then Equations (14.8) and (14.9) provide only apparent lifetimes. If the lifetimes calculated from measurement at a single frequency using the phase and modulation are significantly different, then the presence of species with more than one exponential decay is indicated.

It is possible to resolve the emission from one species in the presence of another by phase-suppression techniques. The emission from a single fluorophore excited by sinusoidally modulated light is described by

$$F(t) = 1 + m_L m \sin(\omega t - \phi) \tag{14.10}$$

where m_L is the modulation of the incident light and m is the modulation of the emitted light. The output of a phase-sensitive detector is proportional to the modulation intensity and to the cosine of the difference between the detector phase ϕ_D and the sample phase angle, resulting in a fluorescence emission spectrum described by

$$F(\lambda, \phi_D) = kF(\lambda) \cos(\phi_D - \phi) \tag{14.11}$$

where $F(\lambda)$ is the fluorescence emission spectrum at a given wavelength, and k represents a constant comprised of instrument factors, sample factors, and the constant m_L. As indicated by Equation (14.11), the observed fluorescence intensity at a given wavelength varies with the detector phase angle, and with the phase angle of the fluorescent species under observation. Selection of the detector phase is arbitrary, and can be referenced incrementally with respect to the phase of the excitation. In a mixture of two fluorescent species with different lifetimes, the emission is the product of two sine waves that have the same period, but different phase angles. Each contributes to the emission spectrum according to

$$F(\lambda, \phi_D) = F_A(\lambda)m_A \cos(\phi_D - \phi_A) + F_B(\lambda)m_B \cos(\phi_D - \phi_B) \tag{14.12}$$

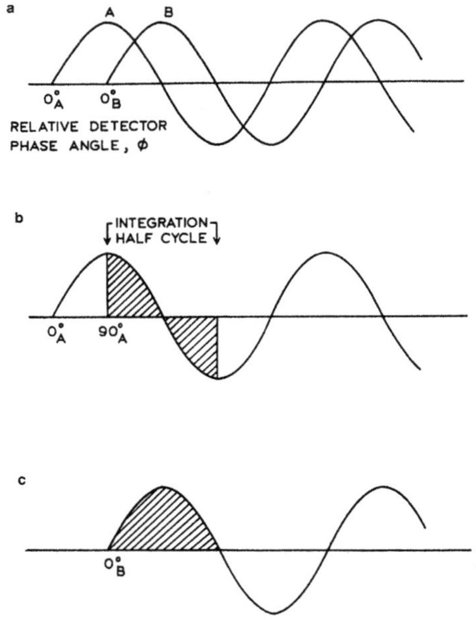

Figure 14.12. Schematic diagram of phase-resolved fluorometry of a mixture of two fluorophores in which the emission of one component is suppressed and the other is quantitated. (a) Emission of components A and B; (b) component A emission is suppressed; (c) component B emission is quantitated.

It is evident from this expression that the contribution of each fluorophore to the total emission varies with the detector phase. The critical property of modulated emission that enables PRFIA is the fact that for each component of a mixture of fluorophores, there is a detector phase angle at which the emission of that component is zero (see Figure 14.12). The detector phase is set at a selected phase angle to span half of a single period, or a π cycle. With reference to the point of the emission at which the intensity for a given component is at a maximum, if the detector phase is set 90° higher or lower than that maximum, then a zero contribution for that component results, and the observed emission is that of the remaining component. In this manner, the emission from either component can be suppressed. Thus in the PRFIA, either the bound or free fluorophore can be measured.

The PRFIA analysis of human lactoferrin, an iron binding glycoprotein with a molecular weight of about 80,000, has been described[(53)] using direct quenching in one mode, and energy transfer in another mode. In both modes, the antigen–antibody binding interaction is transduced into a measurable phase shift by a change in the lifetime of the tracer fluorophore at a modulation frequency of 30 MHz. The binding event alone alters the fluorophore lifetime by direct quenching. In a direct quenching sandwich format, where one antibody of the pair is labeled with either fluorescein, tetramethylrhodamine, or Texas Red, the fluorophores are mildly quenched. Texas Red, with the longest lifetime of the three, is quenched more than the other two fluorophores, changing from 4.06 to 3.68 nsec. In the energy transfer mode, donor fluorophore lifetimes are shortened by both energy transfer and direct quenching effects. The lifetime change therefore is greater for the energy transfer case, resulting in a broader dynamic range for the immunoassay. The competitive format is less effective than the sandwich format, in which one antibody of the pair is labeled with fluorescein and the other with tetramethylrhodamine. With this donor–acceptor pair, the lifetimes of the free and bound forms of the fluorescein tracer are 3.43 and 2.92 nsec, respectively. Lactoferrin detection sensitivity is in the 10^{-9} *M* range with this approach. Energy transfer is less efficient using a donor–acceptor pair comprised of tetramethylrhodamine-labeled antibody and a Texas Red-labeled antibody. Competitive energy transfer and direct quenching formats using a fluorescein–lactoferrin tracer are also less effective. The fluorescence lifetime of the fluorescein label in the unbound lactoferrin tracer is only about one-third that of fluorescein-antibody conjugates. The large reduction in the lifetime of fluorescein on covalent attachment to the lactoferrin in the tracer is attributed to self-quenching, and results in only small lifetime changes on binding to the antibody quencher.

A phenobarbital PRFIA has been implemented using direct quenching of fluorescein-labeled phenobarbital by antiphenobarbital antibody in a homogeneous, competitive format.[(49)] The lifetime change of fluorescein is modest, from 4.04 to 3.94 nsec between free and bound forms. Using averages of five detector phase angles at 30 MHz, the sensitivity of phenobarbital determination is 1 ng/ml. The direct quenching PRFIA method also has been applied to human serum albumin (HSA) measurement[(106)] using Texas Red-labeled HSA. Sensitivity is more than adequate, with six decades of dilution required to execute the assay, thereby also minimizing matrix effects. Three detector phase angles, 10.7°, 55.7°, and 100.7° are used at a modulation frequency of 30 MHz. Results are tabulated for each individual detector phase angle, for the average of two angles, and for the average of three. The average of two detector phase angles, 55.7° and 100.7°, results in significantly lower standard errors than for the other four cases. The factors responsible for the tendency toward cancellation of the errors for the two angle average were not studied.

Since greater dynamic ranges are observed for PRFIAs using energy transfer rather than direct quenching, reagents that lead to greater fluorescence lifetime changes should result in further improvements in sensitivity. In addition to certain limitations that are inherent in homogeneous energy transfer formats (see Section 14.5.3), a limitation of the technology as practiced to date is imposed by use of the Debye–Sears modulator, with its few frequencies and modest modulation efficiency. Investigation of a broader distribution of modulation frequencies should be helpful in finding more robust frequencies that will improve the precision and dynamic range.

14.5.4.3. Phase Fluorescence and Fluorescence Lifetime Immunoassays

PFIAs and fluorescence lifetime immunoassays (FLIAs) are uniquely based on measurement of probe emission properties other than the intensity. The phase and modulation are measured, and they directly reflect the fluorescence lifetime of the fluorophore. This provides a major advantage, since the intensity can vary over a broad range, with only minor effects on the results. Phase-modulation measurements can be less sensitive than intensity measurements to sources of interference such as scatter, photobleaching, stray light, and endogenous fluorescence. Interfering contributions to the results can be identified and compensated for by the use of multifrequency phase-modulation determinations and algorithms to deconvolute the data.

The apparent phase lifetime (τ_p) can be determined at a single modulation frequency (ω) from the phase angle (ϕ) (Figure 14.7) by

$$\tau_p = \frac{1}{\omega} \tan \phi, \tag{14.13}$$

and the apparent modulation lifetime can be determined from the modulation (m) by

$$\tau_m = \frac{1}{\omega}\left(\frac{1}{m^2} - 1\right)^{-1/2} \tag{14.14}$$

The apparent lifetimes calculated by these expressions are the true lifetimes only if the fluorophore obeys single exponential decay kinetics. In the case of a single exponential decay, the apparent lifetimes as determined from the two equations should be the same. If the apparent phase and modulation lifetimes are not equal, more than one decay process is indicated.

The multiexponential decay law of the emission from a mixture of fluorophores can be recovered from phase and modulation measurements over a range of multiple frequencies by

$$I(t) = \sum_{i=1}^{n} \alpha_i \exp(-t/\tau_i) \tag{14.15}$$

Table 14.2. Fluorescence Lifetime Distributions of Some Fluorophore–Protein Conjugates

Conjugate	τ, nsec (fl. yield)	Method	Reference
Sulforhodamine 101-BSA in PBS solution	4.49 (82%) 1.45 (15%) 0.31 (3%)	TCSPC	135
Sulforhodamine 101-BSA, adsorbed to hydrophilic surface	1.55 (27%) 0.61 (42%) 0.18 (31%)	TCSPC	135
Dichlorotriazinofluorescein-goat anti-mouse IgG, in PBS solution	3.15 (91%) 0.62 (9%)	PMFS	107
Dichlorotriazinofluorescein-goat anti-mouse IgG, bound to tetramethylrhodamine in PBS solution	2.58 (76%) 0.63 (24%)	PMFS	107

Abbreviations: BSA, bovine serum albumin; PBS, phosphate buffered saline; PMFS, phase-modulation fluorescence spectroscopy; TCSPC, time-correlated single photon counting.

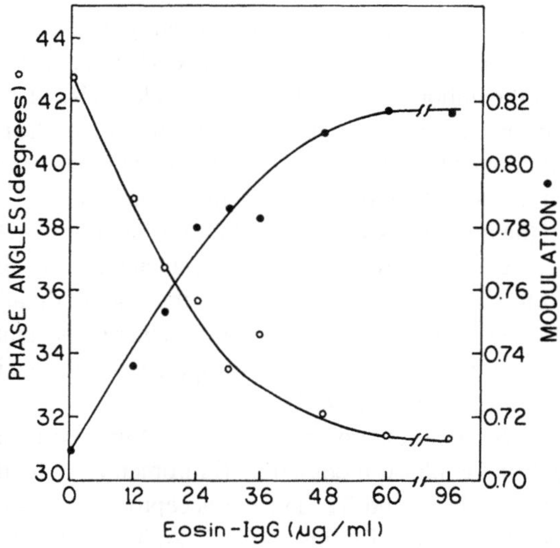

Figure 14.13. Changes in the phase angle and modulation of fluorescein-goat anti-human IgG donor on binding to varying concentrations of eosin-human IgG acceptor at a modulation frequency of 50 MHz. (From Ref. 107 with permission.) [FITC–IgG] = 5×10^{-7} $M = 75$ μg/ml. λ_{ex} = 442 nm.

where τ_i are the decay times associated with the amplitudes α_i.

A solution of a pure fluorophore may reasonably be expected to display a single exponential decay time. The emission from fluorophore-protein conjugates, on the other hand, may be best characterized by two or three exponential decay times (Table 14.2). In labeling proteins with fluorophores, a heterogeneity of labeled sites results in fluorophore populations that have different environments, and hence different lifetimes. The lifetime distribution of a fluorophore-protein conjugate in bulk solution may vary further when immobilized on a solid support (Table 14.2).

The apparent lifetimes are a weighted average based on the modulation frequency, the relative amounts of free donor and of donor–acceptor complex, their fluorescence yields, and lifetimes. To accurately reflect the relative amounts of different donor populations, both bound and free donor must contribute to the signal. While it is advantageous that bound donor fluorescence is highly quenched, the energy transfer must not be so efficient that it entirely extinguishes the fluorescence from any bound donor population. Should this occur, then the phase angle and lifetime contributions for the extinguished population are lost, and the free donor level appears to be greater than it really is.

Since the phase and modulation values change in response to changes in fluorescence lifetimes, a means of transducing the antigen–antibody binding event into a lifetime change enables PFIA measurements. Fluorescence resonance energy transfer (see Section 14.5.3.) provides an excellent transduction mechanism. The first energy transfer, competitive FLIA of human IgG based on phase-modulation measurements has been recently described.[107] The donor is FITC-labeled goat anti-human IgG, and the acceptor antigen is eosin-labeled human IgG. In the absence of acceptor, or with saturating amounts of unlabeled antigen, the free donor has a maximum apparent lifetime, since no binding complex is formed. Hence, at a given modulation frequency for the unbound donor, the phase angle is also at a maximum value. Conversely, in the absence of analyte, when the concentration of the donor–acceptor complex is greatest, the donor lifetime is at a minimum due to energy transfer. Therefore the phase angle of the bound donor is also at a minimum value. This constitutes the full dynamic range of the immunoassay. The modulation has a similar, but opposite response to the lifetime changes. Dose-response curves indicate that the phase angle is indirectly proportional to the analyte concentration and that the modulation is directly proportional (Figure 14.13). As the analyte concentration increases, it interferes with formation of the donor acceptor complex, and the average donor lifetime increases in direct proportion to the analyte concentration (Figure 14.14).

A model energy transfer PFIA of thyroxine has been described in another application of phase-modulation fluorometry.[34] In this study, the donor is BPE with covalently bound T_4, and the acceptor is CY5-labeled antithyroxine antibody. The Förster distance for this pair is 72 Å, indicating exceptionally efficient energy transfer (see Figure 14.6 and Section 14.5.3). Multifrequency measurements of the unbound donor, of donor with a nearly saturating amount of acceptor, and with added T_4 as analyte are shown in Figure 14.15. The phase angle curves increase with increasing frequency, and the modulation curves decrease. The phase and modulation curves for donor alone cross at about 100 MHz, whereas the curves for donor-acceptor complex (triangles) are shifted to higher frequency, crossing at 300 MHz. This shift to higher frequencies indicates that the donor lifetime has been reduced in the donor–acceptor complex. At each individual frequency, the difference between the phase angle of the

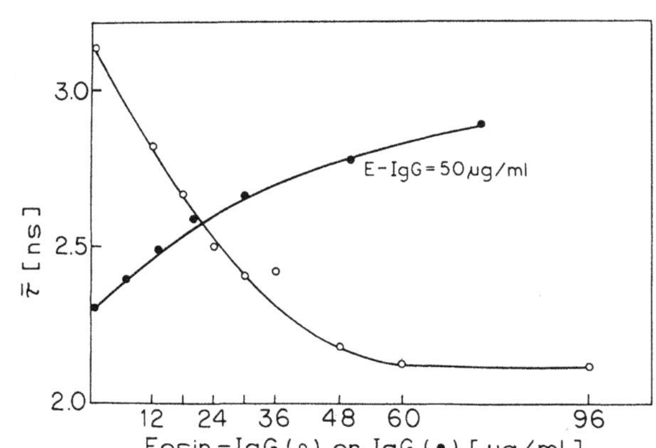

Figure 14.14. Competitive fluorescence lifetime immunoassay dose-response curve (●), in which the analyte, human IgG, competes with eosin-human IgG for fluorescein-anti-human IgG binding sites. Fluorescence lifetime changes of fluorescein-human IgG donor in the presence of increasing amounts of eosin-human IgG acceptor are also shown (○). (From Ref. 107 with permission.)

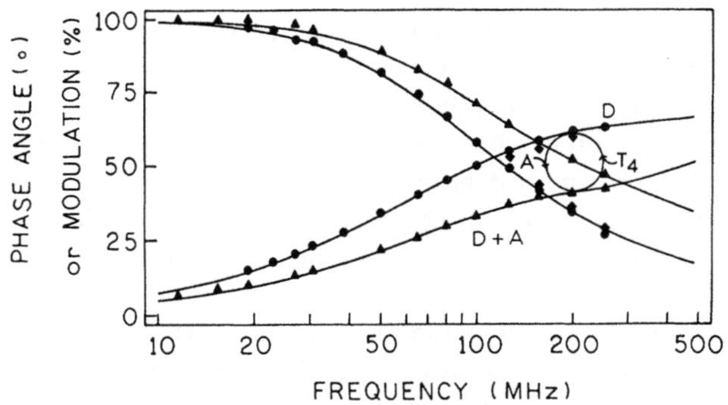

Figure 14.15. Multifrequency phase-modulation curves of 4 n*M* BPE-thyroxine donor conjugate (●); of donor in the presence of CY5-anti-thyroxine IgG acceptor conjugate (▲); and of both donor and acceptor in the presence of a nearly saturating amount of thyroxine (♦). (From Ref. 34 with permission).

donor and of the donor–acceptor complex defines the dynamic range of the assay. Phase-modulation at a single frequency is sufficient to conduct an immunoassay, provided that the donor is previously well characterized by multifrequency determinations. The change in phase angle with respect to T_4 concentration at selected modulation frequencies (Figure 14.16) indicates an acceptable response over the clinically useful range for this analyte, and compares favorably with steady-state dose-response curves.

As a homogeneous technology, PFIAs have a simplified workflow, but are subject to some of the limitations of energy transfer immunoassays (see Section 14.5.3). Nevertheless, PFIA appears to offer a somewhat broader dynamic range and greater sensitivity than steady-state measurements made with the same reagents under comparable conditions. To maximize dynamic range, energy transfer should be as efficient

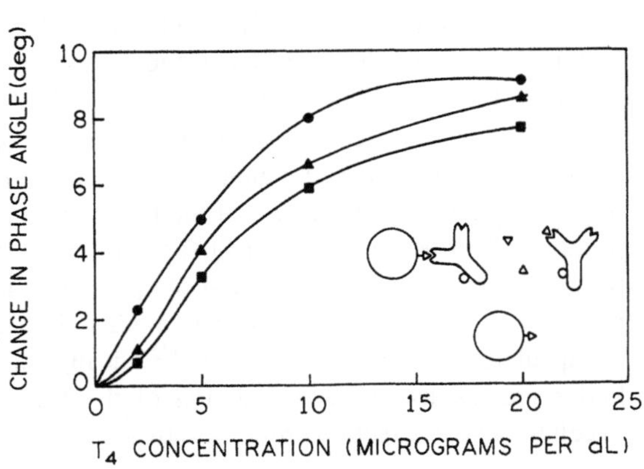

Figure 14.16. Dose-response curves for a phase fluorescence immunoassay of thyroxine at modulation frequencies of 49.5 MHz (■), 64.8 MHz (▲), and 99.1 MHz (○). (From Ref. 34 with permission.)

as possible. In the fluorescence lifetime immunoassay described in this section for human IgG, the dynamic range was about 10° of phase angle. This corresponds to 50 points of resolution at a nominal precision of 0.2° of phase angle, achievable with typical instrumentation. It can be reasonably anticipated that with the appropriate combination of donors and acceptors and optimized reagent chemistries, phase angle changes of 50° or more will be possible, resulting in at least 250 points of resolution. As instrumentation is optimized, the precision can also be expected to improve.

A significant advantage of phase-modulation measurements for immunoassays is that analyte determination is achieved by monitoring fluorescence lifetime changes rather than fluorescence intensity changes. Fluorescence lifetimes, as absolute quantities, are not affected by fluorescence intensity fluctuations. Measurements at as few as three frequencies allow determination of the relative amounts of bound and free donor within a reaction mixture, enabling antibody binding constants to be determined. Kinetics of immunoassay reactions can be easily monitored as well, particularly with the advances that have been made recently in rapid measurement with commercial phase fluorometers (ISS, Inc. and SLM, Inc., both in Champagne-Urbana, Illinois).

14.5.5. Liposome Fluoroimmunoassays

Liposomes that encapsulate reporter groups within their phospholipid bilayer membranes can amplify immunoassay binding interactions by several orders of magnitude over signal levels that are achievable using direct labeling of antibodies or antigens.[108] Individual liposomes can encapsulate up to 10^5 fluorophore molecules.[109] Immunological recognition is mediated by antibodies or antigens immobilized on the liposome surface. Since the external surface of an individual liposome reagent usually is coupled to many molecules of either antigen or antibody, the ratio of liposomes to antibody-antigen binding events in an immunoassay can be anticipated to be lower than the best case 1:1 ratio. Liposomes that are bound in an immunoassay may be quantitated while still intact, although they are generally lysed to release entrapped dye or other reporter groups. Liposomes typically are lysed through the action of detergents[108, 109] or of complement.[110]

Fluorophores are encapsulated within liposomes at concentrations that either optimize the fluorescence signal of the intact liposome, or maximize the amount of the fluorophore at the expense of the fluorescence signal, depending on the intended purpose. The Becton Dickinson Q Test Strep® is a liposome-based rapid manual assay for the determination of group A streptococcal antigen from throat swabs. Visual detection on a membrane device is mediated by antibody bound to liposomes that encapsulate sulforhodamine B dye.[111] The high concentration of encapsulated sulforhodamine results in self-quenching of the fluorescence, but the liposomes retain excellent visual detection properties. In addition to Q Test assays, additional liposome-

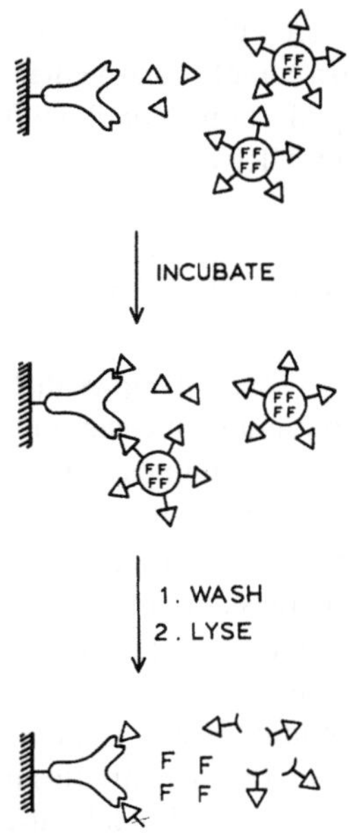

Figure 14.17. Liposome fluoroimmunoassay depiction, in which antigen–liposome conjugates containing fluorescent dye (concentration quenched) compete with analyte antigen for antibody binding sites, followed by wash and detergent lysis, to release the fluorophore for fluorescence measurement.

based rapid manual tests are available in the Becton Dickinson *Direct*igen test system. Recently, the Becton Dickinson IQ® Immunochemistry System has become available, with liposome immunoassays for thyroxine, TSH, and thyroxine uptake. The immunoassays are conducted using antibody-coated tubes and either antigen (T_4 and T uptake) or antibody (TSH) bound to sulforhodamine B-encapsulated liposomes. In these assays, the highly quenched dye is released by detergent lysis from liposomes that become immunologically bound to the coated tube (Figure 14.17). A fluorescence intensity measurement of the dequenched fluorophore is then made.

An immunosensor has been reported that is comprised of carboxyfluorescein-containing liposomes immobilized in an agar gel at the tip of an optical fiber.[112] The dye is quenched due to the high encapsulation concentration. Dinitrophenyl (DNP) moieties immobilized on the liposome surface are bound by anti-DNP antibody, the liposomes are subsequently lysed by complement, and the dequenched dye is quantitated. Liposome flow injection immunoassays were demonstrated for theophylline and for antitheophylline IgG using carboxyfluorescein liposomes that have theophylline immobilized on the surface.[113] For both assays, a regenerable column with antitheophylline IgG bound to silica particles serves as the solid phase. Both bound and unbound liposome fractions are quantitated after surfactant-mediated lysis. Although carboxyfluorescein is negatively charged at physiological pH, it leaks through the membrane at an appreciable rate. This problem can be avoided by using more hydrophilic fluorophores, as demonstrated with novel hydrophilic fluorescein derivatives in a complement-mediated liposome immunolytic assay (LILA) of digoxin.[110] In another application of a carboxyfluorescein LILA, C-reactive protein (CRP) was determined using anti-CRP antibody-coupled liposomes in a homogeneous sandwich format.[114]

14.5.6. Fluoroimmunosensors

Developments in fluoroimmunosensors have been driven by an interest in the continuous measurement of analytes in the homogeneous formats that they enable. Fluoroimmunosensors may be used for remote, on-site sensing of any analyte that is amenable to immunological detection, such as pesticides, toxins, and other pollutants in a water source. The potential for continuous *in vivo* measurement of concentration changes of clinically relevant analytes exists with optical fibers, which would provide primary care physicians the advantage of rapid clinical and therapeutic patient management during a single patient visit. The sensitivity limits of optical fiber-based immunoassays require improvements to make them clinically useful at present. Detection sensitivities using planar waveguides, such as microscope slides, appear to be nearing clinically useful ranges.

Fluoroimmunosensors require waveguides to transmit excitation and emission to and from immunoassay reaction zones that are at or near a surface of the waveguide. Optical fibers and planar waveguides with immobilized reagents can be used in immunoassay formats that distinguish between bound and free reagents, without a requirement for separation and washing steps. This two-dimensional approach may preclude the establishment of true binding equilibrium, but has the advantages of sensing kinetic parameters of the immunoreaction.[7] Immunosensor configurations that do not require immobilized reagents are possible. These devices are capable of direct measurement in solution, and may offer greater sensitivity, although this may be at the expense of device complexity. For example, the tip of an optical fiber may be coupled to a cuvette or reaction chamber in which the immunoreaction occurs. Various immunosensor configurations are capable of measuring analytes remotely, continuously, and directly within the fluid to be analyzed. Direct immunosensors are functionally complete and independent of separate reagents, whereas indirect sensors require reagent additions. The binding surfaces of reversible sensors can be regenerated, and multiple measurements can be made with some irreversible sensors before their surfaces are saturated. Immunosensors are often reviewed as a subgroup of biosensors.[115–117]

14.5.6.1. Fiber Optic Fluoroimmunosensors

A phenytoin energy transfer immunosensor has been reported that is reversible and direct, in which macromolecular reagents are retained within a short length of 200-μm dialysis tubing cemented to the end of an optical fiber (Figure 14.18).[118, 119] A

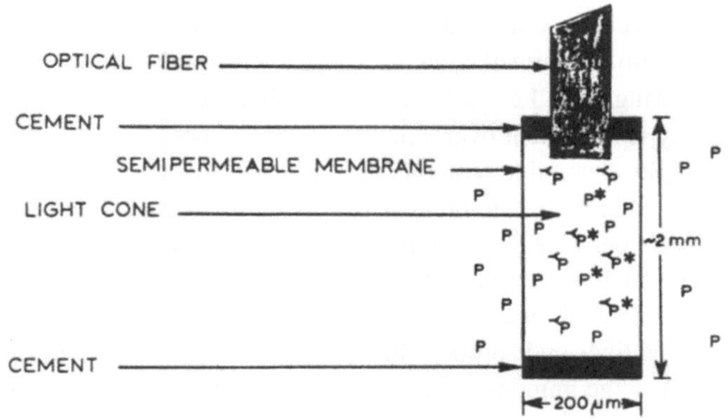

Figure 14.18. Diagrammatic representation of a phenytoin fluorescence energy transfer fiber optic sensor, as described in the text. ≺ = Antibody–Texas Red; P = unlabeled phenytoin; P* = phenytoin-b-phycoerythrin. (Adapted from Ref. 119.)

BPE-phenytoin donor or phenytoin in a sample is bound by biotinylated antiphenytoin antibody. The donor is quenched by a Texas Red-avidin acceptor. The assembly is small enough to fit into a 23-gauge needle. Cycle time is from 5 to 30 min, with excellent reproducibility. The sensor has been further optimized and the reagents have been characterized by the determination of association and dissociation constants.[120, 121] Intact antibody as well as Fab fragments are used, resulting in a sensor which responds well over the clinically useful concentration range of phenytoin.

An energy transfer immunosensor described for IgG measurement exploits controlled release of donor and acceptor from ethylene-vinyl acetate copolymer plugs positioned in opposite ends of a cylindrical reaction chamber (Figure 14.19).[122] Fluorescein-antibody donor and Texas Red-antigen-IgG acceptor are continuously released from the polymeric plugs. The reagents diffuse to the center of the cylindrical reaction chamber, where a perpendicular top-mounted optical fiber delivers the excit-

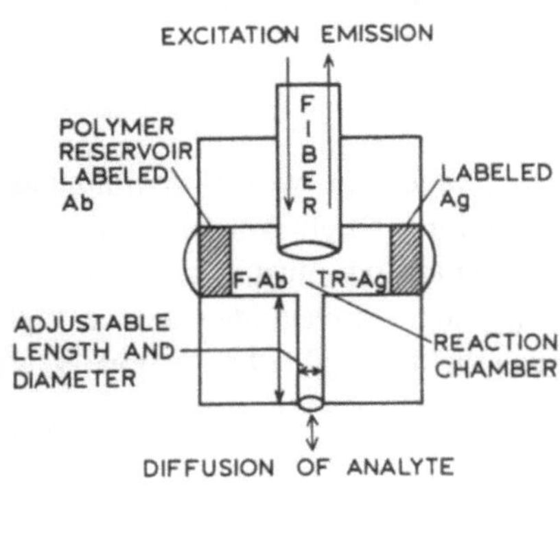

Figure 14.19. Schematic diagram of a continuous-release fluoroimmunosensor, as described in the text. (From Ref. 122 with permission.) (©1991, AAAS.)

ing light, and collects the emission. Analyte diffuses into the reaction chamber via a perpendicular bottom-mounted tube, in contact with bulk solution. The sensor is compact, measuring 15 × 12 × 10 mm. Cycling between samples with and without analyte antibody is consistent over a period of several hours. Continued functioning over several weeks is possible with this sensor.

A reversible, direct fluoroimmunosensor for human serum albumin (HSA) measurement has been described by Bright *et al.*[123] Antibody Fab′ fragments are first immobilized on small quartz plates by hinge-region thiols, and then dansylated. The immunosensor is formed by attaching the quartz plates with bound Fab′ to the distal end of a bifurcated fiber-optic probe, which transmits both the excitation and emission. Binding of HSA to the immunosensor results in a three- to five-fold enhancement of dansyl fluorescence. The sensor can be reused up to 50 times, with a detection limit of about 1.8×10^{-8} *M*, and a somewhat limited dynamic range.

An optical immunosensor for continuous T_4 measurement has been described, in which the fluorescent indicator protein is separated from the sample flow chamber by a dialysis membrane.[124] The indicator is T_4-binding globulin (TBG), the intrinsic fluorescence (ex. 290 nm) of which is quenched by T_4 binding. Due to the high affinity of the TBG for thyroxine, the immunosensor is not reversible, but multiple measurements can be made until the TBG is saturated. Sensitivity is inadequate for clinically useful concentrations of T_4, but suggestions for improvement of the method are made.

Greater analyte detection sensitivities can be achieved using a time-resolved approach. The long lifetimes and large Stokes shifts of lanthanide chelates facilitate rejection of scattered light, which is a significant advantage in fiber optic immunoassay applications, as described in an assay for rabbit IgG.[125] Rabbit IgG in the sample competes with rabbit IgG that is covalently bound to the distal end of a quartz optical fiber for antibody binding sites on the tracer, Eu^{3+} 2-naphthoyltrifluoroacetonate-labeled anti-rabbit IgG. The detection limit is about 7×10^{-10} *M*. Time-resolved measurement minimizes back-scattered light, and results in detection sensitivity that is three orders of magnitude greater than that achievable using a fluorescein isothiocyanate label. However, time-resolved detection using optical fibers in this format is at least two orders of magnitude less sensitive than measurement of comparable immunoassay mixtures in cuvettes.

14.5.6.2. Evanescent Wave Fluoroimmunosensors

Evanescent wave fluoroimmunosensors exploit the properties of total internal reflection of light within a waveguide. The characteristic refractive index of the waveguide, $\eta 1$, must be greater than the refractive index of the aqueous medium, $\eta 2$, at the liquid-waveguide interface. Light is launched into a fiber or a planar waveguide such that the angle of reflection, Θ, is greater than the critical angle, Θ_c, defined as

$$\Theta_c = \sin^{-1}(\eta 2/\eta 1) \tag{14.16}$$

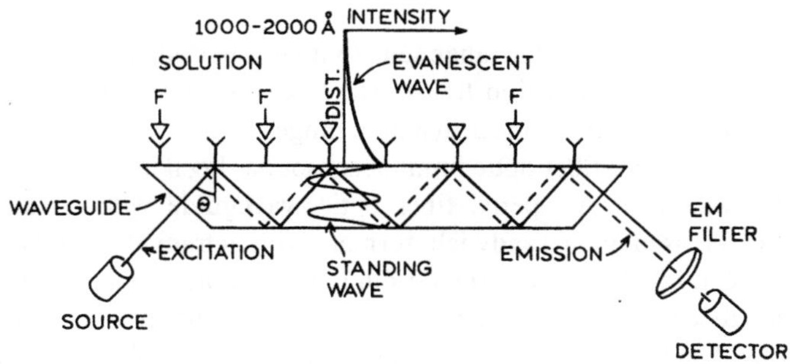

Figure 14.20. Principle of evanescent wave fluoroimmunoassay using a planar waveguide. The principle is comparable using an optical fiber, except that light is launched and collected at the proximal end of the fiber.

resulting in total internal reflection (TIR) of the incident light (Figure 14.20). As a result of total internal reflection of the light, an electromagnetic standing wave is produced perpendicular to the waveguide surface. The standing wave emanates into the liquid with an exponential intensity decay. The depth of penetration (dp) is relatively short and is dependent on Θ, the ratio $\eta\,2/\eta\,1$, and the excitation wavelength. The longer the wavelength, the greater the dp. The dp is just a fraction of a light wavelength, typically about 1,000 to 2,000 Å. Antibodies, with oblate spheroid shapes, have axial diameters of about 5×10 nm. Thus, even on a waveguide surface with a multilayered immunocomplex, bound tracer can be excited by an evanescent wave with an intensity near 90% of the initial intensity. Most of the unbound tracer is outside of the influence of the evanescent wave. The emission tunnels back into the waveguide. The intensity of the emission that is observed in-line with the excitation can be 50-fold greater than the emission intensity that is observed perpendicular to the waveguide.[126–128]

Fused silica slides have been used in a highly sensitive evanescent wave immunosensor for hCG measurement.[129] Biotinylated Fab′ fragments are immobilized on the slides by binding to avidin that is adsorbed on the waveguide surface. A tetramethylrhodamine-hCG-peptide conjugate is bound to the Fab′ binding sites prior to initiating the assay, which is displaced by hCG in the sample in a competitive format. The peptide of hCG is selected for ease of displacement by hCG. In a sandwich format, tetramethylrhodamine-labeled Fab′ fragments are used with the same Fab′ waveguide that bears prebound labeled hCG-peptide. The waveguide is irradiated with an argon ion laser, and the emission is detected by a CCD camera. Detection sensitivity in the sandwich format is 10^{-13} *M*, and 10^{-12} *M* in the competitive format.

An example of an evanescent wave fiber optic immunoassay and the associated optics has been described in detail for measurement of anti-rabbit IgG.[130] Rabbit antibody is immobilized on the distal face of an optical fiber. Unlabeled anti-rabbit IgG competes with fluorescein-labeled anti-rabbit IgG for rabbit antibody binding sites on the fiber surface. A nanomolar-range sensitivity is reported, using volumes as small as 10 μl, with 20-minute reaction times. The fibers are irreversible, and interfiber variation is 10%, with a small linear dynamic range.

An evanescent wave fiber optic immunosensor has been used for the detection of ricin in river water.[131] A tapered fiber optic waveguide with covalently bound anti-ricin IgG is used in a sandwich format, with tetramethylrhodamine-labeled antibody as tracer. In a two-step format, ricin in the sample is bound to the fiber first, and then the fiber is exposed to the tracer antibody. Sensitivity is 1 ng/ml for the two-step assay. The one-step assay, in which the fiber optic probe contacts the sample and labeled antibody simultaneously is less sensitive, but more convenient.

A fluorescence capillary fill device, formed from two glass plates sandwiched together with a small gap between them, has been described in an evanescent wave immunoassay of human IgG (hIgG).[132] Polyclonal anti-hIgG is covalently bound to one of the plates, and FITC-labeled hIgG (FITC-hIgG) is used as the tracer in a competitive format. The sample is excited with light pulses from the top of the device, perpendicular to the flat surface, exciting all of the molecules in solution. Unbound fluorophores fluoresce into the waveguide at relatively large angles, and therefore emerge at large angles at the end of the waveguide. The fluorophore bound at the surface emits at all angles into the waveguide. Light is monitored at smaller angles to quantitate surface-bound fluorescence, to the exclusion of unbound probe. However, unbound probe fluorescence is also measured at larger angles to normalize the total fluorescence, and to correct for background and system fluctuations.

A disposable, patterned, planar waveguide with a number of individual wells has been reported for a one-step homogeneous immunoassay of IgG.[133] The device is fabricated by an ion-exchange process, etching, and covalent reagent immobilization. The sample fills the waveguide by capillary action. The sample well, as well as fluorescent and nonfluorescent control wells are excited by an evanescent field, and individually scanned. The IgG detection limit is in the 10^{-9} *M* range.

In another application of multiple position detection, a cooled CCD detector has been used to detect four spots of fluorescein-labeled HSA, which bind to anti-HSA antibody that is immobilized on a fused silica plate.[134] Irradiation of the waveguide at a 70° angle of incidence on the top surface generates an evanescent wave in the waveguide. The CCD is positioned parallel to the top surface of the waveguide to capture the emission of the labeled HSA. Three-dimensional images of the scans are shown in this study.

14.6. Discussion and Conclusions

Considerable effort is being expended toward the development and understanding of fluorescence-based immunoassay technology. A number of factors can be suggested that tend to drive interest in the pursuit of specific immunoassay methods in general, including sensitivity, accuracy, precision, ease of manufacture, capability for a broad range of analytes, simplicity, originality, and value, to name a few, within the framework of research, clinical, and commercial utility. Several of these factors seem to stand out as major driving forces. The sensitivity and broad analyte capabilities of some fluorescence methods are useful for the most demanding clinical analytes. Most notable, in terms of sensitivity at present, are TRIAs and enzyme-mediated fluorogenic immunoassays. Yet, at the other extreme, FPIAs, which are far less sensitive, and which are limited to small-molecular-weight analytes, are among the more commercially successful immunoassays. As a homogeneous technology, PIAs do not have the manipulation requirements of heterogeneous technologies. Thus, instrumentation and procedures are greatly simplified, contributing to the excellent precision of FPIAs.

The overall factors involved in the development of FPIAs, or any immunoassay technology, are rather complex, but practical analysis emphasizes the importance of simplicity and ease of manufacture. By this reasoning, future focus could well be on homogeneous technologies. However, since homogeneous immunoassays do not currently achieve the sensitivity levels of heterogeneous immunoassays, both strategies may be prominent in future development. Fluorescence resonance energy transfer enables homogeneous immunoassays, although energy transfer immunoassays have not experienced wide utility. Energy transfer has been applied in phase-resolved immunoassays, and has been utilized more recently in reports of fluorescence lifetime and phase fluorescence model immunoassays. The remarkably low degree of dependence of phase-modulation measurements on fluorescence intensity changes indicates the potential for fluorescence lifetime-based immunoassays to fluorish, given that instrumentation for phase-modulation fluorometry has recently made rapid advances in reduction of costs and in simplification of the assay process. It remains to be seen whether phase-modulation fluorometry will ultimately have significant utility in immunoassay development.

Fluoroimmunosensors may well be the "next generation" in immunoassay technology development and commercialization. The novelty and intriguing aspects of the technology notwithstanding, much of the recent activity in fluoroimmunosensor development is dedicated to the achievement of practical devices and detection schemes. Planar waveguides that fill by capillary action seem simple to use, merely requiring that sufficient sample be contacted with the device to fill it. Fiber optic devices can be completely self-contained, suggesting a "dip-and-read" capability analogous to electrode measurements, and they can be compact enough to be used for *in vivo* measurements. The use of probes that excite at near-IR wavelengths with fluoroimmunosensors may facilitate development of *in vivo* applications. One of the challenges in fluoroimmunosensor development at present is to make continuing advancements in detection sensitivity, such that they may become an important addition to clinically useful diagnostics. However, unless fluoroimmunosensors are capable of being manufactured reproducibly and inexpensively, even though the technologies may exceed performance expectations, they will not easily become widely used.

Movement of detection wavelengths for immunoassay purposes into the near-IR is underway, to gain the advantages of measurement at long wavelengths, and the potential for whole blood measurements. Laser diodes are now mass produced at wavelengths as low as 630 nm, and are inexpensively available with associated optics and more than adequate power levels. Various types of detectors are presently available with red-enhanced capabilities at high sensitivities. While a few near-IR probes are readily available, more effort is warranted at developing applications for them and for continuing development of new probes. Progress in application of near-IR probes to immunoassays has been relatively slow, but the momentum is increasing. Thus, with the proper combination of near-IR probes and immunosensing technology, practical whole blood testing and perhaps *in vivo* immunodiagnostics should become possible.

Acknowledgments

Support of this work by Becton Dickinson Advanced Diagnostics is gratefully acknowledged. For insightful suggestions regarding this work, the author thanks Robert W. Rosenstein, Lewis Pollack, Gary Krauth, and Joseph R. Lakowicz.

References

1. I. A. Hemmilä, *Applications of Fluorescence in Immunoassays* (J. D. Winefordner and I. M. Kolthoff, series eds.), John Wiley & Sons, New York (1991).
2. K. Van Dyke and R. Van Dyke, Eds., *Luminescence Immunoassay and Molecular Applications*, CRC Press, Boca Raton, Florida (1990).
3. C. P. Price and D. J. Newman, Eds., *Principles and Practice of Immunoassay*, Stockton Press, New York (1991).
4. T. T. Ngo, *Nonisotopic Immunoassay*, Plenum Press, New York (1988).
5. W. P. Collins, Ed., *Alternative Immunoassays*, John Wiley & Sons, New York (1985).
6. E. P. Diamandis, Immunoassays with time-resolved fluorescence spectroscopy. Principles and applications, *Clin. Biochem. 21*, 139–150 (1988).
7. J. F. Place, R. M. Sutherland, and C. Dähne, Opto-electronic immunosensors: A review of optical immunoassay at continuous surfaces, *Biosensors 1*, 321–353 (1985).
8. J. P. Gosling, A decade of development in immunoassay methodology, *Clin. Chem. 36*, 1408–1427 (1990).
9. V. P. Butler, Jr., D. H. Schmidt, T. W. Smith, E. Haber, B. D. Raynor, and P. Demartini, Effects of sheep digoxin-specific antibodies and their Fab fragments on digoxin pharmacokinetics in dogs, *J. Clin. Invest. 59*, 345–359 (1977).
10. E. Lamoyi and A. Nisonoff, Preparation of $F(ab')_2$ fragments from mouse IgG of various subclasses, *J. Immunol. Methods 56*, 235–243 (1983).
11. P. Parham, On the fragmentation of monoclonal IgG1, IgG2a, and IgG2b from BALB/c mice, *J. Immunol. 131*, 2895–2902 (1983).
12. G. Kohler and C. Milstein, Continuous cultures of fused cells secreting antibody of predefined specificity, *Nature 256*, 495–497 (1975).
13. Linscott's Directory, 4877 Grange Road, Santa Rosa, California 95404.

14. J. H. Howanitz, Immunoassay innovations in label technology, *Arch. Pathol. Lab. Med. 112*, 775–779 (1988).
15. E. Soini and I. Hemmilä, Fluoroimmunoassay: Present status and key problems, *Clin. Chem. 25*, 353–361 (1979).
16. R. S. Davidson and M. M. Hilchenbach, The use of fluorescent probes in immunochemistry, *Photochem. Photobiol. 52*, 431–438 (1990).
17. M. Brinkley, A brief survey of methods for preparing protein conjugates with dyes, haptens, and cross-linking reagents, *Bioconjugate Chem. 3*, 2–13 (1992).
18. S. H. Wong, *Chemistry of Protein Conjugation and Cross-Linking,* CRC Press, Boca Raton, Florida (1991).
19. R. B. Mujumdar, L. A. Ernst, S. R. Mujumdar, C. J. Lewis, and A. S. Waggoner, Cyanine dye labeling reagents: Sulfoindocyanine succinimidyl esters, *Bioconjugate Chem. 4*, 105–111 (1993).
20. T. A. Kelly, C. A. Hunter, D. C. Schindele, and B. V. Pepich, Aluminum phthalocyanine-streptavidin: New, sensitive fluorescent tracer for immunoassay, *Clin. Chem. 37*, 1283–1286 (1991).
21. D. C. Schindele and G. E. Renzoni, Ultra Fluors™: New fluorophores for immunological applications, *J. Clin. Immunoassay 13*, 182–186 (1990).
22. S. A. Soper, Q. L. Mattingly, and P. Vegunta, Photon burst detection of single near-infrared fluorescent molecules, *Anal. Chem. 65*, 740–747 (1993).
23. L. E. Morrison, Time-resolved detection of energy transfer: Theory and application to immunoassays, *Anal. Biochem. 174*, 101–120 (1988).
24. M. N. Kronick and P. D. Grossman, Immunoassay techniques with fluorescent phycobiliprotein conjugates, *Clin. Chem. 29*, 1582–1586 (1983).
25. J. D. Rodwell, V. L. Alvarez, C. Lee, A. D. Lopes, J. W. F. Goers, H. D. King, H. J. Powsner, and T. J. McKearn, Site-specific covalent modification of monoclonal antibodies: *in vitro* and *in vivo* evaluations, *Proc. Natl. Acad. Sci. USA 83*, 2632–2636 (1986).
26. M.-M. Chua, S.-T. Fan, and F. Karush, Attachment of immunoglobulin to liposomal membrane via protein carbohydrate, *Biochim. Biophys. Acta 800*, 291–300 (1984).
27. B. Packard and M. Edidin, Site-directed labeling of a monoclonal antibody: Targeting to a disulfide bond, *Biochemistry 25*, 3548–3552 (1986).
28. F. J. Martin and D. Papahadjopoulos, Irreversible coupling of immunoglobulin fragments to preformed vesicles, *J. Biol. Chem. 257*, 286–288 (1982).
29. E. Ishikawa, M. Imagawa, S. Hashida, S. Yoshitake, Y. Hamaguchi, and T. Ueno, Enzyme-labeling of antibodies and their fragments for enzyme immunoassay and immunohistochemical staining, *J. Immunoassay 4*, 209–327 (1983).
30. D. L. Meadows, J. S. Shafer, and J. S. Schultz, *J. Immunol. Methods 143*, 263–272 (1991).
31. J. R. Lakowicz, *Principles of Fluorescence Spectroscopy*, Plenum Press, New York (1983).
32. E. Amler, L. Mazzanti, E. Bertoli, and A. Kotyk, Lifetime distribution of low sample concentrations: A new cuvette for highly accurate and sensitive fluorescence measurements, *Biochem. Int. 27*, 771–776 (1992).
33. W. Groskopf, B. Green, L. Sohn, and S. Hsu, Furosemide as a displacing agent in assay of total triiodothyronine, *Clin. Chem. 37*, 587–588 (1991).
34. A. J. Ozinskas, H. Malak, J. Joshi, H. Szmacinski, J. Britz, R. B. Thompson, P. A. Koen, and J. R. Lakowicz, Homogeneous model immunoassay of thyroxine by phase-modulation fluorescence spectroscopy, *Anal. Biochem. 213*, 264–270 (1993).
35. R. P. Ekins, Current concepts and future developments, in: *Alternative Immunoassays* (W. P. Collins, ed.), pp. 219–237, John Wiley & Sons, New York (1985).
36. J. El Jabri, S. De Lauzon, and N. Cittanova, Estrogen fluoroimmunoassay with a fluorimeter designed for low-intensity light detection, *Anal. Chim. Acta 227*, 129–134 (1989).
37. W. K. Wang, L. T. Ho, Y. Chiang, and T.C. Chen, A space-resolved fluorometer and its application to immunoassay, *J. Immunol. Methods 112*, 173–176 (1988).

38. W. K. Wang, L. T. Ho, and Y. Chiang, Space-resolved fluoroimmunoassay for quantifying α-fetoprotein in serum, *Clin. Chem. 39*, 1659–1661 (1993).
39. V. M. Bertram, M. P. Bailey, and B. F. Rocks, Multiple releasable fluorescein labels for immunoassay: The principle illustrated by an immunoassay for antibodies to the human immunodeficiency virus, *Ann. Clin. Biochem. 28*, 487–491 (1991).
40. T. L. Keimig and L. B. McGown, Micellar modification of the spectral, intensity and lifetime characteristics of fluorescein-labeled phenobarbital, *Talanta 33*, 653–656 (1986).
41. D. R. Bethell, M. Dawson, and M. J. LaFoe, Characterization of monoclonal antibodies to cell surface antigens by particle concentration fluorescence immunoassay (PCFIA), *BioTechniques 3*, 466–473 (1985).
42. K. Auditore-Hargreaves, R. L. Houghton, N. Monji, J. H. Priest, A. S. Hoffman, and R. C. Nowinski, Phase-separation immunoassays, *Clin. Chem. 33*, 1509–1516 (1987).
43. W. B. Dandliker, R. J. Kelly, J. Dandliker, J. Farquhar, and J. Levin, Fluorescence polarization immunoassay. Theory and experimental method, *Immunochemistry 10*, 219–227 (1973).
44. M. Fiore, J. Mitchell, T. Doan, R. Nelson, G. Winter, C. Grandone, K. Zeng, R. Haraden, J. Smith, K. Harris, J. Leszczynski, D. Berry, S. Safford, G. Barnes, A. Scholnick, and K. Ludington, The Abbott IMx™ automated benchtop immunochemistry analyzer, *Clin. Chem. 34*, 1726–1732 (1988).
45. R. A. A. Watson, J. Landon, E. J. Shaw, and D. S. Smith, Polarisation fluoroimmunoassay of gentamicin, *Clin. Chim. Acta 73*, 51–55 (1976).
46. A. R. McGregor, J. O. Crookall-Greening, J. Landon, and D. S. Smith, Polarisation fluoroimmunoassay of phenytoin, *Clin. Chim. Acta 83*, 161–166 (1978).
47. F. V. Bright, Multifrequency phase fluorescence study of hapten-antibody complexation, *Anal. Chem. 61*, 309–313 (1989).
48. F. Perrin, Polarization de la lumiere de fluorescence. Vie moyenne de molecules dans l'etat excite, *J. Phys. Radium. 7*, 390–401 (1926).
49. F. V. Bright and L. B. McGown, Homogeneous immunoassay of phenobarbital by phase-resolved fluorescence spectroscopy, *Talanta 32*, 15–18 (1985).
50. S. A. Eremin, D. E. Schiavetta, H. Lotey, D. S. Smith, and J. Landon, Design and development of a single-reagent polarization fluoroimmunoassay for methamphetamine, *Ther. Drug Monitoring 10*, 327–332 (1988).
51. T. Uematsu, R. Sato, A. Mizuno, M. Nishimoto, S. Nagashima, and M. Nakashima, A fluorescence polarization immunoassay evaluated for quantifying astromicin, a new aminoglycoside antibiotic, *Clin. Chem. 34*, 1880–1882 (1988).
52. P. Urios, N. Cittanova, and M.-F. Jayle, Immunoassay of the human chorionic gonadotropin using fluorescence polarization, *FEBS Lett. 94*, 54–58 (1978).
53. K. Nithipatikom and L. B. McGown, Homogeneous immunochemical technique for determination of human lactoferrin using excitation transfer and phase-resolved fluorometry, *Anal. Chem. 59*, 423–427 (1987).
54. P. Urios and N. Cittanova, Adaptation of fluorescence polarization immunoassay to the assay of macromolecules, *Anal. Biochem. 185*, 308–312 (1990).
55. S. H. Grossman, Fluorescence polarization immunoassay applied to macromolecules: Creatine kinase-BB, *J. Clin. Immunoassay 7*, 96–100 (1984).
56. M. Tsuruoka, E. Tamiya, and I. Karube, Fluorescence polarization immunoassay employing immobilized antibody, *Biosensors and Bioelectronics 6*, 501–505 (1991).
57. R. P. Fisher and J. D. Winefordner, Pulsed source-time resonance phosphorimetry, *Anal. Chem. 44*, 948–956 (1972).
58. C. G. Barnes and J. D. Winefordner, Optimization of time-resolved phosphorimetry, *Appl. Spectrosc. 38*, 214–228 (1984).
59. E. Soini and H. Kojola, Time-resolved fluorometer for lanthanide chelates—a new generation of nonisotopic immunoassays, *Clin. Chem. 29*, 65–68 (1983).

60. N. Sabbatini, M. Guardigli, A. Mecati, V. Balzani, R. Ungaro, E. Ghidini, A. Casnati, and A. Pochini, Encapsulation of lanthanide ions in calixarene receptors. A strongly luminescent terbium(3+) complex, *J. Chem. Soc. Chem. Commun.* 878–879 (1990).
61. V.-M. Mukkala and J. Kankare, New fluorescent Eu(III) and Tb(III) chelates of 2,2′-bipyridine derivatives, *Eur. J. Solid State Inorg. Chem. 29*, 53–56 (1992).
62. G. F. de Sa, L. H. A. Nunes, and 0. L. Malta, Synthesis, characterization and luminescence of europium(III) and terbium(III) complexes of 3-aminopyrazine-2-carboxylic acid, *J. Chem. Res. (S)*, 78–79 (1992).
63. L. Prodi, M. Maestri, R. Ziessel, and V. Balzani, Luminescent Eu^{3+}, Tb^{3+}, and Gd^{3+} complexes of a branched-triazacyclononane ligand containing three 2,2′-bipyridine units, *Inorg. Chem. 30*, 3798–3802 (1991).
64. V. Balzani and R. Ballardini, New trends in the design of luminescent metal complexes, *Photochem. Photobiol. 52*, 409–416 (1990).
65. M. P. Bailey, B. F. Rocks, and C. Riley, Chelated terbium as a label in fluorescence immunoassay, in: *Nonisotopic Immunoassay* (T. Ngo, ed.), pp. 187–197, Plenum Press, New York (1988).
66. I. A. Hemmilä, S. Dakubu, V.-M. Mukkala, H. Siitari, and T. Lövgren, Europium as a label in time-resolved immunofluorometric assays, *Anal. Biochem. 137*, 335–343 (1984).
67. I. A. Hemmilä, Time-resolved fluorometric determination of terbium in aqueous solution, *Anal. Chem. 57*, 1676–1681 (1985).
68. T. Lövgren, I. Hemmilä, K. Pettersson and P. Halonen, Time-resolved fluorometry in immunoassay, in: *Alternative Immunoassays* (W. P. Collins, ed.) pp. 203–217, John Wiley & Sons, New York (1985).
69. E. Soini, Pulsed light, time-resolved fluorometric immunoassay, in: *Monocolonal Antibodies and New Trends in Immunoassays* (Ch. A. Bixollon, ed.) pp. 197–208, Elsevier Science Publishers, Amsterdam (1984).
70. V.-M. Mukkala, H. Mikoia, and I. Hemmilä, The synthesis and use of activated *N*-benzyl derivatives of diethylenetriaminetetraacetic acids: Alternative reagents for labeling of antibodies with metal ions, *Anal. Biochem. 176*, 319–325 (1989).
71. P. Helsingius, I. Hemmilä, and T. Lövgren, Solid-phase immunoassay of digoxin by measuring time-resolved fluorescence, *Clin. Chem. 32*, 1767–1769 (1986).
72. K.-T. Yeo, T. M. Sioussat, J. D. Faix, D. R. Senger, and T.-K. Yeo, Development of time-resolved immunofluorometric assay of vascular permeability factor, *Clin. Chem. 38*, 71–75 (1992).
73. P. Nuutila, P. Koskinen, K. Irjala, L. Linko, H.-L. Kaihola, J. U. Eskola, R. Erkkola, P. Seppcälä, and J. Viikari, Two new two-step immunoassays for free thyroxine evaluated: Solid-phase radioimmunoassay and time-resolved fluoroimmunoassay, *Clin. Chem. 36*, 1355–1360 (1990).
74. T. Lövgren, I. Hemmilä, K. Pettersson, J. U. Eskola, and E. Bertoft, Determination of hormones by time-resolved fluoroimmunoassay, *Talanta 31*, 909–916 (1984).
75. E. P. Diamandis and R. C. Morton, Time-resolved fluorescence using a europium chelate of 4,7-bis(chlorosulfophenyl)-1,10-phenanthroline-2,9-dicarboxylic acid (BCPDA). Labelling procedures and applications in immunoassays, *J. Immunol. Methods 112*, 43–52 (1988).
76. E. P. Diamandis and T. K. Christopoulos, Europium chelate labels in time-resolved fluorescence immunoassays and DNA hybridization assays, *Anal. Chem. 62*, 1149A–1157A (1990).
77. V. Bhayana and E. P. Diamandis, A double monoclonal time-resolved immunofluorometric assay of carcinoembryonic antigen in serum, *Clin. Biochem.* 22, 433–438 (1989).
78. E. Reichstein, Y. Shami, M. Ramjeesingh, and E. P. Diamandis, Laser-excited time-resolved solid-phase fluoroimmunoassays with the new europium chelate 4,7-bis(chlorosulfophenyl)-1,10-phenanthroline-2,9-dicarboxylic acid as label, *Anal. Chem. 60*, 1069–1074 (1988).
79. M. J. Khosravi and E. P. Diamandis, Immunofluorometry of choriogonadotropin by time-resolved fluorescence spectroscopy, with a new europium chelate as label, *Clin. Chem. 33*, 1994–1999 (1987).

80. M. A. Chan, A. C. Bellem, and E. P. Diamandis, Time-resolved immunofluorometric assay of alpha-fetoprotein in serum and amniotic fluid with a novel detection system, *Clin. Chem. 33*, 2000–2003 (1987).

81. P. Shankaran, E. Reichstein, M. J. Khosravi, and E. P. Diamandis, Detection of immunoglobulins G and M to rubella virus by time-resolved immunofluorometry, *J. Clin. Microbiology 28*, 573–579 (1990).

82. E. P. Diamandis and T. K. Christopoulos, Time-resolved immunofluorometric detection of antigens separated by high-performance liquid chromatography and coated to polystyrene, *BioTechniques 10*, 646–648 (1991).

83. T. K. Christopoulos and E. P. Diamandis, Enzymatically amplified time-resolved fluorescence immunoassay with terbium chelates, *Anal. Chem. 64*, 342–346 (1992).

84. A. Papanastasiou-Diamandi, T. K. Christopoulos, and E. P. Diamandis, Ultrasensitive thyrotropin immunoassay based on enzymatically amplified time-resolved fluorescence with a terbium chelate, *Clin. Chem. 38*, 545–548 (1992).

85. R. A. Evangelista, A. Pollak, and E. F. G. Templeton, Enzyme-amplified lanthanide luminescence for enzyme detection in bioanalytical assays, *Anal. Biochem. 197*, 213–224 (1991).

86. P. L. Khanna, Fluorescence energy transfer immunoassays, in: *Nonisotopic Immunoassay* (T. T. Ngo, ed.), pp. 211–229, Plenum Press, New York (1988).

87. K. Albertsson-Wickland, C. Jansson, S. Rosberg, and Anne Novamo, Time-resolved immunofluorometric assay of human growth hormone, *Clin. Chem. 39*, 1620–1625 (1993).

88. S. E. Kakabakos, T. K. Christopoulos, and E. P. Diamandis, Multianalyte immunoassay based on spatially distinct fluorescent areas quantified by laser-excited solid-phase time-resolved fluorometry, *Clin. Chem. 38*, 338–342 (1992).

89. Y.-Y. Xu, K. Pettersson, K. Blomberg, I. Hemmilä, H. Mikola, and T. Lövgren, Simultaneous quadruple-label fluorometric immunoassay of thyroid-stimulating hormone, 17α-hydroxyprogesterone, immunoreactive trypsin, and creatine kinase, *Clin. Chem. 38*, 2038–2043 (1992).

90. I. Hemmilä, O. Malminen, H. Mikola, and T. Lövgren, Homogeneous time-resolved fluoroimmunoassay of thyroxin in serum, *Clin. Chem. 34*, 2320–2322 (1988).

91. G. Barnard, F. Kohen, H. Mikola, and T. Lövgren, Measurement of estrone-3-glucuronide in urine by rapid, homogeneous time-resolved fluoroimmunoassay, *Clin. Chem. 35*, 555–559 (1989).

92. E. F. Ullman, M. Schwarzberg, and K. E. Rubenstein, Fluorescent excitation transfer immunoassay. A general method for determination of antigens, *J. Biol. Chem. 251*, 4172–4178 (1976).

93. I. B. Berlman, in: *Handbook of Fluorescence Spectra of Aromatic Molecules*, 2nd ed., Academic Press, New York (1971).

94. M. N. Kronick, Phycobiliproteins as labels in immunoassay, in: *Nonisotopic Immunoassay* (T. Ngo, ed.), pp. 163–185, Plenum Press, New York (1988).

95. P. L. Khanna and E. F. Ullman, 4′,5′-Dimethoxy-6-carboxyfluorescein: A novel dipole–dipole coupled fluorescence energy transfer acceptor useful for fluorescence immunoassays, *Anal. Biochem. 108*, 156–161 (1980).

96. E. F. Ullman and P. L. Khanna, Fluorescence excitation transfer immunoassay (FETI), *Methods in Enzymology 74*, 28–60 (1981).

97. I. Wieder and R. L. Hale, PCT Patent Application WO 87/07,955 (1987).

98. M. Genet, V. Brandel, M.-P. Lahalle, and E. Simoni, Electronic energy transfer between coumarin 460 and Eu^{3+} in thorium phosphate xerogel, *C. R. Acad. Sci. Paris 311* (Series II), 1321–1325 (1990).

99. E. Gratton and M. Limkeman, A continuously variable frequency cross-correlation phase fluorometer with picosecond resolution, *Biophys. J. 44*, 665–669 (1983).

100. J. R. Lakowicz and B. P. Maliwal, Construction and performance of a variable-frequency phase-modulation fluorometer, *Biophys. Chem. 21*, 61–78 (1985).

101. J. R. Lakowicz, G. Laczko, and I. Gryczynski, A 2 GHz frequency-domain fluorometer, *Rev. Sci. Instrum. 57*, 2499–2506 (1986).

102. R. B. Thompson, J. K. Frisoli, and J. R. Lakowicz, Phase fluorometry using a continuously modulated laser diode, *Anal. Chem. 64*, 2075–2078 (1992).
103. R. D. Spencer and G. Weber, Measurement of subnanosecond fluorescence lifetimes with a cross-correlation phase fluorometer, *Ann. N.Y. Acad. Sci. 158*, 361–376 (1969).
104. J. R. Lakowicz and S. Keating, Binding of an indole derivative to micelles as quantified by phase-sensitive detection of fluorescence, *J. Biol. Chem.* 5519–5524 (1983).
105. F. V. Bright, T. L. Keimig, and L. B. McGown, Thermodynamic binding parameters evaluated by using phase-resolved fluorescence spectroscopy, *Anal. Chim. Acta 175*, 189–201 (1985).
106. Y. R. Tahboub and L. B. McGown, Phase-resolved fluoroimmunoassay of human serum albumin, *Anal. Chim. Acta 182*, 185–191 (1986).
107. J. R. Lakowicz, B. Maliwal, A. J. Ozinskas, and R. B. Thompson, Fluorescence lifetime energy-transfer immunoassay quantified by phase-modulation fluorometry, *Sensors and Actuators B 12*, 65–70 (1993).
108. J. P. O'Connell, R. L. Campbell, B. M. Fleming, T. J. Mercolino, M. D. Johnson, and D. A. McLaurin, A highly sensitive immunoassay system involving antibody-coated tubes and liposome-entrapped dye, *Clin. Chem. 31*, 1424–1426 (1985).
109. A. L. Plant, M. V. Brizgys, L. Locasio-Brown, and R. A. Durst, Generic liposome reagent for immunoassays, *Anal. Biochem. 176*, 420–426 (1989).
110. M. Fiechtner, M. Wong, C. Bieniarz, and M. T. Shipchandler, Hydrophilic fluorescein derivatives: Useful reagents for liposome immunolytic assays, *Anal. Biochem. 180*, 140–146 (1989).
111. M. A. Gerber, M. F. Randolph, and K. K. DeMeo, Liposome immunoassay for rapid identification of group A streptococci directly from throat swabs, *J. Clin. Microbiol. 28*, 1463–1464 (1990).
112. Y. Tatsu, S. Yamamura, and S. Yoshikawa, Fluorescent fibre-optic immunosensing system based on complement lysis of liposome containing carboxyfluorescein, *Biosensors and Bioelectronics 7*, 741–745 (1992).
113. L. Locascio-Brown, A. L. Plant, V. Horvath, and R. A. Durst, Liposome flow injection immunoassay: Implications for sensitivity, dynamic range, and antibody regeneration, *Anal. Chem. 62*, 2587–2593 (1990).
114. M. Umeda, Y. Ishimori, K. Yoshikawa, M. Takada, and T. Yasuda, Liposome immune lysis assay (LILA), *J. Immunol. Methods 95*, 15–21 (1986).
115. P. Vadgama and P. W. Crump, Biosensors: Recent trends, a review, *Analyst 117*, 1657–1670 (1992).
116. D. G. Buerk, *Biosensors*, Technomic Publishing Co., Lancaster, Pennsylvania (1993).
117. O. S. Wolfbeis, Ed., *Fiber Optic Chemical Sensors and Biosensors*, CRC Press, Boca Raton, Florida (1991).
118. F. P. Anderson and W. G. Miller, Fiber optic immunochemical sensor for continuous, reversible measurement of phenytoin, *Clin. Chem. 34*, 1417–1421 (1988).
119. W. G. Miller and F. P. Anderson, Antibody properties for chemically reversible biosensor applications, *Anal. Chim. Acta 227*, 135–143 (1989).
120. J. R. Astles and W. G. Miller, Reversible fiber-optic immunosensor measurements, *Sensors and Actuators B 11*, 73–78 (1993).
121. W. G. Miller and J. R. Astles, *First European Conference on Optical Sensors and Biosensors*, Graz, Austria, 12–15 April (1992).
122. S. M. Barnard and D. R. Walt, Chemical sensors based on controlled-release polymer systems, *Science 251*, 927–929 (1991).
123. F. V. Bright, T. A. Betts, and K. S. Litwiler, Regenerable fiber-optic-based immunosensor, *Anal. Chem. 62*, 1065–1069 (1990).
124. B. Reck, K. Himmelspach, N. Opitz, and D. W. Lübbers, Possibilities and limitations of continuous thyroxine measurement in an optode using the principle of homogeneous fluoroimmunoassay, *Analyst 113*, 1423–1426 (1988).
125. R. D. Petrea, M. J. Sepaniak, and T. Vo-Dinh, Fiber-optic time-resolved fluorimetry for immunoassays, *Talanta 35*, 139–144 (1988).

126. R. Sutherland, C. Dähne, R. Slovacek, and B. Bluestein, Interface immunoassays using the evanescent wave, in: *Alternative Immunoassays* (W. P. Collins, ed.), pp. 331–357, John Wiley & Sons, New York (1985).
127. E. H. Lee, R. E. Benner, J. B. Fenn, and R. K. Chang, Angular distribution of fluorescence from liquids and monodispersed spheres by evanescent wave excitation, *Appl. Optics 18*, 862–870 (1979).
128. C. K. Carniglia, L. Mandel, and H. Drexhage, Absorption and emission of evanescent photons, *J. Opt. Soc. Am. 62*, 479–486 (1972).
129. J. N. Herron, K. D. Caldwell, D. A. Christensen, S. Dyer, V. Hlady, P. Huang, V. Janatova, H.-K. Wang, and A.-P. Wei, Fluorescent immunosensors using planar waveguides, *Proc. SPIE 1885*, 28–39 (1993).
130. B. J. Tromberg, M. J. Sepaniak, T. Vo-Dinh, and G. D. Griffin, Fiber-optic chemical sensors for competitive binding fluoroimmunoassay, *Anal. Chem. 59*, 1226–1230 (1987).
131. R. A. Ogert, L. C. Shriver-Lake, and F. S. Ligler, Toxin detection using a fiber optic-based biosensor, *Proc. SPIE 1885*, 11–17 (1993).
132. R. A. Badley, R. A. L. Drake, I. A. Shanks, F. R. S., A. M. Smith, and P. R. Stephenson, Optical biosensors for immunoassays: The fluorescence capillary fill device, *Phil. Trans. R. Soc. Lond. B 316*, 143–160 (1987).
133. Y. Zhou, J. V. Magill, R. M. De La Rue, P. J. R. Laybourn, and W. Cushley, Evanescent fluorescence immunoassays performed with a disposable ion-exchanged patterned waveguide, *Sensors and Actuators B 11*, 245–250 (1993).
134. V. Hlady, J. N. Lin, and J. D. Andrade, Spatially resolved detection of antibody-antigen reaction on solid/liquid interface using total internal reflection excited antigen fluorescence and charge-coupled device detection, *Biosens. Bioelectron. 5*, 291–301 (1990).

Index

MIX
Papier aus verantwortungsvollen Quellen
Paper from responsible sources
FSC® C105338

If you have any concerns about our products,
you can contact us on
ProductSafety@springernature.com

In case Publisher is established outside the EU,
the EU authorized representative is:
Springer Nature Customer Service Center GmbH
Europaplatz 3, 69115 Heidelberg, Germany

Printed by Libri Plureos GmbH
in Hamburg, Germany